Soilless Culture, Growing Media and Horticultural Plants

Soilless Culture, Growing Media and Horticultural Plants

Editors

Nazim S. Gruda
Brian Eugene Jackson

Basel • Beijing • Wuhan • Barcelona • Belgrade • Novi Sad • Cluj • Manchester

Editors
Nazim S. Gruda
University of Bonn
Bonn, Germany

Brian Eugene Jackson
NC State University
Raleigh, NC, USA

Editorial Office
MDPI
St. Alban-Anlage 66
4052 Basel, Switzerland

This is a reprint of articles from the Special Issue published online in the open access journal *Agronomy* (ISSN 2073-4395) (available at: https://www.mdpi.com/journal/agronomy/special_issues/Soilless_Media_Horticultural).

For citation purposes, cite each article independently as indicated on the article page online and as indicated below:

Lastname, A.A.; Lastname, B.B. Article Title. *Journal Name* **Year**, *Volume Number*, Page Range.

ISBN 978-3-0365-8760-8 (Hbk)
ISBN 978-3-0365-8761-5 (PDF)
doi.org/10.3390/books978-3-0365-8761-5

Cover image courtesy of Nazim S. Gruda

Contents

About the Editors

Nazim S. Gruda

Nazim S. Gruda, Professor of Horticulture at the University of Bonn in Germany, vice-chair of the ISHS-Division Vegetables, Roots, and Tubers, and chair of one ISHS-Working Group. Professor Gruda is an internationally recognized leading authority in soilless culture, growing media, and controlled environment cultivation. He has been working as a successful researcher, research advisor, project expert and evaluator, member of steering committees and editorial boards, editor, and scholar for more than three decades. He obtained his Doctorate at the Technical University of Munich and Habilitation at the Humboldt University of Berlin, both in Germany. His research focused on the scientific understanding and application of innovative and sustainable horticultural food production. He has achieved an impressive track record of over 300 publications. He has co-edited four Acta Horticulturae for the ISHS. Dr. Gruda is one of the authors and the editor of the successful book "Advances in Horticultural Soilless Culture", from Burleigh Dodds Science Publishing Limited, Cambridge, UK, in 2021. According to the evaluation carried out by Stanford University over the last three years, Prof. Gruda is in the top 2% of the most influential scientists in the world ("World's Top 2% Scientists"). In recognition of his excellent research, Professor Gruda was awarded the "Dr Heinrich-Baur-Prize" 2003 by the Technical University of Munich, Germany, the "National Scientific Prize" 2018 by the Academy of Science of Albania, and was elected as "Distinguished Scientist" 2020 by the Academy of Science of China. In addition, Professor Gruda is a full member, a foreign correspondent member, and an honorary member of three European Academies of Sciences.

Brian Eugene Jackson

Brian Eugene Jackson is a Professor and Director of the Horticultural Substrates Laboratory in the Department of Horticultural Science at North Carolina State University. Brian earned both his graduate degrees (M.S and Ph.D.) in soilless substrates and has continued as a soilless substrate scientist for the past 15 years at NC State University. Brian's major research focus and international expertise are in the engineering and development of alternative substrates in soilless growing systems, primarily wood fiber, and other forest and agricultural-based biomass materials. He has advised or co-advised 5 Ph.D. and 14 M.S. students and has served on the committees of 18 others nationally and internationally. Brian has developed a global reputation as one of the leading voices, educators, researchers, and advocates for the soilless substrate industry. Brian has traveled to over 50 countries to research, collaborate, or consult on soilless substrate-related problems and innovations in the past decade. At home, his career efforts and successes have been recognized by his receipt of the 2020 Outstanding Undergraduate Educator Career Award and the 2021 Outstanding International Horticulturalist Career Award from the American Society of Horticultural Science.

Editorial

Advances in Soilless Culture and Growing Media in Today's Horticulture—An Editorial

Nazim S. Gruda

Division of Horticultural Sciences, Institute of Crop Science and Resource Conservation, University of Bonn, Auf dem Hügel 6, 53121 Bonn, Germany; ngruda@uni-bonn.de

Abstract: The soilless culture system is a promising, intensive, and sustainable approach with various advantages for plant production. The Special Issue "Soilless Culture, Growing Media, and Horticultural Plants" includes 22 original papers and 1 review written by 84 authors from 15 countries. The purpose of this Special Issue was to publish high-quality research articles that address the recent developments in the cultivation of horticultural plants in soilless culture systems and solid growing media. The published articles investigated new developments in simplified and advanced systems; the interaction between soilless and environmental factors with their effects on plant growth and photosynthesis, and the accumulation of secondary metabolites; the analyses of nutrient solution and hydraulic properties of substrates and mixtures; and the microbe–plant growing media interactions. Climate change and environmental and ecological issues will determine and drive the development of soilless culture systems and the choice of growing media constituents in the near future. Bioresources and renewable raw materials have great potential for use as growing medium constituents.

Citation: Gruda, N.S. Advances in Soilless Culture and Growing Media in Today's Horticulture—An Editorial. *Agronomy* **2022**, *12*, 2773. https://doi.org/10.3390/agronomy12112773

Received: 2 November 2022
Accepted: 4 November 2022
Published: 7 November 2022

1. Introduction

Decreasing arable land, rising urbanisation, water scarcity, and climate change have placed pressure on agricultural producers [1]. The soilless culture system (SCS) is a promising approach with different advantages for plant production. As an intensive and sustainable cultivation method, SCSs have rapidly expanded worldwide, particularly in areas close to cities or with a shortage of water supply, poor soil quality, and problems with soil-borne diseases and salinity. These systems produce pot ornamentals, seedlings, and transplants and increase plant metabolites in fruits, vegetables, and medicinal and aromatic plants. Production technology affects plant growth, yield, and overall plant quality, which, in turn, improves the cumulative benefits of plants [2–4].

Horticultural crops, such as vegetables, floral crops, ornamentals, and fruits, have become essential components of aesthetics and nutrition in our daily life. Currently, SCSs have received significant interest and are used for the intensive production of vegetables, floral crops, ornamentals, green roofs, and rain gardens [2,3,5]. Furthermore, because of their lightweight and sustainable resource efficiency, soilless systems are especially suitable for urban areas, including green infrastructure projects and vertical farming [6]. The increased worldwide production of crops in controlled environmental systems has been further accelerated by the increased interest in growing small/soft fruit crops, greens, herbs, and cannabis in soilless container systems. In addition, they are used to increase metabolites in medicinal and aromatic plants and to introduce new crops [2,3]. As a result, the demand for SCS and growing media continues to increase worldwide, as does the need for novel research to address problems and continue creating opportunities for this industry [4].

The purpose of this Special Issue was to publish high-quality research articles that address the recent developments in the cultivation of horticultural plants using SCSs with or without solid growing media. It aims to provide contributions from various currently relevant topics in horticultural sciences, physiology, root medium properties, plant propagation, plant nutrition and chemistry, substrate hydrology and physics, compost and waste management, engineering, and all other research fields familiar with soilless culture and growing media.

The Special Issue "Soilless Culture, Growing Media, and Horticultural Plants" includes 22 original papers and 1 review written by 84 authors from 15 countries. Considering this is just the tip of the iceberg, the remaining papers were rejected during the published review process, showing the great interest in this Special Issue from the scientific community. Writing an editorial after several years allowed us to analyse the papers' importance. Following citations from the publication date until the end of October 2022, papers from this Special Issue were cited 363 times, with an average of 16.5 times per paper, which is relatively high. The review article [1] received the highest number of citations (126) among all the published papers. It should be mentioned that this article received the second-best paper award on the tenth anniversary of the journal, while the Web of Science-Clarivate lists it as a highly cited paper (1% of all papers included in the database). The article from Dou et al. [7] received the highest number of citations among the research papers, with 41 citations.

2. Soilless Culture Systems

Soilless culture is a modern cultivation technology applied mainly in greenhouses, which has developed rapidly during the last 30–40 years [5]. Most SCS plants are grown in high-tech greenhouse structures with fully automatic climate control features [2,3]. This Special Issue focused on new developments in simplified, advanced, and complex SCSs. For instance, Bentrary et al. [8] and Michelon et al. [9] investigated the feasibility of a low-tech SCS for cultivating *Pelargonium zonale* and *Lactuca sativa*, respectively. As a result, the yield for lettuce cultivation in tropical areas was improved by +35% and +72% in Brazil and Myanmar, respectively, and the water-use efficiency (WUE) was 7.7 and 2.7 times higher in Brazil and Myanmar, respectively, compared to traditional on-soil cultivation [9]. The soilless system typology can also significantly affect the rooted cutting growth, commercial features, and WUE. For example, adopting an open-cycle drip system significantly improved all commercial crop characteristics of geranium (*Pelargonium zonale*) compared to a substrate and a nutrient film technique system. The water consumption of this treatment system was higher than that of the other systems. However, it induced the highest fresh weight and, therefore, the highest WUE [8].

Given its flexibility in manipulating the nutrient status and efficient utilisation of nutrient components, SCSs could be used as an efficient tool for producing high-value vegetables and herbs and crucial root vegetables in temperate and tropical zones, such as sweetpotatoes (*Ipomoea batatas*) [10].

In recent years, research on soilless culture has mainly focused on the automation of nutrient and water supply, particularly in closed systems [5]. A closed-loop SCS is an environment-friendly cultivation method. However, variations in nutrients can lead to instability in nutrient management. Ahn and Son [11] analysed nutrient variation in a closed-loop SCS based on a theoretical model and found fluctuations around the target value. However, the total nutrient concentration did not continuously deviate from the target value in the conventional method and showed a tendency to increase. Therefore, the authors concluded that these characteristics of the alternative method could help minimise nutrient and water emissions from the cultivation system.

Theoretical and experimental analyses of nutrient solutions, variations in electrical conductivity, fertiliser selection, and nutrient solution replenishment methods have been discussed in the papers published in this Special Issue. The fertiliser used in the SCS should contain balanced elements without any precipitates [12]. For sweet pepper yields, the commercial fertiliser 5N-4.8P-21.6K was responsible for the highest yield of both cultivars, 'Bentley' and 'Orangella'. Fertilisers and cultivars did not affect the shape index. For eggplant, the shoot fresh weight was greater for 'Angela' than for 'Jaylo' at 5N-4.8P-21.6K and 7N-3.9P-4.1K. Furthermore, both eggplant cultivars were affected by yellowing fruits for all the fertiliser treatments after two months, probably due to the accumulation of nutrients in the closed hydroponic system [12].

3. The Interaction of SCS with Environmental Greenhouse Factors

The effects of the interaction of soilless culture and different environmental greenhouse factors, such as supplemental lighting intensity, UV radiation, and CO_2 enrichment, on biomass accumulation, gas exchange properties, and plant quality are addressed in this Special Issue. For instance, Llewellyn et al. [13] found that increasing levels of supplemental light had only minor effects on vegetative growth (young plants) and the size and quality of harvested flowers (mature plants). However, cut gerbera (*Gerbera jamesonii*) plants grown under higher light intensity produced 10.3 and 7.0 more total and marketable flowers per plant than the lowest light intensity and matured faster [13].

One other factor is the CO_2 concentration in the air. According to Li et al. [14]), the accumulation of cucumber biomass can be significantly increased by elevated CO_2 concentrations and high N supply. In addition, a high N supply can further improve photosynthesis. The authors concluded that if we had a greater understanding of the mechanisms that control mineral concentration changes in cucumber plants in response to elevated CO_2, mineral fertilisation could be optimised to improve the growth of plants under elevated CO_2 conditions. Thus, sustainable vegetable production with higher C and N use efficiency and lower CO_2 emissions and fertiliser input could be achieved [14].

4. SCS and Produce Quality

Using SCSs to control nutrients, the temperature in the root area, and managing environmental and agronomic factors can improve product quality [1,15]. This Special Issue investigated the effects on plant photosynthesis and growth, the accumulation of secondary metabolites, and seasonal antioxidant changes. For instance, Neocleous et al. [16] indicated that lower solar irradiance, ultraviolet radiation, and temperature in Mediterranean greenhouses could be insufficient to stimulate phytochemical production during late autumn and winter in peppers. Thus, plant light interception must be more actively managed. Furthermore, Ellenberger et al. [17] investigated how stress affects the content of secondary metabolites in leaf bell papers. Therefore, high UV stress should be considered a tool for enriching plant leaves with valuable secondary metabolites.

The absence of ultraviolet (UV) radiation and low photosynthetic photon flux density (PPFD) in a controlled environment reduced the phenolic compounds in herbs. Dou et al. [7] investigated green and purple basil to characterise the optimal UV-B radiation dose and PPFD for enhancing the synthesis of phenolic compounds in basil plants (*Ocimum basilicum*). Plants were grown at two PPFDs, 160 and 224 $\mu mol \cdot m^{-2} \cdot s^{-1}$, and treated with five UV-B radiation doses. In purple basil plants, the concentrations of phenolics and flavonoids increased after $2\ h \cdot d^{-1}$ of UV-B treatment. Among all treatments, $1\ h \cdot d^{-1}$ for 2 d of UV-B radiation under a PPFD of 224 $\mu mol \cdot m^{-2} \cdot s^{-1}$ was the optimal condition for green basil production in a controlled environment [7].

Interestingly, Giménez et al. [18] found that compost in growing media boosted the product's final quality, with a higher total phenolic content and antioxidant capacity in the leaves of baby leaf lettuce in a floating system, particularly during the second cut.

5. Growing Media and the Diversity of Inorganic and Organic Substrates

In SCSs, solid inorganic or organic substrates are used for plant cultivation, usually in containers. Therefore, studies submitted to this Special Issue have investigated the physicochemical and hydraulic properties of organic and mineral substrates and mixtures and the substrate volumetric water content to improve water-use efficiency in growing media. Furthermore, the chemical properties and the microbe–plant growing media interactions were investigated.

According to Gohardoust et al. [19], an essential first step towards developing advanced soilless culture management strategies is the comprehensive characterisation of the growing media's hydraulic and physicochemical properties. These parameters can be applied to the engineering of growing media by mixing organic and inorganic constituents at different ratios to meet specific plant physiological demands. Furthermore, these results could also be used to visualise three-dimensional numerical computer codes to simulate water and nutrient dynamics in containerised growth modules.

Moreover, Currey et al. [20] found that the growth of basil, dill, parsley, and sage can be affected by the water supply, with no signs of stress or visual damage resulting from the reduced volumetric water content of the substrate. Therefore, restricting irrigation and substrate volumetric water content is an effective non-chemical growth control method for containerised culinary herbs.

Bacterial enhancement has a significant potential to modulate plant performance in horticultural systems. However, the effectiveness of bacterial amendment regarding plant performance depends on the bacterial source and its interaction with the growth medium. Therefore, an appropriate selection of the plant growth medium composition is critical for the efficacy of bacterial amendments and optimal plant performance in a plant factory with artificial lighting [21].

6. Peat Alternatives in Growing Media Mixtures

Peat is the most commonly used substrate constituent in horticulture. However, the use of peat in horticulture has been strongly criticised because of environmental and climate change concerns [1–3]. Therefore, new peat additives and/or peat alternative growing media, such as biochar, green compost, olive oil-processing waste composts, and vermicompost, were investigated in the Special Issue. In addition, the raw materials used as growing media constituents should be free from phytotoxic compounds [22] and should demonstrate good chemical properties, such as a suitable pH [23,24] and the content of certain elements and/or salt content [18,25].

Composts from different raw materials, such as vineyard waste, tomato waste, leek waste, and olive mill cake, can be alternatives to peat in producing baby leafy vegetables in a floating system. The use of 25% compost as a component of the growing media in the production of baby leafy vegetables in a floating system not only favours crop yield and product quality, but also suppresses *Pythium irregulare* [18].

Tüzel et al. [26] found that compost obtained from two-phase and three-phase olive mill solid wastes and olive oil wastewater sludge that can be used in a ratio of 25% in mixtures with peat was appropriate for most of the measured tomato seedling properties.

Moreover, biochar has been proposed as a soil amendment and a growing medium component that positively affects plant growth and yield [24]. Chrysargyris et al. [25] investigated four types of commercial-grade biochar from wood-based materials used in mixtures with peat for cabbage seedling production. Biochar material had a high K content and a pH $\geq$ 8.64, which increased the growing media's pH. In addition, the leachate pH of all biochar mixes was higher than that of the control [27]. Potassium, phosphorous, copper accumulation and magnesium deficiency in cabbage leaves were related to the presence of biochar. Therefore, wooden biochar from beech, spruce, and pine species and fertilised biochar from fruit trees and hedges is promising for cabbage seedling production [25].

While recent studies on biochar mentioned the importance of the feedstock used, Prasad et al. [24] stated for the first time the need for information on particle size because the fractions from the same biochar can have different levels of total extractable nutrients and pH levels. Particle size could have a profound effect on the nutrient availability of Ca and Mg. This could lead to nutrient imbalances during the cultivation of plants on substrate mixtures. In addition to nutrient ratios, a suitable pH level for a given species should be achieved [24].

Mixes with 80% biochar and vermicompost had lower container capacities than the control. Nevertheless, plants in the BC mixes had similar growth indices and total dry weights concerning those in 100% commercial substrate. Therefore, BC mixed with vermiculite has the potential to replace commercial peat-based substrates for container-grown plants [27].

Yu et al. [28] conducted a greenhouse experiment to evaluate the potential of replacing mixed hardwood biochar with sugarcane bagasse. Both tomato and basil plants grown in biochar-incorporated mixes had a similar or higher growth index, leaf greenness, and yield than bark-based commercial substrates. The authors concluded that hardwood and sugarcane bagasse biochar could replace 50% and 70% of bark-based substrates for tomato and basil plants without adverse growth effects [28].

7. Concluding Remarks and Future Trends

The articles published in this Special Issue stated that climate change and environmental and ecological issues would soon determine and drive the development of soilless cultural systems and the choice of growing media constituents. It is clear that while much has been achieved in this Special Issue, many challenges remain. Understanding the optimisation of root-zone conditions [29] and clarifying the mechanism of interaction between roots and surroundings will contribute to a better understanding of SCS. Advances in soilless culture will be supported by findings from other scientific fields that will contribute to the further development of soilless cultures. In addition, bioresources and renewable raw materials have great potential for use as growing media constituents. We expect these publications to promote further discussion about these two exciting topics.

Funding: This research received no external funding.

Conflicts of Interest: The author declares no conflict of interest.

References

1. Gruda, N.S. Increasing Sustainability of Growing Media Constituents and Stand-Alone Substrates in Soilless Culture Systems. *Agronomy* **2019**, *9*, 298. [CrossRef]
2. Gruda, N.S. Soilless culture systems and growing media in horticulture: An overview. 1–20. In *Advances in Horticultural Soilless Culture*; Gruda, N., Ed.; Burleigh Dodds Science Publishing Limited: Cambridge, UK, 2021. [CrossRef]
3. Hong, J.; Gruda, N.S. The Potential of Introduction of Asian Vegetables in Europe. *Horticulturae* **2020**, *6*, 38. [CrossRef]
4. Gruda, N.S.; Fernández, J.A. Optimising Soilless Culture Systems and Alternative Growing Media to Current Used Materials. *Horticulturae* **2022**, *8*, 292. [CrossRef]
5. Savvas, D.; Gruda, N. Application of soilless culture technologies in the modern greenhouse industry—A review. *Eur. J. Hortic. Sci.* **2018**, *83*, 280–293. [CrossRef]
6. Eigenbrod, C.; Gruda, N. Urban vegetable for food security in cities. A review. *Agron. Sustain. Dev.* **2015**, *35*, 483–498. [CrossRef]
7. Dou, H.; Niu, G.; Gu, M. Pre-Harvest UV-B Radiation and Photosynthetic Photon Flux Density Interactively Affect Plant Photosynthesis, Growth, and Secondary Metabolites Accumulation in Basil (*Ocimum basilicum*) Plants. *Agronomy* **2019**, *9*, 434. [CrossRef]
8. Brentari, L.; Michelon, N.; Gianquinto, G.; Orsini, F.; Zamboni, F.; Porro, D. Comparative Study of Three Low-Tech Soilless Systems for the Cultivation of Geranium (*Pelargonium zonale*): A Commercial Quality Assessment. *Agronomy* **2020**, *10*, 1430. [CrossRef]
9. Michelon, N.; Pennisi, G.; Myint, N.; Dall'Olio, G.; Batista, L.; Salviano, A.; Gruda, N.; Orsini, F.; Gianquinto, G. Strategies for Improved Yield and Water Use Efficiency of Lettuce (*Lactuca sativa* L.) through Simplified Soilless Cultivation under Semi-Arid Climate. *Agronomy* **2020**, *10*, 1379. [CrossRef]
10. Sakamoto, M.; Suzuki, T. Effect of Nutrient Solution Concentration on the Growth of Hydroponic Sweetpotato. *Agronomy* **2020**, *10*, 1708. [CrossRef]
11. Ahn, T.; Son, J. Theoretical and Experimental Analysis of Nutrient Variations in Electrical Conductivity-Based Closed-Loop Soilless Culture Systems by Nutrient Replenishment Method. *Agronomy* **2019**, *9*, 649. [CrossRef]
12. Singh, H.; Dunn, B.; Payton, M.; Brandenberger, L. Selection of Fertilizer and Cultivar of Sweet Pepper and Eggplant for Hydroponic Production. *Agronomy* **2019**, *9*, 433. [CrossRef]
13. Llewellyn, D.; Schiestel, K.; Zheng, Y. Increasing Levels of Supplemental LED Light Enhances the Rate Flower Development of Greenhouse-grown Cut Gerbera but does not Affect Flower Size and Quality. *Agronomy* **2020**, *10*, 1332. [CrossRef]
14. Li, X.; Dong, J.; Gruda, N.; Chu, W.; Duan, Z. Interactive Effects of the CO_2 Enrichment and Nitrogen Supply on the Biomass Accumulation, Gas Exchange Properties, and Mineral Elements Concentrations in Cucumber Plants at Different Growth Stages. *Agronomy* **2020**, *10*, 139. [CrossRef]
15. Gruda, N. Assessing the impact of environmental factors on the quality of greenhouse produce. In *Achieving Sustainable Greenhouse Cultivation*; Marcelis, L., Heuvelink, E., Eds.; Burleigh Dodds Science Publishing Limited: Cambridge, UK, 2019. [CrossRef]

16. Neocleous, D.; Nikolaou, G. Antioxidant Seasonal Changes in Soilless Greenhouse Sweet Peppers. *Agronomy* **2019**, *9*, 730. [CrossRef]
17. Ellenberger, J.; Siefen, N.; Krefting, P.; Schulze Lutum, J.; Pfarr, D.; Remmel, M.; Schröder, L.; Röhlen-Schmittgen, S. Effect of UV Radiation and Salt Stress on the Accumulation of Economically Relevant Secondary Metabolites in Bell Pepper Plants. *Agronomy* **2020**, *10*, 142. [CrossRef]
18. Giménez, A.; Fernández, J.; Pascual, J.; Ros, M.; Saez-Tovar, J.; Martinez-Sabater, E.; Gruda, N.; Egea-Gilabert, C. Promising Composts as Growing Media for the Production of Baby Leaf Lettuce in a Floating System. *Agronomy* **2020**, *10*, 1540. [CrossRef]
19. Gohardoust, M.; Bar-Tal, A.; Effati, M.; Tuller, M. Characterization of Physicochemical and Hydraulic Properties of Organic and Mineral Soilless Culture Substrates and Mixtures. *Agronomy* **2020**, *10*, 1403. [CrossRef]
20. Currey, C.; Flax, N.; Litvin, A.; Metz, V. Substrate Volumetric Water Content Controls Growth and Development of Containerized Culinary Herbs. *Agronomy* **2019**, *9*, 667. [CrossRef]
21. Van Gerrewey, T.; Vandecruys, M.; Ameloot, N.; Perneel, M.; Van Labeke, M.; Boon, N.; Geelen, D. Microbe-Plant Growing Media Interactions Modulate the Effectiveness of Bacterial Amendments on Lettuce Performance Inside a Plant Factory with Artificial Lighting. *Agronomy* **2020**, *10*, 1456. [CrossRef]
22. Gruda, N.; Rau, B.J.; Wright, R.D. Laboratory Bioassay and Greenhouse Evaluation of a Pine Tree Substrate Used as a Container Substrate. *Eur. J. Hortic. Sci.* **2009**, *74*, 73–78.
23. Jackson, B.E.; Wright, R.D.; Gruda, N. Container medium pH in a pine tree substrate amended with peat moss and dolomitic limestone affects plant growth. *Hortscience* **2009**, *44*, 1983–1987. [CrossRef]
24. Prasad, M.; Chrysargyris, A.; McDaniel, N.; Kavanagh, A.; Gruda, N.S.; Tzortzakis, N. Plant nutrient availability and pH of biochars and their fractions, with the possible use as a component in a growing media. *Agronomy* **2020**, *10*, 10. [CrossRef]
25. Chrysargyris, A.; Prasad, M.; Kavanagh, A.; Tzortzakis, N. Biochar Type and Ratio as a Peat Additive/Partial Peat Replacement in Growing Media for Cabbage Seedling Production. *Agronomy* **2019**, *9*, 693. [CrossRef]
26. Tüzel, Y.; Ekinci, K.; Öztekin, G.; Erdal, İ.; Varol, N.; Merken, Ö. Utilization of Olive Oil Processing Waste Composts in Organic Tomato Seedling Production. *Agronomy* **2020**, *10*, 797. [CrossRef]
27. Huang, L.; Gu, M.; Yu, P.; Zhou, C.; Liu, X. Biochar and Vermicompost Amendments Affect Substrate Properties and Plant Growth of Basil and Tomato. *Agronomy* **2020**, *10*, 224. [CrossRef]
28. Yu, P.; Huang, L.; Li, Q.; Lima, I.; White, P.; Gu, M. Effects of Mixed Hardwood and Sugarcane Biochar as Bark-Based Substrate Substitutes on Container Plants Production and Nutrient Leaching. *Agronomy* **2020**, *10*, 156. [CrossRef]
29. Balliu, A.; Zheng, Y.; Sallaku, G.; Fernández, J.A.; Gruda, N.S.; Tuzel, Y. Environmental and Cultivation Factors Affect the Morphology, Architecture and Performance of Root Systems in Soilless Grown Plants. *Horticulturae* **2021**, *7*, 243. [CrossRef]

Review

Increasing Sustainability of Growing Media Constituents and Stand-Alone Substrates in Soilless Culture Systems

Nazim S. Gruda

Department of Horticultural Science, INRES-Institute of Crop Science and Resource Conservation, University of Bonn, 53121 Bonn, Germany; ngruda@uni-bonn.de

Received: 2 May 2019; Accepted: 5 June 2019; Published: 9 June 2019

Abstract: Decreasing arable land, rising urbanization, water scarcity, and climate change exert pressure on agricultural producers. Moving from soil to soilless culture systems can improve water use efficiency, especially in closed-loop systems with a recirculating water/nutrient solution that recaptures the drain water for reuse. However, the question of alternative materials to peat and rockwool, as horticultural substrates, has become increasingly important, due to the despoiling of ecologically important peat bog areas and a pervasive waste problem. In this paper, we provide a comprehensive critical review of current developments in soilless culture, growing media, and future options of using different materials other than peat and rockwool. Apart from growing media properties and their performance from the point of view of plant production, economic and environmental factors are also important. Climate change, CO_2 emissions, and other ecological issues will determine and drive the development of soilless culture systems and the choice of growing media in the near future. Bioresources, e.g., treated and untreated waste, as well as renewable raw materials, have great potential to be used as growing media constituents and stand-alone substrates. A waste management strategy aimed at reducing, reusing, and recycling should be further and stronger applied in soilless culture systems. We concluded that the growing media of the future must be available, affordable, and sustainable and meet both quality and environmental requirements from growers and society, respectively.

Keywords: biochar; compost; climate change; hydroponics; growing medium; life cycle analysis; organic bioresources; peat alternatives; renewable raw materials; rockwool; waste; wood fibers

1. Introduction

According to the United Nations, the current world population of 7.79 billion people will increase to 9.77 billion people by 2050 [1], while the arable land per capita continues to be reduced. This development is following the same pattern in all countries, although the rate varies between countries. For instance, in North America there were 1.06 ha, and in the European Union 0.32 ha, per person available in the year 1961, while in 2015 only 0.55 ha and 0.21 ha per person, respectively. This is nearly to 2× and more than 1.5× reduction for North America and the European Union, respectively (Figure 1) [2].

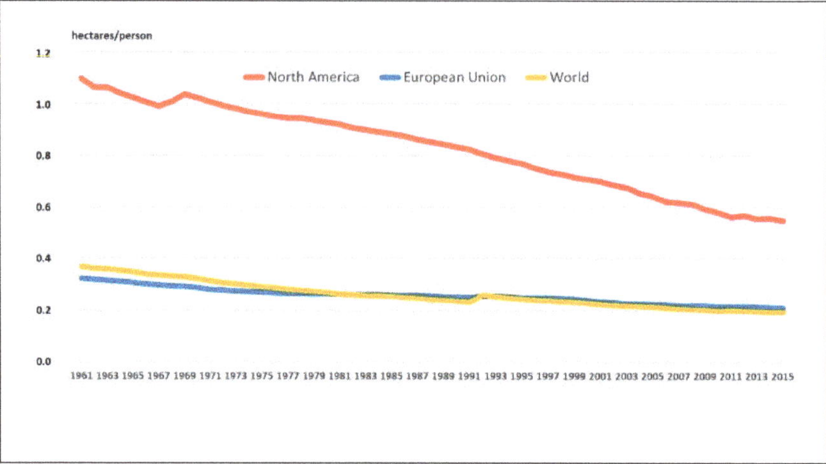

Figure 1. The arable land per person has been continuously reduced in the recent past. Arable land in hectares per person from 1961–2015 for North America, the European Union, and worldwide, according to World Bank [2].

In addition, worldwide urbanization is increasing rapidly. In 2008, the global urban population overtook the rural population for the first time in history. Today, over 50% of the world's population lives in cities; by 2030, this number is projected to increase to 70% [3].

Future climate change scenarios predict more frequent occurrence of extreme conditions, such as drought years and the uneven distribution of precipitation during the year [4]. The possible increase in water shortage and extreme weather events may cause lower yields and higher yield fluctuations [5]. These disadvantages will be predominately in warmer regions worldwide. Therefore, besides securing sufficient water, it will become increasingly important to improve the use efficiency of this resource [6–8]. Water, as a valuable resource, can be used more efficiently in protected vegetable production, which is considered less dependent on weather conditions than open field production, because micro-climates can be manipulated [6,7].

Decreasing arable land, rising urbanization, water scarcity, and climate change exert pressure upon agricultural producers. One of the most promising approaches to tackle this challenge is termed "sustainable intensification", which tries to combine increased production without damaging its supporting ecosystem. Examples for this approach are protected, soilless culture systems (SCS) [9]. "Soilless culture" is defined as the cultivation of plants in systems without soil in situ [10]. The percentage of SCS to the total commercial horticultural protected cultivation area varies from country to country. For instance, in the Netherlands and Almeria, Spain, soilless culture represents the main cultivation system used [11]. In Europe, Canada, and in the large horticultural industry complexes in the US, 95% of greenhouse tomatoes are produced in SCS [12,13].

Growing media, "substrates" or "plant substrates" provide a root environment that is initially free of plant pathogens and properties that ensure an adequate aeration, water, and nutrient supply. In the horticultural industry, generally, mixtures of growing media constituents and additives are used. Organic or inorganic materials can be used as constituents, while additives include fertilizers, liming materials, and bio-control or wetting agents [14–16].

Blok and Urrestarazu [17] estimated an area of more than 10,000 ha cultivated in rockwool slabs worldwide, including 6000 ha greenhouse area in Europe, mainly in Northern Europe. Rockwool has a low volume weight, is inert, and has a buffering capacity, limited to the quantity of nutrients and water held within the pore space of the medium [18]. To feed the plant with water and fertilizer a complete nutrient solution is supplied through the irrigation system (Figure 2).

(a)   (b)

Figure 2. Tomato production in soilless culture with rockwool as a growing medium: (**a**) Transplants in rockwool cubes, shortly before the start of greenhouse cultivation; (**b**) tomato plants in rockwool slabs (photos: Gruda, private collection).

However, it is important to note that the disposal problem for mineral wool has led to criticism of its current usage. Some authors, such as Bussell and Mckennie [19], showed some options to reuse rockwool, but when analyzing the life cycle assessment of horticultural growing media, Quantis [20] reported that mineral wool has the highest negative impacts on human health. In addition, freight costs are relatively high.

Besides rockwool, other inorganic substrates, such as perlite, volcanic rock, tuff, expanded clay granules, vermiculite, zeolite, pumice, sand, and synthetic materials could be used directly or in combination with other materials as a growing medium.

Of all organic materials, peat is the most used substrate constituent in horticulture [7]. The leading peat-production countries are Finland, Ireland, Germany, Sweden, Belarus, Canada, and Russia, which account for 80% of the world's production. Commercial applications include lawn and garden soil amendments, potting soils, and turf maintenance on golf courses [21]. The extensive use of peat as a basic and main component of substrates is due to relatively low costs in these areas, its excellent chemical, biological, and physical properties with low nutrient content, low pH, a unique combination of high water-holding capacity by high air space and drainage characteristics, light weight, and freedom from pests and diseases [14,16,21]. The unique microporous properties of *Sphagnum* peat and its resistance to degradation are matched by few other growing medium constituents [22].

However, peat is a limited resource with a great demand, and the extraction of peat bogs causes negative impacts on environment. Peatlands are areas with a layer of dead plant materials (peat) at the surface. The water-saturated and oxygen-free conditions prevent peat from fully decomposing. Peatlands are a habitat with special ecological value with the most important long-term carbon sinks and one of the most effective eco-systems in the terrestrial biosphere, providing different environmental services, such as biodiversity, carbon (C) storage, regulation of the local water quality, and local hydrology conditions, including flood protection [23–25]. Covering only about 3% of Earth's land area, they may store 21% [26] to 33% [27] of the total world's terrestrial organic carbon. In the long-term, peatlands are the largest stores of organic carbon of all terrestrial ecosystems [28]. However, when peat bogs are drained or destroyed, i.e., used in agriculture, forestry, and/or horticulture, they no longer act as carbon sinks. Degraded peatlands contribute disproportionally to global greenhouse gas emissions, with approximately 25% of all CO_2 emissions from the land use sector [29]. Annual emissions equivalent of 15 million tons of carbon are estimated [23,24,30,31]. In addition, the renewal process of peatlands takes a very long time, and in arid areas peat is imported, with an impact both in environmental and economic terms. Therefore, Quantis [20] indicates that peat has the highest impact on "climate change" and "resources" of all commonly-used substrate materials.

Recently, the energy use and carbon emissions in horticultural production systems have moved into the public spotlight. Thus, retailers increased the pressure and are now requiring not only traceable healthy and safe horticultural products, but also "clean and green" produce with a low carbon footprint. On the other hand, due to limited natural resources and waste recycling issues, environmentally acceptable solutions are needed for materials used as growing media constituents.

The objective of this paper is to critically review and expand the knowledge of impacts of soilless culture and growing media on the environment, targeting an improvement of sustainability of all horticultural systems. First, an overview on the pros and cons of soilless culture and growing media use is provided. Second, different important economic and environmental factors are analyzed. Moreover, different organic materials are explored with the objective to recognize successful alternatives for peat and rockwool.

2. Results and Discussion

2.1. Soilless Culture and Growing Media: Pros and Cons

Soilless culture systems are commonly integrated in controlled environment agriculture, i.e., heated greenhouses, that in turn are associated with environmental concerns and the production of high amounts of greenhouse gases (GHGs). Indeed, major studies conducted showed that from an environmental point of view, plants cultivated directly in soil in tunnels or greenhouses without using auxiliary systems perform better than those with heating in SCS [32–34]. However, even if the heated protected cultivation systems present a good opportunity to move from soil to SCS, we do not have to associate SCS only with heated greenhouses. The specific features along the entire production system in these structures include the large amount of energy consumption for heating during the cold season, artificial lighting, the greenhouse structure itself [35], the use of fertilizer and growing media [7], postharvest transport, and packaging [36]. The equipment of SCS contribute to some degree to an increase of the energy needed together with growing media used in these systems. But, on the other hand, SCS contributes to a reduction of many problems associated with traditional cultivation on soil in situ, such as soil-borne diseases and pests, and to an exact control of water and fertilizer supplies. As a consequence, higher yields at a reasonable production cost and high product quality can be attained in these systems [13,37]. Recently, the greenhouses production is increasingly carried out with machines as an "unmanned working model" in some soilless systems [38].

Moreover, high precision in modulating nutrient solution composition, the exact dosage and controlled exposure, make SCS a good instrument to predict the product supply and enhance the organoleptic plant parameters and bioactive quality components. Moderate salinity and/or nutritional stress and the biofortification of vegetables with beneficial micronutrients to human health, such as iodine, iron, molybdenum, selenium, silicon, and zinc are well known methods that have been successfully used to enhance the health-promoting phytochemicals in vegetables [13,39–42].

Therefore, in general, growing plants in soilless media is a sustainable production manner. This is due to the inherent space, nutrient, and water use efficiencies of this production method; all of which are higher than soil-grown crops [9]. At present, life cycle analysis (LCA) is used for the classification of growing media constituents, based on their environmental impact and sustainability, environmental protection, and the application of "green technologies" for their production [7,16]. Mugnozza et al. [43] determined, using LCA, that soilless cultivation reduced the environmental impact by more than double, due to lower levels of fertilizers and pesticides emitted to the environment, compared to soil cultivation. The total GHG emissions from a tomato rockwool culture averaged 853 g (exp. 1) and 999 g CO_2 equivalent (exp. 2), and from a soil-based production averaged 1303 g (exp. 1) and 1509 g CO_2 equivalent (exp. 2), respectively. In addition, 16S ribosomal ribonucleic acid gene abundance in soil samples was 10-fold higher than in rockwool samples [44].

Every year, the majority of freshwater, approx. 87%, is used worldwide for agricultural production [45]. The lack of freshwater resources is an acute issue for arid and semiarid areas in Africa, the Middle East, Southern Europe, and South America that may not only threaten economic development, but also lead to drastic environmental and social problems. One of the major advantages of using SCS is water economization. For instance, lettuce nutrient film technique (NFT) production in South-East Spain requires 62% less water than soil cultivation [46]. In this context, sometimes a comparison between local and imported products is discussed. Stoessel et al. [47] studied a wide range of vegetables, including tomatoes, and concluded that, from a carbon footprint viewpoint, it is often better to import vegetables produced in warm Southern countries during periods when Northern production requires heating. However, surprisingly, sometimes LCA studies, e.g., for tomato production in different Mediterranean countries, have been carried out without considering the impacts of freshwater use [48–51]. Webb et al. [52] also did not address the impacts of freshwater use in their comparison of locally produced tomatoes in the UK and imported tomatoes from Spain [53].

Tomato is the most important vegetable crop in the world [54] and the most cultivated in SCS. When comparing water consumption and water use efficiency (WUE), defined as the obtained yield per unit of irrigation water, vast improvements in WUE are made, with varying degrees, when moving from traditional, soil-based production to protected SCS cultivation methods (Figure 3). For instance, for one kilogram of tomatoes produced in the field, on average about 200 ± 100 L of water are needed. Using drip irrigation, this amount is reduced to about 60 L/kg [55,56]. Moving from soil to SCS can further improve WUE.

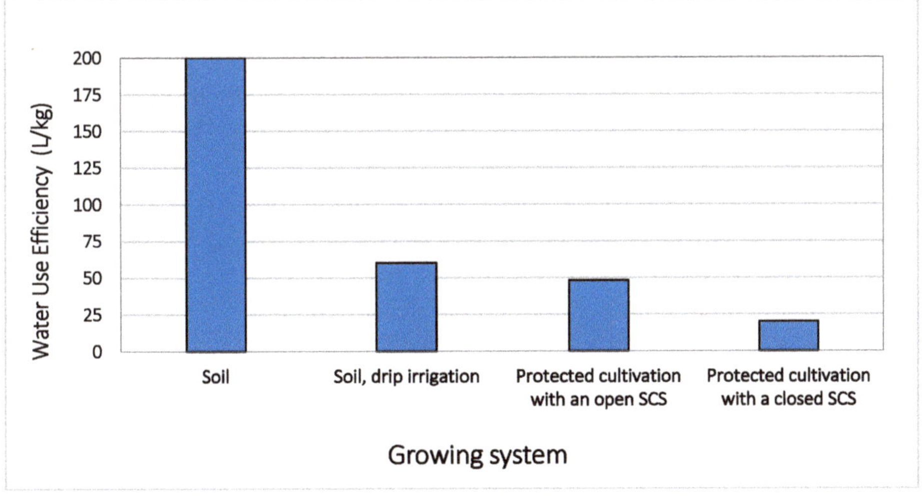

Figure 3. Applying new techniques and new irrigation systems can significantly improve water use efficiency, here calculated as L/kg tomatoes, using different growing systems. Soilless culture system (SCS).

The SCS could be either open-loop or closed-loop cultivation systems. The latter, which involves re-using any drainage solution, can substantially reduce potential pollution of water resources by nitrates and phosphates, while contributing to an appreciable reduction in water and fertilizer consumption [10], even if an ion accumulation (Na^+ and Cl^-) is a challenge in these systems [57]. Comparing data from a commercial tomato farm in Italy and referring to one summer growing season, the savings from a closed irrigation system were 25%, 40%, 24%, and 11% in water, nitrogen (N), phosphorus (P), and potassium (K), respectively [58]. In an open system, where drainage water is not captured and recycled, 10–20% water and fertilizers can be saved, while production and quality can be improved [59]. However, in closed-loop or recirculating water systems that recapture the drainage water for reuse [13], the average use is between 14 and 20 L/kg, i.e., reduced by a factor of up to 5–10 [51,55,60] (Figure 3). By combining a modern irrigation system with modern environmental management, such as the use of closed/semi-closed greenhouses [8] with the regaining and reusing of condensed evaporated water [55], the use of light selective shading and evaporative cooling systems [60,61] make more water savings possible. To come back to our example regarding tomatoes, according to van Kooten et al. [55], it is possible to achieve WUE of 1.5 L water per kg tomato. Under practical conditions, the levels of WUE are rather higher than this. However, these values are possible to achieve and every reduction in water consumption is a step in the right direction. Moreover, under the expected climate change scenarios and water limitation for agriculture, desalinated seawater coupled with hydroponic systems could be a valuable strategy to sustain a high productive agriculture [46].

WUE has a direct economic and environmental effect [8]. Apart from WUE, growing crops with high water requirements in water-scarce areas has important implications. Payen et al. [51] compared the production of tomatoes in Morocco with a production in France. They found that, although the water use efficiency was similar, Moroccan tomato freshwater deprivation was almost four times higher, with 28.0 L H_2O eq kg^{-1} for Moroccan tomatoes and 7.5 L H_2O eq kg^{-1} for French tomatoes. This was explained by the high-water stress index of the cultivation area. Therefore, the authors concluded that, from a water perspective, sourcing vegetables from water-scarce countries is questionable [51].

Because of their light weight and sustainability in terms of resource efficiency, soilless systems are especially suitable for urban areas as well as hobby gardening. These systems allow for an exact dosage and application of nutrients [3,62]. Nowadays, "vertical farming systems" in tower shapes have started to be applied. This system provides 10x more plants per unit area, a 50% reduction in the harvest period, water and fertilizer savings, clean production, and all year-round production [38].

The major disadvantages of soilless cultures are the high investment and energy costs that are required for the initial installation, as well as the increased technical skills that are needed. Other advantages and disadvantages by using certain organic materials as growing media constituents and stand-alone substrates are analyzed below.

2.2. Organic Materials Other than Peat Used as Growing Media Constituents

Different organic materials may play an important role in decreasing the C footprint of the horticultural industry by fully or partly replacing peat-based substrates. Compost, coir, bark, and wood fiber are some organic materials that are already being used in a commercial way as an alternative to peat [23]. In addition, some inorganic materials, such as vermiculite, perlite, clay granules, lava, and pumice are used instead of rockwool or in mixture with peat and other combinations, while new organic materials, such as *Sphagnum* moss, waste and digestates, biochars, and hydrochars are still in their test phase. Below, some of these organic materials and bioresources are briefly described.

2.2.1. Compost, as a Bioresource and Growing Media Constituent

Compost is a general term, describing all organic matter that has undergone thermophilic, aerobic decomposition. It represents a bioresource and a sustainable use case for a potential waste material [9,63]. Several materials are used as growth media after adequate composting. Abad et al. [64] created a database with 105 materials suitable for use as growing media for ornamental potted plant production in Spain. The authors differentiate between urban, sea, agricultural, forest, animal, industrial, and food waste. The disposal needs for waste materials is already an environmental problem, and their recycling in the form of potting media provides a suitable solution. However, some of these materials cannot be used directly. They either contain pathogens, are not stable, or have high water [65] or nutrients content. In some cases, the legal basis needs also be clarified.

Table 1 presents several waste materials used for compost production, which, afterwards, alone or in mixture with other materials, can be used as plant substrates. These include urban and municipal solid wastes, animal manure, grape marc, olive mill, and other food processing wastes; bark, sewage sludge, paper waste, greenhouse waste, pruning waste, spent mushroom compost, and green waste. Different nursery, ornamental, and vegetable plants can grow into these substrates (Table 1). Materials such as bark, wood, several shells or hulls, and coconut coir possess good physical properties after composting. However, being relatively resistant to decomposition, these materials should be subjected to long and well-controlled composting, which may be shortened using N and N-rich organic matter, such as animal manures [66]. According to Raviv [66], high temperatures may cause ashing of these materials, which leads to reduced porosity and increased bulk density. Therefore, temperatures above 65 °C are not desirable.

Table 1. Waste materials used for compost, which, in turn, is used as a plant substrate on its own or in a mixture with other materials.

Feedstock Waste	Use as Growing Medium for Plant Production	Reference(s)
Animal manures	Pot plant production, landscape nurseries, vegetables, and cut flowers production	[67]
Broccoli plants	Lettuce	[68]
Chestnut plants	Lettuce	[68]
Coconut coir dust	Gerbera	[69]
Dredged sediment co-composted with green waste	Ornamental plants	[70]
Corn cobs	*Anthurium*	[71]
Cotton gin	*Azalea*	[72]
Coffee pulp	Tomato seedling	[73]
Farm yard manure	Gerbera	[69]
Grape fruit with coir or vermiculite	Seedlings of lemon basil	[74]
Grapes	Lettuce	[68]
Green waste and sewage sludge	Ornamental bedding plant	[75]
Green/pruning; green/pruning wastes compost, vermicompost, and slumgum compost	Rosemary, Leyland cypress, lettuce, onion, petunia, and pansy	[76]
Olive mill [1], olive [2]	Melon, cress, and tomato plants [1]; lettuces [2]	[77] [1], [68] [2]
Plant leaves	Gerbera	[69]
Posidonia residues	Tomato [1], lettuce seedlings [2], melon, and tomato seedlings [3], pot basil [4], pot sea fennel [5]	[78] [1],[79] [2],[80] [3],[81] [4],[82] [5]
Pruning wastes; pruning waste and municipal solid, or sewage sludge	Ryegrass and cypress [1], Pistacia (nursery) [2]	[83][1], [84][2]
Sewage sludge	Ornamental conifer plants	[84]
Slumgum compost	Rosemary, Leyland cypress, lettuce, onion, petunia, and pansy	[76]
Spent mushroom	Ryegrass and cypress	[85]
Urban solid wastes	Tomato transplant	[86]

Superscripted reference numbers (e.g., [1,2,3,4,5]) link feedstock waste and growing media with the corresponding literature, applicable only within rows, not columns.

Some value-added benefits have to be highlighted here. These are based on specific properties, such as the potential to suppress some diseases and the capacity to control some plant pathogens. Biofertilization and biostimulation could be mentioned as well. However, composts are variable with respect to physical, chemical, and biological properties. Volume weight, air space, water retention, pH, and available plant nutrient elements can vary greatly from batch to batch as well as with the degree of microbiological degradation and primary organic material used. Even within the different green composts there are differences concerning the quality of the compost. For instance, only the use of selected raw material from greenhouse vegetables, nursery shrubs, and green wastes, i.e., plant trimmings, prunings, and crop residues, could contribute to the production of high-quality compost [87]. The selected green compost was found to be a valuable growing medium for peat substitution, while the green compost derived from mixed raw material negatively influenced *Pelargonium* plant nutrition and photosynthesis, thus significantly reducing plant biomass accumulation and quality. Raw material selection increases the production costs of compost. Therefore, according to Massa et al. [87], efforts should involve the adaptation of new technologies for tracking raw materials and supporting sustainable circular chains for compost production at a local level. In addition, strict quality control procedures are essential in preparing composts for use in growing media [22].

Composts produced from so-called green materials, such as prunings, shredded branches, plant debris, and waste from gardens and nurseries, are widely used as components of growing media in the Netherlands, the United Kingdom, Italy, and Germany, primarily in media for the hobby market [22]. However, they can be used as a component of a growing medium up to 50%, but not as stand-alone substrates [88]. The limiting factor regarding the use of composted green waste is its high electrical conductivity (EC) and potassium (K) concentration. There can also be a problem of plant pathogens, human pathogens, and weed contamination if the composting process is not properly conducted, i.e., if the temperature time exposure is not sufficient [14]. Moreover, compost has a low (5–10%) carbon efficiency, which is reflected in material mass and volume reduction and a relatively high pH.

The use of waste as composting material with a further use as growing media and/or growing media constituents is of a dual benefit. For instance, the removal and disposal of large volumes of plant biomasses of *Posidonia*, a marine phanerogam endemic of the Mediterranean Sea, represent, on one hand, a high cost for local administrations [79]. On the other hand, posidonia-based compost, produced from posidonia residues, may have a considerable potential as a peat substitute in horticultural substrates. Several studies evidenced its use for production of tomatoes [78], lettuce transplants [79], melon and tomato seedlings [80], pot basil [81], and pot sea fennel [82].

The same is true for mushroom substrates. Over three million tons of spent mushroom substrates are produced in Europe every year as a by-product of the cultivation of *Agaricus bisporus* [89]. Due to its physical properties and nutrient content, spent mushroom substrate has great potential to be employed as a growing medium in horticulture. However, spent mushroom substrate should be first matured and stabilized through a composting system [89] before being used, e.g., for vegetable production (Figure 4).

(a)   (b)

Figure 4. Spent mushroom substrate used as growing media in simple soilless culture systems (SCS) in Shandong province in China. (**a**) Spent mushroom substrate. Mushroom production is usually placed in the North part of the greenhouse. (**b**) Tomato production in simple SCS in the South part of the greenhouse. Here, the spent mushroom substrate is utilized as a growing medium (Photos: Gruda, private collection).

Compost, when mixed into growing media, is a source of fiber, i.e., a rooting medium, as well as an important source of nitrogen (N), phosphorus (P), and potassium (K). Therefore, the substrate mixtures containing compost required adjusted fertigation due to nutrients supplied by the compost [90]. In addition, the degree of infection with powdery mildew and aphids was strongly positively correlated with the N status of the crop, pointing at the risks of high N supply for the crop [90].

As an alternative to conventional composting, the action of worms and their gut microorganisms can be used to break down organic waste materials to produce vermicompost. Particle-size distribution and fertility were superior in the vermicompost-based media than in the conventional compost-based media. The compost-based media showed an approx. 2.2× higher coarseness index than the vermicompost medium that possessed more fine particles as compost, due to the effect of earthworms [91]. Earthwoms increase the quantity of small particles by ingesting, mixing, grinding, and then egesting organic material [92]. In addition, the nutrient level was higher and the heavy metal concentration was lower in vermicompost [91,93,94]. Moreover, the supplement of additives could counteract some negative aspects of composting processes, such as emissions of GHGs and odorous molecules.

Due to the large range of raw materials used, composting durations and conditions leads to different compost qualities are produced. Concerning the reproducibility, this is a weakness. However, on the other hand, the diversity of final materials may be treated as a force. The use for plant growth and the properties of materials should meet plant biological requirements.

2.2.2. Coir, a Growing Media Constituent and Stand-Alone Substrate

Coir is the material that forms the middle layers or mesocarp of coconut fruits (*Cocos nucifera* L.). Coir pith, coir fibers, and coir chips are some of the most abundant plant-derived organic waste materials in many tropical and subtropical countries, notable as a rapidly renewable resource. The use of coir as and in growing media has vastly increased since 2004, particularly in Europe but also in the western United States [22].

Similar to peat, coir is used in mixtures for the potting industry as it is a lightweight material and has good air and water holding characteristics. Since coir contains more lignin and less cellulose than peat, it is more resistant to microbial breakdown and usually shrinks less; it is also more hydrophilic and easier to re-wet after drying than peat moss and tends to retain its basic structure when wet or dry [18,95,96].

Leaching of nitrogen is marginally higher and the total water-holding capacity is lower than in peat when comparing materials of a similar particle size, and sometimes natural higher total soluble salts, sodium, and chloride levels are found in coir, depending on their origin [96–98].

However, coir pith has the highest impact on "ecosystem quality", which is often due to land occupation during the coconut harvesting stage [20]. Therefore, efforts have been undertaken to investigate and develop growing media from locally sourced materials, such as, for instance, bark or other wood-based materials, co-products from a forest harvest, or wood processing industries [99–102].

2.2.3. Bark and Wood-Based Materials as Bioresources, Growing Media Constituents, and Stand-Alone Substrates

Bark is a major component of growing media, particularly in areas where peat is scarce or expensive [22], due to transportation cost. It is a lightweight material with a bulk density of 0.1–0.3 g cm^{-3} [63]. Similar to coir, bark can be produced in different particle sizes, which makes adjusting the air and water-holding capacities possible by varying the percentage of fine particles [103].

As with coir, pine bark is not produced specifically for use in growing media and tends to have variable physical, chemical, and biological properties [24]. Bark is usually used as a composted or aged material, in order to avoid potential problems with phytotoxicity, since the presence of phenolic compounds, terpenes, and tannins are typical in the chemical composition [30]. High manganese content, especially at low pH could also be a source of potential phytotoxicity [104]. In addition, N deficiency is a common issue, depending on the origin of the material used and the processing method. Recent studies showed that hydrothermal treatments were effective regarding phytotoxicity removal from industrial bark. After this treatment, bark maintains a very high air content that can be a plus in aeration improvement when added to commercial peat-based substrates [31].

Wood fiber, wood chips, and sawdust are renewable resources from the woodworking industry. All these products are characterized by low water retention and good air content. Depending of the initial material, they could sometimes contain phytotoxins that may affect the plant growth at the beginning of cultivation. In this case, a pretreatment with substrate washing would be recommended [105]. Particle-size distribution determines further physical properties, e.g., water retention and water-holding capacity [99,100,106]. A very good correlation was detected between the high percentage of particles <1 mm and max. water holding capacity, and therefore plant growth [101,107].

Wood fibers are further used to optimize the physical properties of other material components, e.g., reducing bulk density, increasing air space, improving re-wetting capacity [24,107,108] and/or as an organic mulch to reduce soil temperature fluctuations, and soil water evaporation and suppress weeds [109,110].

2.2.4. Biochar and Hydrothermal Carbonization Products as Bioresources and Growing Media Constituents

Different investigations have been carried out to search for methods that transform agricultural, industrial, and municipal wastes into materials that can be used in growing media. The benefit of diverting wastes from landfills and providing large quantities of organic growing media in the future is particularly important for arid and semiarid regions of the globe [22,23].

Biochar and hydrothermal carbonization (HTC) might play a more important role as constituents of growing media. Whereas biochar is manufactured by heating organic matter in an anoxic situation (pyrolysis), the HTC process requires only moderate temperatures [31] and pressures and is usually used for materials with high water content, e.g., plants. Both processes, pyrolysis and HTC, show great potential for the production of sustainable CO_2-neutral energy from biomass, because plants capture the sun energy and convert carbon dioxide from the atmosphere into carbohydrates via photosynthesis [23].

Biochar and HTC char have physical and chemical properties that are variable, depending on the raw material used and the carbonization technique. Usually, the electrical conductivity (EC) and pH values are similarly low in peat and HTC and are slightly increased in biochar [25].

Biochars contain various amounts of different micronutrients in addition to P and K. These nutrients are usually slowly available to plants much like slow release fertilizers, rather than being immediately available [65]. However, there are some problems that need to addressed. For instance, biochar usually contains about 1% nitrogen (N). A high N-immobilization occurs in hydrochar as well. This, and the presence of some phytotoxic substances, were the factors that lead to reduced growth of potted basil, even in mixtures of only 30% by volume [111]. After composting, N-immobilization was reduced and phytotoxic substances degraded within a few weeks [111]. However, as mentioned before, low carbon efficiency, high volume reduction, and time needed for composting make this process not particularly economically attractive. Therefore, apart from feedstock choice, carbonization processes seem to be important for future research.

2.2.5. Other Organic Materials as Bioresources and Growing Media Constituents

Apart from materials analyzed above, several more novel materials and bioresources are used at a small scale and/or have the potential to be used as growing media constituents. These include untransformed waste stream materials, which are affordable and available in certain areas. Waste materials can include, e.g., rice hulls [112–114], almond shell waste [115–117], hazelnut husks [118–120], and paper waste [121]. The main disadvantage of using these materials in commercial soilless media is that they are not produced specifically for horticultural applications; they can therefore be highly inconsistent. As such, they are almost always used in conjunction with more traditional materials [24].

Furthermore, peat moss (*Sphagnum*) from paludiculture has recently been used as a sustainable high-quality alternative to fossil white peat, i.e., as a raw material for plant substrates. *Sphagnum* farming refers to the cultivation of *Sphagnum* mosses to produce *Sphagnum* biomass sustainably [122].

Moreover, *Sphagnum* farming is a feasible large-scale, climate-friendly, and sustainable land use option for abandoned cutover bogs and degraded bog grassland [123]. It reduces human pressure on the remaining natural peatlands in surroundings areas [122].

In areas where forestry activity is minimal, but arable farmland is abundant, the development of soilless growing media from crops normally used as biofuels has been investigated [24]. *Miscanthus* is one such fast-growing crop. *Miscanthus* is a renewable raw material and a low-input crop that can be locally produced, providing ecosystem services, such as CO_2 mitigation and biodiversity [124]. Moreover, switchgrass (*Panicum virgatum* L.) [125,126], giant reed (*Arundo donax* L.) [127], reed canary grass (*Phalaris arundinacea* L.) straw [128], and willow (*Salix* spp.) [126] have been used in plant production alone or in mixtures with other materials.

2.3. Growing Medium Choice

The question as to which is the best growing medium does not have a single answer. This will depend on the location, the availability and cost of potential growing medium constituents, and the crop production system envisaged.

The materials for growing media have to fulfil different requirements: First, they should be available consistently from batch to batch and economically feasible, i.e., the materials and the production process should not be very expensive. Second, the physical, chemical, and biological properties of the growing medium should meet the biological plant requirement. However, there is no universal substrate or mixture that is valid for all plant species and in all situations of cultivation [11,14,23]. Gruda et al. [14], Barrett et al. [24], Savvas, and Gruda [16] also speak for the performance of growing media. Here, they included not only substrate properties, but also the ability to perform well in real growing conditions.

Third, the material used for production and growing media itself should be sustainable and environmentally friendly. Carbon footprint analyses show that the largest share of emissions from heated greenhouse farms results from energy consumption, followed by substrate, packing, and containers used [129]. The biodiversity concern and climate change emphasize the significance of peat bogs as carbon sinks. Generally, avoiding or reducing the use of peat as a growing media constituent, can substantially reduce the carbon footprint in horticulture [23,130]. Apart from extraction, processing, manufacturing, and transportation are important business factors to distinguish between materials from specific sources [131]. Therefore, the authors suggested a list of eight criteria that reflect current, and potentially future, social and environmental issues in relation to the use of growing media. These include the energy and water used in previously mentioned business factors, the social compliance, ensuring minimum labor standards, continuity of supply, habitat and biodiversity, pollution, renewability, and resource use efficiency. In order to guarantee a continued growth and sustainable development of soilless cultivation, it is important to identify effective and environmentally sustainable materials for growing media [24].

Selecting growing media is not an easy task because environmental issues and technical and financial implications must be considered [14,20]. The geographical location, the selection of plant cultivation and production types, the substrate cost and performance, as well as other societal concerns, govern which growing media has to be selected. In addition, the evidence indicates that growers and gardeners tend to favor the types of growing media they are accustomed to and know how to manage. Hence, inertia is also a barrier to change [132]. In the following, we identified two perspectives and functions that we found important to consider: Production systems and transportation distances.

2.3.1. Production Systems

2.3.1.1. Nursery Production

Peat-based growing media are mainly used for production of seedlings and transplants for vegetables and ornamental plants. Nowadays, efforts in the substrate industry are made toward peat reduction in the entirety of the components, used for growing media. Even 10% wood fiber mixed in pure black peat would significantly reduce the carbon footprint for lamb's lettuce, grown in 4 cm press pots [133]. Higher percentages of wood fiber can result in additional emission reductions. For instance, Gruda and Schnitzler [107] reported that, from a performance point of view, the optimal percentage of wood fiber for the prevention of considerable degradation of press pots was approximately 30% in volume. Similarly, biochars can be favorably used as an amendment to peat-based substrates for the development of sustainable greenhouse production [134]. The authors evaluated the effects of additional biochars at a rate of 15% (v/v) to a peat-based substrate and found that the biochar addition increased the C, decreased the N availability in fertigated peat-based growing media, and mitigated CO_2, CH_4, and N_2O emissions. To increase microbial activity, compost at a rate of 4% (v/v) was added. This reaction is similar to results reported for agricultural soils by an additional biochar application.

On the other hand, using the large definition of a plant nursery that includes the production of plants for gardens, agriculture, forestry and conservation biology, bark, and wood fiber substrates are the standards in nursery production. This sustainable way of production will remain steady in the near future.

2.3.1.2. Greenhouse Vegetable Production

Growing media have been used traditionally, mostly for plant propagation, bedding, and pot plant production, but this range of use has expanded to include the total production of many food crops, especially high-value crops grown under protection in greenhouses [14]. For instance, stand-alone substrates, such as rockwool and perlite are used for the commercial soilless production of vegetables [15,16].

The use of polythene-wrapped rockwool, originally produced as insulation in the construction industry, aided by its lightweight and ease of handling, has become the dominant soilless culture system for greenhouse vegetables worldwide and especially in Europe [10]. The advantages of rockwool are substrate uniformity, ease of handling, and ease plant production steering.

Materials which can be pressed in slabs, such as coir, can be successfully used instead of rockwool. The water-buffering capacity is lower in coir dust than in rockwool and peat, and the level of air space varies considerably depend on the origin of the material [97]. Hence, mixing different particle sizes and ratios together or adding other materials is recommended to meet crop-specific moisture and aeration requirements in order to use coir products as stand-alone substrates. For instance, adding perlite to coir improved the physical and hydraulic characteristics of the media, such as total porosity and wettability, by manipulating the porosity and capillarity [135]. However, while coir products can make excellent growing media, the long transportation distance makes this alternative less attractive for many areas, such as Northern Europe and North America (see Section 2.3.2. for more information).

White spruce and fir bark alone or mixed with low-grade peat showed high potential for greenhouse tomato production and represented an environmentally sound alternative to rockwool [136]. Moreover, pine bark can be successfully used as a stand-alone substrate for the cultivation of vegetables, such as bell pepper, cucumbers, and muskmelons [137–139]. An economic analysis determined that pine bark was nearly one-eighth the cost of perlite and could be reused for several consecutive crops, resulting in reduced production costs and greater profits. However, bark could become a limited resource due to the changing timber industry and the fact that it is an effective energy source [140], increasingly used as fuel.

Wood chips and fibers are also gaining traction as an alternative to rockwool for slab culture [141]. Depardieu et al. [142], stated that sawdust- and bark-based materials can be used for strawberry soilless culture production, as long as an initial basic fertilization is applied to avoid the initial tie up. Additional N fertilization from the beginning of plant cultivation is recommended to overcome N immobilization in wood fiber substrates [143].

Recently, Kraska et al. [124] found that cucumbers and tomatoes grown on different stand-alone Miscanthus substrates, such as shreds, chips, and fibers, obtained comparable cumulative yields to rockwool. Generally, by using rockwool alternative substrates, the plant cultivation technology has to be adapted to the growing medium's properties [7].

2.3.1.3. Greenhouse Ornamental Production

The standard substrate component used for the production of greenhouse ornamentals is peat moss. Several stand-alone substrates, such as perlite and volcanic lava are used to produce cut ornamentals. If SCS, such as ebb-and-flow bunches or floors, are applied, pot ornamentals could also be cultivated in alternative peat substrates. Other materials, such as bark, wood fibers etc., can be used up to 100% to produce plants. Since nutrient solution is used to supply the plants, the substrate function is vital to keep and support the plants.

However, depending on the crops and technologies used, the portion of usage of growing media constituents other than peat in pot ornamentals varies between 20–50%. Apart from porosity that is much higher in growing media, an important difference between soil and substrate culture is the limited volume of plant roots in a container. This provides a reduced root system for a comparable and sometimes much higher developed aerial part. According to Savvas and Gruda [16], the particle size of the growing media used and the container geometry have to be properly selected to balance water availability and aeration in the root zone. In addition, an adaptation in cultivation methods, mainly in irrigation systems, is required. Furthermore, investing in SCS demands excellent water quality, drainage water collection systems, and an increase in laborers' skills. A soilless crop is much more sensitive to mistakes as there is hardly any buffer [59].

Bark is used as stand-alone substrate in the production of orchids and as a growing media constituent in pot ornamentals, whereas wood fiber substrates are becoming more and more popular in ornamental plant production. Wood chips and sawdust are usually used in the proportion of 20–30% (volume basis) in mixtures with other substrate components. A reduction in particle size, an increase in volume weight, and an increase in the irrigation frequency is recommended [99,100,106]. Furthermore, clay is added, to increase the water holding capacity and nutrient buffer ability of potting mixes.

Álvarez et al. [144] showed that it is possible to grow container plants of geranium (*Pelargonium peltatum* (L.) L'Hér. ex Aiton) and petunia (*Petunia x hybrida* hort. ex E. Vilm.) using a peat-based substrate mixed with biochar and/or vermicompost. Plants in these substrates showed a similar or enhanced physiological response to those grown under control using a commercial peat-based substrate. When compost is used, perlite may be utilized as a growing medium constituent to increase the drainage and air content of the growing media mix.

Several studies reported that biochar in potting media results in the same ornamental plant growth as in peat-based standard substrates [65,145,146]. According to Kern et al. [25], char materials must not necessarily remain on the level of a minor ingredient, but have the potential to be used as major constituents. Furthermore, since they are characterized by a high porosity and a high water-holding capacity, these materials may also be usable as a substitute for constituents, which are already established in the growing media market, but which have a limited supply [25,147,148]. For instance, rice hull-derived biochar would be a practically applicable amendment to improve the properties of growing media, in terms of an increased cation exchange capacity and water content [149]. The typically high porosity and surface area of biochars promote the retention of water and the sorption of nutrients [25].

Non-decomposed *Sphagnum* has been used with great success in the cultivation of orchids as well as together with peat substrates for the cultivation of *Tagetes patula* L. [150]. These results were confirmed by investigations with *Pelargonium* and *Petunia* [151]. Adding *Sphagnum* fibers to peat increased water retention and hydraulic conductivity, but either reduced or had no impact on air-filled porosity. Moreover, the quality of brown peat can be improved by adding a minimum of 30% *Sphagnum* fibers to sieved peat. Therefore, Jobin et al. [151] stated that *Sphagnum* biomass production will most likely continue to develop, offering the growing mix industry an alternative material with a low carbon footprint and a better use of peatlands.

However, the chosen substrate has to be stable enough and possess a good bulk density within the entire cultivation period and after the sale to the end-consumer. For bed, balcony, bowl, and hanging basket plants, the irrigation management of the end-consumer is a challenge. Since the end-consumers are usually inexperienced, mistakes occur. Any incorrectness is frustrating and associated with product rejection. End-consumers think that they do not possess the "green fingers" and this in turn creates a great loss for horticulture, not only from the profit side.

2.3.2. Transportation Distances

The second perspective is a function of growing media use from distances from sources of primary raw materials to growers. Due to transportation ways, the cost of a growing medium is also a function of location. For instance, in peat-rich regions, such as Northern Europe and Canada, where the transportation distances are relatively short, peat may still be an economical option. Similar to peat, coconut coir is produced in specific locations (mainly South-East Asia) and, if not used locally, has to be transported to growers in other parts of the world, with unavoidable costs [9,23]. This is the reason why regional substrates, such as volcanic lava and pumice are and will certainly remain important in the South of Europe in the future. However, location is not only important from an economical point of view, but also from a sustainability perspective, due to the high CO_2 footprint. Therefore, compost, together with biochar and hydrochar, has good chances, since usually they are locally produced. Materials, whether sourced from industrial, agricultural, or municipal waste are being investigated as soilless substrate components [24]. A particular trend has been the use of renewable raw materials locally sourced, natural in occurrence and fast-growing, in particular in industrialized countries [16,30].

2.4. Disposal Concerns and Waste Management

The disposal issue is one of the biggest concerns of using soilless culture and growing media. The question is, what can be done with several fertilizer leachates and water waste during the cultivation period as well as the growing media after its end-of-life?

The generally accepted waste management hierarchy includes the three Rs: Reduce, reuse, and recycle [152]. Reducing the amount of growing medium per plant contributes to reducing CO_2 emissions in the production chain of plants [7].

In the seedling and transplant industry there has recently been a trend among producers towards more cells per tray, which decreases the need for growing medium and increases the number of seedlings or transplants produced per unit area [153]. However, the reduction of growing media amount is not always a viable option, due to a direct influence on yield and product quality parameters [9,13]. For instance, Gruda and Schnitzler [153] reported that a reduction of the pot size decreased the quality of the lettuce seedlings. However, no differences were found in the lettuce yield after transplanting to the field and this is of much importance. Certainly, culture methods, such as irrigation and a good root development of seedlings in wood fiber substrates, have been responsible for these results [153].

On the other hand, using SCS means using a reliabe and precise dosage of both fertilizer and water, and this is one of the advantages of using closed systems, at least theoretically. However, in practice, soilless culture vegetables are usually over-fertilized, and an excessive synthetic N fertilizer is applied to ensure that no nutrient deficiency occurs. Indeed, as Truffault et al. [154] reported, over-fertilized tomatoes provided an accumulation of N in leaves and stems. However, yield, leaf photosynthetic activity, and plant architecture were not significantly improved. In addition, the quality of tomato fruits decreased in terms of their sugar:acid ratio and dramatically decreased in the pericarp, whereas the locular gel composition remained similar [154]. Therefore, the reduction of fertilizer used, first and foremost the N fertilizer, is the first appropriate and sustainable step that should be undertaken. The impacts are not only related to the use of fertilizers itself but also to the amount of energy, materials, and transport processes involved in the production of fertilizers [155] and manufacturing facilities. As Gruda et al. [7,8] reported, the fertilizer reduction is directly linked with a reduction of N-emissions (N_2O, NH_3, and NO_x) that, in turn, have an enormous effect on GHGs.

One way to address the runoff nutrient wastewater pollution in open-loop hydroponic systems is the reuse of drained nutrient solutions to a second greenhouse crop. This system is called the "cascade cropping system" [156,157]. Muñoz et al. [157] reported that the N leachate from a soilless tomato system decreased by more than 60% when the nutrient solution was used in a tomato soil system. Moreover, intense and year-round crop production, high N-fertilizer application, suitable temperatures, and frequent irrigation make the greenhouse system an ideal environment for high N-emissions that are considered to be extremely damaging to the ozone layer [7]. The adoption of the cascade crop system reduced the environmental impact by 21%, but increased the eutrophication category by 10% because of the yield reduction [157]. Similarly, cherry tomatoes may be grown with an exhausted nutrient solution that is flushed out from a culture of a salt-sensitive tomato cultivar in semi-closed soilless systems [156]. Several other studies stated that nutrient solution discharged from hydroponic culturing systems can be reused for the production of several vegetables in indoor or outdoor conditions, such as Chinese cabbage [158], melon, and cucumber [159]. These results are in agreement with the growth promotion of poinsettias (*Euphorbia pulcherrima* Willd. ex Klotzsch) after reusing the waste nutrient solution from rose hydroponic cultures [160].

Growing media can be reused as well. Reuse is the best approach in terms of its environmental impact and the results of LCA [9]. For instance, multiple cucumber cycles can be produced on the same growing media in soilless or substrate culture systems, whereas a reuse of substrates in containers systems is generally not common. However, reusing could be associated with distributions of pathogen infections and the possible deterioration of substrate properties. Therefore, in accordance with the Directive EU2018/851 of the European Parliament and of The Council, "waste management in the European Union should be improved and transformed into sustainable material management, with a view to protecting, preserving, and improving the quality of the environment, protecting human health, ensuring prudent, efficient, and rational utilization of natural resources, promoting the principles of the circular economy … " [161]. The directive further regulates how to reuse and prepare for reuse and recycling, in line with the waste hierarchy. With regards to growing media, the reuse of substrates may induce a higher compaction with increased volume weight (bulk density) and reduction of porosity, due to shrinkage [9,162], with a limited air and low water buffer capacity [101] accompanied by failures and a bottleneck situation of nutrients [163]. On the other hand, the gradual accumulation of nutrients in organic substrates during growing season may have adverse effects on plant development [148], and these effects are further increased by a substrate reuse. Xing et al. [164] identified a total of 358 differentially abundant proteins, including 11 mineral ion binding and transport related proteins, such as a calmodulin-like protein and a nitrate transporter 3.2 under peat-vermiculite and coir tomato cultivation. Xing et al. [164] suggested that these indicators could contribute to a better control of SCS and a waste reduction.

The investigations of crop response to the cultivation in reused growing media compared to virgin substrates show contradictory results: (a) Reduction of crop yield and/or produce quality in reused media, (b) minimal differences between virgin and reused substrates, or (c) even better results in reused materials [165]. Similar to virgin growing media, the reused materials have to possess good physical, chemical, and biological properties. Therefore, generally, some remediation steps are recommended to amend the substrate properties before reusing [9].

First, growing media should be free from any infection with pests and diseases, otherwise a disinfection process has to be undertaken. For instance, cleaning and disinfecting perlite with hot water at a temperature of 96 °C before reuse produced a better marketable tomato yield in comparison to a virgin one, due to the collective effect of salt reduction, medium disinfection, and the optimum level of nutrients [166]. Second, the nutrient level of growing media should be analyzed and eventually adjusted according to crop demands. This step is very important when a nutrient solution is not used in the second crop. Third, physical properties have to be amended by breaking up and sifting growing media as well as by removing older roots [165].

Further, organic substrates with high microbial activities, such as compost, are often added to used peat substrates, because of their suppressive properties against soilborne diseases, such as *Pythium*. In addition, an artificial inoculation with selected microorganisms or the introduction of microbial antagonists, preliminarily isolated from suppressive soils and/or used soilless media, could be used to increase the suppressive properties against root rot diseases [165,167]

Recycling is the final approach in the waste management hierarchy. To recycle something means that it will be transformed again into raw material, which can be shaped into a new item [152] for second or multiple life uses. Until recently, growing media were always the last step of the value chain, and usually it was all about how to dispose of them without further negative impact on the environment and climate. Composting offers a good option to drastically reduce this impact, as shown in Section 2.2.1. Organic substrates can be used immediately or after their composting as soil amendments. This method is highly evaluated in arid and semi-arid areas, increasing not only organic matter in soil but also improving water holding capacity. In addition, composted materials can be used to cultivate less-demanding crops, such as forest tree saplings [9]. Moreover, Kraska et al. [124] opted for a cascade way of recycling and found a subsequent use of *Miscanthus*-based growing medium for combustion feasible, after the production of cucumbers and tomatoes on different stand-alone *Miscanthus* substrates. As mentioned before, *Miscanthus* is a renewable raw material and a low-input crop that can be locally produced.

2.5. Other Factors Having an Impact on Sustainability

In temperate regions, controlled environment systems are characterized by large amounts of energy consumption for heating during the cold season. Large energy consumption is the greatest environmental concern [7,8]. As Eigenbrod and Gruda [3] stated, the motto for future plant production should not be "local at any price," but "as sustainable as possible." Therefore, Gruda et al. [7,8] recommend the implementation of so-called next generation culture methods: Better insulation thanks to double cladding and triple screens, following biological and nature-oriented culture techniques, dehumidifying the blown-in air, and, if necessary, humidifying (rewetting) and "harvesting" greenhouse existing heat amounts. In addition, the use of alternative energy sources can fundamentally increase and improve the sustainability of protected cultivation systems and nursery production. Replacing or recycling rockwool and plastic items are other important factors [7,8].

Plastic containers, pots, bags, and trays have been the predominant containers in greenhouse and nursery production. However, most plastics are derived from petroleum—a nonrenewable resource [168]. Therefore, different examples of alternative containers made from plantable and compostable materials, such as bamboo, coconut or wood pulp fiber, rice hulls, and recycled paper have been developed. The use of these containers will furthermore contribute to sustainable systems along with suitable growing media.

Moreover, the lifetime of structure materials, e.g., plastic covers and auxiliary equipment, e.g., drippers, should be further extended and manufactured out of biodegradable material to reduce waste. Better management of the nutrient supply as well as the reduction of fertilizer use is required [7].

Another way to reduce the amount of peat (not only for SCS), used as soil improvements for acidophilic plants, is the breeding of new varieties that have neutral requirements related to pH in the root zone. In addition, the use of plant biostimulants, such as humic substances, protein hydrolysates, seaweed extracts, and beneficial microorganisms, such as mycorrhizal fungi and nitrogen fixation bacteria [37,167,169], can contribute to improve effectiveness and interaction in the root zone of plants into growing media.

3. Conclusions

In conclusion, soilless culture is one of the best techniques to overcome local water shortages, while also producing high quality produce, even in areas with poor soil structure and problematic conditions. Reduce, reuse, and recycle issues should be more frequently applied in SCS. The application of these systems is likely to increase close to existing cities as well as in mega-cities worldwide in the near future.

In this paper, we reviewed different organic materials and bioresources used or intended to be used as growing media constituents in the future. All of these have their respective advantages and disadvantages. Different areas in the world, with different conditions and requirements, require different crops, different distances to sources of primary raw materials used as growing media components, and different technologies used to produce plants.

However, factors such as climate change, CO_2 emissions, and other ecological issues will determine and drive the adoption and influence of growing media in the near future. Materials that are easily available, financially feasible, environmentally friendly, and that can provide a high-quality growing medium will become replacements for rockwool and peat in the future.

Further research on the innovative approaches in SCS and materials used as growing media components is required.

Acknowledgments: Special thanks to Michael Maher, Dublin, Ireland for language revision and valuable recommendations.

Conflicts of Interest: I declare no conflict of interest.

References

1. UN. Available online: https://esa.un.org/unpd/wpp (accessed on 27 July 2018).
2. World Bank Group. International Development, Poverty, & Sustainability. Available online: www.worldbank.org (accessed on 7 June 2018).
3. Eigenbrod, C.; Gruda, N. Urban vegetable for food security in cities. A review. *Agron. Sustain. Dev.* **2015**, *35*, 483–498. [CrossRef]
4. Abukari, M.K.; Tok, M.E. Protected cultivation as adaptive response in climate change policy: The case of smallholders in northern Ghana. *J. Emerg. Trends Econ. Manag. Sci.* **2016**, *7*, 307–321.
5. Olesen, J.E.; Bindi, M. Consequences of climate change for European agricultural productivity, land use and policy. *Eur. J. Agron.* **2002**, *16*, 239–262. [CrossRef]
6. Bisbis, M.B.; Gruda, N.; Blanke, M. Potential impacts of climate change on vegetable production and product quality—A review. *J. Clean. Prod.* **2018**, *170*, 1602–1620. [CrossRef]
7. Gruda, N.; Bisbis, M.B.; Tanny, J. Impacts of protected vegetable cultivation on climate change and adaptation strategies for cleaner production—A review. *J. Clean. Prod.* **2019**, *225*, 324–339. [CrossRef]
8. Gruda, N.; Bisbis, M.B.; Tanny, J. Influence of climate change on protected cultivation: Impacts and sustainable adaptation strategies—A review. *J. Clean. Prod.* **2019**, *225*, 481–495. [CrossRef]
9. Raviv, M. Can compost improve sustainability of plant production in growing media? *Acta Hortic.* **2017**, *1168*, 119–133. [CrossRef]
10. Gruda, N.; Gianquinto, G.; Tüzel, Y.; Savvas, D. Soilless Culture. In *Encyclopedia of Soil Sciences*, 3rd ed.; Lal, R., Ed.; CRC Press Taylor & Francis Group: Boca Raton, FL, USA, 2016; pp. 533–537.
11. Di Lorenzo, R.; Pisciotta, A.; Santamaria, P.; Scariot, V. From soil to soil-less in horticulture: Quality and typicity. *Ital. J. Agron.* **2013**, *8*, 30. [CrossRef]
12. Peet, M.M.; Welles, G.W.H. Greenhouse tomato production. In *Tomatoes—Crop Production Science in Horticulture*; Heuvelink, E., Ed.; CABI Publishing: Wallingford, UK; Cambridge, MA, USA, 2005; Volume 13, pp. 257–304.
13. Gruda, N. Do soilless culture systems have an influence on product quality of vegetables? *J. Appl. Bot. Food Qual.* **2009**, *82*, 141–147.
14. Gruda, N.; Caron, J.; Prasad, M.; Maher, M.J. Growing media. In *Encyclopedia of Soil Sciences*, 3rd ed.; Lal, R., Ed.; CRC Press Taylor & Francis Group: Boca Raton, FL, USA, 2016; pp. 1053–1058.
15. Savvas, D.; Gianquinto, G.; Tüzel, Y.; Gruda, N. Soilless culture. In *Good Agricultural Practices for Greenhouse Vegetable Crops–Principles for Mediterranean Climate Areas*; Plant Production and Protection Paper 217; Baudoin, W., Ed.; FAO: Rome, Italy, 2013; pp. 303–354.
16. Savvas, D.; Gruda, N. Application of soilless culture technologies in the modern greenhouse industry—A review. *Eur. J. Hortic. Sci.* **2018**, *83*, 280–293. [CrossRef]

17. Blok, C.; Urrestarazu, M. Substrate Growing Developments in Europe 2010–2027. Available online: www.horticom.com (accessed on 12 June 2009).
18. Nichols, M. Coir: Sustainable Growing Media. 2013. Available online: http://cocopeatcompany.com/report/1475602985.pdf (accessed on 7 June 2018).
19. Bussell, W.T.; Mckennie, S. Rockwool in horticulture, and its importance and sustainable use in New Zealand. *N. Z. J. Crop Hortic. Sci.* **2004**, *32*, 29–37. [CrossRef]
20. QUANTIS. Comparative Life Cycle Assessment of Horticultural Growing Media Based on Peat and Other Growing Media Constituents 2012. Available online: http://epagma.eu/evidence-based (accessed on 23 November 2018).
21. Apodaca, L.E. Peat (advance release). In *Minerals Yearbook 2016*; U.S. Department of the Interior: Washington, DC, USA; U.S. Geological Survey: Reston, VA, USA, 2018.
22. Carlile, W.R.; Cattivello, C.; Zaccheo, P. Organic growing media: Constituents and properties. *Vadose Zone J.* **2015**, *14*. [CrossRef]
23. Gruda, N. Current and future perspective of growing media in Europe. *Acta Hortic.* **2012**, *960*, 37–43. [CrossRef]
24. Barrett, G.E.; Alexander, P.D.; Robinson, J.S.; Bragg, N.C. Achieving environmentally sustainable growing media for soilless plant cultivation systems–A review. *Sci. Hortic.* **2016**, *212*, 220–234. [CrossRef]
25. Kern, J.; Tammeorg, P.; Shanskiy, M.; Sakrabani, R.; Knicker, H.; Kammann, C.; Tuhkanen, E.-M.; Smidt, G.; Prasad, M.; Tiilikkala, K.; et al. Synergistic use of peat and charred material in growing media–an option to reduce the pressure on peatlands? *J. Environ. Eng. Landsc. Manag.* **2017**, *25*, 160–174. [CrossRef]
26. Yu, Z.C.; Loisel, J.; Brosseau, D.P.; Beilman, D.W.; Hunt, S.J. Global peatland dynamics since the last glacial maximum. *Geophys. Res. Lett.* **2010**, *37*, L13402. [CrossRef]
27. Weissert, L.F.; Disney, M. Carbon storage in peatlands: A case study on the Isle of Man. *Geoderma* **2013**, *204*, 111–119. [CrossRef]
28. UNEP. *Frontiers 2018/19 Emerging Issues of Environmental Concern*; United Nations Environment Programme: Nairobi, Kenya, 2019.
29. Bonn, A.; Reed, M.S.; Evans, C.D.; Joosten, H.; Bain, C.; Farmer, J.; Emmer, I.; Couwenberg, J.; Moxey, A.; Artz, R. Investing in nature: Developing ecosystem service markets for peatland restoration. *Ecosyst. Serv.* **2014**, *9*, 54–65. [CrossRef]
30. Gruda, N. Sustainable peat alternative growing media. *Acta Hortic.* **2012**, *927*, 973–980. [CrossRef]
31. Chemetova, C.; Fabião, A.; Gominho, J.; Ribeiroa, H. Range analysis of *Eucalyptus globulus* bark low-temperature hydrothermal treatment to produce a new component for growing media industry. *Waste Manag.* **2018**, *79*, 1–7. [CrossRef]
32. Page, G.; Ridoutt, B.; Bellotti, B. Carbon and water footprint tradeoffs in fresh tomato production. *J. Clean. Prod.* **2012**, *32*, 219–226. [CrossRef]
33. Bojacá, C.R.; Wyckhuys, K.A.G.; Schrevens, E. Life cycle assessment of Colombian greenhouse tomato production based on farmer-level survey data. *J. Clean. Prod.* **2014**, *69*, 26–33. [CrossRef]
34. Dias, G.M.; Ayer, N.W.; Khosla, S.; Van Acker, R.; Young, S.B.; Whitney, S.; Hendricks, P. Life cycle perspectives on the sustainability of Ontario greenhouse tomato production: Benchmarking and improvement opportunities. *J. Clean. Prod.* **2017**, *140*, 831–839. [CrossRef]
35. Antón, A.; Torrellas, M.; Montero, J.I.; Ruijs, M.; Vermeulen, P.; Stanghellini, C. Environmental impact assessment of Dutch tomato crop production in a Venlo glasshouse. *Acta Hortic.* **2012**, *927*, 781–791. [CrossRef]
36. Theurl, M.C.; Hörtenhuber, S.J.; Lindenthal, T.; Palme, W. Unheated soil-grown winter vegetables in Austria: Greenhouse gas emissions and socio-economic factors of diffusion potential. *J. Clean. Prod.* **2017**, *151*, 134–144. [CrossRef]
37. Gruda, N.; Savvas, D.; Colla, G.; Rouphael, Y. Impacts of genetic material and current technologies on product quality of selected greenhouse vegetables–A review. *Eur. J. Hortic. Sci.* **2018**, *83*, 319–328. [CrossRef]
38. Fernández, J.A.; Orsini, F.; Baeza, E.; Oztekin, G.B.; Muñoz, P.; Contreras, J.; Montero, J.I. Current trends in protected cultivation in Mediterranean climates. *Eur. J. Hortic. Sci.* **2018**, *83*, 294–305. [CrossRef]
39. Rouphael, Y.; Kyriacou, M.C. Enhancing quality of fresh vegetables through salinity eustress and biofortification applications facilitated by soilless cultivation. *Front. Plant Sci.* **2018**, *9*, 1254. [CrossRef]

40. Gruda, N. Assessing the impact of environmental factors on the quality of greenhouse produce. In *Achieving Sustainable Greenhouse Cultivation*; Marcelis, L., Heuvelink, E., Eds.; Burleigh Dodds Science Publishing Limited: Cambridge, UK, 2019; ISBN 978-1-78676-280-1.

41. Sabatino, L.; D'Anna, F.; D'Anna, F.; Iapichino, G.; Moncada, A.; D'Anna, E.; De Pasquale, C. Interactive effects of genotype and molybdenum supply on yield and overall fruit quality of tomato. *Front. Plant Sci.* **2019**, *9*, 1922. [CrossRef]

42. Sabatino, L.; Ntatsi, G.; Iapichino, G.; D'Anna, F.; De Pasquale, C. Effect of selenium enrichment and type of application on yield, functional quality and mineral composition of curly endive grown in a hydroponic System. *Agronomy* **2019**, *9*, 207. [CrossRef]

43. Mugnozza, G.S.; Russo, G.; De Lucia Zeller, B. LCA methodology application in flower protected cultivation. *Acta Hortic.* **2007**, *761*, 625–632. [CrossRef]

44. Hashida, S.N.; Johkan, M.; Kitazaki, K.; Shoji, K.; Goto, F.; Yoshihara, T. Management of nitrogen fertilizer application, rather than functional gene abundance, governs nitrous oxide fluxes in hydroponics with rockwool. *Plant Soil* **2014**, *374*, 715–725. [CrossRef]

45. Postel, S. Growing more food with less water. *Sci. Am.* **2001**, *284*, 46–51. [CrossRef]

46. Martinez-Mate, M.A.; Martin-Gorriz, B.; Martínez-Alvarez, V.; Soto-García, M.; Maestre-Valero, J.F. Hydroponic system and desalinated seawater as an alternative farm-productive proposal in water scarcity areas: Energy and greenhouse gas emissions analysis of lettuce production in southeast Spain. *J. Clean. Prod.* **2018**, *172*, 1298–1310. [CrossRef]

47. Stoessel, F.; Juraske, R.; Pfister, S.; Hellweg, S. Life cycle inventory and carbon and water food print of fruits and vegetables: Application to a Swiss retailer. *Environ. Sci. Technol.* **2012**, *46*, 3253–3262. [CrossRef]

48. Antón, A.; Montero, J.I.; Muñoz, P.; Castells, F. LCA and tomato production in Mediterranean greenhouses. *Gov. Ecol.* **2005**, *4*, 102–112. [CrossRef]

49. Cellura, M.; Longo, S.; Mistretta, M. Life Cycle Assessment (LCA) of protected crops: An Italian case study. *J. Clean. Prod.* **2012**, *28*, 56–62. [CrossRef]

50. Martínez-Blanco, J.; Muñoz, P.; Anton, A.; Rieradevall, J. Assessment of tomato Mediterranean production in open-field and standard multi-tunnel greenhouse, with compost or mineral fertilizers, from an agricultural and environmental standpoint. *J. Clean. Prod.* **2011**, *19*, 985–997. [CrossRef]

51. Torrellas, M.; Antón, A.; Ruijs, M.; García, N.; Stanghellini, C.; Montero, J.I. Environmental and economic assessment of protected crops in four European scenarios. *J. Clean. Prod.* **2012**, *28*, 45–55. [CrossRef]

52. Webb, J.; Williams, A.G.; Hope, E.; Evans, D.; Moorhouse, E. Do foods imported into the UK have a greater environmental impact than the same foods produced within the UK? *Int. J. Life Cycle Assess.* **2013**, *18*, 1325–1343. [CrossRef]

53. Payen, S.; Basset-Mens, C.; Perret, S. LCA of local and imported tomato: An energy and water trade-off. *J. Clean. Prod.* **2015**, *87*, 139–148. [CrossRef]

54. FAO. 2017. Available online: http//faostat3.fao.org (accessed on 27 May 2019).

55. Van Kooten, O.; Heuvelink, E.; Stanghellini, C. Nutrient supply in soilless culture: On-demand strategies. *Acta Hortic.* **2004**, *659*, 533–540. [CrossRef]

56. Ntinas, G.K.; Neumair, M.; Tsadilas, C.D.; Meyer, J. Carbon footprint and cumulative energy demand of greenhouse and open-field tomato cultivation systems under Southern and Central European climatic conditions. *J. Clean. Prod.* **2017**, *142*, 3617–3626. [CrossRef]

57. Neocleous, D.; Savvas, D. NaCl accumulation and macronutrient uptake by a melon crop in a closed hydroponic system in relation to water uptake. *Agric. Water Manag.* **2016**, *165*, 22–32. [CrossRef]

58. Incrocci, L.; Massa, D.; Pardossi, A.; Bacci, L.; Battista, P.; Rapi, B.; Romani, M. A decision support system to optimise fertigation management in greenhouse crops. *Acta Hortic.* **2012**, *927*, 115–122. [CrossRef]

59. Blok, C.; van Os, E.; Daoud, R.; Waked, L.; Hasan, A. *Hydroponic Green Farming Initiative: Increasing Water Use Efficiency by Use of Hydroponic Cultivation Methods in Jordan*; Report GTB 1447; Wageningen University & Research: Wageningen, The Netherlands; BU Greenhouse Horticulture: Wageningen, The Netherlands, 2017.

60. Gruda, N.; Tanny, J. Protected crops. In *Horticulture: Plants for People and Places*; Dixon, G.R., Aldous, D.E., Eds.; Springer: Dordrecht, The Netherlands, 2014; Volume 1, pp. 327–405.

61. Gruda, N.; Tanny, J. Protected crops–recent advances, innovative technologies and future challenges. *Acta Hortic.* **2015**, *1107*, 271–278. [CrossRef]

62. Rodríguez-Delfín, A.; Gruda, N.; Eigenbrod, C.; Orsini, F.; Gianquinto, G. Soil based and simplified hydroponics rooftop gardens. In *Rooftop Urban Agriculture*; Orsini, F., Dubbeling, M., de Zeeuw, H., Gianquinto, G., Eds.; Springer International Publishing AG: Berlin/Heidelberg, Germany, 2017; pp. 61–81. ISBN 978-3-319-57719-7.

63. Raviv, M.; Wallach, R.; Silber, A.; Bar-Tal, A. Substrates and their analysis. In *Hydroponic Production of Vegetables and Ornamentals*; Savvas, D., Passam, H., Eds.; Embrio publications: Athens, Greece, 2002; pp. 25–102.

64. Abad, M.; Noguera, P.; Burés, S. National inventory of organic wastes for use as growing media for ornamental potted plant production: Case study in Spain. *Bioresour. Technol.* **2001**, *77*, 197–200. [CrossRef]

65. Zulfiqar, F.; Allaire, S.E.; Akram, N.A.; Méndez, A.; Younis, A.; Peerzada, A.M.; Shaukat, N.; Wright, S.R. Challenges in organic component selection and biochar as an opportunity in potting substrates: A review. *J. Plant Nutr.* **2019**, *24*, 1–6. [CrossRef]

66. Raviv, M. Production of high-quality composts for horticultural purposes: A mini-review. *HortTechnology* **2005**, *15*, 52–57. [CrossRef]

67. Raviv, M.; Medina, S.; Krasnovsky, A.; Ziadna, H. Organic matter and nitrogen conservation in dairy manure composting for organic agriculture. *Compost Sci. Util.* **2004**, *12*, 6–10. [CrossRef]

68. Santos, F.T.; Goufo, P.; Santos, C.; Botelho, D.; Fonseca, J.; Queirós, A.; Costa, M.S.; Trindade, H. Comparison of five agro-industrial waste-based composts as growing media for lettuce: Effect on yield, phenolic compounds and vitamin C. *Food Chem.* **2016**, *209*, 293–301. [CrossRef] [PubMed]

69. Riaz, A.; Younis, A.; Ghani, I.; Tariq, U.; Ahsan, M. Agricultural waste as growing media component for the growth and flowering of *Gerbera jamesonii* cv. hybrid mix. *Int. J. Recycl. Org. Waste Agric.* **2015**, *4*, 197. [CrossRef]

70. Mattei, P.; Pastorelli, R.; Rami, G.; Mocali, S.; Giagnoni, L.; Gonnelli, C.; Renella, G. Evaluation of dredged sediment co-composted with green waste as plant growing media assessed by eco-toxicological tests, plant growth and microbial community structure. *J. Hazard. Mater.* **2017**, *333*, 144–153. [CrossRef] [PubMed]

71. Suo, L.N.; Sun, X.Y.; Li, S.Y. Use of organic agricultural wastes as growing media for the production of *Anthurium andraeanum* 'Pink Lady'. *J. Hortic. Sci. Biotechnol.* **2011**, *86*, 366–370. [CrossRef]

72. Cole, D.M.; Sibley, J.L.; Blythe, E.K.; Eakes, D.J.; Tilt, K.M. Effect of cotton gin compost on substrate properties and growth of azalea under differing irrigation regimes in a greenhouse setting. *HortTechnology* **2005**, *15*, 145–148. [CrossRef]

73. Berecha, G.; Lemessa, F.; Wakjira, M. Exploring the suitability of coffee pulp compost as growth media substitute in greenhouse production. *Int. J. Agric. Res.* **2011**, *6*, 255–267. [CrossRef]

74. El-Mahrouk, M.E.; Yaser Hassan Dewir, Y.H.; El-Hendawy, S. Utilization of Grape Fruit Waste-based Substrates for Seed Germination and Seedling Growth of Lemon Basil. *HortTechnology* **2017**, *27*, 523–529. [CrossRef]

75. Grigatti, M.; Giorgioni, M.E.; Ciavatta, C. Compost-based growing media: Influence on growth and nutrient use of bedding plants. *Bioresour. Technol.* **2007**, *98*, 3526–3534. [CrossRef] [PubMed]

76. Morales-Corts, M.R.; Gómez-Sánchez, M.A.; Pérez-Sánchez, R. Evaluation of green/pruning wastes compost and vermicompost, slumgum compost and their mixes as growing media for horticultural production. *Sci. Hortic.* **2014**, *172*, 155–160. [CrossRef]

77. Aviani, I.; Laor, Y.; Medina, S.; Krasnovsky, A.; Raviv, M. Co-composting of solid and liquid olive mill wastes: Management aspects and the horticultural value of the resulting composts. *Bioresour. Technol.* **2010**, *101*, 6699–6706. [CrossRef] [PubMed]

78. Castaldi, P.; Melis, P. Growth and Yield Characteristics and heavy metal content on tomatoes grown in different growing media. *Commun. Soil Sci. Plant* **2004**, *35*, 85–98. [CrossRef]

79. Mininni, C.; Santamaria, P.; Abdelrahman, H.M.; Cocozza, C.; Miano, T.; Montesano, F.; Parente, A. Posidonia-based compost as a peat substitute for lettuce transplant production. *HortScience* **2012**, *47*, 1438–1444. [CrossRef]

80. Mininni, C.; Bustamante, M.; Medina, E.; Montesano, F.; Paredes, C.; Pérez-Espinosa, A.; Moral, R.; Santamaria, P. Evaluation of posidonia seaweed-based compost as a substrate for melon and tomato seedling production. *J. Hortic. Sci. Biotechnol.* **2013**, *88*, 345–351. [CrossRef]

81. Mininni, C.; Grassi, F.; Traversa, A.; Cocozza, C.; Parente, A.; Miano, T.; Santamaria, P. *Posidonia oceanica* (L.) based compost as substrate for potted basil production. *J. Sci. Food Agric.* **2015**, *95*, 2041–2046. [CrossRef] [PubMed]

82. Montesano, F.; Gattullo, C.; Parente, A.; Terzano, R.; Renna, M. Cultivation of potted sea fennel, an emerging mediterranean halophyte, using a renewable seaweed-based material as a peat substitute. *Agriculture* **2018**, *8*, 96. [CrossRef]

83. Benito, M.; Masaguer, A.; De Antonio, R.; Moliner, A. Use of pruning waste compost as a component in soilless growing media. *Bioresour. Technol.* **2005**, *96*, 597–603. [CrossRef] [PubMed]

84. Ostos, J.C.; López-Garrido, R.; Murillo, J.M.; López, R. Substitution of peat for municipal solid waste- and sewage sludge-based composts in nursery growing media: Effects on growth and nutrition of the native shrub *Pistacia lentiscus* L. *Bioresour. Technol.* **2008**, *99*, 1793–1800. [CrossRef]

85. Hernández-Apaolaza, L.; Gascó, AM.; Gascó, JM.; Guerrero, F. Reuse of waste materials as growing media for ornamental plants. *Bioresour. Technol.* **2005**, *96*, 125–131. [CrossRef]

86. Diaz-Perez, M.; Camacho-Ferre, F. Effect of composts in substrates on the growth of tomato transplants. *HortTechnology* **2010**, *20*, 361–367. [CrossRef]

87. Massa, D.; Malorgio, F.; Lazzereschi, S.; Carmassi, G.; Prisa, D.; Burchi, G. Evaluation of two green composts for peat substitution in geranium (*Pelargonium zonale* L.) cultivation: Effect on plant growth, quality, nutrition, and photosynthesis. *Sci. Hortic.* **2018**, *228*, 213–221. [CrossRef]

88. Maher, M.J.; Prasad, M. The effect of N source on the composting of green waste and its properties as a component of a peat growing medium. In *Proceedings of the International Conference Orbit 2001 on Biological Processing of Waste*; Spanish Waste Club: Seville, Spain, 2001; pp. 299–306.

89. Paula, F.S.; Tatti, E.; Abram, F.; Wilson, J.; O'Flaherty, V. Stabilisation of spent mushroom substrate for application as a plant growth-promoting organic amendment. *J. Environ. Manag.* **2017**, *196*, 476–486. [CrossRef] [PubMed]

90. Vandecasteele, B.; Debode, J.; Willekens, K. Recycling of P and K in circular horticulture through compost application in sustainable growing media for fertigated strawberry cultivation. *Eur. J. Agron.* **2018**, *96*, 131–145. [CrossRef]

91. Gong, X.; Li, S.; Sun, X.; Wang, L.; Cai, L.; Zhang, J.; Wei, L. Green waste compost and vermicompost as peat substitutes in growing media for geranium (*Pelargonium zonale* L.) and calendula (*Calendula officinalis* L.). *Sci. Hortic.* **2018**, *236*, 186–191. [CrossRef]

92. Garg, V.K.; Suthar, S.; Yadav, A. Management of food industry waste employing vermicomposting technology. *Bioresour. Technol.* **2012**, *126*, 437–443. [CrossRef] [PubMed]

93. Lazcano, C.; Gómez-Brandón, M.; Domínguez, J. Comparison of the effectiveness of composting and vermicomposting for the biological stabilization of cattle manure. *Chemosphere* **2008**, *72*, 1013–1019. [CrossRef] [PubMed]

94. Wang, L.M.; Zhang, Y.M.; Lian, J.J.; Chao, J.Y.; Gao, Y.X.; Yang, F.; Zhang, L.Y. Impact of fly ash and phosphatic rock on metal stabilization and bioavailability during sewage sludge vermicomposting. *Bioresour. Technol.* **2013**, *136*, 281–287. [CrossRef]

95. Robbins, J.A.; Evans, M.R. Growing Media for Container Production in a Greenhouse or Nursery. Part I (Components and Mixes). Greenhouse and Nursery Series. 2011. Available online: https://www.uaex.edu/publications/PDF/FSA-6097.pdf (accessed on 8 November 2018).

96. Gruda, N.; Qaryouti, M.M.; Leonardi, C. Growing media. In *Good Agricultural Practices for Greenhouse Vegetable Crops–Principles for Mediterranean Climate Areas*; Plant Production and Protection Paper 217; FAO: Rome, Italy, 2013; pp. 271–302.

97. Prasad, M. Physical, chemical and biological properties of coir dust. *Acta Hortic.* **1997**, *450*, 21–30. [CrossRef]

98. Noguera, P.; Abad, M.; Noguera, V.; Puchades, R.; Maquieira, A. Coconut coir waste, a new and viable ecologically-friendly peat substitute. *Acta Hortic.* **2000**, *517*, 279–286. [CrossRef]

99. Gruda, N.; Schnitzler, W.H. The effect of water supply on bio-morphological and plant-physiological parameters of tomato transplants cultivated in wood fiber substrate. *J. Appl. Bot. Food Qual.* **2000**, *74*, 233–239. (In German)

100. Gruda, N.; Schnitzler, W.H. The effect of water supply of seedlings, cultivated in different substrates and raising systems on the bio-morphological and plant-physiological parameters of lettuce. *J. Appl. Bot. Food Qual.* **2000**, *74*, 240–247.

101. Gruda, N.; Schnitzler, W.H. Suitability of wood fiber substrates for production of vegetable transplants. I. Physical properties of wood fiber substrates. *Sci. Hortic.* **2004**, *100*, 309–322. [CrossRef]

102. Jackson, B.E.; Wright, R.D.; Gruda, N. Container medium pH in a pine tree substrate amended with peat moss and dolomitic limestone affects plant growth. *Hortscience* **2009**, *44*, 1983–1987. [CrossRef]
103. Prasad, M.; Chualáin, D.N. Relationship between particle size and air space of growing media. *Acta Hortic.* **2004**, *648*, 161–166. [CrossRef]
104. Maher, M.J.; Thomson, D. Growth and Mn content of tomato (*Lycopersicon esculentum*) seedlings grown in Sitka spruce (*Picea sitchensis* (Bong.) Carr.) bark substrate. *Sci. Hortic.* **1991**, *48*, 223–231. [CrossRef]
105. Gruda, N.; Rau, B.J.; Wright, R.D. Laboratory Bioassay and Greenhouse Evaluation of a Pine Tree Substrate Used as a Container Substrate. *Eur. J. Hortic. Sci.* **2009**, *74*, 73–78.
106. Gruda, N.; Schnitzler, W.H. Suitability of wood fiber substrates for production of vegetable transplants II. The effect of wood fiber substrates and their volume weights on the growth of tomato transplants. *Sci. Hortic.* **2004**, *100*, 333–340. [CrossRef]
107. Gruda, N.; Schnitzler, W.H. Wood fibers as a peat alternative substrate for vegetable production. *Eur. J. Wood Wood Prod.* **2006**, *64*, 347–350. [CrossRef]
108. Jackson, B.E. Substrates on Trial. Nursery Management 10. 2018. Available online: www.nurserymag.com/article/wood-fiber-substrates-trials (accessed on 8 November 2018).
109. Gruda, N. The effect of wood fiber mulch on water retention, soil temperature and growth of vegetable plants. *J. Sustain. Agric.* **2008**, *32*, 629–643. [CrossRef]
110. Gruda, N. Weed suppression in vegetable crops using wood fibre mulch. *Rep. Agric.* **2007**, *85*, 329–334. (In German)
111. Neumaier, D.; Lohr, D.; Voßeler, R.; Girmann, S.; Kolbinger, S.; Meinken, E. Hydrochars as peat substitute in growing media for organically grown potted herbs. *Acta Hortic.* **2017**, *1168*, 377–386. [CrossRef]
112. Tsakaldimi, M. Kenaf (*Hibiscus cannabinus* L.) core and rice hulls as components of container media for growing *Pinus halepensis* M. seedlings. *Bioresour. Technol.* **2006**, *97*, 1631–1639. [CrossRef] [PubMed]
113. Gómez, C.; Robbins, J. Pine bark substrates amended with parboiled rice hulls: Physical properties and growth of container-grown Spirea during long-term nursery production. *HortScience* **2011**, *46*, 784–790. [CrossRef]
114. Bonaguro, J.E.; Coletto, L.; Zanin, G. Environmental and agronomic performance of fresh rice hulls used as growing medium component for *Cyclamen persicum* L. pot plants. *J. Clean. Prod.* **2017**, *142*, 2125–2132. [CrossRef]
115. Urrestarazu, M.; Martínez, G.A.; del Carmen Salas, M. Almond shell waste: Possible local rockwool substitute in soilless crop culture. *Sci. Hortic.* **2005**, *103*, 453–460. [CrossRef]
116. Arvanitoyannis, I.S.; Varzakas, T.H. Vegetable waste treatment: Comparison and critical presentation of methodologies. *Crit. Rev. Food Sci. Nutr.* **2008**, *48*, 205–247. [CrossRef]
117. Valverde, M.; Madrid, R.; García, A.L.; del Amor, F.M.; Rincón, L.F. Use of almond shell and almond hull as substrates for sweet pepper cultivation. Effects of fruit yield and mineral content. *Span. J. Agric. Res.* **2013**, *11*, 164–172. [CrossRef]
118. Özçelik, E.; Pekşen, A. Hazelnut husk as a substrate for the cultivation of shiitake mushroom (*Lentinula edodes*). *Bioresour. Technol.* **2007**, *98*, 2652–2658. [CrossRef]
119. Dede, O.H.; Ozdemir, S. Development of nutrient-rich growing media with hazelnut husk and municipal sewage sludge. *Environ. Technol.* **2018**, *39*, 2223–2230. [CrossRef]
120. Dede, O.H.; Dede, G.; Ozdemir, S.; Abad, M. Physicochemical characterization of hazelnut husk residues with different decomposition degrees for soilless growing media preparation. *J. Plant Nutr.* **2011**, *34*, 1973–1984. [CrossRef]
121. Chrysargyris, A.; Stavrinides, M.; Moustakas, K.; Tzortzakis, N. Utilization of paper waste as growing media for potted ornamental plants. *Clean Technol. Environ. Policy* **2018**. [CrossRef]
122. Pouliot, R.; Hugron, S.; Rochefort, L. Sphagnum farming: A long-term study on producing peat moss biomass sustainably. *Ecol. Eng.* **2015**, *74*, 135–147. [CrossRef]
123. Gaudig, G.; Fengler, F.; Krebs, M.; Prager, A.; Schulz, J.; Wichmann, S.; Joosten, H. Sphagnum farming in Germany–a review of progress. *Mires Peat* **2014**, *13*, 1–11.
124. Kraska, T.; Kleinschmidt, B.; Weinand, J.; Pude, R. Cascading use of *Miscanthus* as growing substrate in soilless cultivation of vegetables (tomatoes, cucumbers) and subsequent direct combustion. *Sci. Hortic.* **2018**, *235*, 205–213. [CrossRef]

125. Altland, J.E.; Krause, C. Use of switchgrass as a nursery container substrate. *HortScience* **2009**, *44*, 1861–1865. [CrossRef]

126. Altland, J. Use of processed biofuel crops for nursery substrates. *J. Environ. Hortic.* **2010**, *28*, 129–134.

127. Andreu-Rodriguez, J.; Medina, E.; Ferrandez-Garcia, M.T.; Ferrandez-Villena, M.; Ferrandez-Garcia, C.E.; Paredes, C.; Bustamante, M.A.; Moreno-Caselles, J. Agricultural and industrial valorization of *Arundo donax* L. *Commun. Soil Sci. Plant Anal.* **2013**, *44*, 598–609. [CrossRef]

128. Kuisma, E.; Palonen, P.; Yli-Hallab, M. Reed canary grass straw as a substrate in soilless cultivation of strawberry. *Sci. Hortic.* **2014**, *178*, 217–223. [CrossRef]

129. Köbbing, J.; Rehme, J.; Röse, D. Identify and evaluate emissions with a carbon footprint. *Gemüse* **2018**, *5*, 26–27. (In German)

130. Martínez-Blanco, J.; Lazcano, C.; Christensen, T.H.; Muñoz, P.; Rieradevall, J.; Møller, J.; Antón, A.; Boldrin, A. Compost benefits for agriculture evaluated by life cycle assessment. A review. *Agron. Sustain. Dev.* **2013**, *33*, 721–732. [CrossRef]

131. Alexander, P.D.; Bragg, N.C. Defining Sustainable Growing Media for Sustainable UK Horticulture. *Acta Hortic.* **2014**, *1034*, 219–225. [CrossRef]

132. Knight, A. Towards Sustainable Growing Media. Chairman's Report and Roadmap, Sustainable Growing Media Task Force; 2012. Available online: https://assets.publishing.service.gov.uk/government/uploads/system/uploads/attachment_data/file/221019/pb13867-towards-sustainable-growing-media.pdf (accessed on 8 November 2018).

133. Köbbing, J.; Rehme, J.; Röse, D. Ways to reduce the emissions: Klimabilanz für Gartenbaubetrieb und Pflanze, Teil 3. *Gemüse* **2018**, *7*, 20–21. (In German)

134. Levesque, V.; Rochette, P.; Ziadi, N.; Dorais, M.; Antoun, H. Mitigation of CO_2, CH_4 and N_2O from a fertigated horticultural growing medium amended with biochars and a compost. *Appl. Soil Ecol.* **2018**, *126*, 129–139. [CrossRef]

135. Ilahi, W.F.F.; Ahmad, D. A study on the physical and hydraulic characteristics of cocopeat perlite mixture as a growing media in containerized plant production. *Sains Malays.* **2017**, *46*, 975–980. [CrossRef]

136. Allaire, S.E.; Caron, J.; Menard, C.; Dorais, M. Potential replacements for rockwool as growing substrate for greenhouse tomato. *Can. J. Soil Sci.* **2005**, *85*, 67–74. [CrossRef]

137. Cantliffe, D.J.; Shaw, N.L.; Saha, S.K.; Gruda, N. Greenhouse cooling for production of peppers under hot-humid summer conditions in a high-roof passively-ventilated greenhouse. *Acta Hortic.* **2007**, *761*, 41–48. [CrossRef]

138. Rodriguez, J.C.; Cantliffe, D.J.; Shaw, N.L.; Karchi, Z. Soilless media and containers for greenhouse production of 'Galia' type muskmelon. *HortScience* **2006**, *41*, 1200–1205. [CrossRef]

139. Shaw, N.L.; Cantliffe, D.J.; Funes, J.; Shine, C. Successful beit alpha cucumber production in the greenhouse using pine bark as an alternative soilless media. *HortTechnology* **2004**, *14*, 289–294. [CrossRef]

140. Owen, J.S., Jr.; Warren, S.L.; Bilderback, T.E.; Cassel, D.K.; Albano, J.P. Physical properties of pine bark 938 substrate amended with industrial mineral aggregate. *Acta Hortic.* **2008**, *779*, 131–138. [CrossRef]

141. Schnitzler, W.H.; Michalsky, F.; Gruda, N. Wood fibre substrate for cucumber in greenhouse cultivation. In Proceedings of the 9th International Congress on Soilless Culture, Saint Helier, Jersey, 12–19 April 1997; pp. 453–463.

142. Depardieu, C.; Prémont, V.; Boily, C.; Caron, J. Sawdust and bark-based substrates for soilless strawberry production: Irrigation and electrical conductivity management. *PLoS ONE* **2016**, *11*, e0154104. [CrossRef]

143. Gruda, N.; Tucher, S.V.; Schnitzler, W.H. N-immobilization of wood fiber substrates in the production of tomato transplants (*Lycopersicon lycopersicum* (L.) Karst. Ex. Farw.). *J. Appl. Bot. Food Qual.* **2000**, *74*, 32–37. (In German)

144. Álvarez, J.M.; Pasian, C.; Lal, R.; López, R.; Díaz, M.J.; Fernández, M. Morpho-physiological plant quality when biochar and vermicompost are used as growing media replacement in urban horticulture. *Urban. For. Urban Green.* **2018**, *34*, 175–180. [CrossRef]

145. Tian, Y.; Sun, X.; Li, S.; Wang, H.; Wang, L.; Cao, J.; Zhang, L. Biochar made from green waste as peat substitute in growth media for *Calathea rotundifola* cv. Fasciata. *Sci. Hortic.* **2012**, *143*, 15–18. [CrossRef]

146. Conversa, G.; Bonasia, A.; Lazzizera, C.; Elia, A. Influence of biochar, mycorrhizal inoculation, and fertilizer rate on growth and flowering of Pelargonium (*Pelargonium zonale* L.) plants. *Front. Plant Sci.* **2015**, *6*, 1–11. [CrossRef] [PubMed]

147. Méndez, A.; Paz-Ferreiro, J.; Gil, E. The effect of paper sludge and biochar addition on brown peat and coir based growing media properties. *Sci. Hortic.* **2015**, *193*, 225–230. [CrossRef]

148. Méndez, A.; Cárdenas-Aguiar, E.; Paz-Ferreiro, J.; Plaza, C.; Gascó, G. The effect of sewage sludge biochar on peat-based growing media. *Biol. Agric. Hortic.* **2017**, *33*, 40–51. [CrossRef]

149. Kim, H.S.; Kim, K.R.; Yang, J.E.; Ok, Y.S.; Kim, W.; Kunhikrishnan, A.; Kim, K.-H. Amelioration of horticultural growing media properties through rice hull biochar incorporation. *Waste Biomass Valorization* **2017**, *8*, 483–492. [CrossRef]

150. Emmel, M. Growing ornamental plants in *Sphagnum* biomass. *Acta Hortic.* **2008**, *779*, 173–178. [CrossRef]

151. Jobin, P.; Caron, J.; Rochefort, L. Developing new potting mixes with *Sphagnum* fibers. *Can. J. Soil Sci.* **2014**, *94*, 585–593. [CrossRef]

152. The 'Reduce, Reuse, Recycle' Waste Hierarchy. Available online: https://www.conserve-energy-future.com/reduce-reuse-recycle.php (accessed on 16 March 2019).

153. Gruda, N.; Schnitzler, W.H. Alternative growing systems for head lettuce. *Rep. Agric.* **2006**, *84*, 469–484.

154. Truffault, V.; Ristorto, M.; Brajeul, E.; Vercambre, G.; Gautier, H. To stop nitrogen overdose in soilless tomato crop: A way to promote fruit quality without affecting fruit yield. *Agronomy* **2019**, *9*, 80. [CrossRef]

155. Stanghellini, C.; Montero, J.I. Resource use efficiency in protected cultivation: Towards the greenhouse with zero emissions. *Acta Hortic.* **2012**, *927*, 91–100. [CrossRef]

156. Incrocci, L.; Pardossi, A.; Malorgio, F.; Maggini, R.; Campiotti, C.A. Cascade cropping systems for greenhouse soilless culture. *Acta Hortic.* **2003**, *609*, 297–300. [CrossRef]

157. Muñoz, P.; Paranjpe, A.; Montero, J.I.; Antón, A. Cascade crops: An alternative solution for increasing sustainability of greenhouse tomato crops in Mediterranean zone. *Acta Hortic.* **2012**, *927*, 801–805. [CrossRef]

158. Choi, B.; Lee, S.S.; Ok, Y.S. Effects of waste nutrient solution on growth of Chinese cabbage (*Brassica campestris* L.). *Korean J. Environ. Agric.* **2011**, *30*, 125–131. (In Korea) [CrossRef]

159. Zhang, C.H.; Kang, H.M.; Kim, I.S. Effect of using waste nutrient solution fertigation on the musk melon and cucumber growth. *J. Bioenviron. Control* **2006**, *15*, 400–405.

160. Kim, J.H.; Kim, T.J.; Kim, H.H.; Lee, H.D.; Lee, J.W.; Lee, C.H.; Paek, K.Y. Growth and development of 'Gutbier V-10 Amy' poinsettia (*Euphorbia pulcherrima* Willd.) as affected by application of waste nutrient solution. *Korean J. Hortic. Sci. Technol.* **2000**, *18*, 518–522.

161. EU (The European Parliament and The Council of the European Communities). Directive (EU) 2018/851 the European Parliament and the Council of 30 May 2018 Amending Directive 2008/98/EC on Waste. *Off. J. Eur. Union* **2018**, *L150*, 109–140.

162. Fonteno, W.C.; Cassel, D.K.; Larson, R.A. Physical properties of three container media and their effect on poinsettia growth. *J. Am. Soc. Hortic. Sci.* **1981**, *106*, 736–741.

163. Gruda, N.; Schnitzler, W.H. Determination of volume weight and water content of wood fiber substrates with different methods. *Agribiol. Res.* **1999**, *53*, 163–170.

164. Xing, J.; Gruda, N.; Xiong, J.; Liu, W. Influence of organic substrates on nutrient accumulation and proteome changes in tomato-roots. *Sci. Hortic.* **2019**, *252*, 192–200. [CrossRef]

165. Diara, C.; Incrocci, L.; Pardossi, A.; Minuto, A. Reusing Greenhouse Growing Media. *Acta Hortic.* **2012**, *927*, 793–800. [CrossRef]

166. Hanna, H.Y. Properly recycled perlite saves money, does nor reduce greenhouse tomato yield and can be re-used for many years. *HortTechnology* **2005**, *15*, 342–345. [CrossRef]

167. Baum, C.; El-Tohamy, W.; Gruda, N. Increasing the productivity and product quality of vegetable crops using arbuscular mycorrhizal fungi: A review. *Sci. Hortic.* **2015**, *187*, 131–141. [CrossRef]

168. Nambuthiri, S.; Fulcher, A.; Koeser, A.K.; Geneve, R.; Niu, G. Moving toward sustainability with alternative containers for greenhouse and nursery crop production: A review and research update. *HortTechnology* **2015**, *25*, 8–16. [CrossRef]

169. Colla, G.; Nardi, S.; Cardarelli, M.; Ertani, A.; Lucini, L.; Canaguier, R.; Rouphael, Y. Protein hydrolysates as biostimulants in horticulture. *Sci. Hortic.* **2015**, *196*, 28–38. [CrossRef]

Article

Pre-Harvest UV-B Radiation and Photosynthetic Photon Flux Density Interactively Affect Plant Photosynthesis, Growth, and Secondary Metabolites Accumulation in Basil (*Ocimum basilicum*) Plants

Haijie Dou [1], Genhua Niu [2,*] and Mengmeng Gu [3]

[1] Department of Horticultural Sciences, Texas A&M University, College Station, TX 77843, USA

[2] Department of Horticultural Sciences, Texas A&M AgriLife Research & Extension, The Dallas Center, Texas A&M University, 17360 Coit Road, Dallas, TX 75252, USA

[3] Department of Horticultural Sciences, Texas A&M AgriLife Extension Service, College Station, TX 77843, USA

* Correspondence: gniu@ag.tamu.edu; Tel.: +1-915-859-9111

Received: 19 July 2019; Accepted: 5 August 2019; Published: 7 August 2019

Abstract: Phenolic compounds in basil (*Ocimum basilicum*) plants grown under a controlled environment are reduced due to the absence of ultraviolet (UV) radiation and low photosynthetic photon flux density (PPFD). To characterize the optimal UV-B radiation dose and PPFD for enhancing the synthesis of phenolic compounds in basil plants without yield reduction, green and purple basil plants grown at two PPFDs, 160 and 224 $\mu mol \cdot m^{-2} \cdot s^{-1}$, were treated with five UV-B radiation doses including control, 1 $h \cdot d^{-1}$ for 2 days, 2 $h \cdot d^{-1}$ for 2 days, 1 $h \cdot d^{-1}$ for 5 days, and 2 $h \cdot d^{-1}$ for 5 days. Supplemental UV-B radiation suppressed plant growth and resulted in reduced plant yield, while high PPFD increased plant yield. Shoot fresh weight in green and purple basil plants was 12%–51% and 6%–44% lower, respectively, after UV-B treatments compared to control. Concentrations of anthocyanin, phenolics, and flavonoids in green basil leaves increased under all UV-B treatments by 9%–18%, 28%–126%, and 80%–169%, respectively, and the increase was greater under low PPFD compared to high PPFD. In purple basil plants, concentrations of phenolics and flavonoids increased after 2 $h \cdot d^{-1}$ UV-B treatments. Among all treatments, 1 $h \cdot d^{-1}$ for 2 days UV-B radiation under PPFD of 224 $\mu mol \cdot m^{-2} \cdot s^{-1}$ was the optimal condition for green basil production under a controlled environment.

Keywords: UVR8; PPFD; dose-dependent; photosynthesis; chlorophyll fluorescence; phenolic compounds

1. Introduction

Decreasing arable land, rising urbanization, water scarcity, and climate change exert pressure on agricultural producers [1]. Conventional food production is severely limited by seasonality, unpredictable weather, pests/diseases, and resources such as land and water. Indoor controlled environment agriculture (CEA) systems, which can be built anywhere, have the potential to be a suitable alternative to open field and greenhouse production [2]. However, crops cultivated in indoor CEA systems using artificial lighting are not exposed to ultraviolet radiation. Ultraviolet (UV) radiation is an important environmental signal that initiates plant responses in photosynthetic function, cell division, plant growth, and development [3,4]. In previous studies, UV-B radiation was mainly considered as a stress factor to plants, focusing on the effects of increasing solar UV-B radiation reaching Earth's surface due to stratospheric ozone depletion [5,6]. Recent studies have highlighted supplemental UV-B radiation as a eustress (i.e., positive stress), and reported that low to moderate

UV-B radiation induces a range of favorable processes in plants, such as synthesis of UV-absorbing compounds (anthocyanin, phenolic acids, and flavonoids) and antioxidants (carotenoids, ascorbate, and glucosinolate) [7–9]. These bioactive compounds represent an important source of antioxidant molecules in human diet reducing the risk of cardiovascular diseases, chronic diseases, and specific forms of cancer [10,11].

Manipulation of secondary metabolites in horticultural crops through supplemental UV-B radiation have demonstrated at least two UV-B signaling pathways, which is determined by UV-B radiation dose [11,12]. Under low UV-B radiation dose, the UV-B specific photoreceptor, UV RESISTANCE LOCUS 8 (*UVR8*), initiates an *UVR8*-dependent pathway [13]. Specifically, *UVR8* stimulates gene expression such as CONSTITUTIVELY PHOTOMORPHOGENIC 1 (*COP1*), ENLONGATED HYPOCOTYL 5 (*HY5*), and HY5 HOMOLOG (*HYH*), which play key roles in the synthesis of phenolic compounds, as well as growth retardation such as the inhibition of hypocotyl elongation [14,15]. Under high UV-B radiation dose, UV-B light acts as a damaging agent inducing formation of reactive oxygen species (ROS), causing damage to plant cells, DNA, proteins, and photosynthesis apparatus and, subsequently, negatively affect plant growth and induces synthesis of antioxidants [16,17].

In addition to being dose-dependent, plant responses to supplemental UV radiation also varied among species and cultivars [18]. For example, anthocyanin concentration of red leaf lettuce (*Lactuca sativa*, 'Red Cross') increased by 11% after 12-days UV-A radiation at 18 $\mu mol \cdot m^{-2} \cdot s^{-1}$ for 16 $h \cdot d^{-1}$ prior to harvest (controlled environment, PPFD of 300 $\mu mol \cdot m^{-2} \cdot s^{-1}$) [19]. Synthesis of anthocyanin and other polyphenols in another red leaf lettuce cultivar ('Red Fire', controlled environment, PPFD of 150 $\mu mol \cdot m^{-2} \cdot s^{-1}$) significantly increased after 3-days UV-B radiation at a much lower dose, 1.5 $\mu mol \cdot m^{-2} \cdot s^{-1}$ for 16 $h \cdot d^{-1}$ prior to harvest [4]. Furthermore, glucosinolate concentration in 7-day-old broccoli (*Brassica oleracea*) sprouts (controlled environment, PPFD not mentioned) was enhanced by 19% after 1-day UV-B radiation at 7.0 $\mu mol \cdot m^{-2} \cdot s^{-1}$ for 2 $h \cdot d^{-1}$, compared to 63% enhancement at 10.3 $\mu mol \cdot m^{-2} \cdot s^{-1}$ for 2 $h \cdot d^{-1}$ [9].

Basil (*Ocimum basilicum*) plants have been considered a source of valuable healthy substances due to their unique flavor and relatively high content of phenolic compounds [20,21]. To improve the yield of high-quality basil, more growers are turning to controlled environment production, which has been proven to be a suitable alternative to open field and greenhouse basil production, due to its high environmental controllability and improved resource utilization efficiency (arable land and clean water) [2,22]. However, crops cultivated in controlled environment systems using artificial lighting are not exposed to UV-B radiation, bearing a direct impact on basil flavor and visual appearance [10]. Meanwhile, considering energy saving, the photosynthetic photon flux density (PPFD) in controlled environment systems is much lower compared to sunlight intensity in open field, resulting in further reduction of secondary plant metabolites [21]. Therefore, there is an increasing interest in the use of supplemental UV-B radiation to enhance the synthesis of health-beneficial phenolic compounds to produce premium quality basil products under controlled environment [3,23,24].

Although some studies investigated the effects of supplemental UV-B radiation on phytochemical accumulation of basil plants, most studies were conducted in the open field or greenhouse using photo-selective film covers, and results varied largely in both biomass production and phenolic contents [25–27]. Meanwhile, most studies only focused on the effects of UV-B radiation on secondary metabolites accumulation, not considering yield reduction caused by UV-B radiation [25,28]. Furthermore, considering the significantly low PPFD used in controlled environment systems, little information is known about the interactive effects between supplemental UV-B radiation and PPFD. Collectively, to identify the optimal combination of UV-B radiation dose and PPFD that enhance concentrations of phenolic compounds without significant yield reduction, further investigation is warranted to characterize the physiological, morphological, and biochemical responses in basil plants to supplemental UV-B radiation and different PPFDs under a controlled environment.

Accordingly, in the present study, we exposed two basil cultivars to five pre-harvest supplemental UV-B radiation doses in order to characterize plant responses to supplemental UV-B radiation under two

PPFDs in a controlled environment system. Photosynthetic photon flux density of 224 $\mu mol \cdot m^{-2} \cdot s^{-1}$ for basil plants was selected according to our previous study [21], and a low PPFD of 160 $\mu mol \cdot m^{-2} \cdot s^{-1}$ was selected to test if UV-B radiation can compensate for the reduced accumulation of phenolic compounds in basil plants grown under low PPFD.

2. Materials and Methods

2.1. Plant Materials and Culture

Experiments were conducted in a walk-in growth room in Texas AgriLife Research Center at El Paso, TX, USA, from 8 August to 15 September 2017 on green basil 'Improved Genovese Compact' and from 5 September to 19 October 2017 on purple basil 'Red Rubin' (Johnny's Selected Seeds, Winslow, ME, USA), respectively. For both experiments, one basil seed per cell was sown in 72 square cell trays (length 3.86 cm; height 5.72 cm; volume 59 cm^3) with Metro-Mix® 360 (peat moss 41%, vermiculite 34%, pine bark 25%, Sun Gro® Horticulture, Bellevue, WA, USA). All trays were put under mist in a greenhouse for germination. Temperature under the mist was maintained at 32.7 °C/22.2 °C day/night. Seedlings were moved out from the mist after the emergence of cotyledons and grown in a greenhouse for two weeks. Temperature and relative humidity in the greenhouse were maintained at 29.1 °C/21.6 °C and 48%/66% day/night, respectively. When one pair of true leaves fully expanded, basil seedlings were transplanted into square pots (length 9.52 cm, height 8.26 cm, and volume 574 cm^3) filled with the Metro-Mix® 360, and uniform plants were selected and moved to the walk-in growth room for different treatments.

After transplanting, multi-layer cultivating shelves were used with mechanical mini fans (LS1225A-X, AC Infinity, City of Industry, CA, USA) circulating air to achieve uniform temperatures across treatments. Plant canopy temperature in each treatment was maintained at 23.9 °C/21.2 °C day/night. All plants were manually sub-irrigated with a nutrient solution containing 1.88 $g \cdot L^{-1}$ (277.5 ppm N) 15N-2.2P-12.5K (Peters 15-5-15 Ca-Mg Special, The Scotts Company, Marysville, OH, USA) as needed. The nutrient solution was mixed and stored in a 100-gallon tank with a lid, and the electrical conductivity (EC) and pH were adjusted to 2.0 $dS \cdot m^{-1}$ and 6.0, respectively, using an EC/pH meter (Model B-173, Horiba, Ltd., Japan).

2.2. Supplemental Ultraviolet B (UV-B) Radiation and Photosynthetic Photon Flux Density (PPFD) Treatments

Uniform green and purple basil plants were grown under two PPFDs of 160 and 224 $\mu mol \cdot m^{-2} \cdot s^{-1}$ with a 16-h photoperiod provided by cool white fluorescent lamps (Philips Lighting, Somerset, NJ, USA). Two or five days prior to harvest (basil plant height reaching about 25 cm), UV-B lamps were switched on and basil plants were treated with one of the five UV-B radiation doses including no supplemental UV-B radiation (control), 1 $h \cdot d^{-1}$ for 2 days (1H2D), 2 $h \cdot d^{-1}$ for 2 days (2H2D), 1 $h \cdot d^{-1}$ for 5 days (1H5D), or 2 $h \cdot d^{-1}$ for 5 days (2H5D) with UV-B light intensity at 16.0 $\mu mol \cdot m^{-2} \cdot s^{-1}$ (equal to 18.7 $kJ \cdot m^{-2} \cdot h^{-1}$). There were a total of 10 treatments created by the combination of two PPFDs and five UV-B radiation doses, and 12 plants per treatment. Supplemental UV-B radiation treatments were applied from 8:00 in the morning and provided by Philips TL 40W/12 and 20W/12 UV-B broadband lamps (wavelength: 270–400 nm, maximum emission wavelength at 315 nm, Svetila.com d.o.o., Domzale, Slovenia, EU). The cool white fluorescent lamps at PPFD of 160 and 224 $\mu mol \cdot m^{-2} \cdot s^{-1}$ radiated low intensity of UV radiation, which was 2.2 and 2.5 $\mu mol \cdot m^{-2} \cdot s^{-1}$, respectively. The UV-B light intensity (including UV radiation provided by broadband UV-B lamps and cool white fluorescent lamps) and PPFD in each treatment were measured at 15 cm underneath the lamps at 9 spots using a MU-200 UV radiation meter (Apogee Instruments, Logan, UT, USA) and PS-100 spectroradiometer (Apogee Instruments, Logan, UT, USA), respectively, before placing the plants. To minimize the disproportionate light distribution within each treatment, all plants were systematically rearranged every 3 days.

2.3. Measurements

2.3.1. Growth Parameters

Growth parameters of basil plants such as plant height, width, the number of internodes, leaf area, and yield including shoot fresh weight (FW) and dry weight (DW) were recorded at harvest (on 15 September and 19 October 2017 for green and purple basil plants, respectively). Plant width was calculated as the average of the widest point and its perpendicular width of plant canopy. A leaf area meter (LI-3100, LI-COR, Lincoln, NE, USA) was used to measure the leaf area. Shoot DW was determined after shoot tissues were dried at 80 °C in an oven (Grieve, Round Lake, IL, USA) for 3 days. Specific leaf area (leaf area per unit leaf dry weight) was calculated as an indicator of leaf thickness.

2.3.2. Gas-Exchange Rate, Relative Chlorophyll Concentration, and Chlorophyll Fluorescence

A portable gas exchange analyzer (CIRAS-3, PP Systems International, Amesbury, MA, USA) was used to measure the gas exchange rate, including net photosynthetic rate (P_n), transpiration rate (E), and stomatal conductance (G_s) of basil leaves at harvest. A PLC3 leaf cuvette with light-emitting diode (LED) light unit (white light, in which the proportions of red, blue, and green light were 38%, 25%, and 37%, respectively) was used. The PPFD, temperature, relative air humidity, and CO_2 concentration inside the leaf cuvette were set at 800 $\mu mol \cdot m^{-2} \cdot s^{-1}$, 25 °C, 50%, and 390 $\mu mol \cdot mol^{-1}$, respectively. The third pair of leaves from the top was used for measuring and measurements were taken until the P_n reached a steady state.

Soil plant analysis development (SPAD) index of basil leaves was recorded on the third pair of leaves from the top at harvest to quantify the relative chlorophyll concentration of basil leaves using a chlorophyll meter SPAD-502 (Konica-Minolta cooperation, Ltd., Osaka, Japan). Three measurements were taken for each leaf and the average was recorded for data analysis.

Chlorophyll fluorescence parameters of basil plants were measured at harvest using a pocket Plant Efficiency Analyzer chlorophyll fluorimeter (PEA, Hansatech Instruments Ltd., Norfolk, UK). The third pair of leaves from the top were dark adapted for at least 30 min prior to the measurement. Minimal fluorescence values (F_0) and maximal fluorescence values (F_m) in the dark-adapted state were measured, and maximum quantum use efficiency of photosystem II (PSII) in the dark-adapted state was calculated as $F_v/F_m = (F_m - F_0)/F_m$. Performance index (PI ABS, where "ABS" specifies that the reaction centers' density is expressed per absorption), dissipation of energy per cross section (DI_0/CS), trapped energy flux per cross section (TR_0/CS), and electron transport flux per cross section (ET_0/CS) parameters were calculated using the PEA Plus software (V1.10, Hansatech Instruments Ltd., Norfolk, UK).

2.3.3. Secondary Plant Metabolites

Five basil plants were randomly selected for the measurement of concentrations of anthocyanin, phenolics, and flavonoids, and antioxidant capacity of basil leaves at harvest. Fresh basil leaves were collected in a cooler and immediately stored in a deep freezer (IU1786A, Thermo Fisher Scientific, Marietta, OH, USA) at −80 °C until phytochemical evaluation.

Extraction. Approximately 2 g fresh basil leaves were ground in liquid nitrogen and extracted with 15 mL 1% acidified methanol at 4 °C in dark. After overnight extraction, the mixture was centrifuged (Sorvall RC 6 Plus Centrifuge, Thermo Fisher Scientific, Madison, WI, USA) at 13,200 rpm (26,669× g) for 15 min, and the supernatant was collected for phytochemical evaluation [29].

Anthocyanin evaluation. Absorbance of the extract was measured at 530 nm using a spectrophotometer (Genesys 10S ultraviolet/Vis, Thermo Fisher Scientific, Madison, WI, USA), and anthocyanin concentration was expressed as mg cyanidin-3-glucoside equivalent per 100 g FW of basil leaves using a molar extinction coefficient of 29,600 [30].

Phenolics evaluation. A modified Folin-Ciocalteu reagent method [29] was used to determine the phenolics concentration of basil leaves: 100 μL extraction sample was added to a mixture of 750 μL

1/10 dilution Folin–Ciocalteau reagent and 150 μL distilled water. After 6 min reaction, 600 μL 7.5% Na_2CO_3 was added and the mixture was incubated at 45 °C in a water bath for 10 min before the absorbance was measured at 725 nm using a microplate reader (EL×800, BioTek, Winooski, VT, USA). Results were shown as mg of gallic acid equivalent per g FW of basil leaves.

Flavonoids evaluation. Flavonoid concentration of basil leaves was determined [21] as the following: 20 μL extraction sample was added to a mixture of 85 μL distilled water and 5 μL 5% $NaNO_2$. After 6 min reaction, a 10 μL of 10% $AlCl_3 \cdot 6H_2O$ was added to the mixture. After another 5 min reaction, 35 μL of 1M NaOH and 20 μL distilled water were added to the mixture and the absorbance was measured at 520 nm using the aforementioned microplate reader. Results were shown as mg of (+)-catechin hydrate equivalent per g FW of basil leaves.

Antioxidant capacity evaluation. A 2,2′-azino-bis (3-ethylbenzothiazoline-6-sulphonic acid) (ABTS) method [31] was used to determine the antioxidant capacity of basil leaves: 150 μL extracted sample was added to 2.85 mL of $ABTS^+$ solution and incubate at room temperature for 10 min. The absorbance of mixed solution was measured at 734 nm using the aforementioned spectrophotometer. Results were shown as mg of Trolox equivalent antioxidant capacity per 100 g FW of basil leaves.

2.4. Statistical Analyses

Experiments were arranged in a two factors factorial design. Five plants per treatment were randomly selected for measurement. After verifying the significance of the two main factors (UV-B and PPFD) and their interaction (PPFD × UV-B), a one-way analysis of variance among 10 treatments was conducted for green and purple basil plants, respectively, according to Student's *t* method ($p < 0.05$). Some data were pooled from two PPFDs because effect of PPFD was not statistically significant. Pairwise correlations method ($p < 0.05$) was used to test correlations between parameters. All statistical analyses were performed using JMP software (Version 13, SAS Institute Inc., Cary, NC, USA).

3. Results

3.1. Gas Exchange Rate, Relative Chlorophyll Concentration, and Chlorophyll Fluorescence

Supplemental UV-B radiation suppressed plant photosynthesis, in which P_n, E, and G_s in both basil cultivars were lower compared to plants grown under control, while PPFD showed no effects (Table 1). In green and purple basil leaves, P_n, E, and G_s was 68%/70%, 55%/68%, and 65%/76% lower under treatment 2H5D compared to plants grown under control, respectively. Relative chlorophyll concentration of green and purple basil plants was 9%–15% and 6%–8% lower under supplemental UV-B radiation compared to plants grown under control, respectively, while PPFD showed no effect on green basil plants but increased relative chlorophyll concentration in purple basil plants (Figure 1).

Supplemental UV-B radiation inhibited plant chlorophyll fluorescence parameters in green basil plants, including F_v/F_m and PI ABS. However, in purple basil plants, F_v/F_m showed no differences between control and 1H2D treatment, and PI ABS was only lower under the highest UV-B radiation dose, 2H5D treatment (Figure 2A,B). Similarly, TR_0/CS and ET_0/CS in green basil plants were lower after UV-B radiation, while they were not affected by UV-B radiation in purple basil plants (Figure 2D,E). On the contrary, DI_0/CS in purple basil plants was significantly higher under treatments 1H5D and 2H5D, while in green basil plants no treatment effect was observed (Figure 2C). Chlorophyll fluorescence parameters in basil plants were not affected by PPFD.

Table 1. Net photosynthetic rate (P_n), transpiration rate (E), and stomatal conductance (G_s) of green basil 'Improved Genovese Compact' and purple basil 'Red Rubin' plants under five supplemental UV-B radiation treatments, including no supplemental UV-B radiation (control), 1 h·d^{-1} for 2 days (1H2D), 2 h·d^{-1} for 2 days (2H2D), 1 h·d^{-1} for 5 days (1H5D), and 2 h·d^{-1} for 5 days (2H5D).

Cultivar	Treatment	P_n (μmol·m^{-2}·s^{-1})		E (mmol·m^{-2}·s^{-1})		G_s (mmol·m^{-2}·s^{-1})	
Green Basil	Control	13.2	a [z]	2.76	a	130	A
	1H2D	7.8	B	1.74	bc	79	B
	2H2D	8.5	B	1.93	b	93	ab
	1H5D	7.4	B	1.82	b	71	B
	2H5D	4.2	C	1.24	c	46	C
Purple Basil	Control	7.4	A	2.73	A	131	A
	1H2D	4.3	B	1.49	B	60	B
	2H2D	3.1	C	1.20	B	42	CD
	1H5D	3.8	BC	1.33	B	49	BC
	2H5D	2.2	D	0.86	C	31	D

Data were pooled from two photosynthetic photon flux density (PPFD) treatments. [z] Means followed by the same lower/upper case letters are not significantly different for green/purple basil plants, according to Student's *t* mean comparison ($p < 0.05$).

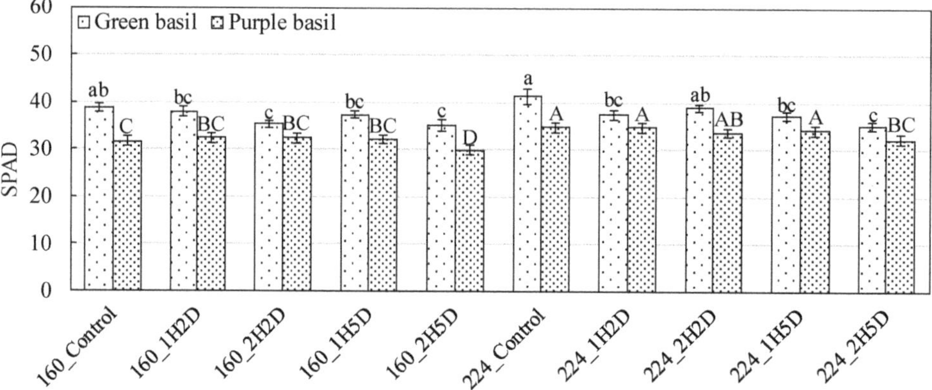

Figure 1. Relative chlorophyll concentration (soil plant analysis development (SPAD) index) of green basil 'Improved Genovese Compact' and purple basil 'Red Rubin' plants at different treatments. There were 10 treatments created by the combination of two photosynthetic photon flux density (PPFD) of 160 and 224 μmol·m^{-2}·s^{-1} and five ultraviolet B (UV-B) radiation treatments, including no supplemental UV-B radiation (control), 1 h·d^{-1} for 2 days (1H2D), 2 h·d^{-1} for 2 days (2H2D), 1 h·d^{-1} for 5 days (1H5D), and 2 h·d^{-1} for 5 days (2H5D). Means followed by the same lower/upper case letters are not significantly different for green/purple basil plants, according to Student's *t* mean comparison ($p < 0.05$). Bars represent standard errors.

3.2. Growth Parameters and Crop Yield

Supplemental UV-B radiation inhibited plant growth in both basil cultivars and performed as lower plant height, width, and leaf area, and the detriment increased with increasing UV-B radiation doses (Table 2). Specifically, under high PPFD (224 μmol·m^{-2}·s^{-1}), plant height of both basil cultivars was the highest under treatments control and 1H2D, followed by treatments 2H2D and 1H5D, and the lowest under treatment 2H5D. Leaf area of green/purple basil plants was 14%/17%, 28%/30%, 28%/34%, and 44%/44% lower, respectively, under treatments 1H2D, 2H2D, 1H5D, and 2H5D compared to control. Specific leaf area (leaf area per unit leaf dry weight) was calculated and used as an indicator of leaf thickness. In the present study, specific leaf area of both basil cultivars was lower under supplemental UV-B radiation, indicating increased leaf thickness after supplemental UV-B radiation (Table 2). Under higher UV-B radiation doses such as 1H5D and 2H5D treatments, basil plants also showed leaf bronze, chlorosis, waxy appearance, and premature leaf defoliation (Figure 3).

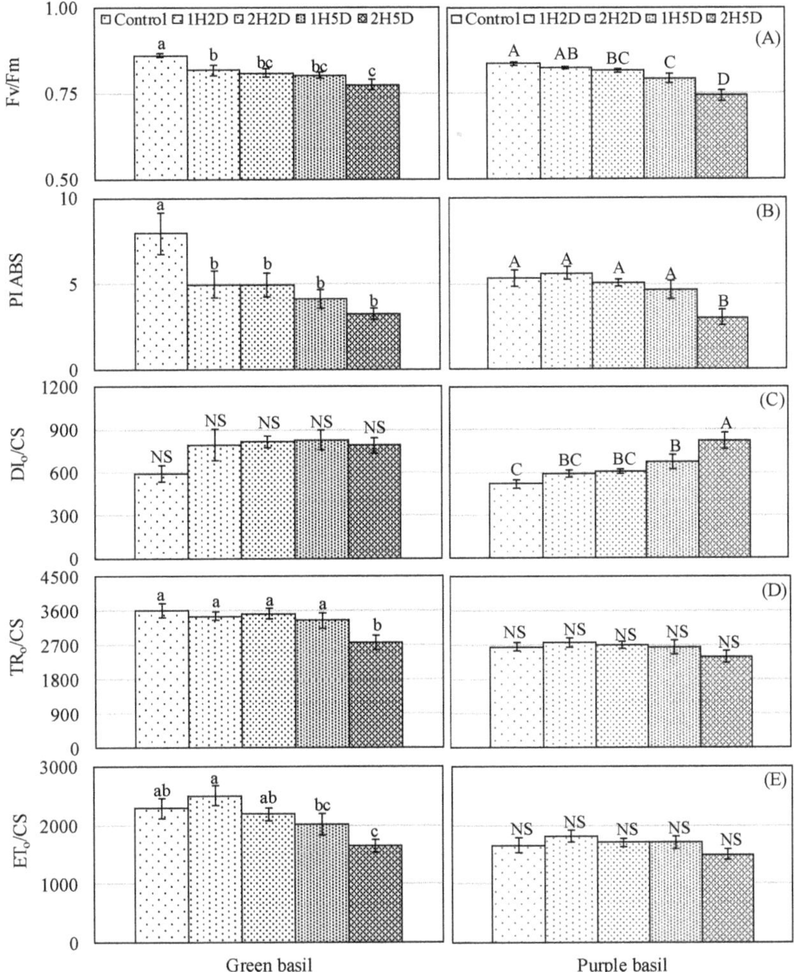

Figure 2. Chlorophyll fluorescence parameters, including maximal photochemical efficiency of Photosystem II (F_v/F_m) (**A**), performance index (PI ABS, where "ABS" specifies that the reaction centers' density is expressed per absorption) (**B**), dissipation of energy per cross section (DI_0/CS) (**C**), trapped energy per cross section (TR_0/CS) (**D**), and electron transport flux per cross section (ET_0/CS) (**E**) of green basil 'Improved Genovese Compact' and purple basil 'Red Rubin' plants under different supplemental UV-B radiation treatments including control, 1H2D, 2H2D, 1H5D, 2H5D. Data were pooled from two photosynthetic photon flux density (PPFD) treatments. Means followed by the same lower/upper case letters are not significantly different for green/purple basil plants, according to Student's *t* mean comparison ($p < 0.05$). Bars represent standard errors.

Table 2. Plant height, width, leaf area, and specific leaf area of green basil 'Improved Genovese Compact' and purple basil 'Red Rubin' plants under different treatments. There were 10 treatments created by the combination of two photosynthetic photon flux density (PPFD) levels of 160 and 224 μmol·m^{-2}·s^{-1} and five UV-B irradiation treatments including control, 1H2D, 2H2D, 1H5D, 2H5D.

Cultivar.	Treatment	Height (cm)		Width (cm)		Leaf Area (cm^2)		Specific Leaf Area (cm^2·g^{-1})	
	160_Control	18.3	bc ^{z}	11.9	ab	520	bc	531	a
	160_1H2D	17.8	Cd	12.1	ab	453	cdef	497	abc
	160_2H2D	16.7	D	11.6	bc	420	ef	528	a
	160_1H5D	17.2	Cd	11.7	b	421	def	502	ab
	160_2H5D	14.4	E	10.6	d	315	g	446	d
Green Basil	224_Control	21.7	A	12.1	ab	687	a	454	d
	224_1H2D	21.3	A	12.3	a	591	b	513	a
	224_2H2D	19.6	B	12.3	a	497	cd	477	bcd
	224_1H5D	19.6	B	12.0	ab	494	cde	466	cd
	224_2H5D	16.7	d	11.0	cd	387	fg	450	d
	PPFD	***		**		***		***	
	UV-B	***		***		***		***	
	PPFD × UV-B	NS		NS		NS		**	
	160_Control	15.6	BC	16.0	A	261	BC	610	A
	160_1H2D	15.4	C	15.5	A	217	DE	553	BC
	160_2H2D	15.1	C	15.4	A	212	DE	575	AB
	160_1H5D	14.6	C	15.2	A	221	D	545	BC
	160_2H5D	12.9	D	13.8	B	176	F	530	C
Purple Basil	224_Control	17.7	A	16.0	A	332	A	558	BC
	224_1H2D	17.2	A	16.1	A	274	B	542	BC
	224_2H2D	16.8	AB	15.7	A	233	CD	531	CD
	224_1H5D	15.5	BC	16.0	A	219	DE	534	C
	224_2H5D	14.6	C	14.0	B	187	EF	490	D
	PPFD	***		NS		***		***	
	UV-B	***		***		***		***	
	PPFD × UV-B	NS		NS		*		NS	

^{z} Means followed by the same lower/upper case letters are not significantly different for green/purple basil plants, according to Student's *t* mean comparison ($p < 0.05$). Asterisks (*) indicate significant differences (* $p < 0.05$; ** $p < 0.01$; *** $p < 0.001$). NS indicates non-significant differences (* $p < 0.05$).

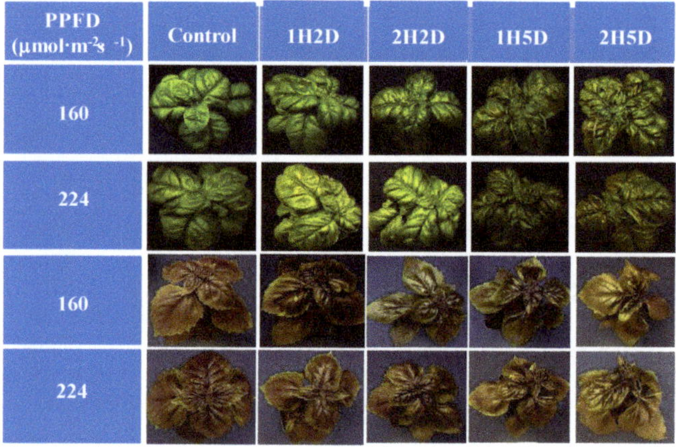

Figure 3. Green basil 'Improved Genovese Compact' and purple basil 'Red Rubin' plants under different treatments at harvest. There were 10 treatments created by the combination of two photosynthetic photon flux density (PPFD) levels of 160 and 224 μmol·m^{-2}·s^{-1} and five UV-B radiation treatments including control, 1H2D, 2H2D, 1H5D, 2H5D.

Shoot FW and DW of green and purple basil plants were generally lower in plants grown under supplemental UV-B treatments, and interactive effects (UV-B × PPFD) were observed on shoot FW ($p = 0.01$) and shoot DW ($p = 0.02$) in purple basil plants, while only interactions in shoot DW were observed in green basil plants ($p = 0.03$). Specifically, under low PPFD (160 $\mu mol \cdot m^{-2} \cdot s^{-1}$), treatment 1H2D showed no effects on shoot FW in green basil plants. So did the 1H2D and 1H5D treatments in purple basil plants, while under high PPFD (224 $\mu mol \cdot m^{-2} \cdot s^{-1}$), shoot FW in both cultivars was lower under UV-B treatments compared to control (Figure 4A,B).

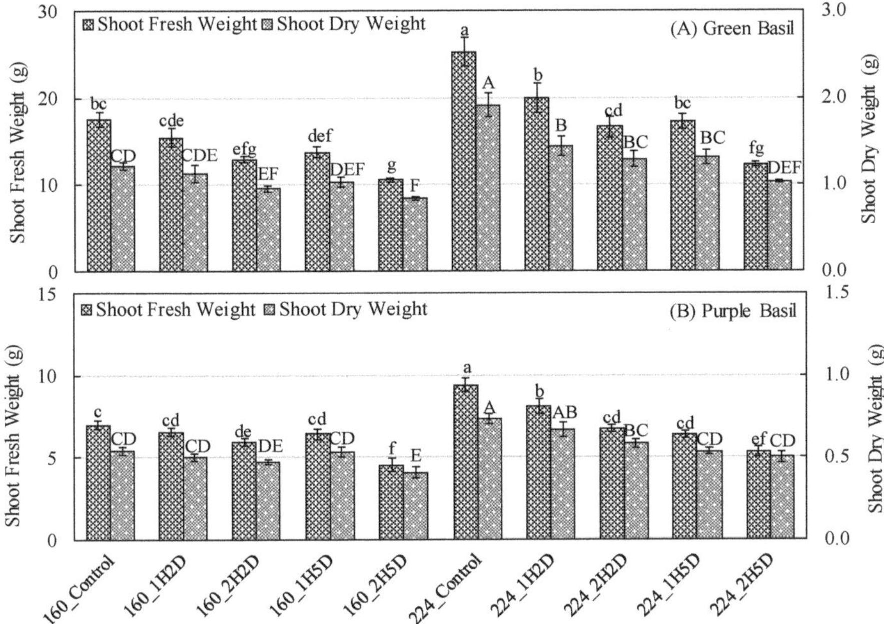

Figure 4. Shoot fresh weight and shoot dry weight of green basil 'Improved Genovese Compact' plants (**A**), and purple basil 'Red Rubin' plants (**B**) under different treatments. There were 10 treatments created by the combination of two photosynthetic photon flux density (PPFD) of 160 and 224 $\mu mol \cdot m^{-2} \cdot s^{-1}$ and five UV-B radiation treatments including control, 1H2D, 2H2D, 1H5D, 2H5D. Means followed by the same lower/upper case letters are not significantly different for green/purple plants, according to Student's *t* mean comparison ($p < 0.05$). Bars represent standard errors.

Plant height, leaf area, leaf thickness, shoot FW, and shoot DW in both basil cultivars were higher under high PPFD (Table 2, Figure 4A,B). Without supplemental UV-B treatments, plant height, leaf area, leaf thickness, shoot FW, and shoot DW in green/purple basil plants were 16%/12%, 24%/21%, 15%/9%, 44%/34%, and 59%/35% higher under high PPFD (224 $\mu mol \cdot m^{-2} \cdot s^{-1}$) compared to plants grown under low PPFD (160 $\mu mol \cdot m^{-2} \cdot s^{-1}$), respectively.

3.3. Secondary Plant Metabolites Accumulation and Antioxidant Capacity

Concentrations of phenolic compounds in green basil plants, including anthocyanin, phenolics, and flavonoids were 9%–23%, 28%–126%, and 80%–169% greater, respectively, after UV-B radiation compared to control (Table 3). Concentrations of anthocyanin and flavonoids in green basil plants were not affected by PPFD, while phenolics concentration was greater under high PPFD (224 $\mu mol \cdot m^{-2} \cdot s^{-1}$). In purple basil plants, only 2 $h \cdot d^{-1}$ UV-B treatments (2H2D and 2H5D) enriched concentrations of phenolics and flavonoids, while UV-B treatments showed no effects on anthocyanin concentration (Table 3). Specifically, under 2H2D and 2H5D treatments, concentrations of phenolics and flavonoids in purple basil plants were 29%–63% and 37%–79% greater, respectively. Concentrations of anthocyanin and phenolics in purple basil plants were greater under high PPFD (224 μmol·m^{-2}·s^{-1}), while flavonoid concentration was not affected by PPFD (Table 3).

Table 3. Anthocyanin concentration (conc.), phenolics conc., and flavonoids conc. of green basil 'Improved Genovese Compact' and purple basil 'Red Rubin' plants under different treatments. There were 10 treatments created by the combination of two photosynthetic photon flux density (PPFD) of 160 and 224 μmol·m^{-2}·s^{-1} and five UV-B radiation treatments including control, 1H2D, 2H2D, 1H5D, 2H5D.

Cultivar	Treatment	Anthocyanin Conc. (mg·100g^{-1} FW)		Phenolics Conc. (mg·g^{-1} FW)		Flavonoids Conc. (mg·g^{-1} FW)	
Green Basil	160_Control	3.19	d [z]	1.10	E	0.45	e
	160_1H2D	3.68	Abcd	1.41	De	0.92	cd
	160_2H2D	3.92	A	1.48	D	0.81	d
	160_1H5D	3.49	Abcd	1.68	Cd	1.00	abcd
	160_2H5D	3.87	Ab	2.49	A	1.21	a
	224_Control	3.29	Cd	1.38	De	0.54	e
	224_1H2D	3.39	Bcd	2.06	B	0.97	bcd
	224_2H2D	3.78	abc	1.95	Bc	0.99	abcd
	224_1H5D	3.35	bcd	2.13	Ab	1.15	abc
	224_2H5D	3.89	ab	2.34	Ab	1.19	ab
	PPFD	NS		***		NS	
	UV-B	**		***		***	
	PPFD × UV-B	NS		NS		NS	
Purple Basil	160_Control	10.63	A	2.06	CD	0.94	CD
	160_1H2D	11.02	A	1.63	E	0.82	D
	160_2H2D	10.84	A	2.66	B	1.41	B
	160_1H5D	10.74	A	2.18	C	1.14	C
	160_2H5D	10.75	A	3.35	A	1.68	A
	224_Control	10.97	A	2.03	CD	1.04	C
	224_1H2D	11.43	A	1.93	CD	1.09	C
	224_2H2D	10.97	A	2.62	B	1.49	B
	224_1H5D	10.85	A	1.85	DE	1.03	C
	224_2H5D	11.07	A	2.85	B	1.42	B
	PPFD	*		*		NS	
	UV-B	NS		***		***	
	PPFD × UV-B	NS		***		**	

[z] Means followed by the same lower/upper case letters are not significantly different for green/purple basil plants, according to Student's *t* mean comparison ($p < 0.05$). Asterisks (*) indicate significant differences (* $p < 0.05$; ** $p < 0.01$; *** $p < 0.001$). NS indicates non-significant differences (* $p < 0.05$).

The total amounts of phytochemicals per plant (i.e., anthocyanin, phenolics, and flavonoids) were calculated by multiplying the phytochemical concentrations by leaf FW per plant (Table 4). Under low PPFD (160 μmol·m^{-2}·s^{-1}), total amount of anthocyanin in green basil plants was 23% lower under treatment 2H5D compared to control, while total amounts of phenolics and flavonoids were 49%–79%% greater (Table 4). Under high PPFD (224 μmol·m^{-2}·s^{-1}), total amounts of anthocyanin and phenolics in green basil plants were 15%–39% lower under supplemental UV-B treatments compared to control, while total amount of flavonoids was 43%–44% higher under treatments 1H2D and 1H5D compared to control (Table 4). In purple basil plants, all supplemental UV-B radiation treatments showed negative or no effects on the total amount of phenolic compounds regardless of PPFD (Table 4).

Table 4. Total amount of anthocyanin, phenolics, and flavonoids per plant of green basil 'Improved Genovese Compact' and purple basil 'Red Rubin' plants under different treatments. There were 10 treatments created by the combination of two photosynthetic photon flux density (PPFD) of 160 and 224 $\mu mol \cdot m^{-2} \cdot s^{-1}$ and five UV-B radiation treatments including control, 1H2D, 2H2D, 1H5D, 2H5D.

Cultivar.	Treatment	Total Amount of Anthocyanin ($mg \cdot plant^{-1}$)		Total Amount of Phenolics ($mg \cdot plant^{-1}$)		Total Amount of Flavonoids ($mg \cdot plant^{-1}$)	
Green Basil	160_Control	0.47	cde [z]	16.0	d	6.6	d
	160_1H2D	0.47	Cde	18.0	d	11.8	b
	160_2H2D	0.42	Def	16.0	d	8.8	cd
	160_1H5D	0.40	Ef	19.2	cd	11.4	bc
	160_2H5D	0.36	F	23.8	bc	11.6	bc
	224_Control	0.67	A	28.4	ab	10.8	bc
	224_1H2D	0.55	Bc	33.2	a	15.4	a
	224_2H2D	0.59	Ab	25.6	b	12.8	ab
	224_1H5D	0.52	Bcd	31.0	a	15.6	a
	224_2H5D	0.41	Ef	24.0	bc	12.2	b
Purple Basil	160_Control	0.63	C	12.0	BC	5.6	DE
	160_1H2D	0.58	D	8.6	E	4.2	F
	160_2H2D	0.51	E	12.6	BC	6.0	CDE
	160_1H5D	0.57	D	11.0	CD	5.2	EF
	160_2H5D	0.38	G	11.4	BC	5.6	DE
	224_Control	0.83	A	15.4	A	8.0	A
	224_1H2D	0.72	B	12.2	BC	7.0	ABC
	224_2H2D	0.57	D	13.0	B	7.2	AB
	224_1H5D	0.54	D	9.4	DE	5.4	DE
	224_2H5D	0.47	F	12.2	BC	6.4	BCD

[z] Means followed by the same lower/upper case letters are not significantly different for green/purple basil plants, according to Student's *t* mean comparison ($p < 0.05$).

Antioxidant capacity in basil plants were not affected by PPFDs. Antioxidant capacity in green basil plants was higher under all supplemental UV-B radiation treatments, while it was only higher under 2 $h \cdot d^{-1}$ UV-B treatments (2H2D and 2H5D) in purple basil plants (Figure 5A). Correlation between antioxidant capacity and UV-B radiation doses was analyzed in three terms according to different UV-B radiation patterns, all UV-B treatments (Figure 5A), 1 $h \cdot d^{-1}$ UV-B treatments (1H2D and 1H5D, Figure 5B), and 2 $h \cdot d^{-1}$ UV-B treatments (2H2D and 2H5D, Figure 5C). Antioxidant capacity in green basil plants were all positively related to UV-B radiation doses regardless of radiation patterns, while antioxidant capacity in purple basil plants showed no correlation with 1 $h \cdot d^{-1}$ UV-B radiation treatments (1H2D and 1H5D, $p = 0.1994$).

Correlation between antioxidant capacity with concentrations of phenolic compounds was analyzed in basil plants. In green basil plants, concentrations of anthocyanin, phenolics, and flavonoids were all positively related to antioxidant capacity (Figure 6A). In purple basil plants, concentrations of phenolics and flavonoids were positively related to antioxidant capacity, while anthocyanin concentration showed no relationship ($p = 0.8812$) (Figure 6B).

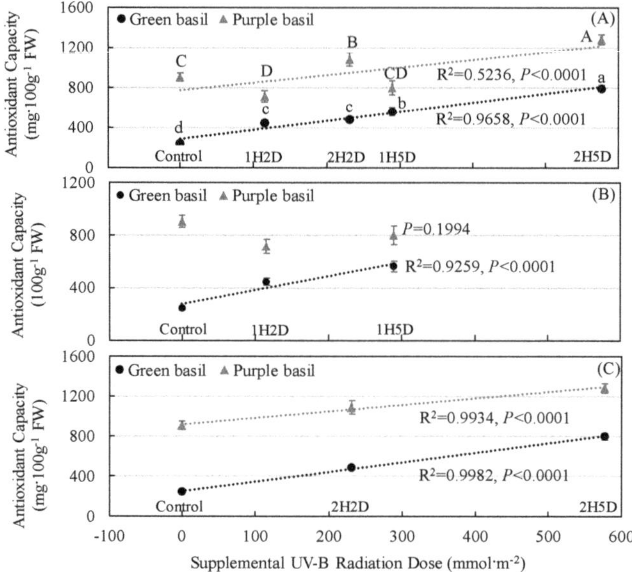

Figure 5. Correlation between antioxidant capacity of green basil 'Improved Genovese Compact' and purple basil 'Red Rubin' plants with UV-B radiation doses. Correlation test was conducted in three terms according to different UV-B radiation patterns, five supplemental UV-B radiation treatments including control, 1H2D, 2H2D, 1H5D, 2H5D (**A**), control and 1 h·d^{-1} UV-B radiation treatments (**B**), and control and 2 h·d^{-1} UV-B radiation treatments (**C**). Data were pooled from two photosynthetic photon flux density (PPFD) treatments. Means followed by the same lower/upper case letters are not significantly different for green/purple basil plants, according to Student's *t* mean comparison ($p < 0.05$). Bars represent standard errors. Dashed lines show the regression between antioxidant capacity with supplemental UV-B radiation dose, according to the pairwise correlation method.

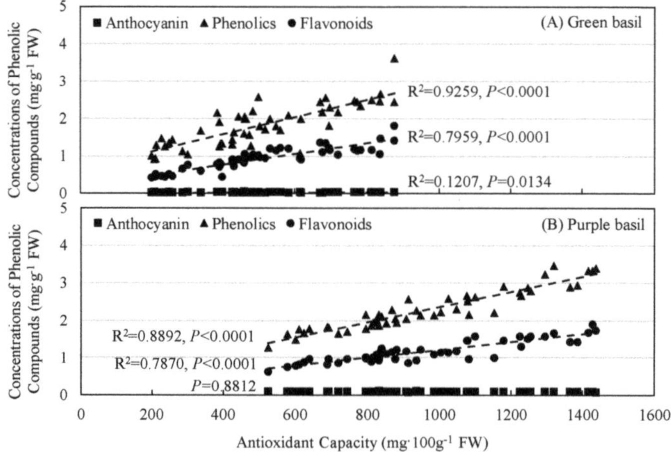

Figure 6. Correlation between antioxidant capacity and concentrations of anthocyanin, phenolics, and flavonoids in green basil plants (**A**), and purple basil plants (**B**). Dashed lines show the regression between concentrations of phenolic compounds with antioxidant capacity according to Pairwise Correlation method.

4. Discussion

4.1. Impacts of UV-B and PPFD on Photosynthesis, Relative Chlorophyll Concentration, and Chlorophyll Fluorescence

Photosynthesis is one of the most sensitive metabolic processes in plants responding to environmental condition changes, such as supplemental UV-B radiation and PPFD. In the present study, P_n in basil leaves was lower after UV-B radiation, which was mainly caused by the direct damage of PSII components and led to reduced photosynthetic capacity, subsequently decreased G_s [32–34]. Meanwhile, relative chlorophyll content in basil leaves was also lower after UV-B radiation, either through degradation or inhibition of enzymes involved in the chlorophyll biosynthetic pathways [34]. However, compared to depressed photosynthesis and reduced chlorophyll content by supplemental UV-B radiation in our study, a meta-analysis of field studies (more than 450 reports from 62 papers) reported unaffected photosynthesis and chlorophyll content after supplemental UV-B radiation [35]. Differences between our study (controlled environment with artificial lighting) from previous field studies (sunlight) probably resulted from significantly low PPFDs and relatively high UV-B proportion used in our study. Firstly, in controlled environment systems, due to the high cost of powering artificial lighting, lower PPFDs are normally used compared to that of sunlight intensity in an open field. Subsequently, lower PPFDs resulted in depressed photochemical protection system of plants, such as decreased photosynthetic capacity, decreased leaf thickness, and reduced concentrations of UV-absorbing agents [21], which aggravated the negative effects caused by UV-B radiation. Secondly, the damage caused by UV-radiation increases with decreasing UV wavelength, since short UV wavelength has more energy than long UV wavelength [36]. The UV component of sunlight consists of 95% UV-A and 5% UV-B, of which the small portion UV-B radiation shows stronger mutagenic and carcinogenic effects compared to UV-A radiation [36,37]. For example, a less prominent and less long-lasting activation of p53 gene ("guardian of the genome") after UV-A radiation compared to UV-B was observed, suggesting stronger effects of UV-B radiation than UV-A [36]. In the present study, the UV radiation provided by broadband UV-B lamps was mainly UV-B radiation with relatively low UV-A radiation, contributing to aggravated negative effects on plant photosynthesis compared to previous field studies, of which mainly consists of UV-A radiation.

Chlorophyll fluorescence parameters provide precise and objective information with regard to photochemical efficiency and non-photochemical de-excitation involved in the conversion of light energy under different conditions [28,38]. The less reduced Fv/Fm, PI ABS, TR_0/CS, and ET_0/CS after UV-B radiation in purple basil plants than green basil plants clearly indicate that purple basil plants are more tolerant to UV-B radiation, resulted from its improved capacity to process excess UV-B energy through PSII [39]. Meanwhile, the uninfluenced DI_0/CS under UV-B treatments in green basil plants suggests its inability to dissipate absorbed UV-B energy in the form of harmless heat, even under the smallest UV-B radiation dose, 16.0 $\mu mol \cdot m^{-2} \cdot s^{-1}$ at 1 $h \cdot d^{-1}$ for 2 days, while purple basil plants coped with excess UV-B energy by increasing heat dissipation. Mosadegh et al. (2018) also reported that the DI_0/CS of green basil plants was not affected after 2-weeks UV-B radiation at 68 and 102 $kJ \cdot m^{-2} \cdot d^{-1}$, confirming that green basil plants failed to dissipate UV-B energy as harmless heat [28]. Differences in chlorophyll fluorescence parameters between green and purple basil plants may be due to the relatively higher concentrations of UV-protective antioxidants in purple basil plants such as anthocyanins, phenolics, and flavonoids, which are known to provide plants with strong protection from excess UV-B energy [40].

In our previous study, the gas exchange rate in green basil plants was positively correlated with PPFD [21], while it was not affected in the present study. This may be due to the large variation of P_n, E, and G_s caused by UV-B radiation at each PPFD. In green basil plants, P_n ranged from 3.7 to 12.6 $\mu mol \cdot m^{-2} \cdot s^{-1}$ at low PPFD (160 $\mu mol \cdot m^{-2} \cdot s^{-1}$), and ranged from 4.8 to 13.8 $\mu mol \cdot m^{-2} \cdot s^{-1}$ at high PPFD (224 $\mu mol \cdot m^{-2} \cdot s^{-1}$). Also, it was observed that the P_n in purple basil plants was much lower compared to the P_n in green basil plants. One hypothesis is that the differences between two cultivars is due to the lower quantum efficiency of photosynthetically active radiation (PAR) in purple basil plants compared to green basil plants. In purple basil plants, the relatively high concentration of anthocyanins and flavonoids absorbs more PAR light, which decreases the absorption of PAR light by chloroplasts and subsequently decreases the photochemistry energy transferred to reaction centers, resulting in decreased P_n in purple basil plants compared to green basil plants [41].

4.2. Impacts of UV-B and PPFD on Growth and Yield

Plant leaf expansion is invariably inhibited by supplemental UV-B radiation and other leaf morphogenesis changes such as reduced leaf area, increased leaf thickness, and accumulation of leaf surface waxes are also observed across a range of plant species [14,42,43]. Internode length is also a very sensitive growth parameter that responds to UV-B radiation [44]. Kaiserli (2018) reported that most cell-wall elongation genes induced by BRI1-EMS-SUPPRESSOR 1 (BES1) are negatively regulated by UV-B radiation [45]. Meanwhile, the biosynthesis and signaling of plant growth hormone auxin, a key regulator of stem elongation, was also suppressed in arabidopsis (*Arabidopsis thaliana*) and coriander (*Coriandrum sativum*) plants after UV-B radiation, thereby reducing plant stem elongation and promoting a compact phenotype [46]. In the present study, similar results such as reduced leaf area, increased leaf thickness, accumulation of leaf surface waxes, and reduced leaf internode length were observed, which are plant acclimation responses to supplemental UV-B radiation. In addition to protecting plants from receiving excess UV-B energy, these acclimation responses also provide plants with improved tolerance to other adverse environmental conditions, such as heat stress and mechanical handling during postharvest [6,47,48].

Reduced gas exchange rate and leaf expansion, and inhibition of stem elongation of basil plants under supplemental UV-B radiation resulted in a reduction in plant size and yield. The greater yield reduction by the UV-B radiation under high PPFD than low PPFD may be due to its taller plants, which shortened the distance between basil plants and UV-B light tube, resulting in increased UV-B radiation intensity sustained by basil plants, and subsequently severer yield reduction.

4.3. Impacts of UV-B and PPFD on Phytochemical Accumulation and Antioxidant Capacity

Across a range of plant species, phenolic compounds, especially flavonoids, act as efficient UV-screening agents to reduce excess UV light received by photosynthetic tissues to protect plants from possible harm [40,49]. Enhanced accumulation of phenolic compounds by supplemental UV-B radiation has been supported by a large body of experimental evidence [50,51], which was confirmed in this study. Ghasemzadeh et al. (2016) reported that total phenolic and flavonoid content in green basil plants increased by 16% and 85%, respectively, after a 13 kJ·m^{-2}·h^{-1} post-harvest UV-B radiation for 4–10 h, but anthocyanin content was not measured [52]. It was also reported that upon supplemental UV-B radiation, the gene expression of phenylalanine ammonia lyase (PAL) and chalcone synthase (CHS), two key molecular markers for phenolic compounds biosynthesis increased significantly [46,53]. Noticeably, in the present study, the enhancement of flavonoids and phenolics by UV-B radiation was much greater than anthocyanin. Consistently, antioxidant capacity was significantly correlated with concentrations of phenolics and flavonoids in both basil cultivars, while marginally or not correlated to anthocyanin concentration. This might be due to the higher ROS-scavenging capacity of phenolics and flavonoids than anthocyanins, resulting in more sensitive reactions of phenolics and flavonoids to UV-B radiation [54]. Csepregi et al. (2017) also reported such differential regulation of different phenolic compounds by UV-B radiation, in which quercetins with additional hydroxyl group on ring-B increased up to 10 folds while kaempferol increased 3–4 fold, due to their different ROS-scavenging capacity [55].

Enhancement of phenolic compounds after UV-B radiation was greater in basil plants grown under low PPFD compared to those grown under high PPFD, indicating basil plants are more sensitive to UV-B radiation under low PPFD. In a similar way, Behn et al. (2010) reported that under low PPFD (550 μmol·m^{-2}·s^{-1}), essential oil quality in peppermint plants was improved in terms of an enhanced menthone to menthol conversion after UV-B radiation, while not affected by UV-B treatment under high PPFD (1150 $\mu mol\cdot m^{-2}\cdot s^{-1}$) [56]. As mentioned, this may be due to a depressed photochemical and biochemical protection system of plants grown under low PPFD, such as lower leaf thickness and reduced concentrations of UV-absorbing agents [21]. As we hypothesized, concentrations of phenolic compounds in basil plants grown under low PPFD with UV-B radiation was significantly higher compared to those of plants grown under high PPFD without UV-B radiation, suggesting that UV-B radiation could be used as a tool to compensate for reduced accumulation of phenolic compounds in basil plants grown under controlled environment.

Similar to plant responses on chlorophyll fluorescence, different responses in phytochemical accumulation between green and purple basil plants were also observed. Specifically, purple basil plants showed fewer biochemical changes than green basil plants after UV-B radiation, which performed as unaffected anthocyanin concentration and less induction of phenolics and flavonoids. Our hypothesis is that the relatively high concentrations of phenolic compounds in purple basil plants act as potent UV-screening agents as well as free-radical scavengers to protect purple basil plants from excess UV-B light. Under high PPFD without UV-B treatment, concentrations of anthocyanin, phenolics, and flavonoids and antioxidant capacity in purple basil leaves were 3.33, 1.47, 1.93, 3.72 times those in green basil leaves, respectively. This hypothesis was confirmed by Tattini et al. (2014), in which he reported that purple basil 'Red Rubin' showed lower metabolic cost of photoprotective mechanisms than green basil 'Tigullio' when being moved from 30% to 100% sunlight condition [57].

4.4. Impacts of UV-B Radiation Doses and Radiation Patterns on Phytochemical Accumulation and Antioxidant Capacity

With the radiation doses and different radiation patterns used in the present study, green basil plants were more dose-dependent, while purple basil plants were both dose-dependent and radiation pattern-dependent. Antioxidant capacity in green basil plants was significantly correlated with the UV-B radiation dose for both 1 $h\cdot d^{-1}$ and 2 $h\cdot d^{-1}$ UV-B radiation patterns, while antioxidant capacity in purple basil plants was not affected by 1 $h\cdot d^{-1}$ UV-B radiation treatments. With the similar UV-B radiation dose (1H5D and 2H2D treatments), after 1 $h\cdot d^{-1}$ UV-B radiation treatments, the recovery time until next day treatment (23 h) allowed purple basil plants' signaling and metabolic adaptation to (at least partially) reset to pre-stress level, without increasing phenolic compounds accumulation, while after 2 $h\cdot d^{-1}$ UV-B radiation (recovery time of 22 h until next treatment), purple basil plants failed to recover from UV-B radiation stress and resulted in an overall increase of phenolic compounds to cope with excess UV-B energy. This indicated that radiation patterns play an important role in regulating purple basil responses to UV-B radiation, while radiation dose is the determining factor in regulating green basil biochemical responses. Mosadegh et al. (2018) also reported that with the same UV-B radiation dose of 102 $kJ\cdot m^{-2}$, phenolics concentration of green basil 'Genovese' was the same level regardless of UV-B radiation pattern, continuous 1-d UV-B radiation or discontinuous 6-d UV-B radiation [28]. However, at lower UV-B radiation doses of 8.5, 34, and 68 $kJ\cdot m^{-2}$, when 'Genovese' green basil plants were treated with the same UV-B radiation dose, continuous 1-d UV-B radiation resulted in significant higher phenolics concentration compared to plants treated with discontinuous 6-d UV-B radiation [28]. Thus, plant responses to UV-B radiation in green basil plants may also depend on radiation patterns, which are affected by the total UV-B radiation dose.

4.5. Implications of Study Findings

Different plant responses to UV-B radiation are observed in studies conducted in the open field with sunlight than in a controlled environment with artificial lighting, due to different PPFDs and components of UV radiation [13,35,58,59]. The novel finding of the present study is that plants grown under a controlled environment with lower PPFDs are more sensitive to UV-B radiation. Therefore, for future studies under a controlled environment, a lower UV-B radiation dose should be applied to reduce its negative effects on plant photosynthesis, growth, or yield. Furthermore, we see differential responses in green and purple basil plants to UV-B radiation doses and radiation patterns. Therefore, to better understand plant responses to supplemental UV-B radiation, more plant species/cultivars, lower radiation doses, and different radiation patterns need to be investigated in future studies.

Plant acclimation responses to supplemental UV-B radiation lead to plant cross-protection against other environmental stresses, through photochemical, morphological, and biochemical mechanisms [60]. For example, *UVR8* was recently shown to be involved in regulating thermomorphogenesis, shade-avoidance responses, and plant immunity, underlining the importance of signaling crosstalk among UV-B radiation, hormone, and defense pathways [47,61]. As a result, supplemental UV-B radiation could be used as a tool to improve plant tolerance to other adverse environmental conditions, and interactions between supplemental UV-B radiation and other key environmental factors still need to be studied.

5. Conclusions

Results of the present study suggest that a short period of pre-harvest supplemental UV-B radiation could significantly improve phytochemical concentrations in basil plants, and plant responses to UV-B radiation vary among plant cultivars, radiation doses, and radiation patterns. Meanwhile, effects of UV-B radiation on basil plants interacted with PPFDs used in the cultivation system, and high PPFD improved plant tolerance to UV-B radiation. Also, supplemental UV-B radiation could compensate for the reduced accumulation of phenolic compounds in basil plants grown under low PPFD. Therefore, combining plant growth performance, yield, and accumulation of health-promoting phenolic compounds, a pre-harvest UV-B radiation of 1 h·d^{-1} for 2 days under a PPFD of 224 μmol·m^{-2}·s^{-1} was recommended for green basil 'Improved Genovese Compact' production under a controlled environment. However, supplemental UV-B radiation doses used in this study decreased the total amount of phenolic compounds in purple basil plants due to yield reduction, and UV-B radiation is not recommended for purple basil 'Red Rubin' production under a controlled environment.

Author Contributions: Conceptualization, H.D., G.N. and M.G.; data analysis, H.D.; methodology, H.D., G.N. and M.G.; writing—original draft preparation, H.D.; writing—review and editing, G.N., M.G.

Funding: This research was partially funded by the USDA National Institute of Food and Agriculture Hatch project TEX090450, Texas A&M AgriLife Research, and Texas A&M AgriLife Extension Service.

Acknowledgments: The authors appreciate the assistance from Youping Sun, Zhanyang Xu, Christina Perez, Triston Hooks, and the student workers at Texas A&M AgriLife Research Center at El Paso, TX.

Conflicts of Interest: The authors declare no conflict of interest.

References

1. Gruda, N.S. Increasing Sustainability of Growing Media Constituents and Stand-Alone Substrates in Soilless Culture Systems. *Agronomy* **2019**, *9*, 298. [CrossRef]
2. Kozai, T.; Niu, G.; Takagaki, M. *Plant Factory—An Indoor Vertical Farming System for Efficient Quality Food Production*, 2nd ed.; Academic Press, Elsevier: New York, NY, USA, 2019; in press.
3. Goto, E.; Hayashi, K.; Furuyama, S.; Hikosaka, S.; Ishigami, Y. Effect of UV Light on Phytochemical Accumulation and Expression of Anthocyanin Biosynthesis Genes in Red Leaf Lettuce. In Proceedings of the VIII International Symposium on Light in Horticulture, East Lansing, MI, USA, 22–26 May 2016; pp. 179–186.
4. Wargent, J.J. UV LEDs in Horticulture: From Biology to Application. In Proceedings of the VIII International Symposium on Light in Horticulture, East Lansing, MI, USA, 22–26 May 2016; pp. 25–32.
5. Caldwell, M.; Flint, S.; Searles, P. Spectral Balance and UV-B Sensitivity of Soybean: A Field Experiment. *Plant. Cell Environ.* **1994**, *17*, 267–276. [CrossRef]
6. Wargent, J.J.; Moore, J.P.; Roland Ennos, A.; Paul, N.D. Ultraviolet Radiation as a Limiting Factor in Leaf Expansion and Development. *Photochem. Photobiol.* **2009**, *85*, 279–286. [CrossRef]

7.   Sun, R.; Hikosaka, S.; Goto, E.; Sawada, H.; Saito, T.; Kudo, T.; Ohno, T.; Shibata, T.; Yoshimatsu, K. Effects of UV Irradiation on Plant Growth and Concentrations of Four Medicinal Ingredients in Chinese Licorice (*Glycyrrhiza uralensis*). In Proceedings of the VII International Symposium on Light in Horticultural Systems, Wageningen, The Netherlands, 15–18 October 2012; pp. 643–648.

8.   Castagna, A.; Dall'Asta, C.; Chiavaro, E.; Galaverna, G.; Ranieri, A. Effect of Post-harvest UV-B Irradiation on Polyphenol Profile and Antioxidant Activity in Flesh and Peel of Tomato Fruits. *Food Bioprocess. Technol.* **2014**, *7*, 2241–2250. [CrossRef]

9.   Moreira-Rodríguez, M.; Nair, V.; Benavides, J.; Cisneros-Zevallos, L.; Jacobo-Velázquez, D.A. UVA, UVB Light Doses and Harvesting Time Differentially Tailor Glucosinolate and Phenolic Profiles in Broccoli Sprouts. *Molecules* **2017**, *22*, 1065. [CrossRef] [PubMed]

10. Rouphael, Y.; Kyriacou, M.C.; Petropoulos, S.A.; de Pascale, S.; Colla, G. Improving Vegetable Quality in Controlled Environments. *Sci. Hort.* **2018**, *234*, 275–289. [CrossRef]

11. Schreiner, M.; Mewis, I.; Huyskens-Keil, S.; Jansen, M.A.K.; Zrenner, R.; Winkler, J.B.; O'Brien, N.; Krumbein, A. UV-B-induced Secondary Plant Metabolites—Potential Benefits for Plant and Human Health. *Crit. Rev. Plant. Sci.* **2012**, *31*, 229–240. [CrossRef]

12. Dotto, M.; Casati, P. Developmental Reprogramming by UV-B Radiation in Plants. *Plant. Sci.* **2017**, *264*, 96–101. [CrossRef]

13. Henry-Kirk, R.A.; Plunkett, B.; Hall, M.; McGhie, T.; Allan, A.C.; Wargent, J.J.; Espley, R.V. Solar UV Light Regulates Flavonoid Metabolism in Apple (*Malus x domestica*). *Plant. Cell Environ.* **2018**, *41*, 675–688. [CrossRef]

14. Jansen, M.A.; Bornman, J.F. UV-B Radiation: From Generic Stressor to Specific Regulator. *Physiol. Plant.* **2012**, *145*, 501–504. [CrossRef]

15. Höll, J.; Lindner, S.; Walter, H.; Joshi, D.; Poschet, G.; Pfleger, S.; Ziegler, T.; Hell, R.; Bogs, J.; Rausch, T. Impact of Pulsed UV-B Stress Exposure on Plant Performance: How Recovery Periods Stimulate Secondary Metabolism While Reducing Adaptive Growth Attenuation. *Plant. Cell Environ.* **2019**, *42*, 801–814. [CrossRef]

16. Brown, B.A.; Jenkins, G.I. UV-B Signaling Pathways with Different Fluence-rate Response Profiles Are Distinguished in Mature *Arabidopsis* Leaf Tissue by Requirement for UVR8, HY5, and HYH. *Plant. Physiol.* **2008**, *146*, 576–588. [CrossRef]

17. Favory, J.J.; Stec, A.; Gruber, H.; Rizzini, L.; Oravecz, A.; Funk, M.; Albert, A.; Cloix, C.; Jenkins, G.I.; Oakeley, E.J. Interaction of COP1 and UVR8 Regulates UV-B-induced Photomorphogenesis and Stress Acclimation in *Arabidopsis*. *EMBO J.* **2009**, *28*, 591–601. [CrossRef]

18. Alexandru Suchar, V.; Robberecht, R. Integration and Scaling of UV-B Radiation Effects on Plants: The Relative Sensitivity of Growth Forms and Interspecies Interactions. *J. Plant. Ecol.* **2017**, *11*, 656–670. [CrossRef]

19. Li, Q.; Kubota, C. Effects of Supplemental Light Quality on Growth and Phytochemicals of Baby Leaf Lettuce. *Environ. Exper. Bot.* **2009**, *67*, 59–64. [CrossRef]

20. Makri, O.; Kintzios, S. *Ocimum* sp. (Basil): Botany, Cultivation, Pharmaceutical Properties, and Biotechnology. *J. Herbs Spices Med. Plants* **2008**, *13*, 123–150. [CrossRef]

21. Dou, H.; Niu, G.; Gu, M.; Masabni, J.G. Responses of Sweet Basil to Different Daily Light Integrals in Photosynthesis, Morphology, Yield, and Nutritional Quality. *HortScience* **2018**, *53*, 496–503. [CrossRef]

22. Liaros, S.; Botsis, K.; Xydis, G. Technoeconomic Evaluation of Urban Plant Factories: The Case of Basil (*Ocimum basilicum*). *Sci. Total Environ.* **2016**, *554*, 218–227. [CrossRef]

23. Hogewoning, S.; Trouwborst, G.; Meinen, E.; van Ieperen, W. Finding the Optimal Growth-Light Spectrum for Greenhouse Crops. In Proceedings of the the VII International Symposium on Light in Horticultural Systems, Wageningen, The Netherlands, 15–18 October 2012; pp. 357–363.

24. Stutte, G.W. Controlled Environment Production of Medicinal and Aromatic Plants. In *Medicinal and Aromatic Crops: Production, Phytochemistry, and Utilization*; Jeliazkov, V.D., Cantrell, C.L., Eds.; ACS Publications: Washington, DC, USA, 2016; pp. 49–63.

25. Johnson, C.B.; Kirby, J.; Naxakis, G.; Pearson, S. Substantial UV-B-mediated Induction of Essential Oils in Sweet Basil (*Ocimum basilicum* L.). *Phytochemistry* **1999**, *51*, 507–510. [CrossRef]

26. Sakalauskaite, J.; Viskelis, P.; Dambrauskiene, E.; Sakalauskiene, S.; Samuoliene, G.; Brazaityte, A.; Duchovskis, P.; Urbonaviciene, D. The Effects of Different UV-B Radiation Intensities on Morphological and Biochemical Characteristics in *Ocimum basilicum* L. *J. Sci. Food Agr.* **2013**, *93*, 1266–1271. [CrossRef]

27. Sakalauskaite, J.; Viskelis, P.; Duchovskis, P.; Dambrauskiene, E.; Sakalauskiene, S.; Samuoliene, G.; Brazaityte, A. Supplementary UV-B Irradiation Effects on Basil (*Ocimum basilicum* L.) Growth and Phytochemical Properties. *J. Food Agr. Environ.* **2012**, *10*, 342–346.

28. Mosadegh, H.; Trivellini, A.; Ferrante, A.; Lucchesini, M.; Vernieri, P.; Mensuali, A. Applications of UV-B Lighting to Enhance Phenolic Accumulation of Sweet Basil. *Sci. Hort.* **2018**, *229*, 107–116. [CrossRef]

29. Xu, C.P.; Mou, B.Q. Responses of Spinach to Salinity and Nutrient Deficiency in Growth, Physiology, and Nutritional Value. *J. Amer. Soc. Hort. Sci.* **2016**, *141*, 12–21. [CrossRef]

30. Connor, A.M.; Luby, J.J.; Tong, C.B. Variability in Antioxidant Activity in Blueberry and Correlations Among Different Antioxidant Activity Assays. *J. Am. Soc. Hort. Sci.* **2002**, *127*, 238–244. [CrossRef]

31. Arnao, M.B.; Cano, A.; Acosta, M. The Hydrophilic and Lipophilic Contribution to Total Antioxidant Activity. *Food Chem.* **2001**, *73*, 239–244. [CrossRef]

32. Sullivan, J.H.; Teramura, A.H. Field Study of the Interaction Between Solar Ultraviolet-B Radiation and Drought on Photosynthesis and Growth in Soybean. *Plant. Physiol.* **1990**, *92*, 141–146. [CrossRef]

33. Lidon, F.J.; Reboredo, F.H.; Silva, M.M.A.; Duarte, M.P.; Ramalho, J.C. Impact of UV-B Radiation on Photosynthesis—An Overview. *Emir. J. Food Agr.* **2012**, *24*, 546–556. [CrossRef]

34. Yadav, S.; Shrivastava, A.K.; Agrawal, C.; Sen, S.; Chatterjee, A.; Rai, S.; Rai, R.; Singh, S.; Rai, L. Impact of UV-B Exposure on Phytochrome and Photosynthetic Machinery: From Cyanobacteria to Plants. In *UV-B Radiation: From Environmental Stressor to Regulator of Plant Growth*; Singh, S., Prasad, S.M., Parhar, P., Eds.; John Wiley & Sons, Ltd.: West Sussex, UK, 2017; pp. 259–277.

35. Searles, P.S.; Flint, S.D.; Caldwell, M.M. A Meta-analysis of Plant Field Studies Simulating Stratospheric Ozone Depletion. *Oecologia* **2001**, *127*, 1–10. [CrossRef]

36. Kappes, U.P.; Luo, D.; Potter, M.; Schulmeister, K.; Rünger, T.M. Short- and Long-wave UV light (UVB and UVA) Induce Similar Mutations in Human Skin Cells. *J. Investig. Dermatol.* **2006**, *126*, 667–675. [CrossRef]

37. Ikehata, H.; Ono, T. The Mechanisms of UV Mutagenesis. *J. Radiat. Res.* **2011**, *52*, 115–125. [CrossRef]

38. Strasser, R.J.; Srivastava, A.; Tsimilli-Michael, M. The Fluorescence Transient as a Tool to Characterize and Screen Photosynthetic Samples. In *Probing Photosynthesis: Mechanisms, Regulation and Adaptation*; Yunus, M., Pathre, U., Mohanty, P., Eds.; Taylor and Francis Inc.: New York, NY, USA, 2000; pp. 445–483.

39. Rai, K.; Agrawal, S. Effects of UV-B Radiation on Morphological, Physiological and Biochemical Aspects of Plants: An Overview. *J. Sci. Res.* **2017**, *61*, 87–113.

40. Takahashi, S.; Badger, M.R. Photoprotection in Plants: A New Light on Photosystem II Damage. *Trends Plant. Sci.* **2011**, *16*, 53–60. [CrossRef]

41. Sun, J.; Nishio, J.N.; Vogelmann, T.C. Green Light Drives CO_2 Fixation Deep Within Leaves. *Plant. Cell Physiol.* **1998**, *39*, 1020–1026. [CrossRef]

42. Cen, Y.P.; Bornman, J.F. The Effect of Exposure to Enhanced UV-B Radiation on the Penetration of Monochromatic and Polychromatic UV-B Radiation in Leaves of *Brassica napus*. *Physiol. Plant.* **1993**, *87*, 249–255. [CrossRef]

43. Wargent, J.J.; Jordan, B.R. From Ozone Depletion to Agriculture: Understanding the Role of UV Radiation in Sustainable Crop Production. *New Phytol.* **2013**, *197*, 1058–1076. [CrossRef]

44. Zhao, D.; Reddy, K.; Kakani, V.; Read, J.; Sullivan, J. Growth and Physiological Responses of Cotton (*Gossypium hirsutum* L.) to Elevated Carbon Dioxide and Ultraviolet-B Radiation under Controlled Environmental Conditions. *Plant. Cell Environ.* **2003**, *26*, 771–782. [CrossRef]

45. Kaiserli, E. Ultraviolet Rays Light Up Transcriptional Networks Regulating Plant Growth. *Dev. Cell* **2018**, *44*, 409–411. [CrossRef]

46. Fraser, D.P.; Sharma, A.; Fletcher, T.; Budge, S.; Moncrieff, C.; Dodd, A.N.; Franklin, K.A. UV-B Antagonises Shade Avoidance and Increases Levels of the Flavonoid Quercetin in Coriander (*Coriandrum sativum*). *Sci. Rep.* **2017**, *7*, 17758. [CrossRef]

47. Teklemariam, T.; Blake, T.J. Effects of UVB Preconditioning on Heat Tolerance of Cucumber (*Cucumis sativus* L.). *Environ. Exp. Bot.* **2003**, *50*, 169–182. [CrossRef]

48. Kakani, V.; Reddy, K.; Zhao, D.; Mohammed, A. Effects of Ultraviolet-B Radiation on Cotton (*Gossypium hirsutum* L.) Morphology and Anatomy. *Ann. Bot.* **2003**, *91*, 817–826. [CrossRef]

49. Logan, B.A.; Stafstrom, W.C.; Walsh, M.J.; Reblin, J.S.; Gould, K.S. Examining the Photoprotection Hypothesis for Adaxial Foliar Anthocyanin Accumulation by Revisiting Comparisons of Green- and Red-leafed Varieties of Coleus (*Solenostemon scutellarioides*). *Photosyn. Res.* **2015**, *124*, 267–274. [CrossRef]

50. Agati, G.; Tattini, M. Multiple Functional Roles of Flavonoids in Photoprotection. *New Phytol.* **2010**, *186*, 786–793. [CrossRef]

51. Hatier, J.H.B.; Clearwater, M.J.; Gould, K.S. The Functional Significance of Black-pigmented Leaves: Photosynthesis, Photoprotection and Productivity in *Ophiopogon planiscapus* 'Nigrescens'. *PLoS ONE* **2013**, *8*, e67850. [CrossRef]

52. Ghasemzadeh, A.; Ashkani, S.; Baghdadi, A.; Pazoki, A.; Jaafar, H.Z.; Rahmat, A. Improvement in Flavonoids and Phenolic Acids Production and Pharmaceutical Quality of Sweet Basil (*Ocimum basilicum* L.) by Ultraviolet-B Irradiation. *Molecules* **2016**, *21*, 1203. [CrossRef]

53. Rodríguez-Calzada, T.; Qian, M.; Strid, A.; Neugart, S.; Schreiner, M.; Torres-Pacheco, I.; Guevara-González, R.G. Effect of UV-B Radiation on Morphology, Phenolic Compound Production, Gene Expression, and Subsequent Drought Stress Responses in Chili Pepper (*Capsicum annuum* L.). *Plant. Physiol. Biochem.* **2019**, *134*, 94–102. [CrossRef]

54. Hideg, E.; Jansen, M.A.K.; Strid, A. UV-B Exposure, ROS, and Stress: Inseparable Companions or Loosely Linked Associations? *Trends Plant. Sci.* **2013**, *18*, 107–115. [CrossRef]

55. Csepregi, K.; Coffey, A.; Cunningham, N.; Prinsen, E.; Hideg, É.; Jansen, M.A. Developmental Age and UV-B Exposure Co-determine Antioxidant Capacity and Flavonol Accumulation in *Arabidopsis* Leaves. *Environ. Exp. Bot.* **2017**, *140*, 19–25. [CrossRef]

56. Behn, H.; Albert, A.; Marx, F.; Noga, G.; Ulbrich, A. Ultraviolet-B and Photosynthetically Active Radiation Interactively Affect Yield and Pattern of Monoterpenes in Leaves of Peppermint (*Mentha × piperita* L.). *J. Agr. Food Chem.* **2010**, *58*, 7361–7367. [CrossRef]

57. Tattini, M.; Landi, M.; Brunetti, C.; Giordano, C.; Remorini, D.; Gould, K.S.; Guidi, L. Epidermal Coumaroyl Anthocyanins Protect Sweet Basil Against Excess Light Stress: Multiple Consequences of Light Attenuation. *Physiol. Plant.* **2014**, *152*, 585–598. [CrossRef]

58. Li, F.R.; Peng, S.L.; Chen, B.M.; Hou, Y.P. A Meta-analysis of the Responses of Woody and Herbaceous Plants to Elevated Ultraviolet-B Radiation. *Acta Oecologica* **2010**, *36*, 1–9. [CrossRef]

59. Wargent, J.J.; Elfadly, E.M.; Moore, J.P.; Paul, N.D. Increased Exposure to UV-B Radiation During Early Development Leads to Enhanced Photoprotection and Improved Long-term Performance in *Lactuca sativa*. *Plant. Cell Environ.* **2011**, *34*, 1401–1413. [CrossRef]

60. Yin, R.; Ulm, R. How Plants Cope with UV-B: From Perception to Response. *Cur. Opin. Plant. Biol.* **2017**, *37*, 42–48. [CrossRef]

61. Schultze, M.; Bilger, W. Acclimation of *Arabidopsis thaliana* to Low Temperature Protects Against Damage of Photosystem II Caused by Exposure to UV-B Radiation at 9 °C. *Plant. Physiol. Biochem.* **2018**, *134*, 73–80. [CrossRef]

Article

Comparative Study of Three Low-Tech Soilless Systems for the Cultivation of Geranium (*Pelargonium zonale*): A Commercial Quality Assessment

Luca Brentari [1], Nicola Michelon [2,*], Giorgio Gianquinto [2], Francesco Orsini [2], Federico Zamboni [3] and Duilio Porro [1]

[1] Edmund Mach Foundation, Technology Transfer Centre, via E. Mach 1, 38098 San Michele (TN), Italy; luca.brentari@fmach.it (L.B.); duilio.porro@fmach.it (D.P.)

[2] Research Centre in Urban Environment for Agriculture and Biodiversity (ResCUE-AB), Department of Agricultural and Food Sciences (DISTAL), Alma Mater Studiorum University of Bologna, Viale Fanin, 44, 40127 Bologna, Italy; giorgio.gianquinto@unibo.it (G.G.); f.orsini@unibo.it (F.O.)

[3] Zamboni Stefano Farm, via Trento 13, 38049 Vigolo Vattaro, Italy; federico.zamboni10@gmail.com

* Correspondence: nicola.michelon@unibo.it; Tel.: +39-051-2096677

Received: 25 August 2020; Accepted: 18 September 2020; Published: 20 September 2020

Abstract: The study evaluated the feasibility of simplified hydroponics for the growth of rooted cuttings of geranium (*Pelargonium zonale*) for commercial purposes in local farms in Northern Italy. Tested systems included a control where soilless system on substrate (peat) (T-1), usually adopted by local farmers, was compared against an open-cycle drip system on substrate (peat) (T-2), and a Nutrient Film Technique system (T-3). For commercial features, assessed parameters included flowering degree (flowering timing, numbers of inflorescences plant^{-1}, and number of flowers inflorescence^{-1}), numbers of leaves plant^{-1}, number of branches plant^{-1}, final height of plant, and the aesthetic-commercial assessment index. Assessed parameters also included fresh and dry weight, SPAD Index, the water consumption, and the water use efficiency (WUE). The soilless systems typology significantly affected rooted cuttings growth, commercial features, and WUE. The adoption of an open-cycle drip system (T-2) resulted in a significant improvement of all the crop commercial characteristics as compared with other treatments, making plants more attractive for the market. The water consumption was higher in T-2 as compared with T-1 and T-3, but it allowed for the highest fresh weight, and therefore also the highest WUE. The results indicate that the typology of soilless system significantly enhances the commercial characteristics of geranium.

Keywords: *Pelargonium zonale*; low-tech soilless cultivation system; commercial quality

1. Introduction

The global gardening pots market size was valued at USD 1.7 billion in 2018. A growing interest in gardening is expected to remain a favorable factor for industry growth [1]. In Italy, the cultivation of cut flowers and potted ornamentals in both greenhouses and open field accounts for a relevant share of the market. In 2017, out of the 2.5 billion euros associated with the national floricultural and ornamental crop sector, about 1.15 billion euros are associated with flower production and potted plants. The Italian cut flowers and potted ornamental sector accounts for 27,000 companies, 100,000 workers, and almost 29,000 ha of farmland. When considering only figures for the ornamental seedling production, 2000 farms for a total area of 1500 ha are also found in Italy [2].

Ornamental plants are typically characterized by a fast growth rate and a large consumption of both nutrients and water, which should be of elevated quality given the limited salt tolerance of these plant species [3]. Furthermore, farmers generally tend to overwater these crops, with the consequence that ornamental plants generally present low water use efficiency (WUE) values. Accordingly, despite the existing variability among ornamental species in terms of water and fertilizer requirements [4], the sector generally accounts for high environmental impact due to the losses of both water and fertilizers [3]. In this scenario, the increasing awareness of environmental pollution caused by agriculture, the scarcity of resources such as water, the need to reduce the production costs, and the growing demand for healthy foods are forcing operators to move towards more sustainable cropping techniques. In greenhouse cultivation, the adoption of soilless culture, coupled with techniques such as fertigation, drip irrigation, integrated plant protection, and climate control, can provide a high-quality product with efficient use of resources, e.g., water, while also increasing the potential yield [5–7] as well as decreasing nutrient losses [8]. Soilless culture can be defined as "any method of growing plants without the use of soil as a rooting medium, in which the inorganic nutrients absorbed by the roots are supplied via the irrigation water" [9]. The soilless systems are classified according to the presence and type of substrate, to the irrigation system, and to the nutrient solution (NS) management, namely the reuse or not of the leaching fraction [9], which results, respectively, in the so-called "closed" or "open" loop systems. In open-loop systems, an excessive amount of NS (120–150% of actual water requirements) is supplied to avoid salt accumulation in the substrate, and the leaching fraction is not reused and commonly released into the environment. In closed soilless systems, on the other hand, water supply is generally higher (150–170% of daily water requirements), but the leaching fraction is reused after being disinfected [10]. Accordingly, water and fertilizer saving (which has both environmental and economic benefits), are the main advantages of closed systems. However, closed systems require more complicated NS management that ultimately results in higher equipment and management costs. Primarily, the higher risks of pest outbreak (mainly root diseases) require the disinfection of the leached fraction. Moreover, controlling nutrients and non-essential ions in the recirculating NS becomes more difficult, especially in the case of saline waters or high concentration of non-essential or scarcely absorbed ions (e.g., Na^+ and Cl^-) [10]. Overall, it is acknowledged that closed systems show a better WUE, at the expenses of possible yield decay in response to salt build-up in the root zone as compared to open systems [5,11–16]. Classification of soilless systems may also be done according to those that feature the presence of a solid inorganic or organic medium, which offers support to the plants and systems without substrate (water based soilless systems), where the bare roots of plants lie directly in the NS [9]. Different features characterize the two groups of systems. Soilless systems on substrate are surely the most popular systems for cut flowers and pot ornamentals [17]. Water-based soilless systems are, on the other hand, associated with reduced environmental impact and costs related to the substrate disposal. However, in water-based soilless systems, the resilience to stresses (e.g., drought) is affected by the absence of a buffer offered by the substrate, and a considerably higher risk of outbreak of root-borne diseases may also be experienced [18].

In soilless systems, fertilization is performed administering a NS containing macro- and micro-nutrients, generally through different types of irrigation systems (drip irrigation, sub-irrigation, or overhead system). Such fertigation can be continuous or discontinuous.

Based on these assumptions, the current research comparatively assessed three low-tech soilless systems for the cultivation of geranium (*Pelargonium zonale*), targeting the identification of the system that would allow for optimal commercial production and improvement of WUE.

2. Materials and Methods

2.1. Location

The experiment was conducted in a greenhouse covered with polyethylene within a commercial farm located in Vigolo Vattaro, Province of Trento, Northern Italy, 46°00′ N, 11°19′ E, at an altitude of 725 m a.s.l. Plants were grown under natural light conditions. The local climate, according to Köppen's classification, is Cfb type [19], which is a mesothermic climate, with the absence of a dry season and cool summer with temperature during the hottest month falling below 22 °C. The experiment was conducted from 25 March 2017 to 2 June 2017.

2.2. Treatments and Experimental Design

Three low-tech soilless systems were compared:

- T-1 (farm system with substrate, Figure 1): 30 rooted cuttings were grown on 0.95 L pot ($\varnothing_1$ 9 cm, $\varnothing_2$ 13 cm, h 10.8 cm), each featuring eight bottom holes, filled with a mixture of two different peats with a 1:1 volume ratio mixture (Peat A: Geotec srl, Adria, Italy, dry bulk density = 0.15 g·cm^{-3}, total porosity = 92%; Peat B: Tercomposti S.p.A., Calvisano, Italy, dry bulk density = 0.10 g·cm^{-3}, total porosity = 95%). The 30 pots were placed on a greenhouse bench, arranged in 3 rows with 10 pots each. Plants were manually watered using a 15 L watering can, as usually done in the local farms, daily supplying 100% of water evapotranspiration. The leaching fraction of water was tending to 0%. The rooted cuttings were fertilized only three times in total with granular fertilizer solubilized in water to have each time a concentration of 2.08 g·L^{-1}.
- T-2 (open-cycle drip system with substrate, Figure 2): 30 rooted cuttings were grown on 0.95 L pot ($\varnothing_1$ 9 cm, $\varnothing_2$ 13 cm, h 10.8 cm), with eight bottom holes, filled with a mixture of two different peats with a 1:1 volume ratio mixture (Peat A: Geotec srl, Adria, Italy, dry bulk density = 0.15 g·cm^{-3}, total porosity = 92%; Peat B: Tercomposti S.p.A., Calvisano, Italy, dry bulk density = 0.10 g·cm^{-3}, total porosity = 95%). The 30 pots were arranged in three rows with 10 pots each. Rows consisted of three plastic troughs measuring 1.50 m in length, 12 cm in width, and 6 cm in depth with a rectangular section and displaying a slope of about 1% to allow collecting the drained NS. The pots were placed inside the plastic troughs. The rooted cuttings were watered only with the NS by the drip irrigation system, daily supplying 130% of daily water requirement (leaching fraction of about 30%). The system was further integrated with a 210 L·NS reservoir tank located at the bottom of the plastic troughs, a submerged pump (Comet Elegance, Germany) with a flow rate of 10 L·min^{-1} and a pressure of 0.5 bar, a 15 L upper tank located at 130 cm high to receive the pumped NS, a drip irrigation system equipped with non-self-compensating emitters (2.4 L·h^{-1} nominal flow rate, one for each pot), and a 30 L reservoir tank for collecting the drainage. The leaching fraction was not reused, and T-2 was managed as an open system. Given that from the upper tank to the drippers the NS descended only by gravity, the actual flow rate of the drippers (0.66 L·h^{-1}, as measured before the experiment started) was lower than their nominal flow rate. Accordingly, the correct amount of NS to be introduced in the system was determined through a programmable electronic timer that activated the pump.
- T-3 (Nutrient Film Technique system, Figure 3): the system adopted the Nutrient Film Technique (NFT) and featured a closed soilless system with a thin layer of around 1–2 mm of NS flowing through sloped watertight troughs that hosted the plant roots. Thirty rooted cuttings were arranged in 3 rows consisting of 3 plastic troughs measuring 1.50 m in length, 12 cm in width, and 6 cm in depth with a rectangular section and a slope of about 1%. The plastic covers featured holes, where the rooted cuttings were placed. T-3 was also composed of a 210 L·NS reservoir located at the bottom of the plastic troughs and a submerged adjustable flow pump (Newa Jet, Italy) that pumped the NS in the plastic troughs.

In each of the three treatments, all 90 rooted cuttings were arranged in rows with 10 plants each, with 15 cm between rooted cuttings and 42 cm between rows, resulting in a planting density of about 16 plants m^{-2}, following common commercial practices. Three replicate plots for each treatment (rows), composed of ten rooted cuttings each (n = 30), were arranged in a randomized complete block design.

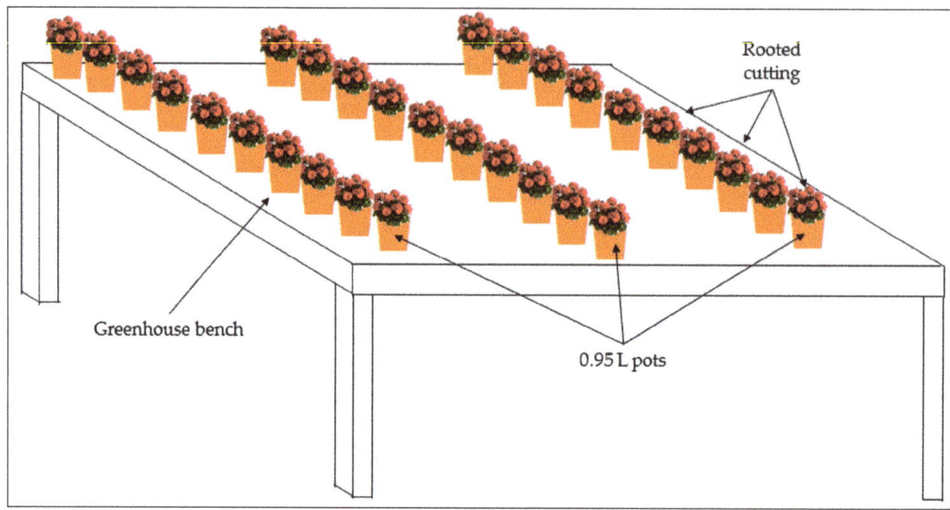

Figure 1. T1, farm system with substrate. Schematic representation of the growing system used.

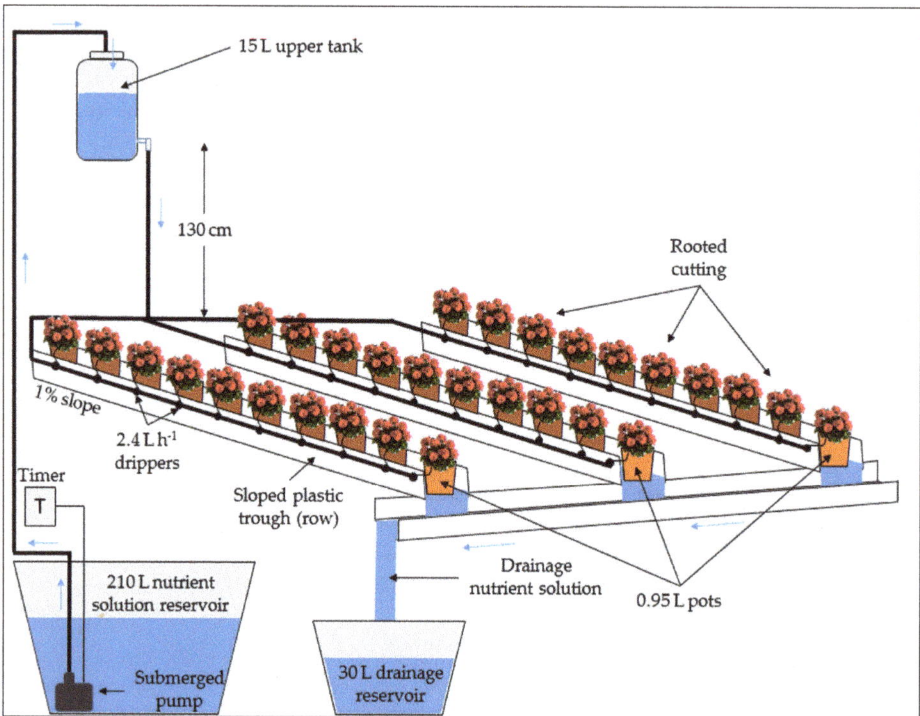

Figure 2. T-2, open-cycle drip system with substrate. Schematic representation of the growing system used.

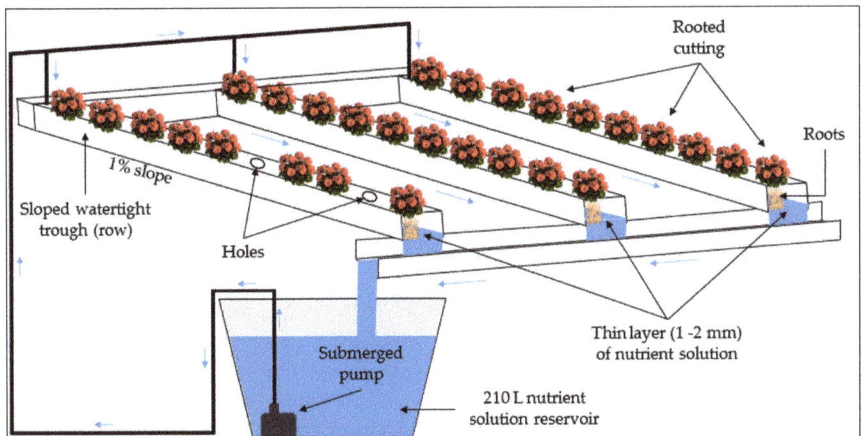

Figure 3. T-3, Nutrient Film Technique system. Schematic representation of the growing system used.

2.3. Plant Material and Crop Management

At the beginning of the trial, rooted cuttings were selected to have uniform plant material (4 cm height and 3 leaves) among the 90 individual plants used for the experimentation. T-1 (control) was managed following traditional practices from local farmers. It was irrigated only with water once every 2 days from the 1st to the 7th week, and once a day from 8th to 10th week, by hand. T-1 was fertilized only three times (discontinuous fertigation, on 4 April, 22 April, and 12 May), with a granular fertilizer (Manna Lin A, Mannafert V., Bolzano, Italy). Granular fertilizer was solubilized in water to have a 2.08 g·L^{-1} concentration for a total amount of 50.50 g applied. Manna Lin A is composed of 7% N-NO$_3$, 13% N-NH$_4$, 5% P$_2$O$_5$, 10% K$_2$O, 2% MgO, 0.025% B, 0.005% Cu, 0.06% Fe, 0.025% Mn, 0.0025% Mo, and 0.02% Zn. The microelements were supplied as chelates.

Unlike T-1, in T-2 and T-3, a continuous fertigation was adopted, using the same NS. The composition of macronutrients of full strength NS was: 10.00 mM NO$_3^-$, 1.00 mM NH$_4^+$, 2.00 mM H$_2$PO$_4^-$, 5.01 mM K$^+$, 4.00 mM Ca^{2+}, 1.50 mM Mg^{2+}, and 3.53 mM SO$_4^{2-}$. A mixed fertilizer for micronutrients was used, with the following full strength NS: 20.00 μM Fe^{3+}, 0.63 μM Cu^{2+}, 4.29 μM Zn^{2+}, 13.88 μM B^{3+}, 19.66 μM Mn^{2+}, and 0.42 μM Mo^{6+}. For all fertigation treatments, NS was prepared using fresh water (pH = 8.00, EC = 359 μS·cm^{-1} at 20 °C). The final EC of full strength NS ranged between 1829 and 1963 μS·cm^{-1} and pH ranged between 5.5 and 6.2. During the first week, in T-2 and T-3, a lower strength NS for macronutrients was used (T-2 top-fertilized by watering can) (EC = 1021 μS·cm^{-1}, pH = 5.5, 5.4 mM NO$_3^-$, 0.50 mM NH$_4^+$, 1.0 mM H$_2$PO$_4^-$, 2.5 mM K$^+$, 2.0 mM Ca^{2+}, 0.97 mM Mg^{2+}, and 0.75 mM SO$_4^{2-}$) to allow the roots to adapt to the new growing environment before using the full strength NS. The EC of leaching fraction was measured every week in both T-2 and T-3 treatment.

T-2 fertigation scheduling took into account the leaching fraction measurement, having drainage around 30% per day as a target. It changed during the crop cycle and ranged from 1 irrigation every 3–5 days at the beginning to 2 irrigations per day at the end of the trial. The NS volume provided for all pots ranged from 3.6 L during the 2nd week to 78.9 L during the 10th week, corresponding to the flowering stage. In T-3, the NS was continuously supplied from sunrise to sunset, by submerged adjustable flow pump with a measured flow rate for every plastic trough of 1.83 L·min^{-1}.

Inside the greenhouse, temperature and relative humidity were monitored every 15 min by GEMINI data logger Tinytag Plus 2. The greenhouse temperature ranged between 12 and 33 °C, and day/night humidity from 30% to 85%, respectively.

2.4. Sampling and Analysis

In the first week, EC, pH, the drained volume of T-2, volume of leftover NS in the 210 L reservoir tank of T-3, its EC and pH were daily measured after the sunset. During the trial, on 22 April and on 13 May, 100 L of fresh NS each were added to the T-2 and T-3 reservoir tanks. EC, pH, and total volume were measured again after the additions. From the 2nd to 10th week, all these parameters were measured weekly. EC was measured by Adwa AD31 Waterproof EC/TDS Tester and pH was measured by Artiglass IP67 pocket pH Tester. All testers were weekly calibrated.

Progressive and final plant heights, determined as the distance from the surface of the medium to the top of the plant, for all 90 rooted cuttings were measured. To evaluate the flowering timing and its quality, the starting date of appearance of inflorescences and their numbers per plant, together with dates of beginning and full flowering, were recorded in ten plants per replicate. Furthermore, in three plants per replicate, weekly counts of the number of fully-grown leaves was performed, as well as counts of the number of flowers of the first inflorescence and number of branches.

The estimation of leaf chlorophyll concentration was performed at the end of the trial through a non-destructive measurement with SPAD-502 (Konica-Minolta, Tokyo, Japan). Measures were taken on the leaf nearby the oldest inflorescence from 10 plants per replicate. The Minolta SPAD meter (Soil Plant Analysis Development) used indirectly measures chlorophyll content in a non-destructive manner. SPAD values were determined by measuring the ratio of light transmitted through the leaf at a red wavelength (650 nm) and an infrared wavelength (940 nm).

At the end of the experiment, all 90 plants were divided into leaves (leaves with petioles), trunks, roots, and inflorescences. Plant organs were weighed for the fresh and dry weight determinations after drying in a ventilated oven at 105 °C for 48 h.

Furthermore, at the end of the experiments, an aesthetic-commercial assessment of 10 plants per replicate was performed. For each rooted cutting, three parameters (vegetative growth, foliage compactness, and general aspect) were evaluated by assigning a score from 1 to 5, with the score 3 being the threshold value for marketability. Whenever at least one of the three parameters received a score below 3, the rooted cutting was evaluated as not marketable. The aesthetic-commercial assessment was performed by the local farmer in a randomized way without being aware of the specific treatments.

2.5. Statistical Analysis

For phenological data regarding the flowering degree (appearance of inflorescences, flowering start, and full flowering) and fully-grown leaves, no statistical analysis was applied, but only kinetic behaviors in relation to the treatment were shown. Analysis of variance (ANOVA) was used to determine the effect of the growing system used on the number of inflorescences plant^{-1}, number of flowers inflorescence^{-1}, final height of the plant, number of branches plant^{-1}, SPAD index, water contents of organs, aesthetic commercial assessments, fresh weight, total water consumption, and water use efficiency. All data were statistically processed using Systat software package (Systat Software 9.0, San Jose, CA, USA).

3. Results

3.1. Climate and Nutrient Solution Monitoring during the Experiment

During the experiment, inside the greenhouse, a data logger was used to measure temperature and humidity every fifteen minutes. Maximum air temperature ranged between 14.7 and 39.0 °C, with an average of 32.2 °C. Minimum air temperature ranged between 8.2 and 20.7 °C, with an average of 12.1 °C. The average daily temperature was 18.7 °C. The daily maximum relative humidity ranged between a minimum of 50.0% and a maximum of 93.3%, with an average of 83.9%. The daily minimum relative humidity ranged between a minimum of 16.5% and a maximum of 85.6%, with an average of 31.7%. The average daily humidity was 65.3%. During the experiment, the mean value of daily global radiation, outside the greenhouse, was 15.87 MJ·m^{-2}·day^{-1} in April and 20.25 MJ·m^{-2}·day^{-1} in May.

A NS was applied only in T-2 and T-3 treatments, with periodical control of both EC and pH. During the first week, in which a lower strength NS (EC = 1021 μS·cm^{-1} and pH = 5.5) was used, EC of the leaching fraction ranged between 1026 and 1040 μS·cm^{-1}, while the pH ranged between 5.5 and 5.8. From the second week, when a full strength NS (EC ranged 1829–1963 μS·cm^{-1} and pH ranged 5.5–6.2) was used, EC of leaching fraction ranged between 2171 and 3923 μS·cm^{-1}, while the pH ranged between 5.8 and 6.5.

3.2. Date of the Appearance of Inflorescences

The starting date of appearance of inflorescences was not affected by the soilless systems (16–17 days after transplanting) (Figure 4a). Concurrently, T-2 and T-3 showed a more extended period (3–4 days) to conclude this phase as compared to T-1 (Figure 4a).

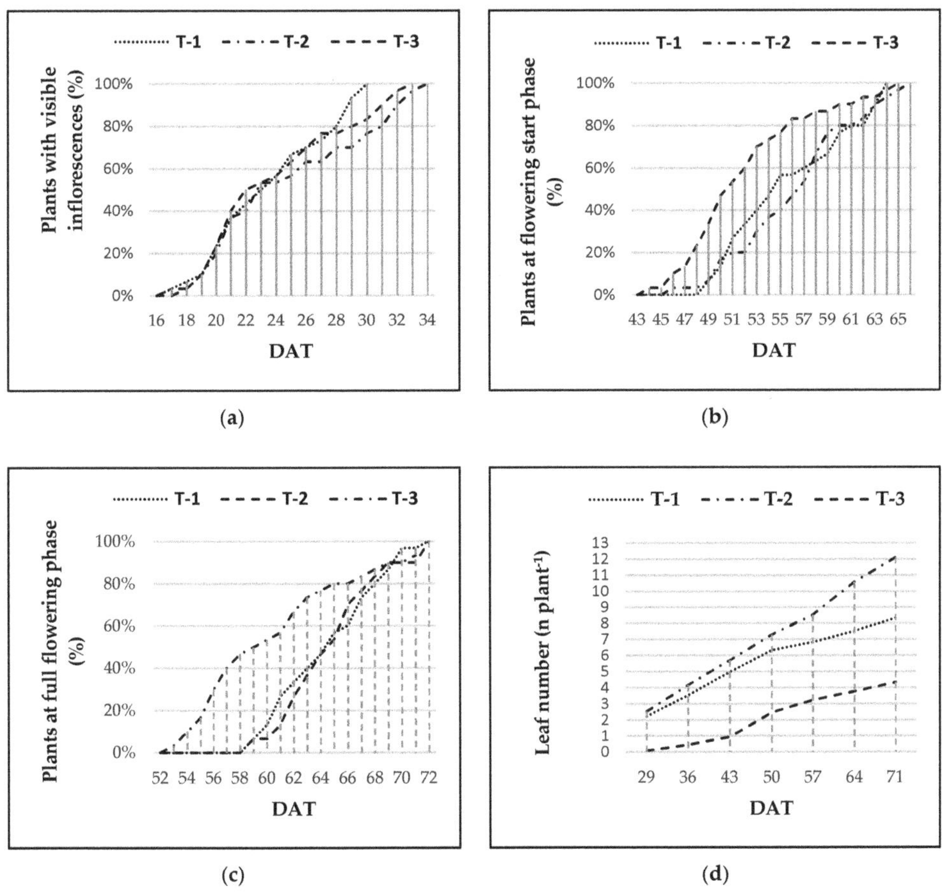

Figure 4. Effect of growing systems on *Pelargonium zonale*: (**a**) plants with inflorescences just visible; (**b**) plants at flowering start phase; (**c**) plants at full flowering phase; and (**d**) leaf number. T-1, farm system with substrate; T-2, open-cycle drip system with substrate; T-3, Nutrient Film Technique; DAT, Days After Transplanting.

3.3. Date of Flowering Start

The three soilless systems affected the date when the flowering started. As compared with T-1, flowering started six and four days earlier, respectively, in T-3 and T-2 (Figure 4b). Flowering was concluded between 64 and 66 days in all treatments, independently from the growing system.

3.4. Date of Full Flowering

There were no differences between T-1 and T-2 treatments in terms of date of starting of the full flowering phase (58 days after transplanting) and flowering duration (14 days) (Figure 4c). Conversely, full flowering was anticipated by about six days and lasted five days longer under the T-3 treatment (Figure 4c).

3.5. Biometrical Parameters

All biometrical parameters were significantly affected by treatments (Table 1), with highest values always associated with T-2 and lowest values found in plants grown under T-3. The plants' height was also affected by treatment (Table 1 and Figure 5): at the final assessment, T-2 had higher values than T-1, which in turn was significantly higher than T-3.

Table 1. Mean *Pelargonium zonale* biometrical responses to growing systems. Within-columns mean values followed by different letters are significantly different by Tukey test.

Treatment	Inflorescences (n·Plant^{-1})	Flowers (n·Inflorescence^{-1})	Branches (n·Plant^{-1})	Plant Height (cm)
T-1	10.63 b	96.33 b	5.33 b	11.66 b
T-2	13.67 a	143.11 a	8.33 a	15.13 a
T-3	6.00 c	64.67 b	2.78 c	9.41 c
Mean	***	***	***	***

With significance (***) for $p \leq 0.001$. T-1, farm system with substrate; T-2, open-cycle drip system with substrate; T-3, Nutrient Film Technique.

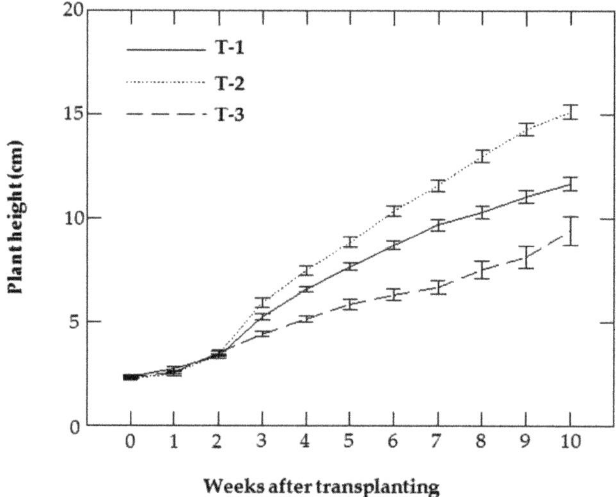

Figure 5. Height of *Pelargonium zonale* plant during growing period in response to the growing system used. Mean values ± standard error. T-1, farm system with substrate; T-2, open-cycle drip system with substrate; T-3, Nutrient Film Technique.

3.6. Number of Leaves

The three soilless systems affected the number of fully-grown leaves, which was the highest in T-2 (Figure 4d). In general, T-1 and T-2 treatments resulted in a different crop kinetic behavior as compared to T-3.

3.7. Leaf Chlorophyll

Leaf greenness of plants (SPAD values) was affected by treatment. Higher SPAD values were detected in rooted cuttings grown in T-3 treatment as compared with plants grown in T-1 and T-2 (Table 2).

Table 2. SPAD values of *Pelargonium zonale* in response to the growing system used. Within-columns mean values followed by different letters are significantly different by Tukey test.

Treatment	SPAD Value
T-1	47.16 c
T-2	54.88 b
T-3	65.63 a
Mean	***

With significance at $p \leq 0.001$ (***). T-1, farm system with substrate; T-2, open-cycle drip system with substrate; T-3, Nutrient Film Technique.

3.8. Fresh and Dry Weight

The three soilless systems affected both fresh (Figure 6a) and dry (Figure 6b) weight, which were the highest in T-2 and the lowest in T-3 and T-1. Leaves, flowers, and branches fresh weights had similar behavior of total biomass, presenting highest values in T-2 as compared with T-1 and T-3, while the fresh weight of roots was not affected by treatment (Figure 6a). Among dry weights (Figure 6b), different behaviors were found across organs. Plants grown under T-2 presented the highest leaf and flower dry biomass as compared to T-1 and T-3. On the other hand, higher dry biomass of both branches and roots was associated with T-1 and T-2 as compared with T-3.

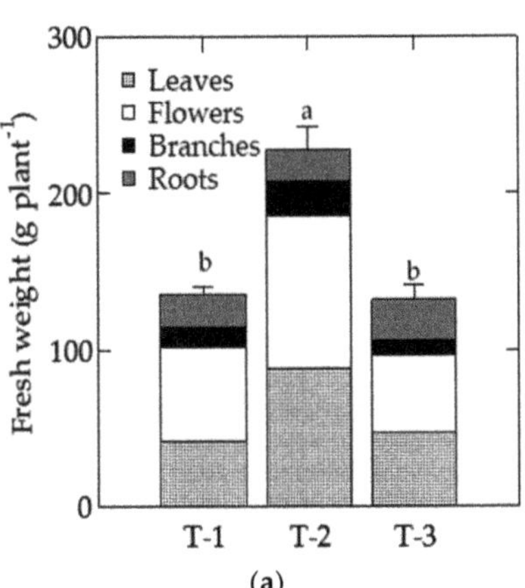

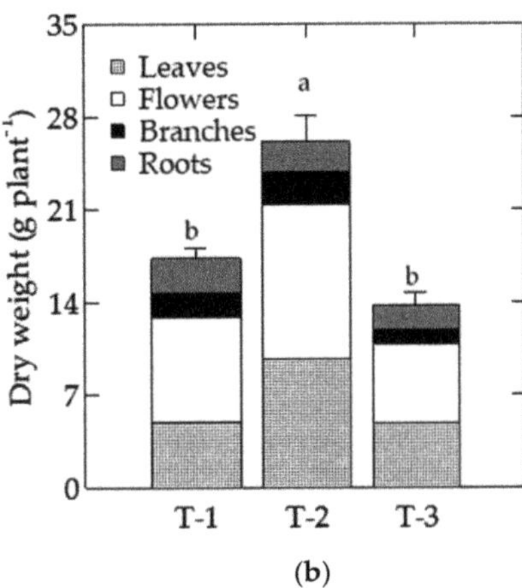

Figure 6. Effect of growing system used on *Pelargonium zonale*. (**a**) Plan fresh weight (g plant^{-1}) in relation to growing system and relative partitioning into different organs. Means values ± standard error for total biomass. (**b**) Plan dry weight (g plant^{-1}) in relation to growing system and relative partitioning into different organs. Mean values ± standard error for total biomass. T-1, farm system with substrate; T-2, open-cycle drip system with substrate; T-3, Nutrient Film Technique.

Treatment affected the water contents of different organs (Table 3): in particular, T-3 showed higher values than T-1, except for flowers where the difference was not significant. T-1 always presented the lowest levels, while T-2 had an intermediate behavior, with high values for both flowers and branches and low value for roots. It is interesting to note that roots of T-3 plants showed the highest values of water contents.

Table 3. Water contents (%) of *Pelargonium zonale* organs in response to the growing system used. Within-columns mean values followed by different letters are significantly different by Tukey test.

Treatment	Leaves (%)	Flowers (%)	Branches (%)	Roots (%)
T-1	88.28 b	86.90 b	84.76 b	87.55 b
T-2	89.06 ab	87.76 a	89.46 a	87.93 b
T-3	89.68 a	87.58 ab	89.26 a	93.22 a
Mean	***	*	***	***

n.s., not significant; * significance for $p \leq 0.050$ and $p \geq 0.010$; *** significance for $p < 0.001$. T-1, farm system with substrate; T-2, open-cycle drip system with substrate; T-3, Nutrient Film Technique.

3.9. Aesthetic-Commercial Assessment

T-3 always had the lowest values for all investigated parameters. T-2 showed the best scores in all the parameters evaluated, except for the vegetative growth, in which T-1 had the highest score as the absolute value, even if statistical analysis did not detect any significant difference as compared to T-2 (Table 4).

Table 4. Aesthetic-commercial assessment of *Pelargonium zonale* in response to the growing system used. Within-columns mean values followed by different letters are significantly different by Tukey test.

Treatment	MV	Vegetative Growth	Foliage Compactness	General Aspect
T-1	4.02 a	4.17 a	3.80 b	4.10 a
T-2	4.27 a	4.13 a	4.53 a	4.13 a
T-3	3.30 b	2.98 b	3.67 b	3.28 b
Mean	***	***	***	***

With significance for $p < 0.001$ (***). MV is the arithmetic mean among vegetative growth, foliage compactness, and general aspect values. T-1, farm system with substrate; T-2, open-cycle drip system with substrate; T-3, Nutrient Film Technique.

3.10. Water Consumption and WUE

The comparison of the three treatments revealed significant differences as regards the biomass produced, the water consumption (leaching fraction included), and the related WUE values (Table 5). In particular, T-2 differed from the other two for its most considerable vegetative development. Total water consumption revealed significantly different values among the three treatments: T-2 showed the highest values, T-3 the intermediate ones, and T-1 the lowest ones. The calculated values of WUE, therefore, showed the highest values in T-2, and the lowest ones in T-3.

Table 5. Total biomass (fresh weight), total water consumption and WUE of *Pelargonium zonale* in response to the growing system used. Within-columns mean values followed by different letters are significantly different by Tukey test.

Treatment	Plant FW (g·Plant^{-1})	TWC (L·Plant^{-1})	WUE (g·FW·L^{-1}·H$_2$O)
T-1	132.22 b	7.50 c	17.63 a
T-2	220.95 a	10.11 a	21.85 a
T-3	113.34 b	8.68 b	13.07 b
Mean	***	***	***

With significance for $p < 0.001$ (***). T-1, farm system with substrate; T-2, open-cycle drip system with substrate; T-3, Nutrient Film Technique; FW, Fresh Weight; TWC, Total Water Consumption; WUE, Water Use Efficiency.

4. Discussion

The adoption of different soilless cultivation systems significantly affected the growth (including flowering, fresh weight, and dry weight) and the commercial characteristics (number of inflorescences per plant, number of flower per inflorescence, number of branches per plant, and number of leaves) of geranium grown in a greenhouse, in Northern Italy. As also reported by Rouphael and Colla [20], the optimal concentration of fertilizer solutions for greenhouse crops may be affected by irrigation method, because it influences the accumulation of nutrients in the growing medium, which in turn affects the nutrient uptake by plants. For example, Cardarelli [21] reported that, when averaged over NS concentration, the number of geranium flowers per plant was significantly (27%) higher with sub-irrigation than with drip-irrigation.

The growth of rooted cuttings of geranium continues until full bloom, when they are ready for sale. Given the characteristics of the market in the area where the trial was conducted, where farmers generally supply local retailers, a gradual flowering could help producers. However, T-1 and T-2 showed no differences (Figure 4), displaying both a 14-day flowering window (from 23 May to 5 June). T-3, on the other hand, showed a much more scalar flowering (20 days, from 17 May to 5 June) and earlier than T-1 and T-2 (Figure 4). Despite this, T-3 did not develop adequate commercial characteristics for the market. Furthermore, some of the rooted cuttings of T-3 treatments highlighted a delayed growth demonstrating stress conditions, which may have caused the observed flowering pattern. In fact, according to Riga [22], stress conditions in geranium can influence the flowering timing, anticipating the opening of the flowers.

The application of different cultivation systems, to which three different fertigation managements are associated, significantly affected all the plant commercial features. T-2 showed the highest number of inflorescences (13.67 inflorescences plant^{-1}), followed by T-1 and T-3, where 10.63 and 6.00 inflorescences plant^{-1} were observed, respectively (Table 1). T-2 showed significant differences from both T-1 and T-3, and T-1 from T-3. T-2 also showed the highest number of flowers per inflorescence (143.11 flowers inflorescence^{-1}), followed by T-1 and T-3, where values of 96.33 and 64.67 flowers inflorescence^{-1} were observed, respectively. No significant differences were observed by comparing T-1 and T-3 (Table 1). Regarding the vegetative behavior (number of fully grown leaves at the end of the experiment, number of branches plant^{-1}, and final height of plants), as well as considered commercial characteristics, T-2 always showed the highest values, demonstrating a better efficiency of this treatment, as also confirmed by fresh and dry weight results (Figure 6a,b). Overall, T-1 and T-2 developed adequate leaf mass for marketing, whereas T-3 showed insufficient development. Cardarelli [21], reported that the net assimilation of CO_2 of geranium was significantly affected by the irrigation systems with the highest values recorded with the drip-irrigation. The mean value of the number of fully-grown leaves (Figure 4d), at the end of the experiment, was 12.10 in T-2, followed by 8.33 in T-1 and 4.33 in T-3. Concerning the number of branches plant^{-1}, T-2 developed 8.33 branches plant^{-1}, followed by T-1 and T-3, where 5.33 and 2.78 branches plant^{-1} were observed, respectively (Table 1). T-2 showed significant differences from both T-1 and T-3, and T-1 from T-3. The final height of plants was 15.13 cm in T-2, followed by T-1 with a value of 11.66 cm and T-3 with a value of 9.41 cm. T-2 showed significant differences from both T-1 and T-3, and T-1 from T-3. Moreover, regarding the growth trend, as reported in Figure 5, it is possible to see that, only two weeks after transplanting, treatment significantly modified the rate of growth until the final assessment. During the first two weeks, plants had a similar trend due to low temperatures registered, which strongly depressed growth.

Regarding the SPAD values, T-1 had the lowest values (47.16) and differed from both T-2 and T-3, with values of 54.88 and 65.63, respectively, which in turn were significantly different (Table 2) [23]. These behaviors reflected management of fertilization: in fact, when only three fertilizer supplies were provided (T-1), the lowest values were recorded, while, for other treatments (T-2 and T-3), in which the concentration of nutrient (nitrogen in particular) was constantly kept, SPAD values were always high. T-3 presumably had too elevated SPAD values, confirmed by the worst performances, while T-2 reached good SPAD levels, suggested to have more equilibrated leaf greenness [23]. Previously, in geranium, SPAD values were linearly correlated with total chlorophyll in fresh tissue. For example, in geranium "Ringo Deep Scarlet", there the following correlation was observe: SPAD = 14.96 + 37.30 *chlorophyll content (mg·g^{-1} of dried tissue), r^2 = 0.95 and p < 0.001 [24]. In this experiment, the high SPAD values of T-2 and T-3 are attributable to the high nutrient concentration of NS provided. This is confirmed by EC values of leaching fraction, always showing higher values compared to EC of NS applied. EC values of leaching fraction fluctuated between 2171 and 3923 µS·cm^{-1}. The fact that the percentage of leaching fraction has always been sufficient (around 30%) [25] and that no particularly high temperatures were experienced during the experiment may have resulted in the use of a too concentrated NS, thus suggesting that the use of a less concentrated NS should be recommended. This was also confirmed by the fact that, during the first week, when a lower concentration of the NS was used, the EC of the leaching fraction did not increase as compared to the EC of NS supplied.

The visual assessment (aesthetic-commercial assessment) confirmed that the rooted cuttings of T-2 reached the best score, except for vegetative growth, and developed the best characteristics for the market (Table 4). Only one rooted cutting of T-2 treatment scored 2, and therefore was considered unmarketable because of an excessive asymmetry of the shape of the canopy. In T-3 treatments, 50% of the plants reached a MV score below 3, mainly since they showed a reduced growth. All of these plants also showed roots darkening. In particular, seven plants scored below 3 in one of the three parameters, six plants in two parameters, and two plants in all parameters. The roots darkening and the stunted growth could be due to the reuse of the non-sterilized leaching fraction, favoring the spread of root pathogens to the whole system. Indeed, spreading of root-borne diseases may occur, thus sterilization of the solution must be provided to avoid pathogens outbreak [9].

The application of different cultivation systems, each featuring a different fertigation management, also affected both water consumption and WUE (Table 5). Water consumption was highest in T-2 (10.11 L·plant^{-1}), followed by T-3 and T-1 with 8.68 and 7.50 L·plant^{-1}, respectively. T-2 showed significant differences from both T-1 and T-3, and T-1 from T-3. Despite these results, T-2 showed the highest value of WUE (21.85 g·L^{-1}), followed by T-1 and T-3 with 17.63 and 13.07 g·L^{-1}, respectively. No significant differences were observed by comparing T-2 and T-1 (Table 5). It should be considered, however, that, when converting the T-2 treatment into a closed system, the values could significantly improve [26–28]. In this case, the consumption of the NS could be lower (thanks to recycling of the drained solution), if compared with the other two treatments [28].

However, it is important to underline that in this scenario (closed system) the changing relationships between the nutrients in the drained solution need to be carefully considered, since they constitute an aspect that could influence the development of the rooted cuttings. Closed systems show a better water use efficiency, despite a slightly lower yield due to salt build-up in the root zone as a consequence of degradation of NS quality compared to the open systems [5,11–16]. In some cases, according to Savvas et al. [9], "switching over to closed cultivation systems does not seem to restrict crop yield or product quality". Given that in the context considered the farms often integrate their income with other crops, the drained fraction can also be used in open-air crops [29].

5. Conclusions

The experiment shows that the adoption of simplified soilless technology may allow enhancing the commercial characteristics of geranium, making it more attractive for the market and ultimately improving water and nutrient management. In particular, the adoption of a cultivation system with continuous fertigation on the substrate (peat) with drip irrigation can enable to obtain more attractive plants for the market. Moreover, this strategy improves water use efficiency, which could also be further improved with the adoption of a closed system. Modernization in the cultivation system and fertigation management may help to improve the commercial features of geranium even without using high technologies currently still not economically sustainable for most of the often family-run farms operating in the cut flowers and pot ornamentals sector in Trento province.

Author Contributions: Conceptualization, L.B. and F.Z.; methodology, L.B.; validation, L.B., F.Z., D.P., N.M. and F.O.; formal analysis, L.B. and D.P.; investigation, L.B. and F.Z.; resources, L.B. and F.Z.; data curation, L.B., F.Z., and D.P.; writing—original draft preparation, L.B.; writing—review and editing, D.P., N.M., F.O., and G.G.; visualization, L.B., D.P., N.M., F.O., and G.G.; supervision, D.P., N.M., F.O., and G.G.; project administration, L.B.; and funding acquisition, L.B. and F.Z. All authors have read and agreed to the published version of the manuscript.

Funding: This research received no external funding.

Acknowledgments: We thank the Zamboni Stefano farm located in Vigolo Vattaro (Northern Italy) for providing the infrastructure and equipment to carry out the experiments. We also thank Maria Vender for her contribution to the revision of the English version of the article.

Conflicts of Interest: The authors declare no conflict of interest.

References

1. Size, E.O.M. Share & Trends Analysis Report by Application (Cleaning & Home, Medical, Food & Beverages, Spa & Relaxation), By Product, By Sales Channel, And Segment Forecasts, 2019–2025. In *Market Analysis Report*; Grand View Research: San Francisco, CA, USA, 2020; 978-1.
2. MiPAAF. Available online: https://www.politicheagricole.it/flex/cm/pages/ServeBLOB.php/L/IT/IDPagina/10906 (accessed on 9 February 2017).
3. Pardossi, A.; Incrocci, L.; Incrocci, G.; Tognoni, F. What limits and how to improve water use efficiency in outdoor container cultivation of ornamental nursery stocks. *Acta Hortic.* **2009**, *843*, 73–80. [CrossRef]
4. Pardossi, A.; Marzialetti, P. La concimazione delle piante in contenitore. In *Uso Razionale delle Risorse nel Forovivaismo: L'acqua*; ARSIA: Florence, Italy, 2004; pp. 183–192.
5. Pardossi, A.; Tognoni, F.; Incrocci, L. Mediterranean greenhouse technology. *Chron. Hortic.* **2004**, *44*, 28–34.
6. Meric, M.K.; Tuzel, I.H.; Tuzel, Y.; Oztekin, G.B. Effects of nutrition system and irrigation programs on tomato in soilless culture. *Agric. Water Manag.* **2011**, *99*, 19–25. [CrossRef]
7. Fernández, J.A.; Orsini, F.; Baeza, E.; Oztekin, G.B.; Muñoz, P.; Contreras, J.; Montero, J.I. Current trends in protected cultivation in Mediterranean climates. *Eur. J. Hortic. Sci.* **2018**, *83*, 294–305. [CrossRef]
8. Van Os, E.A.; Gieling, T.H.; Lieth, J.H. Technical equipment in soilless production systems. In *Soilless Culture: Theory and Practice*; Raviv, M., Lieth, J.H., Eds.; Elsevier Publications: London, UK, 2008; pp. 157–207.
9. Savvas, D.; Gianquinto, G.; Tüzel, Y.; Gruda, N. Soilless culture. In *Good Agricultural Practices Principles for Greenhouse Vegetable Production in the Mediterranean Region*; FAO Paper, AGP series n° 217; FAO: Rome, Italy, 2013; pp. 303–354.
10. Incrocci, L.; Leonardi, C.I. Sistemi di coltivazione fuori suolo a ciclo chiuso. In Proceedings of the La produzione in Serra dopo l'Era del Bromuro di Metile International Workshop, Comiso, Italy, 1–3 April 2004.
11. Hardgrave, M.R. Recirculation system for greenhouse vegetables. *Acta Hortic.* **1993**, *342*, 85–92. [CrossRef]
12. Van Os, E.A. Closed growing system for more efficient and environmental friendly production. *Acta Hortic.* **1994**, *361*, 194–200. [CrossRef]
13. Böhme, M. Influence of Closed Systems on the Development of Cucumber. Proceedings of the ISOSC Proceedings, Berlin, Germany, January 1996. pp. 75–87. Available online: https://www.library.wur.nl/WebQuery/titel/898177 (accessed on 20 September 2020).
14. Tüzel, I.H.; Tüzel, Y.; Gül, A.; Meric, M.K.; Yavuz, Ö.; Eltez, R.Z. Comparison of open and closed systems on yield, water and nutrient consumption and their environmental impact. *Acta Hortic.* **2000**, *554*, 221–228. [CrossRef]
15. Giuffrida, F.; Lipari, V.; Leonardi, C. A simplified management of closed soilless cultivation systems. *Acta Hortic.* **2003**, *614*, 155–160. [CrossRef]
16. Lopez, J.; Perez, J.S.; Trejo, S.L.; Urrestarazu, M. Mineral nutrition and productivity of hydroponically grown tomatoes in relation to nutrient solution recycling. *Acta Hortic.* **2003**, *609*, 219–223. [CrossRef]
17. Malorgio, F. Le colture fuori suolo per le produzioni floricole in serra. In *Uso Razionale Delle Risorse nel Florovivaismo: L'acqua*; ARSIA: Florence, Italy, 2004; pp. 49–57.
18. Wortman, S.E. Crop physiological response to nutrient solution electrical conductivity and pH in an ebb-and-flow hydroponic system. *Sci. Hortic.* **2015**, *194*, 34–42. [CrossRef]
19. Eccel, E.; Cordano, E. Progetto Indiclima: Elaborazione di indici climatici per il Trentino: Relazione finale. 2015. Fondazione Edmind Mach, San Michele all'Adige, Italy. Available online: http://www.hdl.handle.net/10449/34308 (accessed on 19 September 2020).

20. Rouphael, Y.; Colla, G. The influence of drip-irrigation or subirrigation on zucchini squash grown in closed-loop substrate culture with high and low nutrient solution concentrations. *HortScience.* **2009**, *44*, 306–311. [CrossRef]

21. Cardarelli, M.; Rouphael, Y.; Salerno, A.; Rea, E.; Colla, G. Fertilizer concentration and irrigation method affects growth and flowering of two bedding plants. *Acta Hortic.* **2012**, *937*, 689–696. [CrossRef]

22. Riga, P. Avoiding the use of plant growth regulator in geranium production by application of a cyclic deficit irrigation strategy. *JAH* **2012**, *14*, 7–12. [CrossRef]

23. Wang, Y.; Dunn, B.L.; Arnall, D.B.; Mao, P. Use of an active canopy sensor and SPAD chlorophyll meter to quantify geranium nitrogen satus. *HortScience.* **2012**, *47*, 45–50. [CrossRef]

24. Smith, B.R.; Fisher, P.R.; Argo, W.R. Water-soluble fertilizer concentration and pH of a peat-based medium affect growth, nutrient uptake, and chlorosis of container-grown seed geraniums (Pelargonium x hortorum L.H.Bail). *J. Plant. Nutr.* **2004**, *27*, 497–524. [CrossRef]

25. Enzo, M.; Gianquinto, G.; Lazzarin, R.; Pimpini, F.; Sambo, P. *Principi Tecnico-Agronomici Della Fertirrigazione e del Fuori Suolo*; Veneto Agricoltura, ARSAFA: Legnaro (PD), Italy, 2001; pp. 160–181.

26. Voogt, W.; Sonneveld, C. Nutrient management in closed growing system for greenhouse production. In *Plant Production in Closed Ecosystems*; Goto, E., Kurata, K., Hayashi, M., Sase, S., Eds.; Kluwer Academic Publishers: Dordrecht, The Netherlands, 1997; pp. 83–102.

27. Van Os, E.A. Closed soilless growing systems: A sustainable solution for Dutch greenhouse horticulture. *Water Sci. Technol.* **1999**, *39*, 105–112. [CrossRef]

28. Rouphael, Y.; Colla, G.; Battistelli, A.; Moscatello, S.; Rea, E. Yield, water requirement, nutrient uptake and fruit quality of zucchini squash grown in soil and soilless culture. *J. Hort. Sci. Biotech.* **2004**, *79*, 423–430. [CrossRef]

29. Van Os, E.A.; Ruijs, M.N.A.; van Weel, P.A. Closed business system for less pollution from greenhouses. *Acta Hortic.* **1991**, *294*, 49–57. [CrossRef]

Article

Strategies for Improved Yield and Water Use Efficiency of Lettuce (*Lactuca sativa* L.) through Simplified Soilless Cultivation under Semi-Arid Climate

Nicola Michelon [1], Giuseppina Pennisi [1], Nang Ohn Myint [2], Giacomo Dall'Olio [1], Lucrecia Pacheco Batista [3], Adeodato Ari Cavalcante Salviano [4], Nazim S. Gruda [5], Francesco Orsini [1,*] and Giorgio Gianquinto [1]

[1] Department of Agricultural and Food Sciences (DISTAL), Alma Mater Studiorum University of Bologna, Viale Fanin, 44, 40127 Bologna, Italy; nicola.michelon@unibo.it (N.M.); giuseppina.pennisi@unibo.it (G.P.); giacomo.dallolio@alice.it (G.D.); giorgio.gianquinto@unibo.it (G.G.)

[2] Department of Soil and Water Management, Yezin Agriculture University, Yezin 15013, Myanmar; dr.nangohnmyint@yau.edu.mm

[3] Department of Agriculture Sciences, Federal Rural University of Semi-arid, Mossorò-RN (UFERSA), Rio Grande do Norte 59625-900, Brazil; lucreciapbatista@gmail.com

[4] Department of Agricultural Engineering and Soil, Federal University of Piauí (UFPI), Teresina-Piauí 64049-550, Brazil; adeodatosalviano@hotmail.com

[5] Department of Horticultural Sciences, University of Bonn, INRES, 53113 Bonn, Germany; ngruda@uni-bonn.de

* Correspondence: f.orsini@unibo.it; Tel.: +39-0512096677

Received: 14 August 2020; Accepted: 9 September 2020; Published: 12 September 2020

Abstract: Simplified soilless cultivation (SSC) systems have globally spread as growing solutions for low fertility soil regions, low availability of water irrigation, small areas and polluted environments. In the present study, four independent experiments were conducted for assessing the applicability of SSC in the northeast of Brazil (NE-Brazil) and the central dry zone of Myanmar (CDZ-Myanmar). In the first two experiments, the potentiality for lettuce crop production and water use efficiency (WUE) in an SSC system compared to traditional on-soil cultivation was addressed. Then, the definition of how main crop features (cultivar, nutrient solution concentration, system orientation and crop position) within the SSC system affect productivity was evidenced. The adoption of SSC improved yield (+35% and +72%, in NE-Brazil and CDZ-Myanmar) and WUE (7.7 and 2.7 times higher, in NE-Brazil and CDZ-Myanmar) as compared to traditional on-soil cultivation. In NE-Brazil, an eastern orientation of the system enabled achievement of higher yield for some selected lettuce cultivars. Furthermore, in both the considered contexts, a lower concentration of the nutrient solution (1.2 vs. 1.8 dS m^{-1}) and an upper plant position within the SSC system enabled achievement of higher yield and WUE. The experiments validate the applicability of SSC technologies for lettuce cultivation in tropical areas.

Keywords: urban agriculture; simplified soilless culture; hydroponics; conventional agriculture

---

1. Introduction

The detrimental effects of climate change are resulting in dramatic environmental, economic, and social consequences across the world [1]. Current projections show an overall increase in temperatures, with rainfall being irregularly distributed and characterized by heavy downpours [2–5]. Erratic climate can negatively affect natural resources availability (e.g., water and agricultural land), as well as posing severe risks on both ecosystems and human health [2,6]. Many developing countries, also located in tropical areas, are vulnerable to climate change due to their dependence on rain-fed agriculture, widespread poverty, and limited access to innovative technologies and improved agricultural practices [7]. An evident interdependence between climate change, economic vulnerability and migrations exists [8]. Accordingly, climate change is also resulting in a growing rate of migration toward urban and periurban areas of large cities. However, adaptation mechanisms are not yet in place, or are not strong enough, to mitigate the economic vulnerability of the most impoverished strata of the population [8]. Particularly in the tropical areas of Latin America and South-East Asia, health concerns are related to different forms of malnutrition frequently associated with a lack of micronutrients and vitamins in the population diet and low dietary diversification [9].

In Latin America, the Piaui State, located in the north-east area of Brazil (NE-Brazil), is one of the areas most affected by climate change due to its natural resources scarcity and extreme climatic conditions (i.e., semi-dry zone with a rainy season from December to May) [8]. Furthermore, after years of deforestation for agricultural purposes, the soil has a low amount of organic matter which negatively affects agricultural production [10]. Similar to Piaui State, the central dry zone in Myanmar (CDZ-Myanmar) is considered one of the most food-insecure regions of south-east Asia [11]. The climate of CDZ-Myanmar is characterized by a dry season without precipitation from November to March, which compromises and minimizes the agricultural choices of the farmers. Accordingly, climate and water scarcity are considered among the most significant problems of this area [12] and are expected to worsen in the future due to climate change [13].

In both NE-Brazil and CDZ-Myanmar, the introduction of innovative agricultural technologies which allow vegetable production even in urban and periurban areas, while fostering water-saving techniques to improve crop water use efficiency (WUE), is a crucial priority. According to Gianquinto et al. [14], it may be advisable to adopt simplified soilless cultivation (SSC) systems, which are independent of soil fertility and soil-borne diseases, do not require large spaces and intensive work labor and are characterized by high water and nutrient use efficiency thanks to the use of recirculating systems for the nutrient solution [14,15]. SSC systems are adapted from the concept of commercial hydroponics by integrating the advantages of easy construction and maintenance, while also reducing the initial economic investment or input requirements [16]. Different system designs exist for SSC, which mainly differ in the construction material, substrate used for plant growth and management of the nutrient solution [17].

The aim of this study is to assess the viability of an SSC system for the production of lettuce compared to traditional on-soil cultivation techniques in both NE-Brazil and CDZ-Myanmar, considering yield, water use efficiency and the overall physiological plant response. The assumption is that the adoption of SSC can increase yield and reduce water consumption of lettuce as compared to traditional on-soil grown plants also in very highly challenging contexts where soil quality is poor, climate is unfavorable and access to land by many people living in urban and periurban areas of large cities is limited. Moreover, the study integrates figures from different crop features and management strategies, such as crop positioning, garden orientation and cultivar traits to elaborate specific recommendations on the optimal management of the SSC systems proposed.

2. Materials and Methods

2.1. Location

North-East of Brazil (NE-Brazil): The experiments were carried out at the Horticulture Demonstration and Research Centre located at Fazenda Nova Esperança (5°01′ S and 42°46′ W, 87 m a.s.l.), owned by the Foundation Pe. Antonio Dante Civiero, located on the outskirts of the city of Teresina, capital of Piaui. According to Köppen's classification, the local climate is Aw type, with a dry summer and a rainy season between January and May.

Central Dry Zone of Myanmar (CDZ-Myanmar): The experiments were conducted at the Soil and Water Research Station of Yezin Agriculture University (19°83′ N and 96°27′ E, 122 m a.s.l.), located at the university campus in the periurban fringes of the capital NayPyiTaw. According to Köppen's classification, the local climate is Aw type, with a dry summer and a rainy season between June and October.

All experiments were carried out in the open field during the dry season, although the simplified soilless systems were equipped with a shading net (see description in Section 2.3).

2.2. Experimental Design

Four independent experiments were performed with commercial varieties of lettuce. In all experiments, lettuce (*Lactuca sativa* L.) was sown manually in 105 cells plastic seedling trays, and seedlings were transplanted 21 days after sowing (DAS).

- *Experiment 1 (NE-Brazil):* The trial considered a curly green lettuce (cv Isabela). Conventional on-soil cultivation and SSC system were compared. Plants were transplanted on 24 June 2009 and were harvested when reaching full maturity, which occurred at 40 and 31 days after transplanting (DAT) for traditional on-soil cultivation and in the SSC system, respectively. The experimental design was a strip block design with two treatments and three replicates.
- *Experiment 2 (CDZ-Myanmar):* The experiment was carried out with a curly green lettuce (cv Green wave). Conventional on-soil cultivation and SSC systems were compared. Plants were transplanted on 28 December 2018. Harvest occurred at 31 DAT in both systems. The experimental design was a completely randomized block design with two treatments and three replicates.
- *Experiment 3 (NE-Brazil):* Three green curly lettuce cultivars, namely cv Isabela, Veronica and Mimosa verde, and one red curly cultivar, namely cv Banchu Red Fire, were tested on their adaptability to the SSC system. The four cultivars were factorially combined with two different garden/plant row orientation (east and west exposure). Plants were transplanted on 18 July 2009. Harvest occurred at 31 DAT. The experimental design was a completely randomized block design with eight treatments and three replicates.
- *Experiment 4 (CDZ-Myanmar):* The experiment was carried out on two curly green lettuce cultivars, namely cv Green wave and Rapido 344. Plants were tested for their adaptability to SSC, and two different concentrations of nutrient solution salinity, characterized by an electrical conductivity (EC) of 1.2 dS m^{-1} (NS$_{1.2}$) and E.C. 1.8 dS m^{-1} (NS$_{1.8}$), were used. Moreover, the effect of plant growing position (upper position, UP vs lower position, LP) within the garden was evaluated. UP refers to the plant growing in the upper part of the SSC that receives the nutrient solution directly from the nutrient solution tank. LP refers to the plants growing in the lower part of the system, which get the nutrient solution drained from the upper part of the SSC system (Figure 1). Plants were transplanted on 19 February 2019 and were harvest at 31 DAT. The experimental design was a randomized block design with eight treatments and three replicates.

2.3. Simplified Soilless Cultivation System

The SSC system used was the so-called Bottles system (Figure 1), developed and tested in the northeast of Brazil since 2005 [18]. It is composed of a wooden/bamboo frame and a gravity-flow system, where nutrient solution drains from a tank of 310 L volume placed above the system at 2 m height. Hydraulic pipes with an emitter flow rate of 2 L h^{-1} direct the flux into the declined garden with a slope of 24%, which is composed by connecting plastic drinking bottles that host both substrate (rice husk in all experiments) and plants. The excess nutrient solution is then directed through a drainage pipe system to another tank placed below. A 50% shading net was placed above the system, to reduce light intensity. In NE-Brazil, the system used for the experiments was 6 m long and 3 m wide (18 m^2) and accounted for 20 lines of 2 L plastic bottles (8 bottles line^{-1}). Each bottle had two holes for hosting plants. Therefore, at full regime, the system could accommodate 320 plants. In CDZ-Myanmar, the system was tailored to the local context to meet the vegetables production needs of individual households. Accordingly, the system size was reduced (5 m long and 2 m wide, resulting in a garden surface of 10 m^2), and a smaller tank for the nutrient solution (100 L) was adopted. Each module hosted 240 plants. When also considering the surrounding paths allowing for garden access (about half a meter on each side and in internal paths), the net planting density was of 26 plants m^{-2} in both NE-Brazil and CDZ-Myanmar.

In NE-Brazil, a nutrient solution (NS$_{1.6}$) previously adopted for local SSC cultivation was used [18,19]. The NS$_{1.6}$ was prepared with locally available simple mineral salts and soluble fertilizers and was characterized by an electrical conductivity (EC) of 1.6 dS m^{-1} and a pH of 6.5. In CDZ-Myanmar, for both experiments, the NS was prepared by using locally available NPK fertilizer (15-15-15). During experiment 2, the adopted nutrient solution presented an EC of 1.2 dS m^{-1} and a pH of 7.7, while in experiment 4 the nutrient solution was prepared at two concentrations, respectively, 0.6 g L^{-1} in NS$_{1.2}$ (EC = 1.2 dS m^{-1}, pH = 7.3) and 0.8 g L^{-1} in NS$_{1.8}$ (EC = 1.8 dS m^{-1}, pH = 7.5).

Details on macronutrient and micronutrient concentrations of nutrient solutions are reported in Tables 1 and 2.

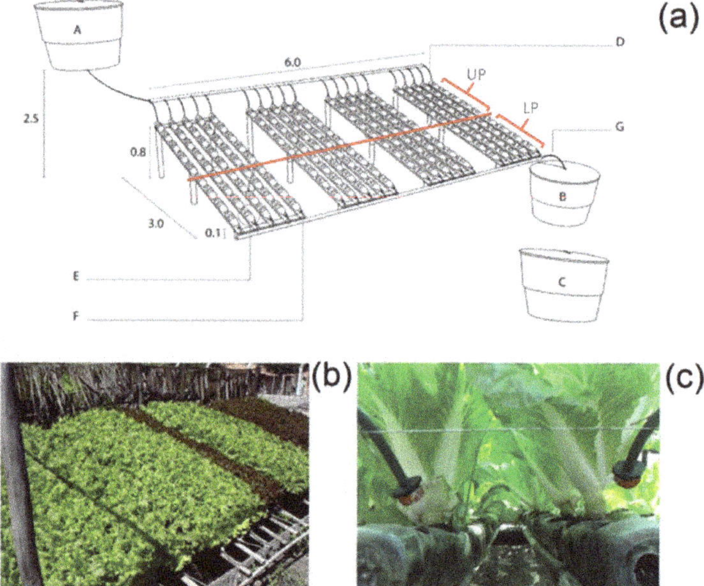

Figure 1. (**a**) Schematic drawing of the growing system used with measurements (in meters) adopted. The system includes a top (A) and a drainage (B) tank, as well as a fresh nutrient solution reservoir (C). The system is fitted with a gravity flow drip-irrigation system (D) that deliver the nutrient solution to 20 lines of recycled plastic bottles (E). Excess nutrient solution is then drained to a re-collection pipe (F) which is connected (G) to the drainage tank (B). UP = Upper position; LP = Lower position. Images of the systems in the cities of (**b**) Teresina (Piaui, Brazil) and (**c**) NayPyiTaw (Myanmar).

Table 1. Macronutrient concentrations in water and nutrient solutions (NS) adopted in the experiments in NE-Brazil and in CDZ-Myanmar.

	Exp.	N (mmol L^{-1})	P (mmol L^{-1})	K (mmol L^{-1})	S (mmol L^{-1})	Ca (mmol L^{-1})	Mg (mmol L^{-1})
			NE-Brazil				
Water	1, 3	1.2	nd	0.6	nd	0.4	0.2
NS$_{1.6}$	1, 3	11.7	0.7	3.4	2.6	3.1	1.7
			CDZ-Myanmar				
Water	2, 4	nd	nd	0.05	nd	0.2	0.06
NS$_{1.2}$	2, 4	6.4	2.9	2.3	3.7	1.0	0.05
NS$_{1.8}$	4	8.6	3.4	3.1	3.1	1.3	0.07

nd = not determined.

Table 2. Micronutrient concentrations in water and nutrient solutions (NS) adopted in the experiments in NE-Brazil and in CDZ-Myanmar.

	Exp.	Fe (µmol L^{-1})	Mn (µmol L^{-1})	Cu (µmol L^{-1})	Zn (µmol L^{-1})	B (µmol L^{-1})	Mo (µmol L^{-1})
			NE-Brazil				
Water	1, 3	nd	0. 5	nd	0.4	2.0	nd
NS$_{1.6}$	1, 3	26.9	12.4	1.6	4.6	21.5	0.5
			CDZ-Myanmar				
Water	2, 4	0.005	0.005	nd	nd	nd	nd
NS$_{1.2}$	2, 4	10.7	0.5	0.3	0.1	nd	0.01
NS$_{1.8}$	4	14.3	0.7	0.4	0.13	nd	0.02

nd = not determined.

2.4. Traditional on-Soil Cultivation

The soil of the two regions had a loamy sand texture with similar hydrological soil parameters (wilting point and field capacity at 6% *v:v* and 13% *v:v*, respectively). The physical and chemical characteristics of the soil in the two locations are described in Table 3. In both NE-Brazil and CDZ-Myanmar, the soil was overturned and dug with a hoe prior to cultivation. Soil fertilization provided a supply of 1.5 kg m^{-2} of cattle manure and 3.75 g m^{-2} of N, P, and K (mineral fertilizer 10-10-10 and Nitrophoska 15-15-15 in NE-Brazil and CDZ-Myanmar, respectively). Fertilizer was manually applied three days before transplanting. No additional fertilizer was applied during the crop cycles. Due to low soil pH in NE-Brazil, 0.15 kg m^{-2} of dolomitic limestone was added into the soil. The plots were raised by 20 cm and a trapezoid shape was developed, ensuring a base 1.2 m wide and a top 1.0 m wide. Finally, each plot was adjusted with a rake. Between the experimental plots, a space of approximately 0.7 m was left to facilitate maintenance, data collection and harvesting process. In both countries, plant spacing was 0.25 m between rows and 0.3 m within rows, resulting in a planting density of 13.3 plant m^{-2}, according to the habits of the local farmers. The elemental unit consisted of a plot of 10 m^2 (133 plants) or 5.4 m^2 (72 plants) in NE-Brazil and CDZ-Myanmar, respectively.

Table 3. Chemical characterization of the soils in the two locations.

OM [1] (%)	pH	EC [2] (dS m^{-1})	N	Available P	K	Ca	Exchangeable Mg	Na	CEC [3] (cmol (+) kg^{-1})
				NE-Brazil (Exp. 1)					
1.01	5.1	nd	nd	9.0	35.2	1.4	0.1	9.2	4.51
				CDZ-Myanmar (Exp. 2)					
0.38	6.2	0.11	54	10.9	25	3.04	0.2	34.8	nd

[1] OM = Organic Matter; [2] EC = Electrical Conductivity; [3] CEC = Cation Exchange Capacity, nd = not determined.

2.5. Irrigation Management

In the SSC system, nutrient solution flux started early in the morning (at 7:00 am) and continued until dusk (6:00 pm). Three times per day (at 7:00 am, 11:00 am, and 3:00 pm), the drained nutrient solution was moved back to the upper tank. The daily nutrient solution consumption was calculated by the difference between the nutrient solution volume between the upper tank (at the beginning of the day) and the bottom tank (at the end of the day). The nutrient solution in the system was refreshed every day by adding new nutrient solution to a set level

When plants were grown on the soil-based system, the irrigation management was different in NE-Brazil and CDZ-Myanmar experiments. In NE-Brazil, irrigation management was carried out based on the traditional local habit of the farmers by using manual irrigation. Water was distributed across experimental plots through manual labor, and a 12 L watering bucket was used. The amount of water distributed in a plot was based on farmers' experience. In CDZ-Myanmar, the irrigation management of soil-based treatments was based on crop evapotranspiration (ETc), restoring 100% of crop ETc by means of a drip irrigation system

ETc was calculated by using the following equation (Equation (1))

$$ET_c = ET_0 \times K_c \tag{1}$$

where ET_c (mm day^{-1}) is the calculated crop evapotranspiration, ET_0 (mm day^{-1}) is the reference evapotranspiration, and K_c is the FAO crop coefficient for lettuce [20].

For the estimation of the reference evapotranspiration (ET_0), the Hargreaves-Samani (HS) equation (Equation (2)) was used,

$$ET_0 = 0.0023 \times (T_{mean} + 17.8) \times (T_{max} - T_{min})^{0.5} \times R_a \tag{2}$$

where ET_0 (mm day^{-1}) is the reference evapotranspiration rate, T_{mean}, T_{max} and T_{min} are the mean, maximum and minimum temperature (°C) of the day, respectively, and R_a (W m^{-2} day^{-1}) is the extraterrestrial solar radiation [20].

The meteorological data for the determination of the reference evapotranspiration were daily downloaded from the website of the Agro-Meteorological Department of Yezin Agriculture University (http://www.yau.edu.mm/), located inside the university campus, excluding extraterrestrial radiation Ra that was calculated according to Duffie and Beckman [21].

The amount of water used for each irrigation was calculated based on plant water balance considering soil properties, root depth, and climate data (including rainfall, if any). Daily ETc was estimated considering the FAO crop coefficient for lettuce crop growth stages. Lettuce cycles were divided into three growth stages, and the Kc used was 0.7, 1.0 and 0.95, respectively. The time of irrigation was determined when readily available soil water (50% available soil water) was depleted.

Sixteen mm diameter drip pipes were used. Drippers had a flow rate of approximately 1.3 L h^{-1}, and each plant was supplied with a single dripper. A flow rate test and calculation of distribution uniformity (DU) were carried out before transplanting. The DU was calculated following the indications from Baum et al. [22]. Irrigation management (time and rate) was manually performed.

2.6. Plant Measurements

At harvest, plants were weighed to determine the fresh weight (g plant^{-1}). Yield (kg m^{-2}) was assessed by excluding external leaves which appeared damaged or wilted. Leaf number was also counted. Water use efficiency (WUE) was determined as the ratio between fresh weight and the volume of water used and was expressed as g FW L^{-1} H$_2$O. In experiment 4, leaf stomatal conductance was also measured using a handheld photosynthesis measurement system model CI-340 (Camas, WA, USA), equipped with 6.25 cm^2 cuvette. Measurements were made at 27 DAT on the upper surface of the canopy on three leaves per each plant from 10:00 to 14:00 taking approximately one hour to complete each replication. All plants were measured on a single day. In the system, EC and pH were constantly monitored using a Combo pH/EC/TDS/Temp tester Model HI98130 (HANNA®, Villafranca Padovana (PD), Italy). In experiments 2 and 4, the nutrient solution temperature was also monitored twice a week.

2.7. Statistical Analysis

Data were collected on 12 plants from the central part of each plot. Data from experiments 1 and 2 were analyzed using one-way ANOVA. Data from experiments 3 and 4 were analyzed by using two- and three-way ANOVA, respectively. Means were separated using the Tukey HSD test at $p \leq 0.05$. Before the analysis, all data were checked for normality and homogeneity of variance. Averages and standard errors (SE) were calculated. Statistical analysis was carried out using R statistical software (version 3.3.2, package "emmeans" and "car").

3. Results

3.1. Climate during the Experiments.

NE-Brazil

During experiment 1, maximum air temperature ranged between 31.4 and 34.7 °C with an average of 33.0 °C. Minimum temperature ranged between 17.8 and 22.4 °C with an average of 19.8°C. The daily relative humidity (RH) ranged between a minimum of 57% and a maximum of 97% (Table 4). Furthermore, 20.3 mm of effective rainfall occurred. During experiment 3, maximum air temperature ranged between 31.7 and 35.4 °C with an average maximum temperature of 34.0 °C. Minimum temperature ranged between 17.8 and 22.4 °C with an average minimum temperature of 19.8 °C. The maximum relative humidity (RH) was 97% and the minimum RH was 55% (Table 4). The growing degree days (GDD) from transplanting to harvest ranged from 710 °C (experiment 3) to 920 °C (experiment 1).

Table 4. Main climatic features during the experiments.

Average Air Temperature (°C)		RH [1] (%)		DLI [2] (mol m^{-2} d^{-1})	Wind Speed (m s^{-1})	GDD [3] (°C)
max	min	max	min			
			NE-Brazil (Exp. 1)			
33.0	19.8	97	57	24.3	0.9	920 */698 **
			CDZ-Myanmar (Exp. 2)			
31.5	16.7	73	48	17.0	1.0	662
			NE-Brazil (Exp. 3)			
34.0	19.8	97	55	24.7	1.0	710
			CDZ-Myanmar (Exp. 4)			
35.6	19.5	59	30	20.9	1.9	731

[1] RH = Relative Humidity; [2] DLI = average daily light integrals; [3] GDD = growing degree days, calculated based on a crop base temperature of 4 °C [23]; * on soil-based system (40 days cropping cycle); ** on simplified soilless system (31 days crop cycle).

CDZ-Myanmar

During the experiment 2, maximum air temperature ranged between 25.0 and 34.0 °C with an average of 31.5°C. Minimum temperature ranged between 14.4 and 20.4 °C, with an average of 16.7 °C. The daily relative humidity (RH) ranged between a minimum of 48% and a maximum of 73% (Table 4). During experiment 4, maximum air temperature ranged between 30.7 and 38.6°C with an average maximum temperature of 35.6 °C. Minimum temperature ranged between 16.0 and 23.0°C with an average minimum temperature of 19.5 °C. The maximum relative humidity (RH) was 59% and the minimum RH was 30%. No rainfall occurred during the experiments. The growing degree days (GDD) from transplanting to harvest ranged from 662 °C (experiment 2) to 731 °C (experiment 4) (Table 4).

3.2. Experiment 1—NE-Brazil

Lettuce yield was higher (+35%) in the SSC system, with a mean value of 2.3 kg m^{-2}, as compared to 1.7 kg m^{-2} achieved on soil (Figure 2a). This was mainly due to larger size of the leaves (data not shown), while leaf number was higher in plants grown on soil (Figure 2b). The increased yield was obtained with a daily water use (L m^{-2} d^{-1}) approximately four times lower in the SSC system, as compared to conventional on-soil cultivation (1.8 vs. 7.5 L m^{-2} d^{-1}) (data not shown). As a consequence, WUE in SSC system was 7.7 times higher, as compared to the conventional on-soil system, with mean values of 43.7 and 5.6 g L^{-1} H$_2$O, respectively (Figure 2c).

3.3. Experiment 2—CDZ-Myanmar

During experiment 2 the average minimum temperature of the nutrient solution was 19.6 ± 1.64 °C while the average maximum temperature was 29.2 ± 1.69 °C. The average pH was 7.7, ranging from 7.4 to 8.1. The average EC was 1.28, ranging from 1.12 to 1.46 dS m^{-1}.

Yield (kg m^{-2}) was increased by 72% (3.1 vs. 1.8 kg m^{-2}) in SSC in comparison to the soil treatment (Figure 2d) and the leaf number was significantly higher in soil-grown lettuce compared to soilless-grown plants (Figure 2e). Daily water use (L m^{-2} d^{-1}) was approximately two times lower in the SSC system (2.66 L m^{-2} d^{-1}) as compared to conventional on soil production (4.07 L m^{-2} d^{-1}). WUE in the SSC system was found to be 2.7 times higher than that obtained with conventional cultivation, with average values of 37.1 and 13.7 g L^{-1} H$_2$O, respectively (Figure 2f).

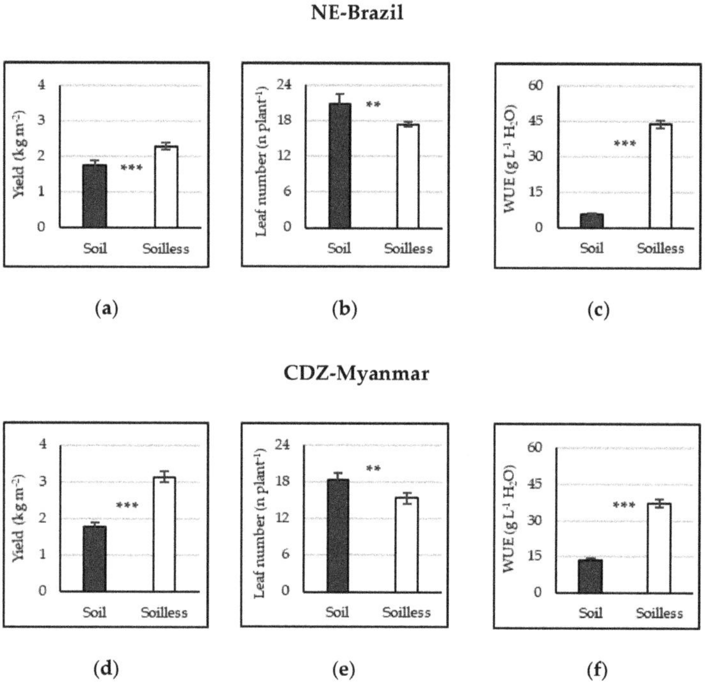

NE-Brazil

CDZ-Myanmar

Figure 2. Results from experiments 1 (Brazil, top row) and 2 (Myanmar, bottom row). Lettuce yields (**a**,**d**), leaf number (**b**,**e**), and water use efficiency (WUE, **c**,**f**). Vertical bars represent standard errors. Significant differences at $p \leq 0.01$ (**), $p \leq 0.001$ (***).

3.4. Experiment 3—NE-Brazil

Considering the system orientation, significant differences for yield were found only in Veronica and Banchu cultivars, for which the west-oriented system showed a reduction in yield of 10 and 44%, respectively, as compared to the east-oriented one. In contrast, yields of cv Isabela and cv Mimosa were not affected by the SSC system orientation (Figure 3). Daily water use was about 1.8 L m^{-2} d^{-1}, as for experiment 1.

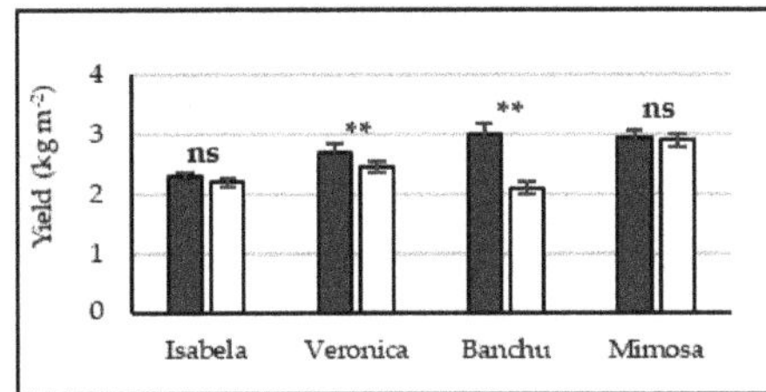

Figure 3. Results from experiments 3 (NE-Brazil). Yield response to the simplified soilless system orientation (east, grey columns; west, white columns) in four lettuce cultivars (Isabela, Veronica, Banchu, and Mimosa). Vertical bars represent standard errors. Significant differences at $p \leq 0.01$ (**), ns = not significant differences.

3.5. Experiment 4—CDZ-Myanmar

The average minimum temperature of the nutrient solution was 20.7 ± 1.1 °C while the average maximum temperature was 39.5 ± 0.87 °C. Daily water use was 2.50 L m^{-2} d^{-1} for NS$_{1.2}$ and 2.23 L m^{-2} d^{-1} for NS$_{1.8}$. The average pH was 7.3 and 7.5 for NS$_{1.2}$ and NS$_{1.8}$, respectively. pH ranged from 6.4–8.7 for the former, and 6.6–8.9 for the latter. Average EC was 1.25 and 1.83 dS m^{-1} for NS$_{1.2}$ and NS$_{1.8}$, respectively, ranging from 1.14–1.48 dS m^{-1} for solution NS$_{1.2}$ and 1.59–2.06 dS m^{-1} for solution NS$_{1.8}$ (data not shown). Results of analysis of variance in Table 5 show that the EC of the nutrient solution (EC), lettuce cultivar (Cv), and plants position (P) significantly affected plant morphological and productive parameters, as well as WUE and the crop physiological response.

Table 5. Results from the ANOVA on experiment 4 (CDZ-Myanmar). Effect of EC of the nutrient solution (EC), cultivar (Cv), and position within the garden (P) on lettuce yield, leaf number, and water use efficiency (WUE). Significant differences at $p \leq 0.05$ (*), $p \leq 0.01$ (**) and $p \leq 0.001$ (***), ns = not significant differences.

	Yield (kg m^{-2})	Leaf Number (n plant^{-1})	g$_s$ (mmol m^{-2} s^{-1})	WUE (g L^{-1})
EC of nutrient solution (EC)	***	***	***	***
Cultivar (Cv)	**	***	***	*
Position (P)	***	ns	***	***
EC × Cv	*	**	**	*
EC × P	***	*	**	***
Cv × P	ns	ns	ns	ns
EC × Cv × P	**	ns	*	**

Yield, stomatal conductance and WUE were affected by EC, Cv, and P—wherein a significant interaction between the three factors was noted—while leaf number was only affected by EC and Cv, with a significant interaction between the two factors (Table 5). Yield of plants placed in the lower position (LP) was not affected by Cv and EC, while for both cultivars the plants in the upper position (UP) yielded more when NS$_{1.2}$ was used (Table 6). The yield of plants belonging to cv Thai and grown by using NS$_{1.2}$ was four times higher, as compared to yield of Thai lettuce supplied with NS$_{1.8}$ and placed in the same position within the system (Table 6). The increased yield was mainly due to leaf number, as Thai plant grown adopting $NS_{1.2}$ showed the highest number of leaves (12.9 leaves plant^{-1}) while no differences were observed between the other treatments (data not shown). Stomatal conductance was highest (212 mmol m^{-2} s^{-1}) in Thai lettuce grown on the upper part of the system by using $NS_{1.2}$ (Table 6). For plants grown in the lower part of the system (LP), stomatal conductance was only affected by CV, and was higher in cv Thai for both considered EC (Table 6). Leaf temperature was only affected by the position (P) in the system (data not shown), and was lowest in plants grown on the top of the system (28.8 °C compared to 29.8 °C measured in plants grown at the bottom of the system). In cv Thai, WUE was highest in plants fed with $NS_{1.2}$ and grown in the upper position (UP) of the system, while for cultivar Rapido 344 the only statistically significant difference was evidenced between plants grown on the upper position and fed with $NS_{1.2}$ and plants grown in the lower position of the system and fed with $NS_{1.8}$ (Table 6).

Table 6. Results from experiment 4 (CDZ-Myanmar). Effects of factorial combination of EC of the nutrient solution (EC, 1.2 vs. 1.8 dS m^{-1}), cultivar (Cv, Thai vs EW) and position (P, upper position, UP vs lower position, LP) within the garden on lettuce yield, stomatal conductance (g_s)and water use efficiency (WUE). Different letters indicate significant differences at $p \leq 0.05$.

EC		Yield		g_s		WUE	
(dS m^{-1})	Cultivar	(kg m^{-2})		(mmol m^{-2} s^{-1})		(g L^{-1} H$_2$O)	
		UP	LP	UP	LP	UP	LP
1.2	Thai	2.88 (a)	1.18 (bc)	212 (a)	118 (b)	38.4 (a)	15.7 (bc)
1.2	Rapido 344	1.79 (b)	1.14 (c)	100 (b)	44 (de)	24.0 (b)	15.2 (bc)
1.8	Thai	0.71(c)	0.65 (c)	121 (b)	91 (bc)	10.57 (c)	9.71 (c)
1.8	Rapido 344	0.82 (c)	0.57 (c)	78 (c)	35 (e)	12.3 (c)	8.54 (c)

4. Discussion

The application of different cropping systems significantly affected yield, physiological response and water use efficiency of lettuce grown in both NE-Brazil and CDZ-Myanmar.

Water availability is one of the major constraints for agricultural development and food production. The first and second experiments aimed to determinate whether SSC lettuce production is a suitable and sustainable alternative to conventional on-soil production in both locations. Barbosa et al. [23], when comparing commercial (high-tech) hydroponic greenhouses against on-soil lettuce production, found that hydroponics could increase yield by 11-folds, thanks to improved nutrition and environmental control. According to our results obtained in both experiments, the use of a simplified (low tech) soilless system allowed increase in the yield of lettuce but to a lesser extent (+35% in NE-Brazil and +72% in CDZ-Myanmar, (Figure 2a,d)). Yield increase can be the result of higher planting density (26 vs. 13 plants m^{-2}, on SSC and on-soil cultivation respectively), fast plant growth and precocity of production (31 vs 40 DAT according to experiment 1) and the improved environmental conditions maintained within the SSC system, including plant nutrition, uniform and constant irrigation, as well as the shading cover integrated in the SSC system. According to Zhao et al., the adoption of a shading net as a cover for lettuce production in the summer season in Kansas led to a slightly lower daily maximum air temperature relative to the open field, with an average reduction of $\approx$0.4 °C [24]. Moreover, Zhao et al. reported that the shading net has a significant impact on soil temperature and leaf temperature [24]. Indeed, in comparison with open field conditions, when shading net is adopted, a considerable reduction of leaf surface temperature, by 1.5 to 2.5 °C, was observed [24], thus affecting the plants' capacity to absorb water and nutrients [25]. In the SSC system, the higher fertigation frequency probably affected production capacity. Silber et al. experimented on the effect of fertigation frequency on yield, water and nutrient uptake of lettuce [26], finding that high fertigation frequency (from 2 to 10 events a day) induced a significant increase (13–15%) in lettuce fresh weight (FW) [26].

Furthermore, SSC systems are also considered water-saving technologies, thanks to the capability to deliver water directly to the plant root [27,28]. Despite limited soil exploration by the shallow rooting system of lettuce, in NE-Brazil, when a conventional growing system was adopted, irrigation water was applied by means of a bucket or can on the entire soil surface and consequently a significant amount of water is lost through evaporation and percolation into the sandy soil. Increase in the use of low-flow and more targeted irrigation techniques, such as the adoption of a drip irrigation system, could lower the overall water use of conventional farming [23]. As a matter of fact, drip irrigation was used as a control treatment in the CDZ-Myanmar experiments. Accordingly, the adoption of an SSC system enabled a reduction of water use by 76% and 59% in NE-Brazil and in CDZ-Myanmar, respectively, as compared to on-soil production. The observed water savings are consistent with previous literature, e.g., when a SSC system was adopted in Colombia, water use was reduced by 90% as compared to the traditional on-soil cropping system [29].

A consequence of higher yield and lower water use was an increased WUE in the SSC systems. In NE-Brazil and CDZ-Myanmar, WUE was, respectively, 7.7 and 2.7-fold higher, as compared to conventional on-soil production (Figure 2c,f). Similarly, WUE for lettuce in hydroponics was previously found in the range of 2.9 g of dry mass per L^{-1} H_2O [30], or 41 g of fresh mass per L^{-1} H_2O [31]. Lettuce grown in high-tech hydroponic conditions showed a reduction in water use by 13-fold, as compared with traditional on-soil cultivation [23]. Under the expected climate change scenarios and water limitation for agriculture, SSC systems could be a valuable strategy to sustain highly productive agriculture where the adoption of high-technology systems is not affordable [3,5,28,32].

In the third experiment, the adaptability of four lettuce cultivars to two different garden orientations was addressed to have a deeper understanding of the SSC system management. It emerged that the response of plant growth to the garden orientation was cultivar dependent (Figure 3). Accordingly, Veronica and Banchu achieved a higher yield with the eastern garden exposure (Figure 3), which received a lower amount of solar radiation in the afternoon mitigating the high air temperatures. Wheeler et al. [33] indicated 23 °C as the optimal daily temperature for growing lettuce, a condition that is far below the mean temperatures observed in NE-Brazil during the experiment. Heat stress could also result in a greater osmotic stress caused by the nutrient solution, resulting in lower water uptake and reduced plant growth [27]. Moreover, due to elevated temperatures in the root zone environment, hypoxia may occur, inhibiting root respiration, mineral uptake and water movement into the roots [34].

In the fourth experiment, the growth response of plants grown in different positions within the system was addressed as a function of both plant cultivar and the nutrient solution concentration (Table 5). Accordingly, the highest yield was associated with Thai cultivar supplied with $NS_{1.2}$ (Table 6). Possibly, under the local climate, plants preferred a nutrient solution with lower EC, and the Thai cultivar better responded to reduced osmotic stress. Furthermore, yield was the highest when Thai cultivar was grown in the upper part of the system. (Table 6). As reported by Gianquinto et al. [14], this aspect could be due to the increased temperature reached by the nutrient solution during the flow between the top and the bottom tank. It would suggest that by the time the nutrient solution reached the lower section of the SSC, it was significantly warmed up by irradiance in the plastic bottles, although this statement should further be confirmed by determination of nutrient solution and substrate temperatures in the different positions. Thompson et al. [35] showed that a 24 °C root temperature in hydroponic systems is the ideal temperature whereby lettuce growth can be maximized, even with elevated air temperature. Different studies also reported that high nutrient solution temperature depresses water and nutrient uptake through reduced oxygen availability, also affecting physiological processes such as root browning and active transport in membranes [36]. Moreover, it was also observed that high solution temperature might decrease nutrient concentration (particularly of N, K and Ca) in the root, which may ultimately decrease crop growth [25]. It should be further studied, however, whether this may be associated with increased nutrient solution temperature as water flows through the system, or with selective absorption of specific nutrients from those plants that receive the nutrient solution first. In this regard, it could also be considered to add additional hydraulic pipes with emitters in the middle of the system.

5. Conclusions

The study addressed the application of SSC technologies [37–39] for lettuce production in tropical wet and dry climates. Elevated potentialities, in terms of both yield increase and improved water use efficiency in comparison with traditional on-soil cultivation technologies, were evidenced in both locations. Furthermore, the study explored alternative crop management strategies evidencing differences in cultivar adaptability and potential productivity. For instance, garden orientation was shown to affect crop productivity on a cultivar-dependent basis. Finally, under the elevated temperatures that are locally experienced, it is advisable to reduce the concentration of the nutrient solution (with EC of 1.2 dS m^{-1} providing better results than EC of 1.8 dS m^{-1}). Interestingly, the yield was also improved when plants were located in the upper positions of the garden. Government and local support services could influence the future of soilless farming, as subsidies could be used to offset the relatively high initial cost of SSC infrastructure. We conclude that a simplified soilless system could become one of the efficient strategies for contributing to sustainably feeding the world's growing population, especially in challenging areas such as the north east of Brazil and the central dry zone of Myanmar.

Author Contributions: Conceptualization, N.M., A.A.C.S., F.O., G.G.; methodology, N.M.; validation, N.M., G.G., G.P., F.O.; formal analysis, N.M., G.P., G.G.; investigation, N.M., L.P.B., G.D., N.O.M.; resources, F.O., G.P.; data curation, N.M., G.P.; writing—original draft preparation, N.M.; writing—review and editing, G.P., G.G., F.O., N.S.G.; visualization, N.M., G.P., G.G.; supervision, F.O., G.P.; project administration, N.M.; funding acquisition, F.O., G.G. All authors have read and agreed to the published version of the manuscript.

Funding: This research received no external funding.

Acknowledgments: We thank Padre Humberto Pietrogrande, former President of Pe. Antonio Dante Civiero Foundation, Nang Hseng Hom, Rector of Yezin agriculture University of Myanmar and Terre des Hommes Italia INGO for providing the area, infracture and equipment to carry out the experiments.

Conflicts of Interest: The authors declare no conflict of interest.

References

1. Eileen, E.B. Impact of Climate Change and Local Adaptation Strategies of Various Socio-economic Groups in Isabela, Northern Philliphines. *Clim. Issue* **2009**, *22*, RN15.
2. Bisbis, M.B.N.; Gruda, M. Blanke 2018: Potential impacts of climate change on vegetable production and product quality—A review. *J. Clean. Prod.* **2018**, *170*, 1602–1620. [CrossRef]
3. Gruda, N.; Bisbis, M.B.; Tanny, J. Impacts of protected vegetable cultivation on climate change and adaptation strategies for cleaner production—A review. *J. Clean. Prod.* **2019**, *225*, 324–339. [CrossRef]
4. Audu, E.B. Gas Flaring: A Catalyst to Global Warming in Nigeria. *Int. J. Sci. Technol.* **2013**, *3*, 6–10.
5. Gruda, N.; Bisbis, M.B.; Tanny, J. Influence of climate change on protected cultivation: Impacts and sustainable adaptation strategies—A review. *J. Clean. Prod.* **2019**, *225*, 481–495. [CrossRef]
6. Intergovernmental Panel on Climate Change (IPCC). Climate Change: Impacts, Adaptation and Vulnerability. In *Exit Epa Disclaimer Contribution of Working Group II for the Fourth Assessment Report of the Intergovernmental Panel on Climate Change*; Cambridge University Press: Cambridge, UK, 2007.
7. Dowuona Nii Nortey, M.; Kwaghe, P.V.; Abdulsalam, B.; Aliyu, H.S.; Dahiru, B. Review of Farm Level Adaptation Strategies to Climate Change in Africa. *Greener J. Agron. For. Hortic.* **2014**, *2*, 038–043.
8. Barbieri, A.F.; Domingues, E.; Queiroz, B.L.; Ruiz, R.M.; Rigotti, J.I.; Carvalho, J.A.; Resende, M.F. Climate change and population migration in Brazil's Northeast: Scenarios for 2025–2050. *Popul. Environ.* **2010**, *31*, 344–370. [CrossRef]
9. Orsini, F.; Kahane, R.; Nono-Womdim, R.; Gianquinto, G. Urban agriculture in the developing world: A review. *Agron. Sustain. Dev.* **2013**, *33*, 695–720. [CrossRef]

10. Souza, E.D.; Carneiro, M.A.C.; Paulino, H.B.; Silva, C.A. Frações do carbono orgânico, biomassa e atividade microbiana em um Latossolo Vermelho sob cerrado submetido a diferentes sistemas de manejos e usos do solo. *Acta Sci. Agron.* **2006**, *28*, 323–329. [CrossRef]

11. Cho, A.; Tun Oo, A.; Speelman, S. Assessment of household food security through crop diversification in Natmauk township, Magway Region, Myanmar. In Proceedings of the Tropentag 2016: Solidarity in a Competing World: Fair Use of Resource, Vienna, Austria, 18–21 September 2016; pp. 1–5.

12. Pavelic, P.; Sellamuttu, S.S.; Johnston, R.; McCartney, M.; Sotoukee, T.; Balasubramanya, S.; Latt, K. *Integrated Assessment of Groundwater Use for Improving Livelihoods in the Dry Zone of Myanmar*; International Water Management Institute (IWMI): Colombo, Sri Lanka, 2015; Volume 164.

13. Aoki, Y. A Problem of Climate Change as Seen by a Pharmaceutical Researcher. *J. Health Sci.* **2009**, *55*, 857–859. [CrossRef]

14. Gianquinto, G.; Orsini, F.; Michelon, N.; da Silva, D.F.; De Faria, F.D. Improving yield of vegetables by using soilless micro-garden technologies in peri-urban area of north-east Brazil. *Acta Hortic.* **2007**, *747*, 57–65. [CrossRef]

15. Savvas, D.; Gruda, N. Application of soilless culture technologies in the modern greenhouse industry—A review. *Eur. J. Hortic. Sci.* **2018**, *83*, 280–293. [CrossRef]

16. Orsini, F.; Morbello, M.; Fecondini, M.; Gianquinto, G. Hydroponic gardens: Undertaking malnutrition and poverty through vegetable production in the suburbs of Lima, Peru. *Acta Hortic.* **2010**, *881*, 173. [CrossRef]

17. Izquierdo, J. Simplified hydroponics: A tool for food security in Latin America and the Caribbean. *Acta Hortic.* **2007**, *742*, 67–74. [CrossRef]

18. Gianquinto, G.; Michelon, N.; Orsini, F. Idroponia in un'area povera del nord est del Brasile. In *Un Esempio di Cooperazione Decentrata: Simplified Soilless Cultivation in a Marginal Area of North-East Brazil: An Example of Decentralised Cooperation*; Franceschetti, G., CLEUP. Regione Veneto-FAO, Eds.; Agricoltura e Ruralità nei Paesi ad Economia Povera: Padova, Italy, 2006; pp. 95–106.

19. Rodríguez-Delfín, A.; Hoyos, M.; Chang, M. *Soluciones Nutritivas en Hidroponía: Formulación y Preparación*; Centro de Investigación de Hidroponía y Nutrición Mineral, Universidad Nacional Agraria La Molina: Lima, Peru, 2001; p. 100.

20. Trajkovic, S. Hargreaves versus Penman-Monteith under humid conditions. *J. Irrig. Drain. Eng.* **2007**, *133*, 38–42. [CrossRef]

21. Duffie, J.A.; Beckman, W.A.; Blair, N. *Solar Engineering of Thermal Processes, Photovoltaics and Wind*, 5th ed.; John Wiley & Sons: Hoboken, NJ, USA, 2020.

22. Baum, M.C.; Michael, D.D.; Grady, L.M. Analysis of residential irrigation distribution uniformity. *J. Irrig. Drain. Eng.* **2005**, *131*, 336–341. [CrossRef]

23. Barbosa, G.L.; Gadelha, F.D.A.; Kublik, N.; Proctor, A.; Reichelm, L.; Weissinger, E.; Halden, R.U. Comparison of Land, Water, and Energy Requirements of Lettuce Grown Using Hydroponic vs. Conventional Agricultural Methods. *Int. J. Environ. Res. Public Health* **2015**, *12*, 6879–6891. [CrossRef]

24. Zhao, X.; Carey, E.E. Summer production of lettuce, and microclimate in high tunnel and open field plots in Kansas. *HortTechnology* **2009**, *19*, 113–119. [CrossRef]

25. Falah, M.A.F.; Wajima, T.; Yasutake, D.; Sago, Y.; Kitano, M. Responses of root uptake to high temperature of tomato plants (*Lycopersicon esculentum* Mill.) in soil-less culture. *J. Agric. Sci. Technol.* **2010**, *6*, 543–558.

26. Silber, A.; Xu, G.; Levkovitch, I.; Soriano, S.; Bilu, A.; Wallach, R. High fertigation frequency: The effects on uptake of nutrients, water and plant growth. *Plant Soil.* **2003**, *253*, 467–477. [CrossRef]

27. Sanchez, S.V. *Avaliação de Alface Crespa Produzidas em Hidropônia Tipo NFT em Dois Ambientes Protegidos em Ribeirão Preto (SP)*; Paulista State University, College of Agricultural and Veterinary Science: Sao Paulo, Brazil, 2007; Available online: http://depositorio.unesp.br/handle/11449/96944 (accessed on 12 August 2020).

28. Gruda, N.S. Increasing Sustainability of Growing Media Constituents and Stand-Alone Substrates in Soilless Culture Systems. *Agronomy* **2019**, *9*, 298. [CrossRef]

29. Bradley, P.; Marulanda, C. Simplified hydroponics to reduce global hunger. *Acta Hortic.* **2000**, *554*, 289–296. [CrossRef]

30. Wheeler, R.M.; Sager, J.C.; Berry, W.L.; Mackowiak, C.L.; Stutte, G.W.; Yorio, N.C.; Ruffe, L.M. Nutrient, acid and water budgets of hydroponically grown crops. *Acta Hortic.* **1997**, *481*, 655–662. [CrossRef]

31. Michelon, N.; Pennisi, G.; Myint, N.O.; Orsini, F.; Gianquinto, G. Strategies for Improved Water Use Efficiency (WUE) of Field-Grown Lettuce (*Lactuca sativa* L.) under a Semi-Arid Climate. *Agronomy* **2020**, *10*, 668. [CrossRef]

32. Martínez-Alvarez, V.; Soto-García, M.; Maestre-Valero, J.F. Hydroponic system and desalinated seawater as an alternative farm-productive proposal in water scarcity areas: Energy and greenhouse gas emissions analysis of lettuce production in southeast Spain. *J. Clean. Prod.* **2018**, *172*, 1298–1310. [CrossRef]

33. Wheeler, T.R.; Hadley, P.; Morison, J.I.L.; Ellis, H. Effects of temperature on the growth of lettuce (*Lactuca sativa* L.) and the implications of assessing the impacts of potential climate changes. *Eur. J. Agron.* **1993**, *2*, 305–311. [CrossRef]

34. Ehret, D.L.; Edwards, D.; Helmer, T.; Lin, W.; Jones, G.; Dorais, M.; Papadopoulos, A.P. Effects of oxygen-enriched nutrient solution on greenhouse cucumber and pepper production. *Sci. Hortic.* **2010**, *125*, 602–607. [CrossRef]

35. Thompson, H.C.; Langhans, R.W.; Both, A.J.; Albright, L.D. Shoot and root temperature effects on lettuce growth in a floating hydroponic system. *J. Am. Soc. Hortic. Sci.* **1998**, *123*, 361–364. [CrossRef]

36. Trejo-Téllez, L.I.; Gómez-Merino, F.C. Nutrient solutions for hydroponic systems. In *Hydroponics-A Standard Methodology for Plant Biological Researches*; Martinez-Mate, M.A., Martin-Gorriz, B., Eds.; InTechOpen: Rijeka, Croatia, 2012; pp. 1–22.

37. Fecondini, M.; Damasio De Faria, A.C.; Michelon, N.; Mezzetti, M.; Orsini, F.; Gianquinto, G. Learning the value of gardening: Results from an experience of community based simplified hydroponics in north-east Brazil. *Acta Hort.* **2010**, *881*, 111–116. [CrossRef]

38. Fecondini, M.; Casati, M.; Dimech, M.; Michelon, N.; Orsini, F.; Gianquinto, G. Improved cultivation of lettuce with a low cost soilless system in indigent areas of northeast brazil. *Acta Hort.* **2009**, *807*, 501–507. [CrossRef]

39. Orsini, F.; Michelon, N.; Scocozza, F.; Gianquinto, G. Farmers-To-Consumers Pipeline: An Associative Example of Sustainable Soil-Less Horticulture in Urban And Peri-Urban Areas. *Acta Hort.* **2009**, *809*, 209–220. [CrossRef]

Article

Effect of Nutrient Solution Concentration on the Growth of Hydroponic Sweetpotato

Masaru Sakamoto * and Takahiro Suzuki

Faculty of Biology-Oriented Science and Technology, Kindai University, Wakayama 649-6493, Japan; tksuzuki@waka.kindai.ac.jp
* Correspondence: sakamoto@waka.kindai.ac.jp; Tel.: +81-0736773888

Received: 24 September 2020; Accepted: 3 November 2020; Published: 4 November 2020

Abstract: Nutrient solution concentration (NSC) is a critical factor affecting plant growth in hydroponics. Here, we investigated the effects of hydroponic NSC on the growth and yield of sweetpotato (*Ipomoea batatas* (L.) Lam.) plants. First, sweetpotato cuttings were cultivated hydroponically in three different NSCs with low, medium, or high electrical conductivity (EC; 0.8, 1.4, and 2.6 dS m^{-1}, respectively). Shoot growth and storage root yield increased at 143 days after plantation (DAP), depending on the NSC. Next, we examined the effect of NSC changes at half of the cultivation period on the growth and yield, using high and low NSC conditions. In plants transferred from high to low EC (HL plants), the number of attached leaves increased toward the end of the first half of the cultivation period (73 DAP), compared with plants transferred from low to high EC (LH plants). Additionally, the number of attached leaves decreased in HL plants from 73 DAP to the end of the cultivation period (155 DAP), whereas this value increased in LH plants. These changes occurred due to a high leaf abscission ratio in HL plants. The storage root yield showed no significant difference between HL and LH plants. Our results suggest that the regulation of hydroponic NSC during the cultivation period affects the growth characteristics of sweetpotato.

Keywords: nutrient solution concentration; hydroponics; sweetpotato; storage root; leaf abscission

---

1. Introduction

Sweetpotato (*Ipomoea batatas*) is an important root vegetable cultivated in temperate and tropical zones, especially in Asia and Africa [1]. Storage roots of sweetpotato contain relatively high amounts of carbohydrates that support the demand for food in developing countries [2,3]. In recent years, sweetpotato has been also evaluated as a candidate for biofuel production [4,5]. Sweetpotato could potentially be used as an alternative to corn-based ethanol production to reduce fertilizer, water, and pesticide inputs and to utilize its ability to fix relatively large amounts of solar energy into starch in storage roots [6,7]. Several efficient methods of extracting biofuels and their residues (hydrogen, ethanol, and methane) from sweetpotato have been reported to date [5,8–11]. Because the demand for sweetpotato is gradually increasing worldwide [12], it is necessary to establish an efficient and cheap cultivation method with low fertilizer requirement.

Fertilizers are widely used in agriculture to increase crop production. In sweetpotato, soil amendment using manure and inorganic fertilizers has a significant impact on plant growth and storage root development [13]. Among chemical fertilizers, nitrogen (N), phosphorus (P), and potassium (K) are the major elements required for supporting shoot and root growth in sweetpotato [14,15]. Under N deficient conditions, the application of N fertilizers significantly increases the storage root weight [16–18].

Agronomy **2020**, *10*, 1708; doi:10.3390/agronomy10111708   www.mdpi.com/journal/agronomy

The relative proportion of N and K fertilizers applied also affects the storage root yield, photosynthesis product distribution, and leaf enzyme activities in field-grown sweetpotato [19–21]. Administration of an adequate quantity of K fertilizer (K_2 was supplied to 300 kg ha^{-1}) has shown to increase the ratio of storage root yield relative to the total yield [22]. Furthermore, N application rate influences the lateral root development at the early growth stage, with 50 kg ha^{-1} application being the most developed [23]. These early root developments are thought to the initiation of storage root formation [24].

In hydroponics, fertilizers are supplied as ions in the nutrient solution [25]. Several formulations of essential macro- and micronutrients have been developed to enhance nutrient uptake and plant growth [26]. Because the nutrient solution is the only source of mineral nutrients in hydroponically-grown plants, extremely low concentrations of nutrients generally leads to growth inhibition [27]. On the other hand, extremely high nutrient solution concentration (NSC) causes osmotic stress, ionic toxicity, and growth restriction [27]. Several studies have demonstrated that NSC influences the growth and components of spinach, tomato, cucumber, salvia, bean, artichoke, wasabi, and lettuce plants [28–37]. In a hydroponic NSC with high electrical conductivity (EC), the growth of tomato plants was restricted, whereas the level of sugars and lycopene in tomato fruits, and consequently fruit quality, were enhanced [30]. In strawberry, flower bud initiation was promoted by treatment with low NSC [38–40].

Sweetpotato plants fail to develop storage roots under continuous waterlogging conditions [41–43]. Therefore, several studies have established hydroponic methods of sweetpotato cultivation to avoid soaking the hypertrophic sites of roots in water [41–49]. Substrates that ensure proper root aeration, such as rockwool, vermiculite, and sand, have been used for the hydroponic cultivation of sweetpotato [43–45]. Additionally, rockwool slab-based hydroponic systems have been demonstrated to produce thickened sweetpotato storage roots between the hydroponic solution surface and rockwool slabs [46]. Similarly, the nutrient film technique and modified deep flow technique have been shown to induce storage root formation at an area where roots are not continuously immersed in the hydroponic solution [41,47–49]. Although some hydroponic methods for sweetpotato have been developed to date, studies on sweetpotato hydroponic NSC are limited. Here, using previously developed vermiculite-based hydroponic methods [43], we investigated the effect of NSC on the growth and yield of sweetpotato.

2. Materials and Methods

2.1. Experimental Conditions

Sweetpotato (*Ipomoea batatas*) cultivar "Narutokintoki" was used in this study. The hydroponics system for sweetpotato was prepared as described previously [43]. Briefly, this system consisted of vermiculite-filled vinyl pots (4.5 L) placed in nutrient solution-filled containers (59 cm × 39 cm × 18 cm). Storage roots developed in vinyl pots, and fibrous roots of sweetpotato plants extended from the upper vinyl pots into the containers placed below. Plants could absorb the nutrient solution from the vermiculite, as the bottom of each pot was in direct contact with the water absorption sheet that extended into the nutrient solution below. Therefore, vermiculite in vinyl pots remained saturated with the nutrient solution throughout the cultivation period. The surface of pots and bottom containers was covered with insulation sheets to maximize the utilization of sunlight by reflection for photosynthesis. The nutrient solution was prepared by mixing OAT house 1: OAT house 2 (OAT Agrio Co., Ltd., Tokyo, Japan) at a ratio of 3:2, and the NSC was adjusted to obtain the target EC (described below). The mixed nutrient solution contains N, 260 mg L^{-1}, P_2O_5, 120 mg L^{-1}; K_2O, 405 mg L^{-1}; CaO, 230 mg L^{-1}; MgO, 60 mg L^{-1}; MnO, 1.5 mg L^{-1}; B_2O_3, 1.5 mg L^{-1}; Fe, 2.7 mg L^{-1}; Cu, 0.03 mg L^{-1}; Zn, 0.09 mg L^{-1} and Mo, 0.03 mg L^{-1}. Reduction in the water level in the tank due to evaporation was compensated by adding more water to the maximum level. Therefore, the EC of the NSC in each tank gradually decreased over time (Supplementary Figure S1). The nutrient solution was renewed approximately once every 30 days.

Two separate experiments were conducted (experiment 1 in 2018, and experiment 2 in 2019) to examine the effect of NSC on the growth and yield of sweetpotato (Figure 1). In experiment 1, three NSCs, each with low (0.8 dS m^{-1}), medium (1.4 dS m^{-1}), or high (2.6 dS m^{-1}) EC, were used throughout the cultivation period, and the effect of each NSC on plant growth was observed. The initial pH of high, medium, and low NSCs were 6.14, 6.46, and 6.72, respectively. In experiment 2, effects of changes in NSC on plant growth were examined using nutrient solutions with low EC (0.8 dS m^{-1}) and high EC (2.6 dS m^{-1}). Four treatments were conducted in experiment 2: (1) LL, plants were grown in low EC nutrient solution throughout the cultivation period; (2) LH, plants were grown in low EC nutrient solution until the end of the first half of the cultivation period, and then transferred to high EC nutrient solution and maintained until the end of the cultivation period; (3) HL, plants grown in high EC nutrient solution were transferred to low EC nutrient solution at the end of the first half of the cultivation period and maintained thereafter; (4) HH, plants were grown in high EC nutrient solution throughout the cultivation period.

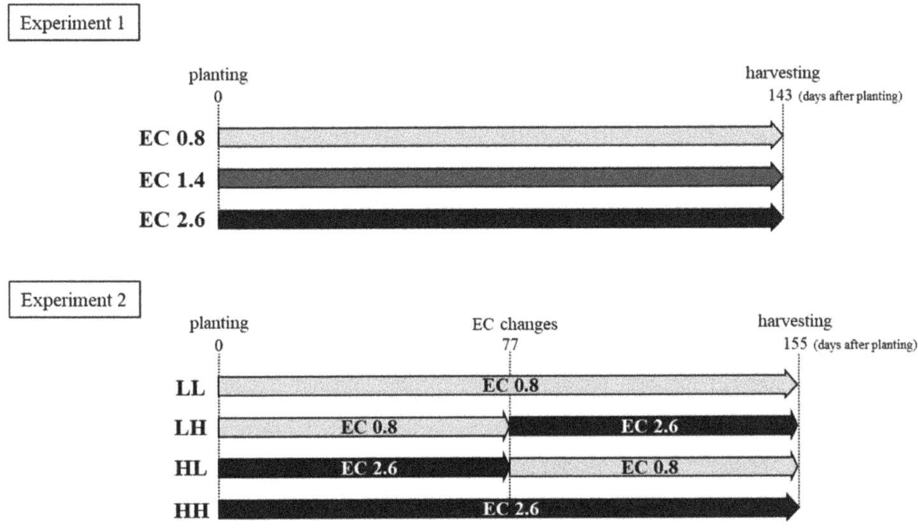

Figure 1. Experimental design. EC: electrical conductivity.

In experiment 1, sweetpotato stem cuttings were planted in vermiculite-filled vinyl pots and grown for 18 days by drenching in nutrient solution with medium EC (1.4 dS m^{-1}). The experiment was started by transferring the pots to the hydroponic system with different NSCs (four pots per container). Plants were then cultivated for 143 days (from 29 May to 19 October in 2018) at the open experimental field of Kindai University (Faculty of Biology-Oriented Science Technology, Wakayama, Japan). The average temperature of Wakayama city in 2018 were 19.7 °C, 23.2 °C, 28.8 °C, 29.1 °C, 24.3 °C, and 19.5 °C in May, June, July, August, September, and October, respectively, according to the website of Japan Meteorological Agency [50]. The average relative humidity was 69%, 77%, 74%, 69%, 78%, and 67% in May, June, July, August, September, and October, respectively [50]. The experimental field is about 17 km away from the meteorological station in Wakayama City, and 90 m higher than the station. The nutrient solution was renewed on 26 June, 31 July, and 30 August. In experiment 2, stem cuttings were planted in pots and grown for 25 days under the same growth conditions as those used for experiment 1. Pots were then transferred to the hydroponic system and cultivated for 155 days (from 18 May to 20 October in 2019) at the open experimental field of Kindai University. The average temperature of Wakayama city in 2019 was 20.1 °C, 23.5 °C, 26.3 °C, 28.5 °C, 26.4 °C, and 20.7 °C in May, June, July, August, September, and October, respectively [51]. The average relative humidity was 60%, 72%, 79%, 76%, 71%, and 73% in May, June, July, August, September, and October, respectively [51]. The nutrient solution was renewed on 22 June, 12 July, 3 August, 31 August, and 28 September. The EC of the nutrient solution was changed for HL and LH plants on 3 August 2019.

2.2. Measurement of Plant Growth and Yield

In experiments 1 and 2, shoot and storage root fresh weight (FW) and storage root number were measured at 143 and 155 days after planting (DAP), respectively. Enlarged roots weighing more than 20 g were included as storage roots. In experiment 2, the number of attached leaves and the maximum length of the stem were measured at 3, 73, and 155 DAP. Total leaf number, abscised leaf ratio, stem number, and stem diameter were measured at 155 DAP. The number of total leaves was counted by adding the number of attached leaves and leaf petioles (without leaves). The abscised leaf ratio was calculated by dividing the number of total leaves with the number of attached leaves. In experiments 1 and 2, all measurements were recorded as the average of eight plants.

2.3. Measurement of Chlorophyll Contents

Relative chlorophyll contents were measured via a nondestructive assay using the Soil and Plant Analyzer Development (SPAD) chlorophyll meter (SPAD-502; Konica Minolta, Tokyo, Japan). Measurements were conducted at 3, 73, and 153 DAP using the second young fully expanded leaf of each plant.

2.4. Data Analysis

Data were analyzed using the JMP statistical package (SAS Institute, Cary, NC, USA). Significant differences among treatments were determined by one-way analysis of variance, followed by the Tukey–Kramer honest significant difference test for pairwise comparisons at $p < 0.05$.

3. Results

3.1. Effect of NSC on the Growth of Hydroponic Sweetpotato (Experiment One)

Experiment one was conducted to examine the effect of NSC on the growth of sweetpotato in a hydroponic system. At 143 DAP, the shoot FW was the highest in the nutrient solution with high EC, followed by medium EC, and low EC (Figure 2A). The storage root FW showed the same trend as that described above (Figure 2B). Additionally, the number of storage roots showed no significant difference among the three treatments (Figure 2C).

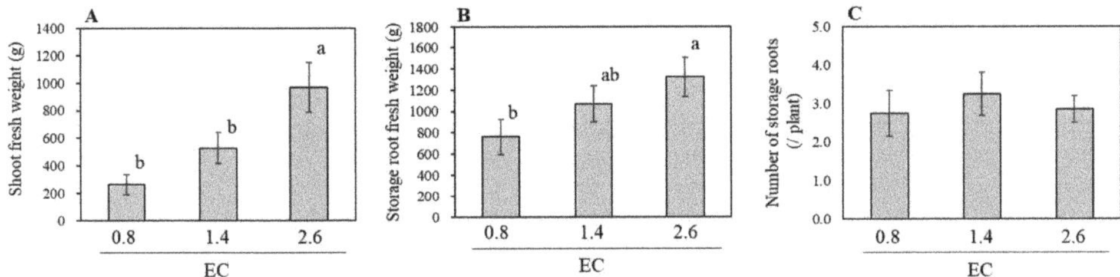

Figure 2. Effects of nutrient solution concentration on shoot fresh weight (**A**), storage root fresh weight (**B**), and number of storage roots (**C**) of sweetpotato at 143 days after plantation in experiment 1. Vertical bars represent the means ± SE (n = 8). Different letters indicate significant differences among the treatments at $p < 0.05$ by Tukey–Kramer's test.

3.2. Effect of Variation in NSC on the Growth of Hydroponic Sweetpotato (Experiment Two)

Next, we examined whether changes in NSC affect the growth of hydroponic sweetpotato. Plant shoot growth was measured at three time points: 3 DAP, 73 DAP (4 days before changing the NSC), and 155 DAP (harvest day). The leaf chlorophyll content increased from 3 to 73 DAP in all plants, reaching similar levels in all treatments (Figure 3A). No significant differences were detected among treatments at each time point, although the leaf chlorophyll contents of HH and LH plants at 155 DAP tended to be higher than that of HL and LL plants (Figure 3A). The number of attached leaves was higher in HH and HL plants than in LH and LL plants at 73 DAP (Figure 3B). Compared with 73 DAP, the number of attached leaves at 155 DAP was approximately 1.51-fold change in HH plants, 0.58-fold change in HL plants, 2.84-fold change in LH plants, and 1.09-fold change in LL plants (Figure 3B). Reduction in the number of attached leaves during cultivation suggests the induction of leaf abscission. This coincides with the pictures of shoots of HL plants at 155 DAP showing that leaves were rarely attached to the petiole at the bottom and middle sections of the stem (Figure 4). To examine leaf abscission, we counted the number of total leaves, including previously abscised leaves, at 155 DAP. HH plants showed the highest number of total leaves, followed LH, HL, and LL plants (Table 1). The abscised leaf ratio was the highest in HL plants, followed by LL, HH, and LH plants (Table 1).

Table 1. Effects of nutrient solution concentration on number of total leaves and abscised leaf ratio of sweetpotato at 155 days after plantation in experiment 2. Different letters indicate significant differences among the treatments at $p < 0.05$ by Tukey–Kramer's test.

Treatment	Number of Total Leaves	Abscised Leaf Ratio (%)
LL	277 c	63.6 ab
LH	545 ab	58.9 c
HL	417 bc	72.9 a
HH	761 a	66.9 ab

LL: plants were grown in low EC nutrient solution throughout the cultivation period; LH: plants were grown in low EC nutrient solution until the end of the first half of the cultivation period, and then transferred to high EC nutrient solution and maintained until the end of the cultivation period; HL: plants grown in high EC nutrient solution were transferred to low EC nutrient solution at the end of the first half of the cultivation period and maintained thereafter; HH: plants were grown in high EC nutrient solution throughout the cultivation period.

At 73 DAP, the maximum shoot length was higher in HH and HL plants than in LH and LL plants (Figure 3C). At 155 DAP, shoot length was the highest in HH plants and lowest in LL plants, while HL and LH plants showed similar intermediate shoot lengths (Figure 3C). The number of stems was significantly higher in HH plants compared with plants in the other treatments (Figure 5A). Stem diameter was the highest in HH plants, followed by HL and LH plants, and the lowest in LL plants (Figure 5B).

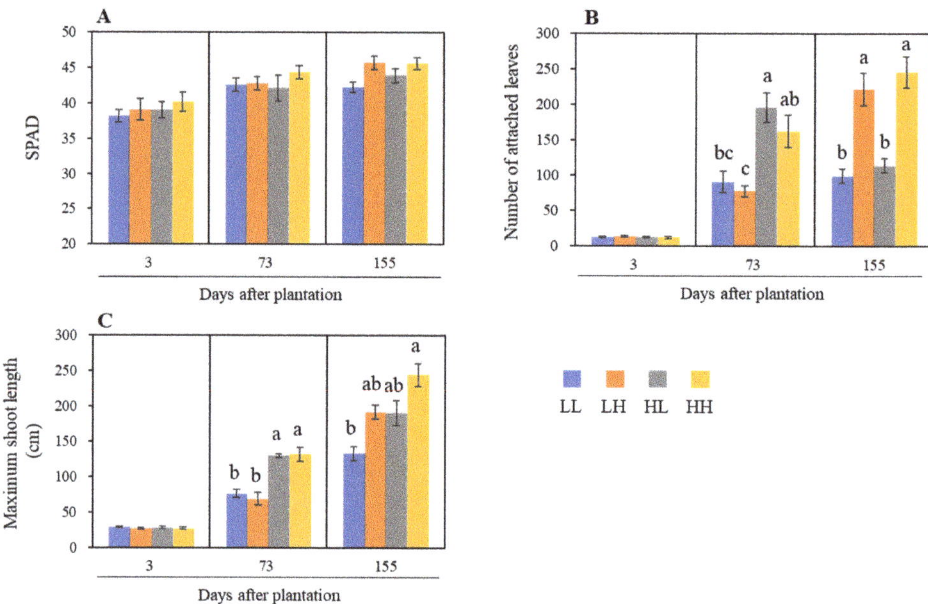

Figure 3. Effects of nutrient solution concentration on number of attached leaves (**A**), maximum shoot length (**B**), and the Soil and Plant Analyzer Development (SPAD) (**C**) of sweetpotato in experiment 2. These parameters were examined after 3, 73, and 155 days after plantation. Vertical bars represent the means ± SE ($n = 8$). Different letters indicate significant differences among the treatments at $p < 0.05$ by Tukey–Kramer's test.

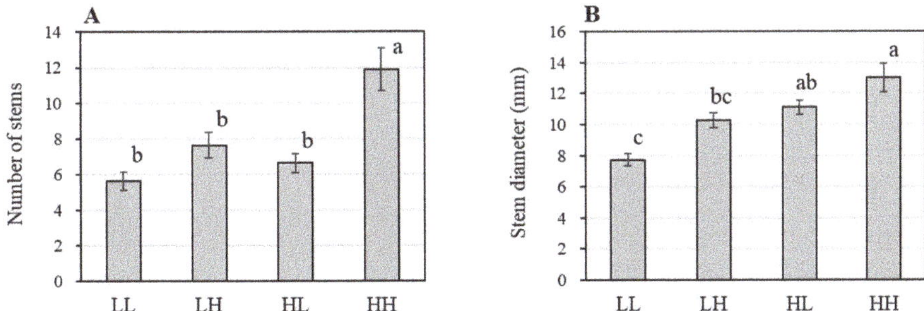

Figure 4. Effects of nutrient solution concentration on the shoot morphology of sweetpotato at 155 days after plantation in experiment 2. Scale bars = 47 cm.

Figure 5. Effects of nutrient solution concentration on number of stems (**A**) and stem diameter (**B**) of sweetpotato at 155 days after plantation in experiment 2. Vertical bars represent the means ± SE ($n = 8$). Different letters indicate significant differences among the treatments at $p < 0.05$ by Tukey–Kramer's test.

The biomass of shoots and storage roots was measured at 155 DAP. Shoot FW was the highest in HH plants, followed by LH, HL, and LL plants (Figure 6A). Storage root FW was the highest in HH plants, followed by HL and LH plants, and the lowest in LL plants (Figure 6B). The number of storage roots showed no significant difference among treatments (Figure 6C). Storage roots developed within vinyl pots in all treatments. Storage roots were round in shape, with a short length and partially undeveloped parts (Figure 7), consistent with previous observations [43]. These morphological characteristics of storage roots exhibited no variation among the different treatments (Figure 7).

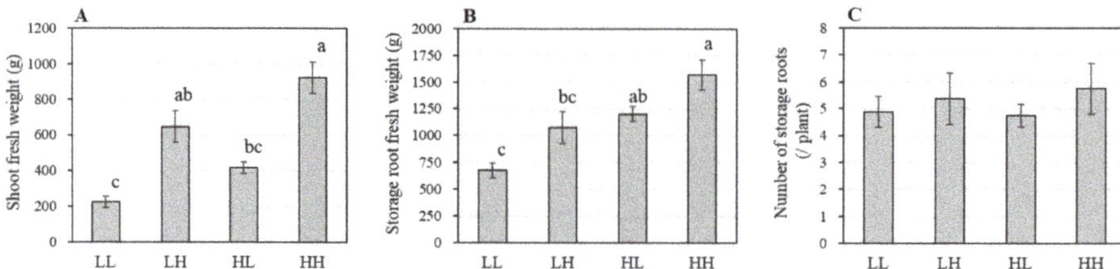

Figure 6. Effect of nutrient solution concentration on shoot fresh weight (**A**), storage root fresh weight (**B**), and number of storage roots (**C**) of sweetpotato at 155 days after plantation in experiment 2. Vertical bars represent the means ± SE (*n* = 8). Different letters indicate significant differences among the treatments at $p < 0.05$ by Tukey–Kramer's test.

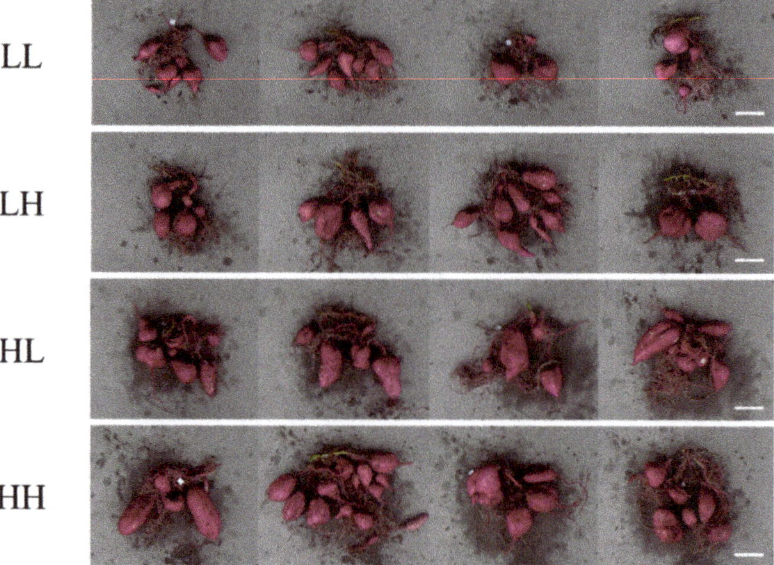

Figure 7. Effect of nutrient solution concentration on the storage root morphology of sweetpotato at 155 days after plantation in experiment 2. Scale bars = 10 cm.

4. Discussion

In hydroponics, the optimal NSC varies among plant species, with EC ranging from 1.5 to 2.5 dS m^{-1} [52,53]. Several studies have shown that high NSCs reduce the growth and photosynthetic parameters of hydroponically-grown plants [30,32,35,54–56]. High NSC-dependent growth restrictions are observed at EC > 1.8 dS m^{-1} in peace lily and at EC > 2.8 dS m^{-1} in peppermint and lettuce [37,54,55]. By contrast, hydroponically-grown bush snap beans can tolerate EC up to 3.6 dS m^{-1} [33]. In the current study, the growth of shoots and storage roots of hydroponic sweetpotatoes increased in an NSC-dependent manner up to an EC of 2.6 dS m^{-1} (Figure 2). Given that continuous growth of sweetpotato plants in a nutrient solution with an EC of 2.6 dS m^{-1} did not influence the leaf chlorophyll content (Figure 3A), this NSC appears to be more favorable for plant growth and storage root development rather than an osmotic stress condition that would deter growth and photosynthetic activity.

Plants sense the nutrient dose and alter the biomass partitioning accordingly [57]. In sweetpotato, the dose of N fertilizer alters the biomass partitioning of storage roots and shoots [58,59]. Increasing the N fertilizer dose from 0 to 1.2 g N per plant increases the biomass partitioning to storage roots [58]. However, at a higher N dose, the storage root biomass decreases, whereas the shoot biomass increases [58]. In experiment one, the growth of shoots and storage roots were enhanced as the NSC increased to an EC of 2.6 dS m^{-1} (Figure 2A,B). Considering the reports that hydroponic plants have different growth characteristics compared with soil grown plants [60,61], the responsiveness of hydroponically-grown sweetpotato to the nutrient dose might be different from that of soil grown sweetpotato plants. In hydroponic potatoes, shoot growth was enhanced when NSC was increased up to EC 2.4 dS m^{-1}, whereas tuber biomass was not affected by the EC of the nutrient solution [56]. Therefore, NSC-dependent partitioning of biomass may differ among plant species in hydroponics. The number of storage roots tend to be higher in experiment two (Figure 6C) compared to experiment one (Figure 2C). This may be caused by the different cultivation periods between experiments. Because two experiments were conducted only one time, the data may be influenced by the environmental condition.

The nutritional requirements of plants vary with the developmental stage [62]. Several studies have shown that changes in NSC influence plant growth characteristics [29,63–66]. In strawberry, restriction of N application at an early developmental stage increased the fruit biomass by enhancing reproductive growth [63]. In hydroponic tomato, increasing the NSC during fruit development reduces the fruit size and increases the sugar content [66]. Nutrient solution formulations have been developed for various growth stages in hydroponic tomato [67]. In sweetpotato, the timing of N fertilizer application influences plant growth and storage root yield [68–71]. Split application of N fertilizer could increase the storage root yield of sweetpotato by improving the efficiency of N uptake [69,70]. The timing of N fertilizer application is also important for increasing the marketable sweetpotato yield [71]. In experiment two, the storage root FW showed no significant difference between HL and LH plants (Figure 6B). This suggests that the timing of high-dose N application is not important for storage root development in hydroponically-grown sweetpotato. On the other hand, the shoot biomass and total leaf number were higher in LH plants compared with HL plants (Figure 6A, Table 1). These results implicate that higher dose of nutrient application at the storage root hypertrophic stage (the second half of growth stage) may enhance the development of shoots as well as storage roots. It should also take into account for the high abscised leaf ratio in HL plants (Table 1) because the abscised leaves, which did not contribute to shoot biomass, were partly responsible for the low shoot FW of HL plants. In general, at the late stage of sweetpotato cultivation, the storage root growth is enhanced, whereas shoot growth is retarded [59]. Thus, nutrient limiting condition at the hypertrophic stage of storage roots in HL plants may represent the field-grown sweetpotato characteristics in the shoot and root development. Because the amount of photosynthetic products translocated to storage roots partly depends on the shoot biomass, modifying the timing of NSC changes might improve the storage root yield.

Leaf abscission occurs during the senescence process and is induced by various stress responses [72]. Before the onset of leaf cell separation, the abscission zone encounters the repression of auxin biosynthesis and enhancement of ethylene production and sensitivity, resulting in the activation of cell wall degradation enzymes [73–75]. In experiment two, leaf abscission was induced at the late stage of cultivation in all NSC treatments (Table 1). This growth stage-dependent leaf abscission in sweetpotato has also been observed in open field conditions [76–78], suggesting a consistent senescence related phenomenon. N or P limitation is known to induce leaf abscission by enhancing ethylene production and sensitivity [79]. Sweetpotato leaves also abscise at the late growth stage under low N or P condition [80]. In HL plants, plant shoot biomass (the number of attached leaves and maximum shoot length) increased during the first half of the cultivation period in the nutrient solution with high EC (Figure 3B,C); however, these shoots grew in relatively poor nutrient conditions during the second half of the cultivation period. These nutrient poor conditions might trigger the high ratio of leaf abscission associated with N or P deficiency. Leaf senescence is accompanied by the breakdown of chlorophyll [81]. HL and LL plants showed a higher abscised leaf ratio and lower relative chlorophyll content (Figure 3A, Table 1), suggesting accelerated leaf abscissions by the progression of senescence. N deficiency also causes oxidative stress to the leaf [82]. Given that oxidative stress could trigger leaf abscission [83–86], it is possible that HL plants exhibit leaf abscission during the second half of the cultivation period due to oxidative stress triggered by N deficiency. On the other hand, LH plants were relatively nutrient-rich condition at the late cultivation period. Therefore, leaf senescence and abscission were thought to be suppressed by relatively rich-N supplement.

5. Conclusions

Compared to traditional soil culture systems, sweetpotato hydroponics saves absorbent material (soil) and can be used anywhere exposed to sunlight. In addition, hydroponics can efficiently utilize nutrient components as supplied components are not dispersed to the soil. In fact, almost all nutrients were absorbed in plants grown on EC 0.8 and 1.4 in this study (Supplementary Figure S1). Here, we presented NSC-dependent storage root yield in hydroponic sweetpotato (experiment one). Although the timing of high and low NSC did not have a significant impact on the storage root yield, shoot growth was apparently increased by high NSC (experiment two). A more precise adjustment of the NSC may increase the yield of storage roots relative to the fertilizer input. Thus, given its flexibility in manipulating the nutrient status, hydroponics could be used as an efficient tool for sweetpotato production.

Supplementary Materials: The following are available online at http://www.mdpi.com/2073-4395/10/11/1708/s1, Figure S1: Time-course changes of nutrient solution EC in experiment 1. The nutrient solution was renewed on June 26.EC was measured two containers of each experimental plot.

Author Contributions: Conceptualization, M.S. and T.S.; formal analysis, M.S.; investigation, M.S.; writing—original draft preparation, M.S.; writing—review and editing, M.S. and T.S. All authors have read and agreed to the published version of the manuscript.

Funding: This research was partly funded by the Japan Society for the Promotion of Science (JSPS), KAKENHI Grant Number 20K12247 to T.S. and 20K06329 to M.S.

Acknowledgments: The authors thank Kokoro Kubota, Takashi Tobe, Kaito Nakano, Daisuke Fukada, Hyuki Kishida, and Rihito Yamabe for supporting the field experiments.

Conflicts of Interest: The authors declare no conflict of interest.

References

1. Petsakos, A.; Prager, S.D.; Gonzalez, C.E.; Gama, A.C.; Sulser, T.B.; Gbegbelegbe, S.; Kikulwe, E.M.; Hareau, G. Understanding the consequences of changes in the production frontiers for roots, tubers and bananas. *Glob. Food Secur.* **2019**, *20*, 180–188. [CrossRef]
2. Woolfe, J. *Sweet Potato: An Untapped Food Resource*, 1st ed.; Cambridge University Press: New York, NY, USA, 1992; p. 643.
3. Pimentel, D.; Doughty, R.; Carothers, C.; Lamberson, S.; Bora, N.; Lee, K. Energy inputs in crop production in developing and developed countries. In *Food Security and Environmental Quality in the Developing World*; CRC Press: Boca Raton, FL, USA, 2002; pp. 129–151.
4. Koçar, G.; Civaş, N. An overview of biofuels from energy crops: Current status and future prospects. *Renew. Sustain. Energy Rev.* **2013**, *28*, 900–916. [CrossRef]

5. Lareo, C.; Ferrari, M.D. Sweet Potato as a Bioenergy Crop for Fuel Ethanol Production: Perspectives and Challenges. In *Bioethanol Production from Food Crops*; Academic Press: New York, NY, USA, 2019; pp. 115–147.

6. Ziska, L.H.; Runion, G.B.; Tomecek, M.; Prior, S.A.; Torbet, H.A.; Sicher, R. An evaluation of cassava, sweet potato and field corn as potential carbohydrate sources for bioethanol production in Alabama and Maryland. *Biomass Bioenergy* **2009**, *33*, 1503–1508. [CrossRef]

7. e Silva, J.O.V.; Almeida, M.F.; da Conceição Alvim-Ferraz, M.; Dias, J.M. Integrated production of biodiesel and bioethanol from sweet potato. *Renew. Energy* **2018**, *124*, 114–120. [CrossRef]

8. Chu, C.Y.; Sen, B.; Lay, C.H.; Lin, Y.C.; Lin, C.Y. Direct fermentation of sweet potato to produce maximal hydrogen and ethanol. *Appl. Energy* **2012**, *100*, 10–18. [CrossRef]

9. Lay, C.H.; Lin, H.C.; Sen, B.; Chu, C.Y.; Lin, C.Y. Simultaneous hydrogen and ethanol production from sweet potato via dark fermentation. *J. Clean. Prod.* **2012**, *27*, 155–164. [CrossRef]

10. Kobayashi, T.; Tang, Y.; Urakami, T.; Morimura, S.; Kida, K. Digestion performance and microbial community in full-scale methane fermentation of stillage from sweet potato-shochu production. *J. Environ. Sci.* **2014**, *26*, 423–431. [CrossRef]

11. Wang, F.; Jiang, Y.; Guo, W.; Niu, K.; Zhang, R.; Hou, S.; Wang, M.; Yi, Y.; Zhu, C.; Jia, C.; et al. An environmentally friendly and productive process for bioethanol production from potato waste. *Biotechnol. Biofuels* **2016**, *9*, 50. [CrossRef]

12. Monteiro, R.L.; de Moraes, J.O.; Domingos, J.D.; Carciofi, B.A.M.; Laurindo, J.B. Evolution of the physicochemical properties of oil-free sweet potato chips during microwave vacuum drying. *Innov. Food Sci. Emerg. Technol.* **2020**, *63*, 102317. [CrossRef]

13. Nedunchezhiyan, M.; Byju, G.; Jata, S.K. Sweet potato agronomy. *Fruit Veg. Cereal Sci. Biotechnol.* **2012**, *6*, 1–10.

14. Halavatau, S.; O'Sullivan, J.N.; Asher, C.J.; Blamey, F.P.C. Better nutrition improves sweet potato and taro yields in the South Pacific. *Trop. Agric.* **1998**, *75*, 6–12.

15. Laxminarayana, K.; John, K.S.; Ravindran, C.S.; Naskar, S.K. Effect of lime, inorganic, and organic sources on soil fertility, yield, quality, and nutrient uptake of sweet potato in Alfisols. *Commun. Soil Sci. Plant Anal.* **2011**, *42*, 2515–2525. [CrossRef]

16. Sawahata, H. Studies on the Characteristics of the Thickening of Storage Root of Sweet Potato: II. Influence of the supply of mineral nutrients on the thickening of storage root. *Jpn. J. Crop Sci.* **1989**, *58*, 290–296. [CrossRef]

17. Osaki, M.; Ueda, H.; Shinano, T.; Matsui, H.; Tadano, T. Accumulation of carbon and nitrogen compounds in sweet potato plants grown under deficiency of N, P, or K nutrients. *Soil Sci. Plant Nutr.* **1995**, *41*, 557–566. [CrossRef]

18. Taranet, P.; Harper, S.; Kirchhof, G.; Fujinuma, R.; Menzies, N. Growth and yield response of glasshouse-and field-grown sweetpotato to nitrogen supply. *Nutr. Cycl. Agroecosyst.* **2017**, *108*, 309–321. [CrossRef]

19. Jia, Z.D.; Ma, P.Y.; Bian, X.F.; Guo, X.D.; Xie, Y.Z. The effects of different N and K fertilizer ratio and planting density on yield and dry matter accumulation of sweetpotato. *Acta Agric. Boreali-Sin.* **2012**, *27*, 321–327.

20. Wang, S.Y.; Li, H.; Liu, Q.; Shi, Y.X. Interactive effects of nitrogen and potassium on root growth and leaf enzyme activities of sweet potato. *Acta Agric. Boreali-Sin.* **2015**, *30*, 167–173.

21. Wang, S.; Liu, Q.; Shi, Y.; Li, H.; Wang, S.; Liu, Q.; Shi, Y.; Li, H. Interactive effects of nitrogen and potassium on photosynthesis product distribution and accumulation of sweetpotato. *Sci. Agric. Sin.* **2017**, *50*, 2706–2716.

22. George, M.S.; Lu, G.; Zhou, W. Genotypic variation for potassium uptake and utilization efficiency in sweet potato (*Ipomoea batatas* L.). *Field Crop. Res.* **2002**, *77*, 7–15. [CrossRef]

23. Villordon, A.; LaBonte, D.; Firon, N.; Carey, E. Variation in nitrogen rate and local availability alter root architecture attributes at the onset of storage root initiation in 'Beauregard' sweetpotato. *HortScience* **2013**, *48*, 808–815. [CrossRef]

24. Villordon, A.Q.; Ginzberg, I.; Firon, N. Root architecture and root and tuber crop productivity. *Trends Plant Sci.* **2014**, *19*, 419–425. [CrossRef] [PubMed]

25. Savvas, D. Hydroponics: A modern technology supporting the application of integrated crop management in greenhouse. *J. Food Agric. Environ.* **2003**, *1*, 80–86.

26. Jones, J.B., Jr. Hydroponics: Its history and use in plant nutrition studies. *J. Plant Nutr.* **1982**, *5*, 1003–1030. [CrossRef]
27. Savvas, D.; Adamidis, K. Automated management of nutrient solutions based on target electrical conductivity, pH, and nutrient concentration ratios. *J. Plant Nutr.* **1999**, *22*, 1415–1432. [CrossRef]
28. Öztekin, G.B.; Uludağ, T.; Tüzel, Y. Growing spinach (*Spinacia oleracea* L.) in a floating system with different concentrations of nutrient solution. *Appl. Ecol. Environ. Res.* **2018**, *16*, 3333–3350. [CrossRef]
29. Sakamoto, Y.; Watanabe, S.; Nakashima, T.; Okano, K. Effects of salinity at two ripening stages on the fruit quality of single-truss tomato grown in hydroponics. *J. Hortic. Sci. Biotechnol.* **1999**, *74*, 690–693. [CrossRef]
30. Wu, M.; Kubota, C. Effects of electrical conductivity of hydroponic nutrient solution on leaf gas exchange of five greenhouse tomato cultivars. *HortTechnology* **2008**, *18*, 271–277. [CrossRef]
31. Wanzheng, M.; Hanping, M.; Jing, H.; Jiheng, N.; Zhongfang, L. Effects of different nutrient solution concentration on the growth and development of cucumber in greenhouse. In Proceedings of the 2011 International Conference on New Technology of Agricultural, Zibo, China, 27–29 May 2011; IEEE: New York, NY, USA, 2011; pp. 277–281.
32. Kang, J.G.; van Iersel, M.W. Nutrient solution concentration affects shoot: Root ratio, leaf area ratio, and growth of subirrigated salvia (*Salvia splendens*). *HortScience* **2004**, *39*, 49–54. [CrossRef]
33. Valdez, M.T.; Ito, T.; Shinohara, Y.; Maruo, T. Effects of nutrient solution levels on the growth, yield and mineral contents in hydroponically-grown bush snap bean. *Environ. Control Biol.* **2002**, *40*, 167–175. [CrossRef]
34. Rouphael, Y.; Cardarelli, M.; Lucini, L.; Rea, E.; Colla, G. Nutrient solution concentration affects growth, mineral composition, phenolic acids, and flavonoids in leaves of artichoke and cardoon. *HortScience* **2012**, *47*, 1424–1429. [CrossRef]
35. Hoang, N.N.; Kitaya, Y.; Shibuya, T.; Endo, R. Development of an in vitro hydroponic culture system for wasabi nursery plant production—Effects of nutrient concentration and supporting material on plantlet growth. *Sci. Hortic.* **2019**, *245*, 237–243. [CrossRef]
36. Shinohara, Y.; Suzuki, Y. Effects of light and nutritional conditions on the ascorbic acid content of lettuce. *J. Jpn. Soc. Hortic. Sci.* **1981**, *50*, 239–246. [CrossRef]
37. Fallovo, C.; Rouphael, Y.; Rea, E.; Battistelli, A.; Colla, G. Nutrient solution concentration and growing season affect yield and quality of *Lactuca sativa* L. var. acephala in floating raft culture. *J. Sci. Food Agric.* **2009**, *89*, 1682–1689. [CrossRef]
38. Sarooshi, R.A.; Cresswell, G.C. Effects of hydroponic solution composition, electrical conductivity and plant spacing on yield and quality of strawberries. *Aust. J. Exp. Agric.* **1994**, *34*, 529–535. [CrossRef]
39. Lieten, P. The effect of nutrition prior to and during flower differentiation on phyllody and plant performance of short day strawberry Elsanta. *Acta Hortic.* **2002**, *567*, 345–348. [CrossRef]
40. Gallace, N.; Boonen, M.; Lieten, P.; Bylemans, D. Electrical conductivity of the nutrient solution: Implications for flowering and yield in day-neutral cultivars. *Acta Hortic.* **2017**, *1156*, 223–228. [CrossRef]
41. Eguchi, T.; Yoshida, S. A cultivation method to ensure tuberous root formation in sweetpotatoes (*Ipomoea batatas* (L.) Lam.). *Environ. Control Biol.* **2004**, *42*, 259–266. [CrossRef]
42. Eguchi, T.; Moriyama, S.; Miyajima, I.; Yoshida, S.; Chikushi, J. A hydroponic method suitable for tops production of a sweetpotato [*Ipomoea batatas*] cultivar 'Suioh'. *J. Sci. High Technol. Agric.* **2007**, *19*, 167–174. [CrossRef]
43. Sakamoto, M.; Suzuki, T. Effect of pot volume on the growth of sweetpotato cultivated in the new hydroponic system. *Sustain. Agric. Res.* **2018**, *7*, 137–145. [CrossRef]
44. Kitaya, Y.; Hirai, H.; Endo, R.; Shibuya, T. Effects of water contents and CO_2 concentrations in soil on growth of sweet potato. *Field Crops Res.* **2013**, *152*, 36–43.
45. Eguchi, T.; Ito, Y.; Yoshida, S. Periodical wetting increases α-tocopherol content in the tuberous roots of sweetpotato (*Ipomoea batatas* (L.) Lam.). *Environ. Control Biol.* **2012**, *50*, 297–303. [CrossRef]
46. Kitaya, Y.; Hirai, H.; Wei, X.; Islam, A.F.M.S.; Yamamoto, M. Growth of sweetpotato cultured in the newly designed hydroponic system for space farming. *Adv. Space Res.* **2008**, *41*, 730–735. [CrossRef]
47. Mortley, D.G.; Loretan, P.A.; Bonsi, C.K.; Hill, W.A.; Morris, C.E. Plant spacing influences yield and linear growth rate of sweetpotatoes grown hydroponically. *HortScience* **1991**, *26*, 1274–1275. [CrossRef]

48. Almazan, A.M.; Zhou, X. Biomass yield and composition of sweetpotato grown in a nutrient film technique system. *Plant Foods Hum. Nutr.* **1997**, *50*, 259–268. [CrossRef]

49. Uewada, T. The solution culture of sweet potatoes. *Environ. Control Biol.* **1990**, *28*, 135–140. [CrossRef]

50. Japan Meteorological Agency Official Website. Past Weather Data of Wakayama City in 2018. Available online: http://www.data.jma.go.jp/obd/stats/etrn/view/monthly_s1.php?prec_no=65&block_no=47777&year=2018&month=&day=&view=p1 (accessed on 23 September 2020).

51. Japan Meteorological Agency Official Website. Past Weather Data of Wakayama City in 2019. Available online: http://www.data.jma.go.jp/obd/stats/etrn/view/monthly_s1.php?prec_no=65&block_no=47777&year=2019&month=&day=&view=p1 (accessed on 23 September 2020).

52. Sonneveld, C.; Voogt, W. *Plant Nutrition of Greenhouse Crops*; Springer: Dordrecht, The Netherlands, 2009; pp. 393–403.

53. Trejo-Téllez, L.I.; Gómez-Merino, F.C. Nutrient solutions for hydroponic systems. In *Hydroponics: A Standard Methodology for Plant Biological Researches*; Asao, T., Ed.; InTechOpen: Rjeka, Croatia, 2012; pp. 1–22.

54. Dewir, Y.H.; Chakrabarty, D.; Ali, M.B.; Hahn, E.J.; Paek, K.Y. Effects of hydroponic solution EC, substrates, PPF and nutrient scheduling on growth and photosynthetic competence during acclimatization of micropropagated *Spathiphyllum* plantlets. *Plant Growth Regul.* **2005**, *46*, 241–251. [CrossRef]

55. Tabatabaie, S.J.; Nazari, J. Influence of nutrient concentrations and NaCl salinity on the growth, photosynthesis, and essential oil content of peppermint and lemon verbena. *Turk. J. Agric. For.* **2007**, *31*, 245–253.

56. Chang, D.C.; Cho, I.C.; Suh, J.T.; Kim, S.J.; Lee, Y.B. Growth and yield response of three aeroponically grown potato cultivars (*Solanum tuberosum* L.) to different electrical conductivities of nutrient solution. *Am. J. Potato Res.* **2011**, *88*, 450–458. [CrossRef]

57. Hermans, C.; Hammond, J.P.; White, P.J.; Verbruggen, N. How do plants respond to nutrient shortage by biomass allocation? *Trends Plant Sci.* **2006**, *11*, 610–617. [CrossRef] [PubMed]

58. Kelm, M.; Brück, H.; Hermann, M.; Sattelmacher, B. *Plant Productivity and Water Use Efficiency of Sweetpotato (Ipomoea batatas) as Affected by Nitrogen Supply*; CIP Program Report 1999–2000; International Potato Center: Lima, Peru, 2000; pp. 273–279.

59. Osaki, M.; Ueda, H.; Shinano, T.; Matsui, H.; Tadano, T. Accumulation of carbon and nitrogen compounds in sweet potato plants grown under different nitrogen application rates. *Soil Sci. Plant Nutr.* **1995**, *41*, 547–555. [CrossRef]

60. Manzocco, L.; Foschia, M.; Tomasi, N.; Maifreni, M.; Dalla Costa, L.; Marino, M.; Cortella, G.; Cesco, S. Influence of hydroponic and soil cultivation on quality and shelf life of ready-to-eat lamb's lettuce (*Valerianella locusta* L. Laterr). *J. Sci. Food Agric.* **2011**, *91*, 1373–1380. [CrossRef]

61. Miller, M.H.; Walker, G.K.; Tollenaar, M.; Alexander, K.G. Growth and yield of maize (*Zea mays* L.) grown outdoors hydroponically and in soil. *Can. J. Soil Sci.* **1989**, *69*, 295–302. [CrossRef]

62. Fageria, N.K.; Baligar, V.C. Enhancing nitrogen use efficiency in crop plants. *Adv. Agron.* **2005**, *88*, 97–185.

63. Yoshida, Y.; Fujime, Y.; Chujo, T. Effects of nitrogen nutrition on flower bud development and fruit malformation in 'Ai-Berry' strawberry. *J. Jpn. Soc. Hortic. Sci.* **1992**, *60*, 869–879. [CrossRef]

64. Dittakit, P.; Thongket, T. Increased nutrient solution concentration during early fruit development stages enhances pungency and phenylalanine ammonia-lyase activity in hot chili (*Capsicum annuum* L.). *Am. J. Agric. Biol. Sci.* **2014**, *9*, 72. [CrossRef]

65. Maruyama, S.; Ishigami, Y.; Goto, E. Effect of nutrient solution concentration at the heading time on the growth, development, and seed storage protein content of rice plants in a controlled environment. *Environ. Control Biol.* **2010**, *48*, 17–24. [CrossRef]

66. Wu, M.; Kubota, C. Effects of high electrical conductivity of nutrient solution and its application timing on lycopene, chlorophyll and sugar concentrations of hydroponic tomatoes during ripening. *Sci. Hortic.* **2008**, *116*, 122–129. [CrossRef]

67. Hochmuth, G.J.; Hochmuth, R.C. *Nutrient Solution Formulation for Hydroponic (Perlite, Rockwool, NFT) Tomatoes in Florida*; HS796; University of Florida Cooperative Extension Service: Gainesville, FL, USA, 2001.

68. Ankumah, R.O.; Khan, V.; Mwamba, K.; Kpomblekou-A, K. The influence of source and timing of nitrogen fertilizers on yield and nitrogen use efficiency of four sweet potato cultivars. *Agric. Ecosyst. Environ.* **2003**, *100*, 201–207. [CrossRef]
69. Du, X.; Xi, M.; Kong, L. Split application of reduced nitrogen rate improves nitrogen uptake and use efficiency in sweetpotato. *Sci. Rep.* **2019**, *9*, 1–11. [CrossRef]
70. Du, X.; Zhang, X.; Xi, M.; Kong, L. Split application enhances sweetpotato starch production by regulating the conversion of sucrose to starch under reduced nitrogen supply. *Plant Physiol. Biochem.* **2020**, *151*, 743–750. [CrossRef] [PubMed]
71. Phillips, S.B.; Warren, J.G.; Mullins, G.L. Nitrogen rate and application timing affect 'Beauregard' sweetpotato yield and quality. *HortScience* **2005**, *40*, 214–217. [CrossRef]
72. Patharkar, O.R.; Walker, J.C. Connections between abscission, dehiscence, pathogen defense, drought tolerance, and senescence. *Plant Sci.* **2019**, *284*, 25–29. [CrossRef]
73. Liljegren, S.J. Organ abscission: Exit strategies require signals and moving traffic. *Curr. Opin. Plant Biol.* **2012**, *15*, 670–676. [CrossRef] [PubMed]
74. Patterson, S.E.; Bolivar-Medina, J.L.; Falbel, T.G.; Hedtcke, J.L.; Nevarez-McBride, D.; Maule, A.F.; Zalapa, J.E. Are we on the right track: Can our understanding of abscission in model systems promote or derail making improvements in less studied crops? *Front. Plant Sci.* **2016**, *6*, 1268. [CrossRef]
75. Patharkar, O.R.; Walker, J.C. Advances in abscission signaling. *J. Exp. Bot.* **2018**, *69*, 733–740. [CrossRef]
76. Somda, Z.C.; Kays, S.J. Sweet potato canopy morphology: Leaf distribution. *J. Am. Soc. Hortic. Sci.* **1990**, *115*, 39–45. [CrossRef]
77. Somda, Z.C.; Mahomed, M.T.M.; Kays, S.J. Analysis of leaf shedding and dry matter recycling in sweetpotato. *J. Plant Nutr.* **1991**, *14*, 1201–1212. [CrossRef]
78. McLaurin, W.J.; Kays, S.J. Substantial leaf shedding—A consistent phenomenon among high-yielding sweetpotato cultivars. *HortScience* **1993**, *28*, 826–827. [CrossRef]
79. Lynch, J.; Brown, K.M. Ethylene and plant responses to nutritional stress. *Physiol. Plant.* **1997**, *100*, 613–619. [CrossRef]
80. Bouwkamp, J.C. *Sweet potato products: A Natural Resource for the Tropics*; CRC Press: Boca Taton, FL, USA, 1985.
81. Matile, P.; Hortensteiner, S.; Thomas, H.; Krautler, B. Chlorophyll breakdown in senescent leaves. *Plant Physiol.* **1996**, *112*, 1403–1409. [CrossRef]
82. Huang, Z.A.; Jiang, D.A.; Yang, Y.; Sun, J.W.; Jin, S.H. Effects of nitrogen deficiency on gas exchange, chlorophyll fluorescence, and antioxidant enzymes in leaves of rice plants. *Photosynthetica* **2004**, *42*, 357–364. [CrossRef]
83. Sakamoto, M.; Munemura, I.; Tomita, R.; Kobayashi, K. Involvement of hydrogen peroxide in leaf abscission signaling, revealed by analysis with an in vitro abscission system in *Capsicum* plants. *Plant J.* **2008**, *56*, 13–27. [CrossRef]
84. Sakamoto, M.; Munemura, I.; Tomita, R.; Kobayashi, K. Reactive oxygen species in leaf abscission signaling. *Plant Signal. Behav.* **2008**, *3*, 1014–1015. [CrossRef] [PubMed]
85. Bar-Dror, T.; Dermastia, M.; Kladnik, A.; Žnidarič, M.T.; Novak, M.P.; Meir, S.; Burd, S.; Philosoph-Hadas, S.; Ori, N.; Sonego, L.; et al. Programmed cell death occurs asymmetrically during abscission in tomato. *Plant Cell* **2011**, *23*, 4146–4163. [CrossRef]
86. Liao, W.; Wang, G.; Li, Y.; Wang, B.; Zhang, P.; Peng, M. Reactive oxygen species regulate leaf pulvinus abscission zone cell separation in response to water-deficit stress in cassava. *Sci. Rep.* **2016**, *6*, 1–17. [CrossRef] [PubMed]

Publisher's Note: MDPI stays neutral with regard to jurisdictional claims in published maps and institutional affiliations.

 *agronomy*

Article

Theoretical and Experimental Analysis of Nutrient Variations in Electrical Conductivity-Based Closed-Loop Soilless Culture Systems by Nutrient Replenishment Method

Tae In Ahn and Jung Eek Son *

Department of Plant Science and Research Institute of Agriculture and Life Sciences, Seoul National University, Seoul 08826, Korea; genisleaf@outlook.com
* Correspondence: sjeenv@snu.ac.kr; Tel.: +82-2-880-4564

Received: 11 September 2019; Accepted: 16 October 2019; Published: 17 October 2019

Abstract: In closed-loop soilless culture systems, variation in nutrients can lead to instability in the nutrient management and forced discharge of nutrients and water. Total nutrients absorbed by plants are replenished in an electrical conductivity-based closed-loop system, and fluctuation in electrical conductivity within a certain range around the initial value can be expected. However, this is not always observed in systems using conventional nutrient-replenishment methods. The objectives of this study were to analyze nutrient variation in a closed-loop soilless culture system based on a theoretical model and derive an alternative nutrient-replenishment method. The performance of the derived alternative method was compared with a conventional nutrient-replenishment method through simulation analysis. A demonstration experiment using sweet peppers was then conducted to confirm whether the theoretical analysis results can be reproduced through actual cultivation. The average amounts of injected nutrients during the experimental period of four months in the conventional and alternative methods were 2257 and 1054 g, respectively. There was no significant difference in the yield of sweet peppers between the two methods. The substrate electrical conductivity in the alternative method was maintained at 2.7 dS·m^{-1} ± 0.5 within the target electrical conductivity value, while that in the conventional method gradually increased to 5.0 dS·m^{-1} ± 1.2. In a simulation study, results similar to the demonstration experiment were predicted. Total nutrient concentrations in the alternative method showed fluctuations around the target value but did not continuously deviate from the target value, while those in the conventional method showed a tendency to increase. As a whole, these characteristics of the alternative method can help in minimizing nutrients and water emissions from the cultivation system.

Keywords: growing medium; nutrient uptake; nutrient variation; simulation model; sweet pepper

1. Introduction

Closed-loop nutrient-management techniques are essential for sustainable soilless cultures with resource savings [1]. Nutrients in soilless culture systems are managed primarily with an open-loop nutrient supply [2,3]. Open-loop soilless culture systems are easier to implement, but resource losses are inevitable. Moreover, due to the intensive use of fertilizers, the threat posed to aquatic environments by repeated discharging a certain ratio of drainage is serious enough to warrant regulation by national governments [3–6]. Since a closed-loop soilless culture system reuses its drainage, the resulting variation in nutrient concentration can significantly affect the plant growth as the reuse period becomes longer [5,7–9]. It is therefore difficult to intuitively explain or interpret nutrient-variation management techniques, unlike open-loop systems. In order to appropriately apply those techniques, theoretical models are required and the problems should be precisely defined [10].

In both closed-loop and open-loop soilless culture systems, the electrical conductivity (EC) of the nutrient solution in the mixing tank is adjusted to a target value before the solution is applied to the plant [9,11,12]. However, unlike an open-loop system, the mixing ratio of tap water to stock solution in a closed-loop system is adjusted by considering the change in nutrient concentration due to the inflow of drainage [12]. Alternatively, in simple systems, a premixed standard nutrient solution of a certain EC is supplied based on the difference between initial and current water levels in the circulation tank, which simultaneously performs drainage collection and nutrient-solution feeding [13,14]. In an EC-based closed-loop soilless culture system, supply of stock solution or standard nutrient solution and tap water is intended to replenish nutrients and water consumed in the system [12]. For a single system in which the plants are grown directly in a nutrient solution container, the nutrients and water consumed due to absorption of plants in the system can be estimated almost exactly [15]. However, errors may occur in systems in which the root zone and nutrient supply are separate from drainage collection. Both elements are widely used in commercial farming conditions.

Considering the functional objective of nutrient and water replenishment in a closed-loop soilless culture system, relatively stable fluctuations within a certain range around the initial EC value should be observed. However, EC changes far exceeding the initial values in the system have generally been observed [13,14,16]. In addition, the effects of these fluctuations are linked to forced discharge of recirculated nutrient solution outside of the system [13,14]. The problems associated with variations in nutrient concentration or EC observed in soilless culture systems are presumed to be inevitable due to the nutrient uptake concentration affected by the environment [5,14]. The experimental results are interpreted depending on the responses of the system according to the treatment application [9,14,16–18], and these have proven difficult to interpret in an integrated way. As a result, technical approaches to managing nutrient variation and the design of experiments are limited. To block nutrient emissions from a soilless culture system, nutrient reuse practices must be standardized, which requires a precise problem definition based on variations of nutrient concentrations or EC.

The objectives of this study were to analyze the cause of EC variation in closed-loop soilless culture systems based on a theoretical model, to derive an alternative nutrient-replenishment method for managing nutrient fluctuation, and to evaluate the performance through theoretical and experimental analyses.

2. Materials and Methods

2.1. Soilless Culture System Model

The model used in this study simulated nutrient changes in a soilless culture system with an automated nutrient-mixing system (Figure 1). The basic structures of the soilless culture system and plant growth models were constructed by referring to the nutrient transport model in a substrate condition [6,19–21]. The measured data of incident radiation intensity in the greenhouse from 10 September 2011 to 9 March 2012 were used as an input variable of transpiration and irrigation control in the simulation. Some units of parameters and variables were converted from the references for simulating the minute-based time scale of the automated soilless culture system. Values and description of the parameters used in the simulation were summarized in Table 1.

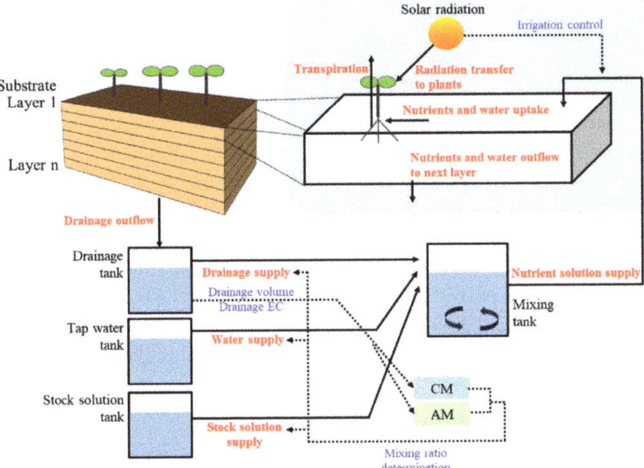

Figure 1. Schematic description of a closed-loop soilless culture system in a simulated condition. Solid lines indicate water and nutrient flow, and dotted lines indicate data flow for nutrient solution mixing and irrigation control. CM and AM mean conventional and alternative nutrient-replenishment methods, respectively.

Table 1. Parameters used for the simulations of soilless culture system.

Symbol	Description	Value	Reference Number
a_{LAI}	Leaf area index parameter	3.5	[22]
b_{LAI}	Leaf area index parameter	13.2	[22]
x_0	Leaf area index parameter	37.2	[22]
a_T	Evapotranspiration parameter	0.98	[22]
b_T	Evapotranspiration parameter	2.08×10^{-4}	[22]
λ	Latent heat of vaporization	2.45	[22]
k	Light extinction coefficient	0.84	[22]
RLD_{max}	Maximal root length density	50,000 m m^{-3}	[20]
K_1	Coefficient of the root growth function	770	[20]
k_1	Coefficient of the root growth function	500	[20]
J_{max}^{K}	Maximum absorption rate	2.89×10^{-3}	[20]
J_{max}^{Ca}	Maximum absorption rate	3.54×10^{-4}	[20]
J_{max}^{Mg}	Maximum absorption rate	4.20×10^{-4}	[20]
K_m^{K}	Michaelis-Menten constant	0.0127	[20]
K_m^{Ca}	Michaelis-Menten constant	0.039	[20]
K_m^{Mg}	Michaelis-Menten constant	0.015	[20]
C_{min}^{K}	Minimal concentration for uptake	0.002	[19]
C_{min}^{Ca}	Minimal concentration for uptake	0.002	[19]
C_{min}^{Mg}	Minimal concentration for uptake	0.002	[19]
C_T	Target total equivalent concentration	15	
C_W	Total equivalent concentration in tap water	1	
F	Field capacity	0.74	[4]
W_{DAW}	Difficult available water	0.0068	[4]
$S_{sub,n}$	Volume of substrate layer n	1.35	

2.2. Water and Nutrient Transport in a Substrate

According to standard practices for automated irrigation of a soilless culture system, the mixing process for a nutrient solution is initiated in the mixing tank, and the nutrient solution is supplied to the substrate after mixing. The target nutrients for the simulation were selected as macronutrient cations (K^+, Ca^{2+}, and Mg^{2+}).

$$\frac{dV_n}{dt} = Q_{n-1} - Q_n - T_n \tag{1}$$

The volume of water in a substrate layer (V_n, L) was calculated depending on the flow rate of the water from the former (Q_{n-1}, L min^{-1}) and to the next layer (Q_n, L min^{-1}) and the evapotranspiration rate (T_n, L min^{-1}). The flow rate of the water to the first substrate layer (V_1) was the irrigation flow rate (Q_0). Q_n is the difference between the flow rate of the water from the former layer and the evapotranspiration rate ($Q_{n-1} - T_n$) or the difference between the irrigation rate and the evapotranspiration rate in the first substrate layer ($Q_0 - T_1$) [23]. The field capacity (F, dimensionless) and difficult available water (W_{DAW}, dimensionless), respectively restrict Q_n and T_n. The flow for Q_n occurs only when $V_n > S_{sub,n}F$, and T_n flows only when $V_n > S_{sub,n}W_{DAW}$. $S_{sub,n}$ (L) is volume of the substrate layer n.

The flow of nutrients in the medium is generated by the flow rate of water.

$$V_n \frac{dC_n^I}{dt} = Q_{n-1}C_{n-1}^I - Q_n C_n^I - P_{RSA}J_n^I \tag{2}$$

C is the molar concentration of nutrient (mM), superscript I is the type of ions (K^+, Ca^{2+}, Mg^{2+}), J_n^I is the uptake rate of nutrients (mmol m^{-2} min^{-1}), and P_{RSA} is the specific root surface area (m^2), which is described as the root length density and specific root surface area.

2.3. Plant Variables and Growth Parameters

In this simulation, evapotranspiration and nutrient uptake rates were applied as plant variables in the substrate. In general, the plant parameters relate to changes in evapotranspiration and nutrient uptake rates with plant growth. The relationship between solar radiation and evapotranspiration is adjusted by the leaf area index [24]. The parameters related to the nutrient uptake rate are derived from the characteristics of the plant ion transporters and are modeled as increasing with growth of the root surface area [25].

2.3.1. Leaf Area Index

The Boltzmann sigmoid equation was used to apply changes in the leaf area index to the evapotranspiration rate:

$$P_{LAI,t} = \frac{a_{LAI}}{\left[1 + e^{\frac{x_0 - t}{b_{LAI}}}\right]} \tag{3}$$

where a_{LAI}, b_{LAI}, and x_0 are constants, and t is time.

2.3.2. Evapotranspiration

The evapotranspiration rate was modeled using the simplified Penman-Monteith equation by Baille et al. (1994) [24].

$$T_n = a_T\left[1 - e^{-k_{LAI}P_{LAI}}\right]\frac{R}{\lambda} + b_T \tag{4}$$

T_n (L min^{-1}, numbers were converted from kg min^{-1}) was calculated depending on the radiation for a minute (R, MJ m^{-2} min^{-1}), the latent heat of vaporization (λ, MJ kg^{-1}), the light extinction coefficient (k_{LAI}), and the leaf area index (P_{LAI}). a_T (dimensionless) and b_T (kg m^{-2} min^{-1}) are regression parameters.

2.3.3. Root Length Density and Specific Root Surface Area

Root length density was used to calculate the specific root surface area and modeled using a logistic function of time [20,26]:

$$P_{len,t} = \frac{RLD_{max}}{1 + K_1 e^{-k_1 t}} \tag{5}$$

$$P_{RSA,t} = 2\pi r_0 P_{len,t} \tag{6}$$

where RLD_{max} (m m^{-3}) is the maximal root length density, and K_1 and k_1 are coefficients. r_0 is the mean root radius (m). Root length density was set to start at the top layer of the substrate and be sequentially assigned to the subsequent layer as the value increased. The allocation of root length density for each layer was calculated by dividing RLD_{max} by the total number of layers.

2.3.4. Nutrient Uptake

The nutrient uptake rate of the plant in the substrate was simulated as a function of Michaelis–Menten:

$$J_n^I = \frac{J_{max}^I \left(C_n^I - C_{min}^I \right)}{K_m^I + \left(C_n^I - C_{min}^I \right)} \tag{7}$$

where J_{max}^I (mmol m^{-2} min^{-1}) is the maximum absorption rate of nutrient I, K_m^I (mM) is the Michaelis-Menten constant, and C_{min}^I (mM) is the minimal concentration at which $J_n^I = 0$.

2.4. Mixing of Nutrient Solutions

The conventional mixing process for stock solution, tap water, and drainage under the automated closed-loop soilless culture system is performed in the mixing tank [11,12]. When the system receives an irrigation command, the entire volume of collected drainage is diluted with tap water within the range of the irrigation volume, and the stock solution is added to the target EC.

However, because drainage is included in the automated mixing process in closed-loop soilless culture systems, the Equation needs to solve for target EC with mixing stock solution, drainage, and water [12]. The nutrient solution mixing process occurs intermittently according to the irrigation interval, and the basic Equation for conventional nutrient replenishment can be summarized based on the dilution Equation:

$$V_T C_T = V_D C_D + V_W C_W + V_S C_S \tag{8}$$

$$V_w = V_T - V_D - V_S \tag{9}$$

$$V_S = \frac{C_T V_T - C_w V_T + C_W V_D - C_D V_D}{C_S - C_W} \tag{10}$$

where V_T (L) is the target irrigation volume per event, C_T (mEq L^{-1}) is the target total equivalent concentration, V_D is the drainage volume, C_D (mEq L^{-1}) is the total equivalent concentration in drainage, V_W (L) is the amount of tap water input to the mixing tank, V_S (L) is the amount of stock solution input to the mixing tank, C_w (mEq L^{-1}) is the total equivalent concentration in tap water, and C_S (mEq L^{-1}) is the total equivalent concentration of the stock solution. Equation (8) can be summarized as Equation (10) by substituting Equation (9) for V_W. Equation (10) is calculating the amount stock solution input based on the total equivalent concentration. In this simulation, we assumed the total equivalent concentration as EC based on the linear relationship between EC and the total equivalent concentration of nutrient solution presented by Savvas and Manos (1999) [27].

The amount of stock solution input to the mixing tank was calculated through this process, and when the irrigation control command was generated during the simulation, the mixing process began based on the volume of drainage stored in the drainage tank at that moment. If the calculated value of the Equation (10) was less than zero, dilution using tap water could not be adjusted to the target concentration within the range of irrigation amount. In this case, the amount of tap water required for diluting the drainage to target total equivalent concentration (C_T) was calculated, and then the ratio between the drainage and calculated tap water was multiplied by V_T. When the doses of V_D, V_W, and V_S were determined through the abovementioned calculation, a flow rate was generated until the corresponding amount was transferred to V_M according to Q_{drg}, Q_{wtr}, and Q_{stk}, respectively. In the simulation, irrigation was controlled by a radiation integral method, which is conventionally used in automated irrigation control [28]. 140 mL of nutrient solution per plant in the mixing tank were supplied whenever the accumulated radiation reached 100 J m^{-2}.

2.5. Experimental Analysis

2.5.1. Cultivation Conditions

Three sweet pepper (*Capsicum annuum* L. "Derby") plants were grown in a rockwool slab, and seven slabs were used per row. Four cultivation lines were installed in a Venlo-type greenhouse at the experimental farm of Seoul National University (Suwon, Korea, Lat. 37.3° N, long. 127.0° E). Each line was an independent closed-loop soilless culture system with a mixing tank, drainage tank, and stock solutions. The stock solution was prepared based on the PBG nutrient solution of the Netherlands. In the greenhouse, daytime temperature was maintained at 25–35 °C and nighttime temperature at 17–22 °C. The solar radiation-based irrigation control was applied; when the cumulative radiation measured by a pyranometer (SP-110-L10, Apogee Instruments, Logan, Utah, USA) reached 100 J cm^{-2}, 150 mL of the nutrient solution was supplied to each plant. However, the irrigation amounts were adjusted according to meteorological conditions to maintain a drainage ratio of approximately 30%. The composition of nutrient solution was 14.17 mM of NO_3^-, 1.14 mM of H_2PO^-, 5.92 mM of K^+, 4.43 mM of Ca^{2+}, 1.59 mM of Mg^{2+}, and 1.6 mM of SO_4^{2-} as macro-elements; and 0.019 mM of Fe^{2+}, 0.01 mM of Zn^{2+}, 0.002 mM of Cu^{2+}, 0.01 mM of Mn^{2+}, and 0.0005 mM of MoO_4^{2-} as micro-elements. After an irrigation event, the drainage solution was returned to the drainage tank (11.7 L). The EC and pH of tap water were 0.17 dS·m^{-1} and 7.11, respectively, and contained 0.21 mM of Na^+, 0.29 mM of Cl^-, 0.04 mM of K^+, 0.36 mM of Ca^{2+}, 0.11 mM of Mg^{2+}, 0.10 mM of SO_4^{2-}, 0.39 mM of NO_3^-, and 0.0 mM of PO_4^{3-}.

2.5.2. Measurement of Fruit Yield and Analyses of Nutrient Content in Leaves and Substrate

The total yield and average fruit weight during the experimental period were measured. The proportion of blossom-end rot (BER) fruits on a sweet pepper plant was measured. At the end of the experiment, 18 leaves (including petiole) from the middle to the top nodes of a sweet pepper were randomly collected from each treatment. Leaves were washed in tap water and dried for 48 h at 70 °C in an oven. The dried leaves were ground, and 0.5 g of each ground sample was digested using concentrated nitric acid. Next, 1 mL of concentrated perchloric acid was added to maintain a set solution temperature of 180 °C, and the digestion process was accelerated on a hot plate at 90 °C for approximately one h, until a clear-colored solution was obtained. After digestion, the tube was cooled, filled with 25 mL deionized water, and the total contents of K^+, Ca^{2+}, and Mg^{2+} in the leaves were determined with an inductively coupled plasma-optical emission spectrometer (ICP-730ES, Varian, Mulgrave, Australia). To determine the nutrient concentrations in the rockwool substrate, samples of nutrient solution in the rockwool slabs were extracted using a syringe. The collection points of the nutrient solution in the rockwool slab were randomly selected to ensure representative samples of the overall concentration in the rockwool slabs. Five 10 mL samples of a rockwool slab nutrient solution were collected for each extraction, for a final volume of 50 mL sample. Four 50 mL samples per treatment were collected every week. SAS (version 9.2, SAS Institute, Cary, NC, USA) was used for statistical analysis.

2.5.3. Nutrient-Replenishment Method

A conventional nutrient-replenishment method (CM) and an alternative nutrient-replenishment method (AM) derived from the theoretical analysis in this study were performed in the mixing tank with two applied nutrient solution mixing modules. In the CM, as explained in Section 2.4, when the system received an irrigation command, the entire drainage volume was diluted with tap water within the range of irrigation volume, and the stock solution was added to match the fixed target EC [12]. In the case when the calculated volume of the diluted drainage exceeds the irrigation volume, the injection ratio of drainage and water was multiplied by the irrigation volume, and the converted drainage and water volumes were injected into the mixing tank without injection of the stock solution. In the AM, the additional volume of the stock solution was determined by the equation derived from the simulation study at every irrigation event (Equation (14)).

2.5.4. Nutrient Solution Mixing Module and Data Collection

The ECs of the nutrient solutions in the mixing tank and drainage tank were measured by EC sensors (SCF-01A, DIK, Chuncheon, Korea). Light intensity in the greenhouse was measured with a pyranometer (SP-110, Apogee, Logan, UT, USA) and used for input data for solar radiation-based irrigation control. Data were measured every 10 s from 15 October to 31 December 2014. Mean values for every hour were used. A datalogger (CR1000, Campbell Scientific, Logan, UT, USA) was used to measure and control the drainage and nutrient mixing process. Water levels of the stock solution tanks and the drainage tanks were monitored by ultrasonic sensors (UHA-300, Unics, Daegu, Korea) and used to estimate the stored volume changes of stock and drainage solutions. ECs in the substrates were measured at intervals of two to five days using a multimeter (Multi 3420 SET C, WTW, Weilheim, Germany).

3. Results and Discussion

3.1. Theoretical Analysis: Reconsideration of Problem and Derivation of Possible Solution

The total concentration of nutrients in the system using CM for nutrient replenishment gradually increased with diurnal level fluctuations, and after approximately 60 days, the total concentration showed repeated fluctuations within a certain range (Figure 2a). The changes with an increasing tendency in total nutrient concentrations relative to initial values have been typically reported in most EC-based closed-loop, semi-closed-loop, and open-loop soilless culture systems [8,13,14,16,29]. Theoretically, the concentration of nutrient solutions in the substrates can be explained by the difference between the concentration of irrigated solution and the concentration of nutrient uptake when the boundary area is limited to a substrate [5]. This can simply explain the nutrient variations in open-loop soilless culture systems. In closed-loop soilless culture systems, on the other hand, the concentration of irrigated solution is also affected by the drained solution, but most of studies on nutrient variations in closed-loop systems have been carried out with a premise that nutrient variations are the result of the changing dynamics of uptake concentrations [5,14,16,30–32].

The total amount of nutrients in the system using CM also increased with time (Figure 2b). In a closed-loop system, the changes in the total amount of nutrients can be interpreted more straightforwardly. The increasing tendency in the total amount of nutrients indicates the accumulation of surplus nutrients supplies. However, most of the previous studies did not attempt to interpret the fluctuations from the perspective of total amount of nutrients. Thus, our theoretical analyses reconsider problems for the nutrient concentration changes in the closed-loop soilless culture system; the nutrient fluctuation with increasing tendency is mainly caused by the accumulated difference between nutrient uptake and replenishment.

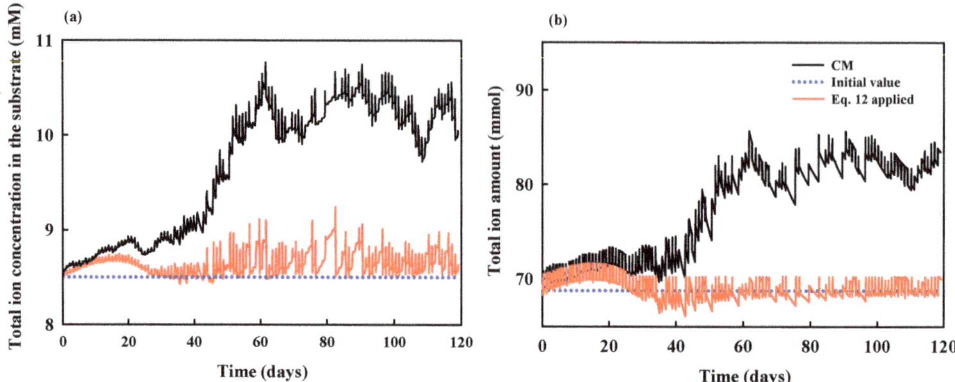

Figure 2. Changes in total ion concentration in the substrate (**a**) and total ions in the system (**b**) in the closed-loop soilless culture system using conventional (CM) and alternative (Equation (12) applied) nutrient-replenishment methods.

In a simple cultivation system sharing root-zone and nutrient solutions in a single container, measurement of changes in nutrient concentration, and water volume in the container corresponds to the actual amount of consumed nutrients in the system [15]. On the other hand, a typical soilless culture system consists of subsystems, including mixing tank, drainage tank, and substrates [6,12]. However, in the conventional nutrient solution mixing method of closed-loop soilless culture, the amount of nutrients replenishment has been mainly determined as a function of EC and volume of irrigation water and drainage [12]. Thus, to remove the errors between the actual nutrient consumption by plants and nutrients supplies in the closed-loop soilless culture system, the determination of the replenishment amount should consider the system-wide nutrients and water. We summarized equations for the estimation of nutrient consumption in the typical soilless culture system as Equation (11) and for the determination of nutrient replenishment as Equation (12)

$$V_{init}C_{init} = V_{drg}C_{drg} + V_{sub}C_{Sub} + V_{mix}C_{mix} + V_{U}C_{U} \tag{11}$$

$$V_{stk} = \frac{C_{init}V_{init} - C_{drg}V_{drg} - C_{mix}V_{mix} - C_{sub}V_{sub}}{C_{stk}} \tag{12}$$

where V_{init} is the initial volume of water in the system; V_{U} is the amount of water absorbed by the plant; C_{init} is the initial total concentration of the system; C_{U} is the average total nutrient uptake concentration; V_{drg}, V_{sub}, V_{mix}, and V_{stk} are the volumes of water stored in the drainage tank, substrate, mixing tank, and the input volume of stock solution, respectively; and C_{drg}, C_{sub}, C_{mix}, and C_{stk} are the total nutrient concentrations in the drainage tank, substrate, mixing tank, and the stock solution concentration, respectively.

In the calculation using Equation (12) for nutrient replenishment by stock solution, the total nutrient concentration showed repeated fluctuations near the initial concentration (Figure 2a). The amount of total nutrients in the system also stayed near the initial value without any apparent increasing or decreasing tendency (Figure 2b).

Precise measurements for the variables in Equation (12) in a real cultivation system have technical limitations. In particular, the amounts of total nutrients C_{sub} and V_{sub} in the substrate are difficult to estimate. In a soilless culture system, the field capacity (*F*) of a substrate corresponds to the parameters of the system, and the volume of water cannot exceed the volume of the substrate multiplied by the field capacity. The EC of the drainage (C_{drg}) can be indicative of a change in concentration of substrate.

Considering this, we can modify Equation (12) as follows for an alternative nutrient-replenishment method (AM):

$$V_{stk} = \frac{C_{init}V_{init} - C_{drg}V_{drg} - C_{mix}V_{mix} - C_{drg}F}{C_{stk}} \tag{13}$$

When the EC of the drainage (C_{drg}) and the field capacity (F) are substituted for C_{sub} and V_{sub}, respectively, errors may occur. However, in this case, total ion concentration fluctuated around the initial concentration (Figure 3).

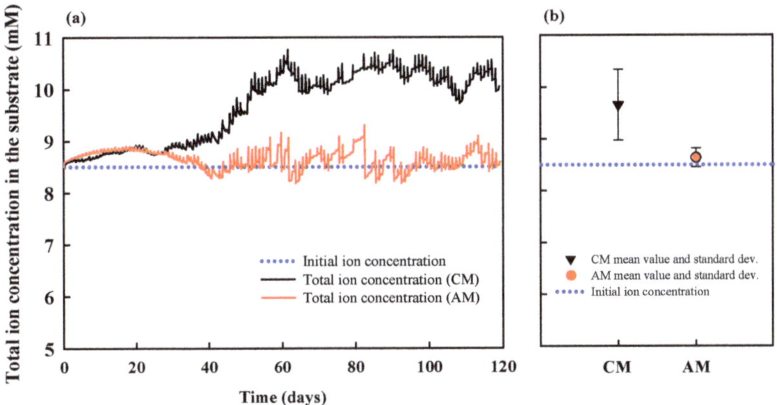

Figure 3. Changes in total ion concentration in the substrate according to the nutrient-replenishment method (**a**) and mean and standard deviation of total ion concentration in the substrate according to the nutrient-replenishment method (**b**). CM is the conventional nutrient-replenishment method and AM is alternative nutrient-replenishment method (Equation (13) applied).

In the existing problem definition, the EC variation in the closed-loop soilless culture system was derived from the dynamic change in nutrient uptake concentration [14,16,30–32]; thus, there were restrictions on active control and interpretation. However, a series of analysis steps leading to Equation (13) makes it possible to convert EC control in the closed-loop soilless culture system to the problem of proper gain search through arbitrary adjustment of system parameters. That is, in Equation (13), all but C_{drg} can be viewed as parameters and the process of calculating the difference between $C_{init}V_{init}$ and the product of the parameters and C_{drg} is performed in every mixing process.

3.2. Experimental Analysis: Demonstration Experiment for the Theoretical Analysis

The AM showed stable changes in the EC control of substrate and drainage against the CM (Figure 4). While the EC of substrate and drainage in the AM was maintained near the initial value of the system, an increasing tendency in stored drainage volume in the drainage tank was not observed (Figure 5). The average level of stored drainage level in the CM was higher than in the AM, and the range of variation was relatively wider (Figure 5).

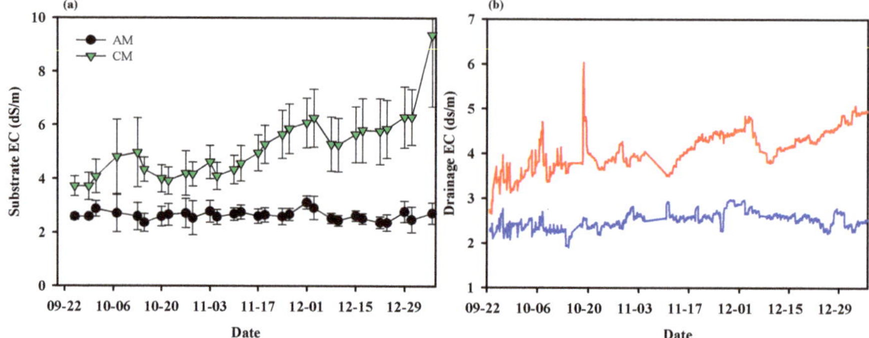

Figure 4. Comparison of electrical conductivity (mean ± SD) in the rockwool substrate (**a**) and the drainage (**b**) of the closed-loop soilless culture system during the experimental period between conventional (CM) and alternative (AM) nutrient-replenishment methods.

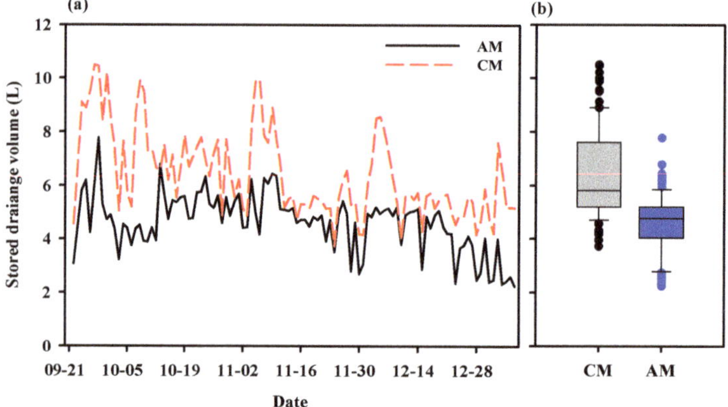

Figure 5. Changes in stored drainage volume in the drainage tank (**a**) and box-plot comparison between conventional (CM) and alternative (AM) nutrient-replenishment methods (**b**) during the experimental period.

The mixing ratio of drainage, water, and stock solution in the conventional nutrient solution mixing process depends on the target EC for the irrigation solution. However, this aspect could generate significant fluctuations in the stored volume of drainage. No increasing or decreasing trend in EC or stored drainage volume can be inferred over the entire experimental period in the closed-loop system, meaning that total nutrient input to the system adequately followed total nutrient uptake by the plant. In the CM, the EC of the rockwool substrate was relatively higher, and gradual increase was observed. The EC in the substrate can eventually be reflected in the EC of the drainage. A high EC value in a closed-loop soilless culture system where concentration control of the recycled nutrient solution is carried out can lead to an increase in the volume of stored drainage solution and subsequently to discharge of drainage when it exceeds system capacity [13,14]. This can be a factor in system instability. The EC changes in the rockwool substrate of the AM applied system indicate a normal effect of the proportional gain adjustment, as in the theoretical analysis in this study.

The cumulative amount of nutrients supplied to the system using the AM increased at a low rate in comparison with the CM, and the final amount of supplied nutrients was also lower than that of the CM; 1054 g for AM and 2257 g for CM, respectively (Figure 6). The AM appeared to work normally, and a reduction in fertilizer input compared with the CM was also observed. In addition, measurement of cumulative amount of nutrients in a state with no overall increases in EC and stored drainage volume were not observed indicates that the system can detect the total nutrient requirement of a plant. This measure could be used as an as index for plant nutritional status, one that is not provided in the CM.

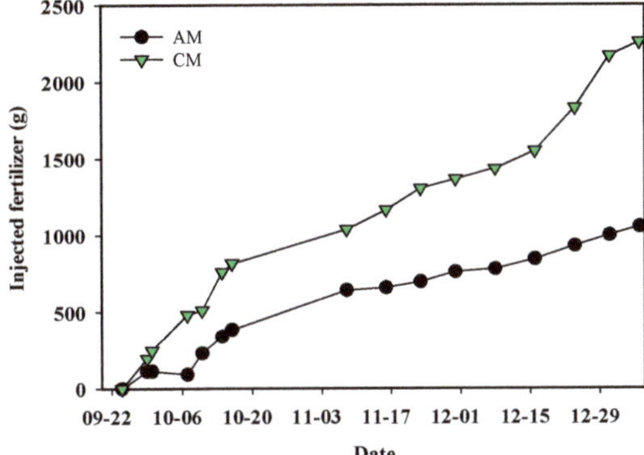

Figure 6. Accumulated amounts of fertilizers injected into the soilless culture systems with conventional (CM) and alternative (AM) nutrient-replenishment methods.

In the case of stock solution input volume change, it was confirmed that the input amount of the AM was relatively evenly distributed during the cultivation period (Figure 7b). On the other hand, in the case of the CM, a concentrated period of nutrient solution injection occurred, and relatively long periods during which the input of stock solution was blocked were observed (Figure 7a). The irregular feeding rate of the stock solution could be an adverse factor in nutrient-balance control when nutrient correction in the system is performed by input of stock or standard nutrient solution [13,14,32].

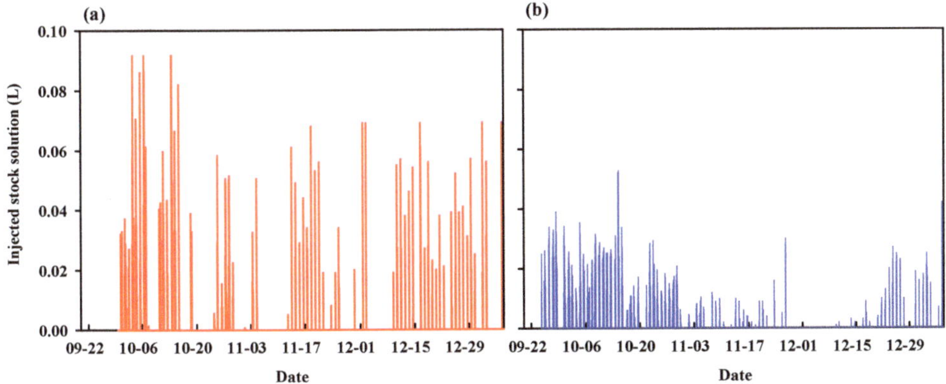

Figure 7. Changes in volume of injected stock solution with conventional (CM, **a**) and alternative (AM, **b**) nutrient-replenishment methods.

In the CM, overall tendencies of increasing Ca^{2+} and Mg^{2+} and decreasing K^+ were observed (Figure 8). In the AM, Ca^{2+} and Mg^{2+} concentrations were stable at a level relatively close to the initial value, but K^+ values showed a rapid decline and then fluctuated at a low concentration (Figure 8a–c). For CM, variations in nutrient concentrations similar to those reported in previous studies were observed [9,14,16]. Previous research on closed-loop soilless culture systems has determined that nutrient variations are a result of dynamic changes in nutrient uptake concentrations, and following those changes is challenging. [5,14,30,33]. However, Figure 8 indicates that a more deterministic change occurred in the system when nutrient replenishment was synchronized with total nutrient uptake through the AM system.

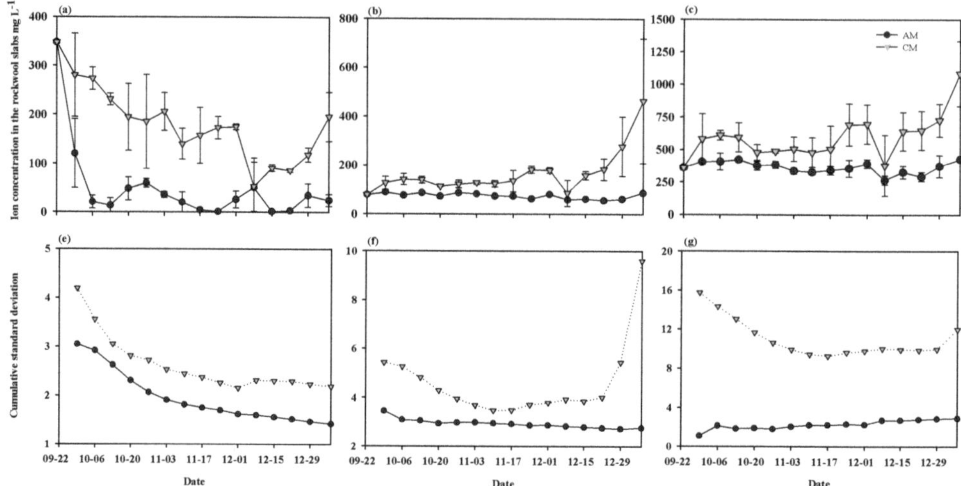

Figure 8. Changes in nutrient concentrations (mean ± SD) of K^+ (**a**), Mg^{2+} (**b**), and Ca^{2+} (**c**) and changes in cumulative standard deviation of nutrient concentrations of K^+ (**e**), Mg^{2+} (**f**), and Ca^{2+} (**g**) in the rockwool substrates using the conventional (CM) and alternative (AM) nutrient-replenishment methods, respectively.

The cumulative standard deviations of the AM were maintained at a lower level than those of the CM during the entire experimental period, and gradually decreasing tendencies were observed in K^+ and Mg^{2+} for the AM (Figure 8e–g). This means that the changes in nutrient concentration in the AM applied system were maintained close to the average concentration values during the experimental period compared with the CM. Considering the nutrient variations of the AM system itself, there may be a limit to defining it as steady state in the strict sense. However, in the actual cultivation conditions in this experiment, input of nutrients and water into the root zone by irrigation occurs intermittently, and the variation in the section where no input occurs cannot be controlled until the next input event. Furthermore, the frequency of changes of such input can affect system fluctuations [34–36], and the AM applied system is also under this influence. Considering these constraints and the CM changes, it can be assumed that the AM entered an average steady state that fluctuated within a certain range. The nutrient concentration control in the soilless culture system can therefore be seen as shifting the fluctuation range of the average steady state to the target range through a compositional change in the stock nutrient solution.

However, because the K^+ concentration of the AM was maintained at a very low level in this study, the impacts on sweet pepper productivity need to be considered [37]. Total sweet pepper yields during the experiment were compared (Figure 9). The average total yield was 827.5 g per plant (standard deviation [SD] ±106.5) in the CM and 838.8 g per plant (±109.8) in the AM, and statistically significant differences were not observed (*t*-test, $P > 0.05$; $n = 10$ per treatment). The average fruit weights were 133.7 g (±35.2) and 137.8 g (±38.6) for the CM and AM, respectively, but no significant effect was observed.

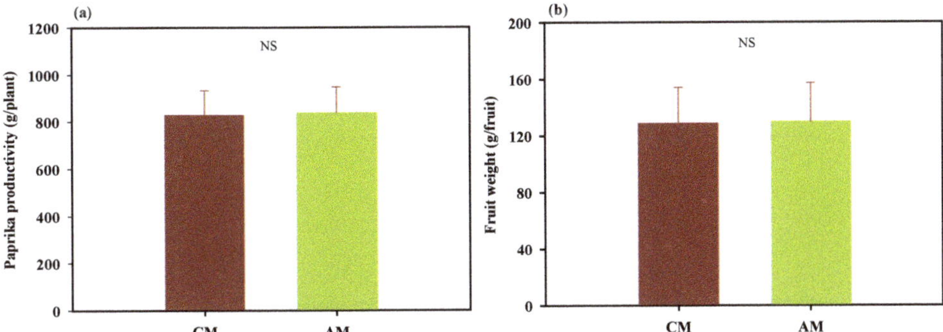

Figure 9. Comparisons (mean ± SD) of total yield (**a**) and fruit weight (**b**) of sweet pepper during the experimental period between conventional (CM) and alternative (AM) nutrient-replenishment methods (*t*-test). NS: Not significant ($P > 0.05$); $n = 10$ per treatment.

In the case of blossom-end rot, the mean value was low in the AM but not by a significant difference (Figure 10). This is considered to be due to the difference in concentration of the root zone when considering the characteristics of sweet pepper responses to root zone nutrient concentration [38].

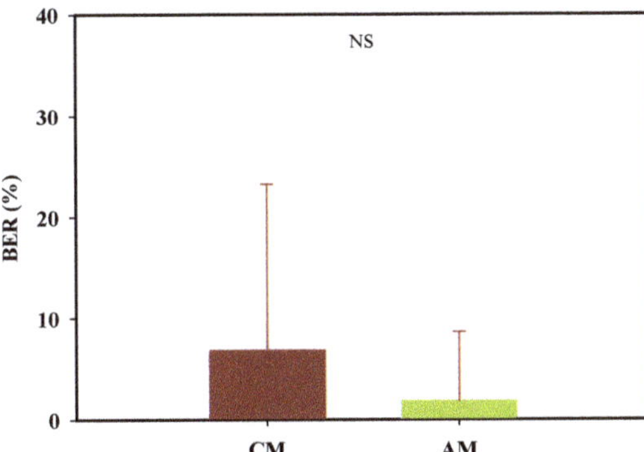

Figure 10. Comparison of blossom-end rot (mean ± SD) of sweet pepper between conventional (CM) and alternative (AM) nutrient-replenishment methods (*t*-test). NS: Not significant.

When comparing the changes in the nutrient ratio in the substrate during the experiment, the AM showed a tendency to accumulate calcium (Figure 11), but it was not in the range of physiological limitations of Steiner's standard [39]. Leaf analysis confirmed that absorption selectivity is maintained by achieving the ratio range of standard nutrient solutions, unlike the ratio of nutrients in the substrate nutrient solution (Figure 11). In the AM, the concentration of K^+ was maintained at a low level, but the supply interval of the stock solution was relatively uniformly distributed, resulting in a periodic supply. That could correspond to the prevention effect of nutrient deficiency through the constant feeding rate of nutrients even at lower concentrations [40].

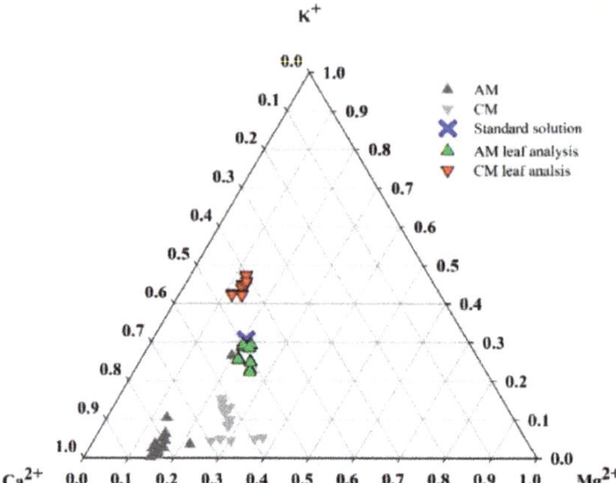

Figure 11. Nutrient balance changes in the rockwool substrates and dried leaves using the conventional (CM) and alternative (AM) nutrient-replenishment methods.

Previous studies and techniques for the EC-based closed-loop soilless culture systems interpreted the nutrient variations mainly focused on the discrepancies between supplied nutrient concentrations and uptake concentrations. Due to the dynamic features in the uptake concentrations and seemingly complex changes of each nutrient in the substrate, this has been a limiting factor in the systematic approach and the development of appropriate technologies so far. Therefore, most of the studies have been carried out through relative comparison by controlled experiments. However, there was no proper theoretical platform for nutrients variation in the closed-loop soilless culture system, so the stability of the cultivation has been verified by changing the terminal factors such as the irrigation, composition of the nutrient solution, and reuse period [9,14,16–18,41,42]. Our study redefined the problem of nutrient variation control in the EC-based closed-loop soilless culture system in the whole system perspective through the theoretical analysis and deduced the proper solution. The experimental results showed theoretically-predicted behaviors in the EC variation control. In addition, the ion concentrations showed convergent changes, which are providing a basis for future studies for technical advancement.

4. Conclusions

The effects of synchronized total nutrient supply on total nutrient uptake by the alternative nutrient-replenishment method (AM) were confirmed and compared with those of the conventional nutrient-replenishment method (CM) in the soilless culture system for sweet pepper cultivation. In the AM, electrical conductivity (EC) was maintained close to the initial value, and the use of fertilizers was reduced by about 45% without significant yield losses compared with the CM. This could mean that a closed-loop soilless culture system, showing complicated nutrient variations, can be stably controlled. Through this study, the problem of EC variation in closed-loop soilless cultures was theoretically analyzed. In addition, more advanced and sustainable control techniques could be applied based on the problem definition provided by this study and repeated experiments for other crops are required to ensure the on-site feasibility.

Author Contributions: Formal analysis, T.I.A.; investigation, T.I.A. and J.E.S.; methodology, T.I.A. and J.E.S.; supervision, J.E.S.; writing—original draft, T.I.A.; writing—review & editing, J.E.S.

Funding: This work was supported by the Korea Institute of Planning and Evaluation for Technology in Food, Agriculture, Forestry and Fisheries (IPET) through the Agriculture, Food and Rural Affairs Research Center Support Program funded by the Ministry of Agriculture, Food and Rural Affairs (MAFRA; 717001-07-1-HD240).

Conflicts of Interest: The authors declare no conflict of interest.

References

1. Gruda, N.S. Increasing Sustainability of Growing Media Constituents and Stand-Alone Substrates in Soilless Culture Systems. *Agronomy* **2019**, *9*, 298. [CrossRef]
2. Bouchaaba, Z.; Santamaria, P.; Choukr-Allah, R.; Lamaddalena, N.; Montesano, F.F. Open-cycle drip vs. closed-cycle subirrigation: Effects on growth and yield of greenhouse soilless green bean. *Sci. Hortic.* **2015**, *182*, 77–85. [CrossRef]
3. Olympios, C.M. Overview of soilless culture: Advantages, constraints and perspectives for its use in Mediterranean countries. *Cah. Options Méditerr.* **1999**, *31*, 307–324.
4. Beerling, E.A.M.; Blok, C.; van der Maas, A.A.; van Os, E.A. Closing the water and nutrient cycles in soilless cultivation systems. *Acta Hortic.* **2014**, *1034*, 49–55. [CrossRef]
5. Van Noordwijk, M. *Synchronisation of Supply and Demand is Necessary to Increase Efficiency of Nutrient Use in Soilless Horticulture*; Plant Nutrition-Physiology and Applications; Kluwer Academic Publishers: Dordrecht, The Netherlands, 1990.
6. Raviv, M.; Lieth, J.H. *Soilless Culture: Theory and Practice*, 1st ed.; Elsevier: Amsterdam, The Netherlands; Boston, MA, USA; Heidelberg, Germany, 2008; ISBN 978-0-444-52975-6.
7. Kläring, H.-P. Strategies to control water and nutrient supplies to greenhouse crops. A review. *Agronomie* **2001**, *21*, 311–321. [CrossRef]
8. Moon, T.; Ahn, T.I.; Son, J.E. Forecasting root-zone electrical conductivity of nutrient solutions in closed-loop soilless cultures via a recurrent neural network using environmental and cultivation Information. *Front. Plant Sci.* **2018**, *9*, 859. [CrossRef]
9. Zekki, H.; Gauthier, L.; Gosselin, A. Growth, productivity, and mineral composition of hydroponically cultivated greenhouse tomatoes, with or without nutrient solution recycling. *J. Am. Soc. Hortic. Sci.* **1996**, *121*, 1082–1088. [CrossRef]
10. Eggert, R.J. *Engineering Design*; Pearson/Prentice Hall: Upper Saddle River, NJ, USA, 2005; ISBN 978-0-13-143358-8.
11. Putra, P.A.; Yuliando, H. Soilless culture system to support water use efficiency and product quality: A review. *Agric. Agric. Sci. Procedia* **2015**, *3*, 283–288. [CrossRef]
12. Savvas, D.; Manos, G. Automated composition control of nutrient solution in closed soilless culture systems. *J. Agric. Eng. Res.* **1999**, *73*, 29–33. [CrossRef]
13. Massa, D.; Incrocci, L.; Maggini, R.; Bibbiani, C.; Carmassi, G.; Malorgio, F.; Pardossi, A. Simulation of crop water and mineral relations in greenhouse soilless culture. *Environ. Modell. Softw.* **2011**, *26*, 711–722. [CrossRef]
14. Signore, A.; Serio, F.; Santamaria, P. A Targeted management of the nutrient solution in a soilless tomato crop according to plant needs. *Front. Plant Sci.* **2016**, *7*, 391. [CrossRef] [PubMed]
15. Anpo, M.; Fukuda, H.; Wada, T. *Plant Factory Using Artificial Light: Adapting to Environmental Disruption and Clues to Agricultural Innovation*, 1st ed.; Elsevier Science: Amsterdam, The Netherlands, 2018; ISBN 978-0-12-813974-5.
16. Hao, X.; Papadopoulos, A.P. Growth, photosynthesis and productivity of greenhouse tomato cultivated in open or closed rockwool systems. *Can. J. Plant Sci.* **2002**, *82*, 771–780. [CrossRef]
17. Incrocci, L.; Malorgio, F.; Della Bartola, A.; Pardossi, A. The influence of drip irrigation or subirrigation on tomato grown in closed-loop substrate culture with saline water. *Sci. Hortic.* **2006**, *107*, 365–372. [CrossRef]
18. Rouphael, Y.; Raimondi, G.; Caputo, R.; De Pascale, S. Fertigation strategies for improving water use efficiency and limiting nutrient loss in soilless *Hippeastrum* Production. *Hortscience* **2016**, *51*, 684–689. [CrossRef]
19. Silberbush, M.; Ben-Asher, J.; Ephrath, J.E. A model for nutrient and water flow and their uptake by plants grown in a soilless culture. *Plant Soil* **2005**, *271*, 309–319. [CrossRef]
20. Silberbush, M.; Ben-Asher, J. Simulation study of nutrient uptake by plants from soilless cultures as affected by salinity buildup and transpiration. *Plant Soil* **2001**, *233*, 59–69. [CrossRef]
21. Snape, J.B.; Dunn, I.J.; Ingham, J.; Prenosil, J.E. *Dynamics of Environmental Bioprocesses: Modelling and Simulation*; VCH: New York, NY, USA, 1995; ISBN 978-3-527-61538-4.
22. Ta, T.H.; Shin, J.H.; Ahn, T.I.; Son, J.E. Modeling of transpiration of paprika (*Capsicum annuum* L.) plants based on radiation and leaf area index in soilless culture. *Hortic. Environ. Biotechnol.* **2011**, *52*, 265–269. [CrossRef]

23. Shin, J.H.; Park, J.S.; Son, J.E. Estimating the actual transpiration rate with compensated levels of accumulated radiation for the efficient irrigation of soilless cultures of paprika plants. *Agric. Water Manag.* **2014**, *135*, 9–18. [CrossRef]

24. Baille, M.; Baille, A.; Laury, J.C. A Simplified model for predicting evapotranspiration rate of nine ornamental species vs. climate factors and leaf-area. *Sci. Hortic.* **1994**, *59*, 217–232. [CrossRef]

25. Bassirirad, H. Kinetics of nutrient uptake by roots: Responses to global change. *New Phytol.* **2000**, *147*, 155–169. [CrossRef]

26. Barber, S.A. *Soil Nutrient Bioavailability: A Mechanistic Approach*; John Wiley and Sons: New York, NY, USA, 1995; ISBN 0-471-58747-8.

27. Savvas, D.; Adamidis, K. Automated management of nutrient solutions based on target electrical conductivity, pH, and nutrient concentration ratios. *J. Plant Nutr.* **1999**, *22*, 1415–1432. [CrossRef]

28. Shin, J.H.; Son, J.E. Application of a modified irrigation method using compensated radiation integral, substrate moisture content, and electrical conductivity for soilless cultures of paprika. *Sci. Hortic.* **2016**, *198*, 170–175. [CrossRef]

29. Lee, J.Y.; Rahman, A.; Azam, H.; Kim, H.S.; Kwon, M.J. Characterizing nutrient uptake kinetics for efficient crop production during *Solanum lycopersicum var. cerasiforme* Alef. growth in a closed indoor hydroponic system. *PLoS ONE* **2017**, *12*, e0177041. [CrossRef] [PubMed]

30. Carmassi, G.; Incrocci, L.; Maggini, R.; Malorgio, F.; Tognoni, F.; Pardossi, A. An aggregated model for water requirements of greenhouse tomato grown in closed rockwool culture with saline water. *Agric. Water Manag.* **2007**, *88*, 73–82. [CrossRef]

31. Carmassi, G.; Incrocci, L.; Maggini, R.; Malorgio, F.; Tognoni, F.; Pardossi, A. Modeling salinity build-up in recirculating nutrient solution culture. *J. Plant Nutr.* **2005**, *28*, 431–445. [CrossRef]

32. Savvas, D. Automated replenishment of recycled greenhouse effluents with individual nutrients in hydroponics by means of two alternative models. *Biosyst. Eng.* **2002**, *83*, 225–236. [CrossRef]

33. Savvas, D.; Stamati, E.; Tsirogiannis, I.L.; Mantzos, N.; Barouchas, P.E.; Katsoulas, N.; Kittas, C. Interactions between salinity and irrigation frequency in greenhouse pepper grown in closed-cycle hydroponic systems. *Agric. Water Manag.* **2007**, *91*, 102–111. [CrossRef]

34. Ta, T.H.; Shin, J.H.; Noh, E.H.; Son, J.E. Transpiration, growth, and water use efficiency of paprika plants (*Capsicum annuum* L.) as affected by irrigation frequency. *Hortic. Environ. Biotechnol.* **2012**, *53*, 129–134. [CrossRef]

35. Xu, G.; Levkovitch, I.; Soriano, S.; Wallach, R.; Silber, A. Integrated effect of irrigation frequency and phosphorus level on lettuce: P uptake, root growth and yield. *Plant Soil* **2004**, *263*, 297–309. [CrossRef]

36. Savvas, D.; Gruda, N. Application of soilless culture technologies in the modern greenhouse industry—A review. *Eur. J. Hortic. Sci.* **2018**, *83*, 280–293. [CrossRef]

37. Gruda, N.; Savvas, D.; Colla, G.; Rouphael, Y. Impacts of genetic material and current technologies on product quality of selected greenhouse vegetables—A review. *Eur. J. Hortic. Sci.* **2018**, *83*, 11. [CrossRef]

38. Tadesse, T.; Nichols, M.A.; Fisher, K.J. Nutrient conductivity effects on sweet pepper plants grown using a nutrient film technique. *N. Z. J. Crop Hortic.* **1999**, *27*, 239–247. [CrossRef]

39. Steiner, A.A. The selective capacity of plants for ions and its importance for the composition and treatment of the nutrient solution. *Acta Hortic.* **1980**, *98*, 87–97. [CrossRef]

40. Ingestad, T. Relative addition rate and external concentration; Driving variables used in plant nutrition research. *Plant Cell Environ.* **1982**, *5*, 443–453. [CrossRef]

41. Ko, M.T.; Ahn, T.I.; Shin, J.H.; Son, J.E. Effects of renewal pattern of recycled nutrient solution on the ion balance in nutrient solutions and root media and the growth and ion uptake of paprika (*Capsicum annuum* L.) in closed soilless cultures. *Korean J. Hortic. Sci. Technol.* **2014**, *32*, 463–472. [CrossRef]

42. Ehret, D.L.; Menzies, J.G.; Helmer, T. Production and quality of greenhouse roses in recirculating nutrient systems. *Sci. Hortic.* **2005**, *106*, 103–113. [CrossRef]

 *agronomy*

Article

Selection of Fertilizer and Cultivar of Sweet Pepper and Eggplant for Hydroponic Production

Hardeep Singh [1,*], Bruce L. Dunn [1], Mark Payton [2] and Lynn Brandenberger [1]

[1] Department of Horticulture and Landscape Architecture, Oklahoma State University, Stillwater, OK 74078-6027, USA

[2] Department of Statistics, Oklahoma State University, Stillwater, OK 74078-6027, USA

* Correspondence: hardeep.singh@okstate.edu

Received: 6 July 2019; Accepted: 5 August 2019; Published: 7 August 2019

Abstract: Dutch bucket hydroponic trials were conducted with the aim to evaluate the effects of different hydroponic fertilizers (5N-4.8P-21.6K, 5N-5.2P-21.6K, and 7N-3.9P-4.1K) on growth, fruit production, and the fruit quality (fruit shape index) parameters of two cultivars of sweet pepper (*Capsicum annuum* L.) and on two cultivars of eggplant (*Solanum melongena* L.). For sweet pepper yield, the 5N-4.8P-21.6K fertilizer was responsible for the greatest yield for both cultivars. For sweet pepper fresh and dry shoot weight interaction, the 'Orangella' cultivar had greater growth in 5N-4.8P-21.6K and 5N-5.2P-21.6K fertilizers, whereas there was no difference among cultivars in 7N-3.9P-4.1K. Shape index was not affected by fertilizers or cultivars. For the eggplant yield, there was no main effect nor interaction between fertilizers and cultivars for fruit yield, while the interaction between fertilizers and cultivars was significant for shoot fresh weight production. Shoot fresh weight was greater for 'Angela' than 'Jaylo' in 5N-4.8P-21.6K and 7N-3.9P-4.1K. Furthermore, both eggplant cultivars were affected with yellowing of fruits in all fertilizer treatments after 2 months, which was probably due to the accumulation of nutrients in the closed hydroponic system. Therefore, hydroponic producers could select 5N-4.8P-21.6K and 5N-5.2P-21.6K fertilizers for the cultivation of the 'Orangella' cultivar of sweet pepper based on yield. It is important to evaluate more fertilizers and cultivars for eggplant hydroponic cultivation.

Keywords: soilless culture; water soluble fertilizers; vegetables; *Capsicum annuum* L.; *Solanum melongena* L.; nutrients; shape index

1. Introduction

Problems such as soil salinity, lack of fertile soil, and soil-borne diseases are causes of hindrance for vegetable production in soil. Therefore, to overcome these problems, soilless culture was developed [1]. Involving growing plants without soil, soilless culture is considered a sustainable method for the cultivation of various greenhouse vegetable crops such as tomatoes (*Solanum lycopersicum* L.), cucumbers (*Cucumis sativus* L.), peppers (*Capsicum* L.), lettuce (*Lactuca sativa* L.), Swiss chard (*Beta vulgaris* L.), and eggplant (*Solanum melongena* L.). It is considered good for increasing agriculture sustainability as well as improving environmental health [2]. Soilless culture has various classification systems and methods such as hydroponics, aeroponics, gravel culture, and rockwool culture [3–5]. Dutch bucket system was introduced in the early 1980s by Dutch and Belgian growers and is defined as a container-type hydroponics system filled with substrates to provide support to the plant and nutrient solution supplied by drippers to each container [6].

The most important factor affecting crop yield and quality in hydroponics is nutrient solution [7]. The fertilizer used in hydroponic production should have balanced amounts of essential elements and should not form any precipitates during its use [8]. In most studies, nutrient solutions such as Copper's, Hoagland and Arnon's and Yamazaki's solution, which require self-preparation, have been evaluated in the hydroponic production of various crops. The self-preparation of nutrient solution for hydroponic production is effective for large-scale growers, whereas small scale growers face difficulties in managing nutrient concentration [9]. Therefore, commercially prepared, also known as one or two bag approach fertilizers, are gaining popularity. According to Mattson and Peters [9], a single bag fertilizer performed well for the production of peppers, cucumbers, and tomatoes at the University of Arizona Controlled Agriculture Center greenhouse. One of the reasons for the importance of a suitable fertilizer selection in hydroponics is that under field conditions, plants can influence nutrient availability by releasing root exudates or exploring new soil regions by growing their roots, while in hydroponics, it is not possible for plant roots to expand because of the confined area for root growth and the low buffering capacity of roots [10]. Furthermore, the accumulation of nutrients into plant structures may occur if nutrients are supplied in excess, posing health risks when plant products are consumed [11]. In addition, if a food product high in nitrate content is ingested, it is transformed into nitrite and subsequently nitrite, and in combination with amines, may form some carcinogenic compounds [12].

Soilless culture not only offers the possibility of growing crops with considerable savings of water and fertilizers, it is also considered as an easy and rapid method for screening cultivars of different crops for production, drought tolerance, and for physiological disorders [13]. Moreover, cultivar selection for hydroponics is not comparable to cultivar selection for field production. The data derived from field experiments for cultivar selection cannot be directly applied for hydroponic production due to the great difference in growth conditions between the two systems [14]. Some studies have evaluated the performance of sweet pepper cultivars for different objectives. Twelve sweet pepper cultivars were evaluated using a hydroponic system and it was concluded that 'Special' and 'Cupra' for red, 'Boogie', 'Fellini', and 'President' for orange, and 'Fiesta' and 'Derby' for yellow color had greater yields compared to other cultivars [15]. Mineral nutrition has the greatest impact on some physical and quality characteristics of sweet pepper, which include soluble solids, pH, fruit shape index, firmness, and pulp thickness [9]. It has also been also suggested that fruit weight and fruit shape index are two important characteristics of sweet peppers determining consumer preference and acceptability [16].

Various cultivars are available in the market for each crop, but for hydroponic cultivation, it is also necessary that the cultivar have a high economic value due to high input costs [5,8]. The yellow and orange colored cultivars of sweet peppers have a higher economic value than green colored cultivars. Therefore, 'Orangella' and 'Bentley' are orange and yellow colored cultivars of sweet pepper, respectively. Among the eggplant cultivars, 'Angela' and 'Jaylo' have been reported to have higher economic values due to their greater fruit size and white stripped fruits, respectively. Due to their high economic value, these cultivars has been tested with different objectives in hydroponic production. 'Bentley' has been tested for susceptibility to *Fusarium* spp. and other water-borne diseases in hydroponic cultivation [17]. Nevertheless, scientific literature evaluating these sweet pepper and eggplant cultivars using different commercially available hydroponic fertilizers is still lacking. Therefore, the objectives of our study were to evaluate the effect of three different commercial hydroponic fertilizers on growth, fruit production, and fruit quality (fruit shape index) parameters of different cultivars of sweet pepper and eggplant in the Dutch bucket system.

2. Materials and Methods

2.1. Plant Materials and Growth Conditions

Seeds of sweet peppers 'Bentley' and 'Orangella' and eggplants 'Angela' and 'Jaylo' were obtained from Johnny's Selected Seeds (Winslow, ME, USA) and sown on 12 February 2016. The seeds were sown in 1.5 cm^3 rockwool starter cubes with a sheet of 98 cubes (Grodan, Milton, ON, Canada) and transplanted into a Dutch bucket system on 20 March 2016 at the Department of Horticulture and Landscape Architecture Research Greenhouses in Stillwater, OK, USA. The average daily temperature, measured using a data logger (T & D Corporation, Nagano, Japan), was 27.2 °C. Light was measured using the same sensor and the daily light integral (DLI) was calculated from this data by multiplying 7992.48 lux by 0.0185 (standard conversion factor for sunlight to convert lux to PPFD), then multiplying 172.9 μmol m^{-2} s^{-1} by 0.0864 (standard conversion based on the total number of seconds in a day divided by 1 million) to obtain a DLI average of 12.8 mol m^{-2} d^{-1} for sweet pepper production [8]. For eggplant production, the average lux for the growth period was 8701.23 lux, therefore, DLI was equal to 13.90 mol m^{-2} d^{-1}. No nutrition was provided during nursery production. Seeds for the second replication were sown on 15 February 2017 and transplanted into the system on 23 March 2017. A single plant was transplanted into each bucket. The Dutch buckets were placed 50 cm apart and the rows were 100 cm apart and arranged on the opposite side of the irrigation and drainage pipes. Water was provided to each plant by a drip emitter, which supplied 3.75 L of water per h. Buckets were filled with expanded clay pebbles (Mother Earth Hydroton, National Garden Wholesale Sunlight Supply, Vancouver, WA, USA). The water that drained away was recirculated from a 150 L capacity storage tank using an electric pump.

2.2. Fertilizers

Both crops were fertilized by 5N-4.8P-21.6K (Jack's, J.R. Peters, Allentown, PA, USA), 5N-5.2P-21.6K (Peters, J.R. Peters, Allentown, PA, USA), and 7N-3.9P-4.1K (Dyna Gro, Richmond, CA, USA). The fertilizers used in this experiment had different elemental compositions (Table 1). Fertilizers 5N-4.8P-21.6K and 5N-5.2P-21.6K did not contain calcium (Ca) in their formulation, therefore, it was recommended by the manufacturer to add calcium nitrate (CaNO$_3$) (Haifa North America, Inc., Altamonte Spring, FL, USA) to supply Ca and a fraction of nitrogen (N). Fertilizer 7N-3.9P-4.1K contained all the recommended dosages of nutrients in one formulation. Tap water with an electrical conductivity (EC) of 0.5 dS m^{-1} and a pH of 7.8 was used to prepare the nutrient solution.

Table 1. Nutrient concentrations (in ppm) of hydroponic fertilizers when 3.69 kg were dissolved in 3785.4 L of water (as suggested by manufacturer).

Nutrients	Concentrations for 5N-4.8P-21.6K	Concentrations for 5N-5.2P-21.6K	Concentrations for 7N-3.9P-4.1K
Nitrogen	150.00	150.00	188.00
Phosphorus	39.00	48.00	41.00
Potassium	216.00	216.00	134.00
Calcium	139.00	139.00	53.00
Magnesium	47.00	31.00	13.44
Iron	2.30	3.00	2.68
Manganese	0.38	0.05	1.34
Zinc	0.11	0.15	1.34
Boron	0.38	0.50	0.53
Copper	0.13	0.15	1.34
Molybednum	0.07	0.10	0.02

2.3. EC, pH, and Data Collection

Sweet pepper fruits were harvested when 80% color (yellow or orange) development occurred and eggplants were harvested when they reached full size (i.e., weighing 250–400 g). Harvesting was carried out once or twice a week depending on number and maturity stage of fruits. The EC of all the nutrient solutions was maintained at 2.5–3.5 dS m^{-1}. If EC was higher than the recommended limit, then water was added and if EC was lower, then some fertilizer was added. The pH was maintained at 5.5–6.5 for eggplants and 5.5–6 for peppers. The commercially available product pH down (General Hydroponics, Santa Rosa, CA, USA) were used to adjust pH. This product was reported to be best among different organic and inorganic products used for pH maintenance in hydroponics [18]. The pH and EC of the solution was checked every alternate day.

At each harvest, data were collected on fruit weight and fruit shape index (for sweet pepper). Shape index was defined by the equatorial to longitudinal length ratio and calculated by dividing the maximum height (H) of fruit to the maximum width (W) of fruit (H/W) [19]. The height and width of each fruit were measured from randomly selected fruit. Nutrient analysis was conducted for the leaves of sweet peppers and eggplants. The nutrient analysis data for sweet pepper are not presented because there were no nutritional disorders in sweet pepper and nutrient concentration were within recommended limits. At the end of the trial, data were collected on fresh shoot weight, dry shoot, and root weight (shoots and roots dried for 2 days at 56 °C). Nutrient analysis of leaf samples was analyzed by the Soil, Water and Forage Analytical Laboratory at Oklahoma State University, using a nutrient analyzer (TruSpec Elemental Analyzer; LECO Corp, St. Joseph, MI, USA).

2.4. Experimental Setup and Data Analyses

The experimental design was a split plot design with two replications over time. The factors were fertilizer (main plots, three levels) and cultivars (sub plots, two levels for each crop). The experimental unit for the fertilizer was 18 plants, while the experimental unit for the cultivar was nine plants of each crop. Therefore, for each fertilizer treatment, there were nine replicas of each cultivar. Tests of significance were performed at the 0.05, 0.001, and 0.0001 levels. Least significance difference (LSD) method was used for comparing differences between treatment means. Data analysis was generated using SAS/STAT software (version 9.4) [20].

3. Results

3.1. Sweet Pepper

Interactions between fertilizer and sweet pepper cultivars occurred for shoot fresh and dry weight, and average fruit weight (Table 2). Shoot fresh weight and dry weight were significantly greater for 'Orangella' as compared to 'Bentley' when fertilized with 5N-4.8P-21.6K and 5N-5.2P-21.6K (Figures 1 and 2). There was no significant difference between shoot fresh and dry weight between sweet pepper cultivars when fertilized with 7N-3.9P-4.1K (Figures 1 and 2). Average fruit weight was significantly greater for 'Orangella' as compared to 'Bentley' when fertilized with 5N-4.8P-21.6K, whereas there was no significant difference between two cultivars when fertilized with 5N-5.2P-21.6K and 7N-3.9P-4.1K (Figure 3). Average fruit weight ranged from 122–172 g.

Table 2. Interaction and main effect for sweet pepper ('Bentley' and 'Orangella'), eggplant ('Angela' and 'Jaylo'), and hydroponic fertilizers (5N-4.8P-21.6K, 5N-5.2P-21.6K, and 7N-3.9P-4.1K).

	Crops	Cultivar	Fertilizer	Cultivar * Fertilizer
Yield	Sweet pepper	NS [z]	*	NS
	Eggplant	NS	NS	NS
Shoot fresh weight	Sweet pepper	***	NS	**
	Eggplant	**	***	*
Root weight	Sweet pepper	NS	***	NS
	Eggplant	NS	NS	NS
Shoot dry weight	Sweet pepper	***	NS	*
	Eggplant	***	***	NS
Shape index	Sweet pepper	NS	NS	NS
Average fruit weight	Sweet pepper	NS	NS	*

[z] NS, *, **, *** indicates non-significant or significant at $p \leq 0.05$, $p \leq 0.001$, or $p \leq 0.0001$, respectively.

For fruit yield and root weight, there was a significant fertilizer effect, while there was no fertilizer or cultivar effect for shape index (Table 2). The fruit yield of sweet pepper was significantly greater in 5N-4.8P-21.6K and 5N-5.2P-21.6K as compared to 7N-3.9P-4.1K (Table 3). The root weight of sweet pepper was significantly greater in 5N-4.8P-21.6K as compared to 5N-5.2P-21.6K and 7N-3.9P-4.1K (Table 3).

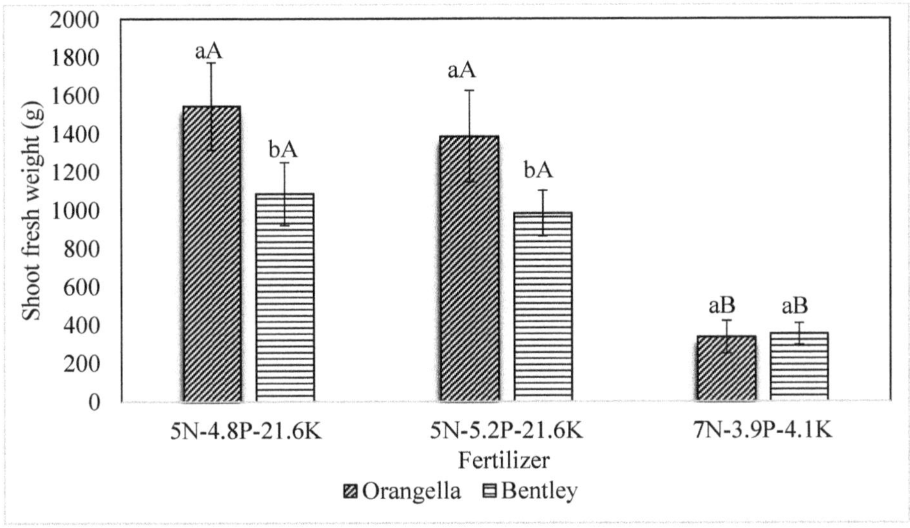

Figure 1. Interaction between sweet pepper cultivars ('Bentley' and 'Orangella') and hydroponic fertilizers (5N-4.8P-21.6K, 5N-5.2P-21.6K, and 7N-3.9P-4.1K) for shoot fresh weight (g) per plant (n = 9). Data are presented as means ± SEM. Means with same lowercase letter are not significantly different by LSD ($p \leq 0.05$) between cultivars within fertilizers. Means with same uppercase letter are not significantly different by LSD ($p \leq 0.05$) among fertilizers within each cultivar.

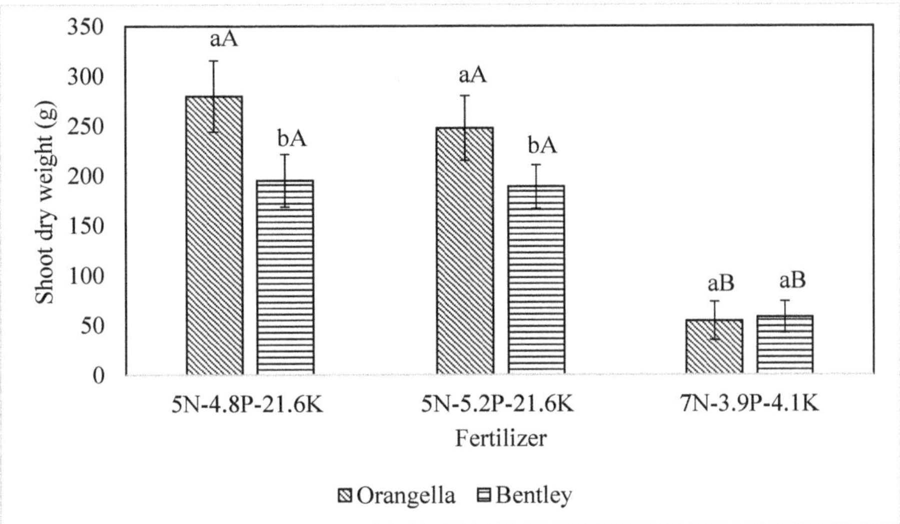

Figure 2. Interaction between sweet pepper cultivars ('Bentley' and 'Orangella') and hydroponic fertilizers (5N-4.8P-21.6K, 5N-5.2P-21.6K, and 7N-3.9P-4.1K) for shoot dry weight (g) per plant (n = 9). Data are presented as means ± SEM. Means with same lowercase letter are not significantly different by LSD ($p \leq 0.05$) between cultivars within fertilizers. Means with same uppercase letter are not significantly different by LSD ($p \leq 0.05$) among fertilizers within each cultivar.

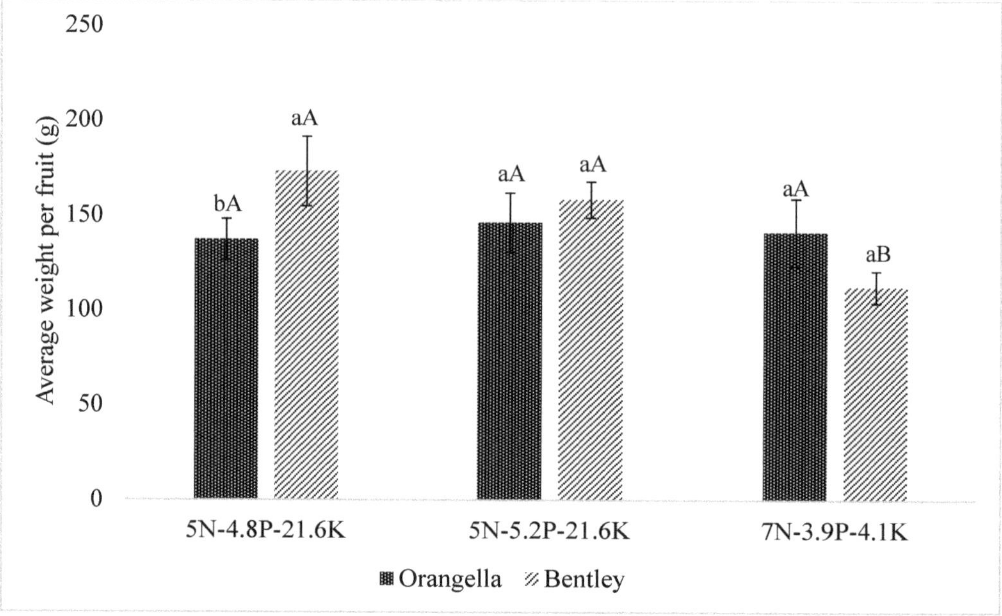

Figure 3. Interaction between sweet pepper cultivars ('Bentley' and 'Orangella') and hydroponic fertilizers (5N-4.8P-21.6K, 5N-5.2P-21.6K, and 7N-3.9P-4.1K) for average fruit weight (g) per plant (n = 9). Data are presented as means ± SEM. Means with same lowercase letter are not significantly different by LSD ($p \leq 0.05$) between cultivars within fertilizers. Means with same uppercase letter are not significantly different by LSD ($p \leq 0.05$) among fertilizers within each cultivar.

Table 3. Main effect of hydroponic fertilizers (5N-4.8P-21.6K, 5N-5.2P-21.6K, 7N-3.9P-4.1K) pooled across cultivars for per plant sweet pepper fruit yield and root weight. (n = 18).

Fertilizer	Yield (g)	Root Weight (g)
5N-4.8P-21.6K	3697.76 ± 352.42a [z]	104.47 ± 12.80a
5N-5.2P-21.6K	3080.97 ± 489.84a	86.60 ± 4.56b
7N-3.9P-4.1K	1378.47 ± 375.41b	39.11 ± 8.96c

[z] Means within a column followed by same lowercase letter are not significantly different by LSD ($p \leq 0.05$). Data are presented as means ± SEM.

3.2. Eggplant

Interactions between fertilizer and eggplant cultivars occurred for shoot fresh weight. The shoot fresh weight was significantly greater for 'Angela' as compared to 'Jaylo' when fertilized with 5N-4.8P-21.6K and 7N-3.9P-4.1K. There was no significant difference between the shoot fresh weights of eggplant cultivars when fertilized with 5N-5.2P-21.6K (Figure 4).

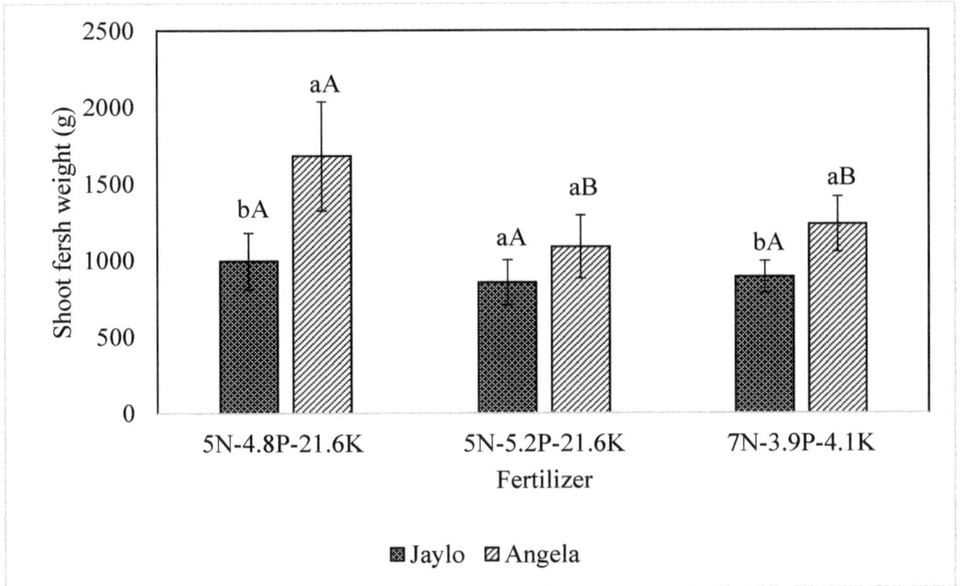

Figure 4. Interaction between eggplant cultivars ('Angela' and 'Jaylo') and hydroponic fertilizers (5N-4.8P-21.6K, 5N-5.2P-21.6K, and 7N-3.9P-4.1K) for shoot fresh weight (g) (n = 9). Data are presented as means ± SEM. Means with the same lowercase letter are not significantly different by LSD ($p \leq 0.05$) between cultivars within fertilizers. Means with same uppercase letter are not significantly different by LSD ($p \leq 0.05$) among fertilizers within each cultivar.

A fertilizer effect was found on the shoot dry weight, while a cultivar main effect was only found for the shoot dry weight of eggplant (Table 2). There was no significant difference for yield and root weight among different fertilizer treatments (Table 2). The shoot dry weight of eggplant was significantly greater in 5N-4.8P-21.6K as compared to 5N-5.2P-21.6K and 7N-3.9P-4.1K when pooled across cultivars (Table 4). The shoot dry weight of 'Angela' was significantly greater than 'Jaylo' when pooled across fertilizers (Table 4).

Table 4. Main effect of hydroponic fertilizers (5N-4.8P-21.6K, 5N-5.2P-21.6K, 7N-3.9P-4.1K) pooled across cultivars (n = 18) and eggplant cultivars ('Angela' and 'Jaylo') pooled across fertilizers for shoot dry weight. (n = 9).

Fertilizer	Shoot Dry Weight (g)	Cultivar	Shoot Dry Weight(g)
5N-4.8P-21.6K	204.07 ± 5.50a [z]	Jaylo	184.30 ± 10.52b
5N-5.2P-21.6K	191.26 ± 3.07b	Angela	248.20 ± 8.96a
7N-3.9P-4.1K	193.42 ± 4.52b		

[z] Means within a column followed by same lowercase letter are not significantly different by LSD ($p \leq 0.05$). Data are presented as means ± SEM.

Eggplant fruits developed an abnormal color after 2 months of production in both years. The fruits of the 'Jaylo' cultivar turned brownish-purple in color, while the 'Angela' cultivar fruits developed a yellow color. Foliar analysis found that the concentration of all nutrients was above the recommended upper limit except N (Table 5). A pairwise comparison was performed between the recommended foliar nutrient concentration by Flores et al. [21] and foliar nutrient concentration of plants grown in different fertilizers was observed.

Table 5. Average foliar nutrient concentration for eggplant in comparison to recommended nutrient level by Flores [21] (n = 18).

Nutrients	Recommended	5N-4.8P-21.6K	5N-5.2P-21.6K	7N-3.9P-4.1K
Nitrogen (%)	4.20a [z]	3.64 ± 0.64a	3.55 ± 0.70a	3.62 ± 0.60a
Phosphorus **,[y] (%)	0.30b	0.39 ± 0.04a	0.42 ± 0.10a	0.38 ± 0.05a
Potassium ** (%)	3.50b	3.96 ± 0.25a	3.95 ± 0.29a	3.72 ± 0.26b
Calcium *** (%)	0.80b	3.82 ± 0.90a	4.02 ± 0.95a	3.63 ± 1.05a
Magnesium ** (%)	0.25b	0.96 ± 0.42a	1.02 ± 0.56a	0.92 ± 0.47a
Manganese *** (ppm)	50.00c	130.25 ± 26.89a	156.30 ± 45.85a	144.80 ± 62.53a
Iron *** (ppm)	50.00b	155.02 ± 39.01a	144.05 ± 52.65a	158.10 ± 55.12a
Boron *** (ppm)	20.00b	80.50 ± 37.02a	88.89 ± 51.46a	96.30 ± 24.05a
Zinc *** (ppm)	20.00b	76.60 ± 16.92a	72.08 ± 18.56a	75.42 ± 25.09a

[z] Means within a row followed by same letter are not significantly different by paired t test ($p \leq 0.05$). [y], *, **, *** indicates non-significant or significant at $p \leq 0.05$, $p \leq 0.001$, or $p \leq 0.0001$, respectively. Data are presented as means ± SEM.

4. Discussion

Fruit weight and fruit shape index are two important characteristics of sweet peppers, determining the fruit quality [22]. For sweet pepper, fruits weighing less than 100 g are considered to be unmarketable [22]; in the current trial, the sweet pepper average fruit weight ranged from 122–172 g (Figure 3). Rubio et al. [22] also looked for the response of Ca and K on the yield and fruit quality of sweet pepper and found that adequate management of Ca and K fertilization could help improve yield and fruit quality (fruit shape index) of sweet pepper in hydroponics. The findings from the current experiment for sweet pepper fruit yield support the results from Rubio et al. [22], as high yielding fertilizers 5N-4.8P-21.6K and 5N-5.2P-21.6K were high in Ca and K as compared to 7N-3.9P-4.1K, whereas there was no effect on fruit quality (fruit shape index). Fertilizer 5N-4.8P-21.6K has been recommended for hydroponic production of tomatoes, cucumbers, and peppers and was found to be similar in nutrient content with the hydroponic recipe prepared by the University of Arizona, which provided remarkable results [9].

Another study evaluated the effect of nutrition and irrigation on sweet pepper production in hydroponics and concluded that in a closed system, the fertilization of nitrogen (N) 240, phosphorus (P) 60, (K) 300, magnesium (Mg) 50, ferrous (Fe) 6, manganese (Mn) 3, boron (B) 1.6, zinc (Zn) 2, (Ca) 90, copper (Cu) 0.8 and molybdenum (Mo) 0.12 (mg L^{-1}) was appropriate for sweet pepper production [23]. Therefore, there is a possibility of a further increase in fruit yield for current sweet pepper cultivars because all the nutrient levels of the current fertilizers were lower than the levels recommended by Gul et al. [23] (Table 1). Adding potassium peroxide at a rate of 1 g L^{-1} has also been reported to result in a 20% increase in sweet pepper yield in hydroponics [24].

For hydroponic eggplant production, we did not find any recommendations of specific fertilizers in the literature other than self-preparation of Hoagland's solution [25]. However, since the manufacturer recommended that the fertilizers tested in the current trial were suitable for fruiting vegetable crops, they were tested for eggplants. Both the form and quantity of N play important roles in hydroponic as well as field vegetable production. The nitrate form of N should dominate in the nutrient solution, while the ammoniacal form should be lower [25]. In the current study, fertilizers 5N-4.8P-21.6K and 5N-5.2P-21.6K had a total N in nitrate form while 7N-3.9P-4.1K had 2.6% as ammoniacal form and 4.4% as nitrate form. In terms of the quantity of N, the recommendation of total N for hydroponic production of eggplants was 120–170 ppm, which was satisfied by the fertilizers used in our study [26]. There was limited literature providing information regarding micronutrient requirements of eggplant in soilless culture. It has been reported that eggplants need 15, 10, 5, 0.75 and 0.5 µM of Fe, Mn, Zn, Cu, and Mo, respectively [26].

The yellowing of eggplant leaves and fruits was initially suspected to be caused by a deficiency of some nutrients. Eggplant is susceptible to boron deficiency and young fully developed leaves turn yellow at the distal end [27]. However, foliar analysis of eggplant revealed that the concentration of all the nutrients was above the recommended limit except N (Table 5). Therefore, the yellowing in plants was more likely due to the toxicity of nutrients. A possible reason explaining this nutrient toxicity in hydroponic eggplant production is the use of expanded clay balls as a stand-alone substrate. Some substrates may have a higher cation-exchange capacity, thereby leading to the localization of some nutrients in root zones and to the toxicity of nutrients. Pine bark has been suggested to be the best stand-alone substrate for fruit vegetable production [28]. Another reason explaining nutrient toxicity could be the higher accumulation of macro and micronutrients in closed hydroponic systems reported in some studies [29]. Therefore, the selection of an adequate stand-alone substrate is important for hydroponic vegetable production to avoid yield loss due to nutrient toxicity [30]. Moreover, there is need for an appropriate method to monitor nutrient concentration in solution during growing cycles.

Many studies have reported different EC ranges for the hydroponic cultivation of eggplants. The response of eggplant to salinity in a recirculating hydroponics system was studied by Savvas et al. [26], who found that high salinity significantly affected osmotic potential due to reduced water uptake leading to less water being directed towards fruit development and they recommended an EC of 1.5 dS m^{-1}. Moazed et al. [31] and Mahjoor et al. [32] recommend an EC of 2.5 dS m^{-1}. According to the foliar nutrient concentrations, by maintaining the EC in the recommended range (2.5–3.5 dS m^{-1}), plants were not able to maintain nutrient concentration in required limits as the concentration of all the nutrients except N was higher than the recommended range. Therefore, some researchers have reported that the EC is not a good indicator for estimating the nutrient concentration of solution, as EC indicates total dissolved ion concentrations only and cannot be used directly to determine individual ion concentrations. Thus, controlling nutrients based on EC in hydroponics may lead to excess or deficiency of some nutrients [33]. Periodic tissue sampling is reported to be the best way to evaluate if the nutrients provided are adequate for the growth stage and growing conditions [9]. Furthermore, some other non-destructive precision agriculture tools, such as mobile phone plant nitrogen applications, can be used to monitor nutrient concentrations in greenhouse production [34].

5. Conclusions

From the results of the present experiment, 5N-4.8P-21.6K and 5N-5.2P-21.6K can be recommended for sweet pepper production in hydroponics because fruit yield was not significantly different between these fertilizers, whereas it was significantly greater than with 7N-3.9P-4.1K. For cultivar evaluation, 'Orangella' produced significantly greater shoot fresh and dry weight in 5N-4.8P-21.6K and 5N-5.2P-21.6K. Nevertheless, vegetable producers are more interested in fruit yield and quality. The average fruit weight of 'Orangella' was significantly lower than 'Bentley' when grown in 5N-4.8P-21.6K. Moreover, some other factors needed to be evaluated to recommend these cultivar for hydroponic production because some studies reported 'Bentley' to be susceptible to *Fusarium* and to water borne disease [17]. Two months data for eggplants showed that there was no effect of cultivar or fertilizer on eggplant yield, while the main effects of cultivar and fertilizer were observed for shoot dry weight, with 'Angela' producing significantly greater results than 'Jaylo' and 5N-4.8P-21.6K producing significantly greater results among the three fertilizers. An interaction among fertilizers and eggplant cultivars was observed for eggplant shoot fresh weight, with 'Angela' producing significantly greater weight in 5N-4.8P-21.6K and 7N-3.9P-4.1K. Based on the results of the current study, it is not possible to recommend either fertilizer or cultivar for the hydroponic production of eggplant, as after 2 months, almost all the fruits were non-marketable due to yellowing. Therefore, future studies are needed to investigate the physiology behind the yellowing of eggplant fruits, and to identify a better indicator of nutrient concentration than EC. Different recycling rates of nutrient solutions and alternatives for stand-alone substrates for eggplant hydroponic production should be evaluated in future studies because this will also affect nutrient accumulation into plant parts.

Author Contributions: Conceptualization, B.L.D.; H.S.; methodology, H.S.; formal analysis, M.P.; investigation, H.S.; writing—original draft preparation, H.S.; writing—review and editing, B.L.D.; H.S.; and L.B.; supervision, B.L.D.

Funding: This research was funded by ODAFF Specialty Block grant #G00000475.

Conflicts of Interest: We declare no conflict of interest.

References

1. Savvas, D.; Gruda, N. Application of soilless culture technologies in the modern greenhouse industry—A review. *Eur. J. Hortic. Sci.* **2018**, *83*, 280–293. [CrossRef]
2. Gruda, N.; Savvas, D.; Colla, G.; Rouphael, Y. Impacts of genetic material and current technologies on product quality of selected greenhouse vegetables—A review. *Eur. J. Hortic. Sci.* **2018**, *83*, 319–328. [CrossRef]
3. Van, E.; Gieling, T.; Lieth, H. Technical Equipment in Soilless Production Systems. In *Soilless Culture Theory and Practice*; Elsevier: London, UK, 2008; pp. 147–207.
4. Atzori, G.; Mancuso, S.; Masi, E. Seawater potential use in soilless culture: A review. *Sci. Hortic.* **2019**, *249*, 199–207. [CrossRef]
5. Lennard, W.; Leonard, B. A comparison of three different hydroponic sub-systems (gravel bed, floating and nutrient film technique) in an aquaponic test system. *Aquac. Int.* **2006**, *14*, 539–550. [CrossRef]
6. Roberto, K. *How to Hydroponics*; The Future Garden Press: Farmingdale, NY, USA, 2003; p. 22.
7. Savvas, D.; Gizas, G. Response of hydroponically grown gerbera to nutrient solution recycling and different nutrient cation ratios. *Sci. Hortic.* **2002**, *96*, 267–280. [CrossRef]
8. Singh, H.; Dunn, B.; Payton, M.; Brandenberger, L. Fertilizer and cultivar selection of lettuce, basil, and swiss chard for hydroponic production. *HortTechnology* **2019**, *29*, 50–56. [CrossRef]
9. Mattson, N.S.; Peters, C. *A Recipe for Hydroponic Success, Inside Grower*; Ball Publishing: Chicago, IL, USA, 2014; pp. 16–19.
10. Savvas, D.; Ntatsi, G.; Rodopoulou, M.; Goumenaki, F. Nutrient uptake concentrations in a cucumber crop grown in a closed hydroponic system under Mediterranean climatic conditions as influenced by irrigation schedule. *Acta Hortic.* **2014**, *1034*, 545–552. [CrossRef]
11. Cavarianni, R.L.; Cecílio Filho, A.B.; Cazetta, J.O.; May, A.; Corradi, M. Nutrient contents and production of rocket as affected by nitrogen concentrations in the nutritive solution. *Sci. Agric.* **2008**, *65*, 652–658. [CrossRef]
12. Elwan, M.; El-Hamed, K. Influence of nitrogen form, growing season and sulfur fertilization on yield and the content of nitrate and vitamin C of broccoli. *Sci. Hortic.* **2011**, *127*, 181–187. [CrossRef]
13. Ogbonnaya, C.; Sarr, B.; Brou, C.; Diouf, O.; Diop, N.; Roy-Macauley, H. Selection of cowpea genotypes in hydroponics, pots, and field for drought tolerance. *Crop Sci.* **2003**, *43*, 1114–1120. [CrossRef]
14. Molders, K.; Quinet, M.; Decat, J.; Secco, B.; Dulière, E.; Pieters, S.; van der Kooij, T.; Lutts, S.; Van Der Straeten, D. Selection and hydroponic growth of potato cultivars for bioregenerative life support systems. *Adv. Space Res.* **2012**, *50*, 156–165. [CrossRef]
15. Won, J.-H.; Gangwon Provincial, A.; Jeong, B.; Gangwon Provincial, A.; Kim, J.; Jeon, S.; Gangwon Provincial, A. Selection of suitable cultivars for the hydroponics of sweet pepper (*Capsicum annuum* L.) in the alpine area in summer. *J. Bio-Environ. Control* **2009**, *18*, 425–430.
16. Navarro, J.; Garrido, C.; Carvajal, M.; Martinez, V. Yield and fruit quality of pepper plants under sulphate and chloride salinity. *J. Hortic. Sci. Biotechnol.* **2002**, *77*, 52–57. [CrossRef]
17. Cerkauskas, R.F. Etiology and management of Fusarium crown and root rot (*Fusarium oxysporum*) on greenhouse pepper in Ontario, Canada. *Can. J. Plant Pathol.* **2017**, *39*, 121–132. [CrossRef]
18. Singh, H.; Dunn, B.; Payton, M. Hydroponic pH modifiers affect plant growth and nutrient content in leafy greens. *J. Hortic. Res.* **2019**, in press.
19. Brewer, M.; Lang, L.; Fujimura, K.; Dujmovic, N.; Gray, S.; van der Knaap, E. Development of a controlled vocabulary and software application to analyze fruit shape variation in tomato and other plant species. *Plant Physiol.* **2006**, *141*, 15–25. [CrossRef]
20. SAS Institute. *Base SAS 9.4 Procedures Guide: Statistical Procedures*; SAS Institute: Cary, NC, USA, 2017.
21. Flores, R.A.; Borges, B.; Almeida, H.; Prado, R. Growth and nutritional disorders of eggplant cultivated in nutrients solutions with suppressed macronutrients. *J. Plant Nutr.* **2015**, *38*, 1097–1109. [CrossRef]

22. Rubio, J.; García-Sánchez, F.; Flores, P.; Navarro, J.; Martínez, V. Yield and fruit quality of sweet pepper in response to fertilisation with Ca^{2+} and K^+. *Span. J. Agric. Res.* **2010**, *8*, 170–177. [CrossRef]

23. Gul, A.; Tuzel, Y.; Tuzel, I.; Irget, M.; Kidoglu, F.; Tepecik, M. Effects of nutrition and irrigation on sweet pepper production in volcanic tuff. *Span. J. Agric. Res.* **2011**, *9*, 221–229. [CrossRef]

24. Urrestarazu, M.; Mazuela, P. Effect of slow-release oxygen supply by fertigation on horticultural crops under soilless culture. *Sci. Hortic.* **2005**, *106*, 484–490. [CrossRef]

25. Savvas, D.; Leneti, H.; Mantzos, N.; Kakarantza, L.; Barouchas, P. Effects of enhanced ammonium N supply and concomitant changes in the concentrations of other nutrients needed for ion balance on the growth, yield, and nutrient status of eggplants grown on rockwool. *J. Hortic. Sci. Biotechnol.* **2010**, *85*, 355–361. [CrossRef]

26. Savvas, D.; Ntatsi, G.; Passam, H. Plant nutrition and physiological disorders in greenhouse grown tomato, pepper and eggplant. *Eur. J. Plant Sci. Biotechnol.* **2008**, *2*, 45–61.

27. De Kreij, C.; Başar, H. Leaf tip yellowing in eggplant is caused by boron deficiency. *J. Plant Nutr.* **1997**, *20*, 47–53. [CrossRef]

28. Gruda, N. Increasing sustainability of growing media constituents and stand-alone substrates in soilless culture systems. *Agronomy* **2019**, *9*, 298. [CrossRef]

29. Rouphael, Y.; Colla, G.; Battistelli, A.; Moscatello, S.; Proietti, S.; Rea, E. Yield, water requirement, nutrient uptake and fruit quality of zucchini squash grown in soil and closed soilless culture. *J. Hortic. Sci. Biotechnol.* **2004**, *79*, 423–430. [CrossRef]

30. Dorai, M.; Papadopoulos, A.; Gosselin, A. Influence of electric conductivity management on greenhouse tomato yield and fruit quality. *Agronomie* **2001**, *21*, 367–383. [CrossRef]

31. Moazed, H.; Ghaemi, A.; Rafiee, M. Evaluation of several reference evapotranspiration methods: A comparative study of greenhouse and outdoor conditions. *Iran. J. Sci. Technol.* **2014**, *38*, 421.

32. Savvas, D.; Adamidis, K. Automated management of nutrient solutions based on target electrical conductivity, pH, and nutrient concentration ratios. *J. Plant Nutr.* **1999**, *22*, 1415–1432. [CrossRef]

33. Lee, J.Y.; Rahman, A.; Azam, H.; Kim, H.S.; Kwon, M. Characterizing nutrient uptake kinetics for efficient crop production during *Solanum lycopersicum* var. cerasiforme Alef. growth in a closed indoor hydroponic system. *PLoS ONE* **2017**, *12*, e0177041. [CrossRef]

34. Swearengin, L.; Dunn, B.; Singh, H.; Goad, C. Evaluation of a mobile phone plant nitrogen recommendation application in the greenhouse. *J. Plant Nutr.* **2018**, *41*, 2615–2625. [CrossRef]

 *agronomy*

Article

Increasing Levels of Supplemental LED Light Enhances the Rate Flower Development of Greenhouse-grown Cut Gerbera but does not Affect Flower Size and Quality

David Llewellyn, Katherine Schiestel and Youbin Zheng *

School of Environmental Sciences, University of Guelph, 50 Stone Road East, Guelph, ON N1G 2W1, Canada; dllewell@uoguelph.ca (D.L.); katie.schiestel@gmail.com (K.S.)
* Correspondence: yzheng@uoguelph.ca

Received: 13 August 2020; Accepted: 3 September 2020; Published: 4 September 2020

Abstract: To investigate the influence of supplemental lighting intensity on the production (i.e., rate of flower development, flower quality, and yield) of cut gerbera during Canada's supplemental lighting season (November to March), trials were carried out at a research greenhouse. Five supplemental light emitting diode (LED) light intensity (LI) treatments provided canopy-level photosynthetic photon flux densities (PPFD) ranging from 41 to 180 μmol m^{-2} s^{-1}. With a 12-h photoperiod, the treatments provided 1.76 to 7.72 mol m^{-2} d^{-1} of supplemental light. Two cultivars of cut gerbera (*Gerbera jamesonii* H. Bolus ex Hook.f) were used to evaluate vegetative growth and flower production. Plugs of 'Ultima' were assessed for vegetative growth and rate of flower development. There were minor LI treatment effects on number of leaves and chlorophyll content index and flowers from plants under the highest versus the lowest LI matured 10% faster. Reproductively mature 'Panama' plants were assessed for flower yield and quality. 'Panama' flowers from the highest LI treatment had shorter stems than the three lowest LI treatments, and flowers from the middle LI treatment had larger diameter than the other treatments. Flowers from the lowest LI treatment had lower fresh mass than the three highest LI treatments. There were linear relationships between LI and numbers of flowers harvested, with the highest LI treatment producing 10.3 and 7.0 more total and marketable flowers per plant than the lowest LI treatment. In general, increasing levels of supplemental light had only minor effects on vegetative growth (young plants) and size and quality of harvested flowers (mature plants), but flowers from plants grown under higher LIs were more numerous and matured faster.

Keywords: flower bud development; flower number; flower quality; *Gerbera jamesonii*; growth; DLI

1. Introduction

In greenhouses at higher latitude regions, such as northern USA and Canada, it is often considered necessary for growers of year-round commodities (e.g., cut flowers) to use supplemental lighting to meet the crops' economic minimum lighting requirements during the darker months, due to low natural light conditions and short daylengths. While many economic (e.g., capital cost of fixtures and electricity prices) and practical (e.g., fixture positioning and capacity of electrical supply infrastructure) elements are considered when outfitting a greenhouse with supplemental lighting systems, the response of the crop(s) to additional lighting is a key factor which can only be evaluated through careful production trials.

The photosynthetic responses of plants to increasing levels of photosynthetically active radiation (PAR), generally described in terms of photosynthetic photon flux density (PPFD, μmol m^{-2} s^{-1}), have been well established for many plant species and environments. When considering supplemental lighting in greenhouse production scenarios, crops are generally subjected to light intensities (LI) that are on the linear portion of the photosynthetic light response curve (i.e., far lower than LIs needed to saturate the photosynthetic machinery). By extension, the yield responses of many greenhouse crops are commonly generalized as being directly proportional to the levels of light provided, with every 1% increase in lighting resulting in concomitant 0.5% to 1% increases in production [1]. This relationship has borne out for some economically relevant production indices in various floriculture commodities, such as cut gerbera [2,3], potted begonias [4], and cut roses [5,6].

For optimization of commercial greenhouse production scenarios which utilize supplemental lighting, it is necessary to determine the impact supplemental daily light integral (DLI), supplied within practical constraints of intensity and photoperiod, has on economically relevant production indices for target commodities. This may be especially relevant for cut flower production, such as cut gerbera, where the harvestable product represents a relatively minor component of total plant biomass production. Therefore, yield may be less directly related to rates of photosynthesis or carbon assimilation, while other crop x lighting interactions, such as photomorphogenic effects and flower bud development, may become more relevant with increasing levels of supplemental DLI [7]. We are unaware of any other references which have directly investigated the effect of DLI on vegetative growth, days between transplant and first visible flower bud, rate of flower development from visible bud to harvest, or fresh harvest metrics of cut gerbera flowers.

Auito [2] investigated cut gerbera production under a range of supplemental PPFD and photoperiods provided by high pressure sodium (HPS) lights. Their results showed that 12-h photoperiod maximized flower production for a given supplemental DLI. Conversely, Pettersen and Gislerød [8] found that a 20-h photoperiod had a higher production of cut gerbera than a 10-h photoperiod. However, the trials were done in a growth chamber, with a fixed PPFD, thus confounding the effects of photoperiod and DLI (i.e., the 20-h photoperiod had twice the DLI as the 10-h photoperiod), making it difficult to extrapolate their results to greenhouse environments. In a parallel study, Auito [2] also found linear or near linear relationships between supplemental light intensity and cut gerbera production in a greenhouse, using PPFD levels ranging from 75 to 300 μmol m^{-2} s^{-1} with a 12-h photoperiod (i.e., DLIs of 3.2 to 13.0 mol m^{-2} d^{-1}). However, natural lighting was only reported as seasonal mean values for outdoor DLI throughout the 6-month trial period ($\approx$ 7.4 mol m^{-2} d^{-1}, November to April) with an (estimated) greenhouse transmission value of 50%. Therefore, it is not possible to draw conclusions based on the absolute light levels (i.e., natural + supplemental) within the treatment plots. Approximate values for natural DLI at crop level in this study would probably have averaged between 3 and 4 mol m^{-2} d^{-1} (based on 50% of 7.4 mol m^{-2} d^{-1}), which is similar to the winter lighting conditions in the research greenhouse facility used in the present study [9]. Spanomitsios et al. [3] found a positive linear relationship between mean daily solar radiation and rate of cut gerbera flower production. In this study, the slope of the relationship between light and production (slope = 0.47) indicated that $\approx$ 0.5% increase in flower production could be expected for every 1% increase in total light. However, it takes approximately four weeks for a cut gerbera flower to mature from visible bud to harvestable stage. Therefore, the reported relationship between daily net radiation and flower yield would have been more realistically portrayed if harvest data had been related to the average DLI for the four weeks prior to each harvest. Further, it is not possible to infer PPFD or DLI at crop level in this study as it is not clear how or where light data were collected or how the data were processed. Mustapić-Karlić et al. [10] found a positive influence of supplemental lighting on flower yield of two cut gerbera cultivars. They compared treatments of natural lighting with natural + supplemental HPS lighting providing $\approx$ 3 mol m^{-2} d^{-1} of additional PAR (i.e., PPFD of $\approx$ 70 μmol m^{-2} s^{-1} with a 12-h photoperiod). However, the DLI at crop level is unknown because the natural light levels at crop level were not reported. Similarly, Gagnon and Dansereau [11] found increases in potted gerbera productivity and reductions in time to flowering with increasing levels of supplemental HPS lighting (ranging from 1.7 to 5.2 mol m^{-2} d^{-1}). However, the authors also did not report natural light levels, making it impossible to draw conclusions about the absolute influence of the lighting treatments on production. While these trials clearly indicate positive relationships between increasing levels of supplemental lighting and production of cut gerbera, insufficient information on canopy-level lighting conditions make it difficult for readers to critically evaluate the total amount of PAR received by the plants in these trials [7].

With respect to the quality of supplemental light, research has shown that at similar PPFD, supplemental PAR from light-emitting diode (LED) technologies have resulted in similar crop production metrics as traditional HPS in greenhouse commodities, such as leafy vegetables [12], fruiting vegetables [13–15], ornamentals [16–18], and cut flowers [19]. While the capital costs of LED technologies are still considerably higher than HPS, LEDs have many advantages over HPS. LEDs can provide narrow wavebands of light specifically targeted at the maximum absorption bands of photosynthetic machinery. LEDs are touted to have greater than twice the lifetime as HPS and also have the potential to achieve higher efficacies (i.e., conversion of electricity into PAR). Moreover, LEDs are naturally dimmable, providing the capacity to adjust intensity according to natural lighting conditions, as well as on-demand customization of spectral recipes, providing greater plasticity for photoperiod and photomorphological control within a single fixture [20,21]. Accordingly, leading researchers and industry professionals consider it only a matter of time before LED technologies replace HPS as the benchmark technology for supplemental lighting in greenhouse applications [22].

The objectives of this study were to evaluate the relationships between increasing levels of supplemental lighting from LEDs during the darker months in Canada on the growth, flower development, yield, and quality of greenhouse grown cut gerbera.

2. Materials and Methods

2.1. Location, Trial Bench, and Greenhouse

The study took place at the University of Guelph in Guelph, ON, Canada, (43.55 ° N, 80.25 ° W) beginning on 9 November 2015 and ending on 25 February 2016 (i.e., 107 d). The study was set up within a single 7.2 × 7.2-m glass-clad research greenhouse compartment, containing four 4.57 × 1.07 m benches, with 0.91 m spacing between them. The long sides of the benches were positioned in an east-west direction (i.e., parallel with the track of the sun).

2.2. Lighting Treatments and Plant Distribution

There were five PPFD treatments, two pots of plants (i.e., two subsamples) for each of two cultivars under each PPFD treatment on each bench, as well as four replicates (i.e., benches) within the greenhouse compartment.

There were four LED fixtures (Pro 325; LumiGrow, Novato, CA, USA) per bench, located 30 and 100 cm (measured on-center of each fixture's LED array) from both ends of each bench. The lights were centered along the long axis of the bench and fixed with the LED arrays 140 cm above pot level. Each fixture was affixed with shrouds arranged parallel with the long sides of the benches made of white vinyl siding (Cedar Creek D4D; Abtco, Milton, ON, Canada) to reduce stray lighting from adjacent benches. The fixtures were set with an area-averaged photon flux ratio of blue (B, 400 to 500 nm) to red (R, 600 to 700 nm) of B22:R78. Fixture positioning and mapping light distribution patterns were done at night using a radiometrically-calibrated spectrometer (USB2000+; Ocean Optics, Dunedin, FL, USA) coupled to a 400-μm diameter UV-VIS optical fiber with a CC-3 cosine corrector (Ocean Optics, Dunedin, FL, USA). Light distribution (intensity and quality) was measured at pot level on a 2 × 12 rectangular grid (i.e., 24 specific locations), centered on the geometric center of the bench, with 30 cm separating adjacent measurement locations. For the trial, individual cut gerbera pots were centered on each of these bench locations and remained there for the duration of the trial. In this configuration, the supplemental light treatment at pot level of each plant was kept at a constant, known value. This design resulted in five unique supplemental PPFD treatment levels on each bench (labeled T1 to T5).

Two cut gerbera (*Gerbera jamesonii* H. Bolus ex Hook.f) cultivars, 'Panama' and 'Ultima', were used for this trial. 'Panama' plants were sourced from an active production environment ($\approx$ 5 months of active flower production) from a local grower (Bayview Flowers Ltd., Lincoln, ON, Canada). Flower stems longer than 2.5 cm were removed from 'Panama' plants at the beginning of the study. 'Ultima' plants came from the supplier, Florist Holland B.V. (De Kwakel, The Netherlands), as 'Jiffy 4' plugs.

On 8 October 2015, the plugs were transplanted into round 19 cm diameter × 19 cm tall pots; filled with coarse coir mix typically used by and obtained from a local cut gerbera grower. 'Ultima' plants began the trial in the vegetative stage, with no visible flower buds. Equal numbers of plants from each cultivar were positioned on the benches such that the cultivars were arranged in an alternating fashion. This arrangement resulted in two plants of each cultivar per treatment per bench, plus two border plants on the ends of each bench. The planting density was $\approx$ 7 plants m^{-2}, which was consistent with local commercial cut gerbera greenhouses. Although the location of each plant was fixed, the plants were rotated one-quarter turn weekly to reduce pot-location effects.

2.3. Environmental Management

The greenhouse environment parameters were set at similar levels to those used by local cut gerbera producers. Supplemental LED lighting was turned on daily 12 h before dusk and turned off at dusk, resulting in a constant 12-h photoperiod. Day and night temperature setpoints were 21 and 14 °C, respectively. Relative humidity (RH) was maintained at 70% using an aerial fogger system located at gutter level. Temperature and humidity dataloggers (HOBO U12-013; Onset Computer Corporation, Bourne, MA, USA) were located at canopy level in the center of each bench. PAR sensors (SQ-110; Apogee Instruments Inc., Logan, UT, USA) were located 1.75 m above the center of each bench (i.e., just above the top of the LED fixtures) and connected to the HOBO dataloggers. Temperature, RH, and PPFD were logged every 120 s throughout the study. Previous light uniformity data, collected by simultaneously logging the natural PPFD at fixture-level and bench-level (supplemental light fixtures present but left off) during a prior supplemental lighting season (i.e., November to March), indicated strong correlations in DLI measured between bench- and fixture-level locations on each bench. Coefficients relating natural DLI at fixture-level to bench-level derived from these data (not shown), were applied to the fixture-level PPFD data collected during the present trial to determine natural DLIs at canopy level on each bench.

2.4. Irrigation Management

Plants were drip irrigated using 20N-3.5P-16.6K All Purpose water soluble fertilizer (250 ppm N, pH 5.5; Plant Products Co. Ltd., Brampton, ON, Canada) with temporary substitutions of well water (pH and EC of 7.9 and 1000 µS cm^{-1}, respectively), when necessary, to maintain an approximate root zone pH of 5.5 and EC of 2500 µS cm^{-1}. Pulse irrigation occurred every second day, at 0915 and 1315 HR for 180 s each time. This irrigation protocol was aimed at producing approximately 10% to 25% leachate. Hand-watering was used as needed to supplement the drip irrigation.

2.5. Plant Growth, Leaf Chlorophyll Content Index, Flower Quality, and Yield Metrics

The number of leaves and chlorophyll content index (CCI) were measured approximately monthly on each 'Ultima' plant using a chlorophyll meter (CCM-200 Plus; Opti-Sciences, Hudson, NH, USA). CCI measurements were taken (three measurements per leaf with the average CCI value recorded), near the leaf margin (i.e., avoiding larger venation) of the youngest fully-expanded leaf of each plant. 'Ultima' plants were also checked twice weekly for the development of flower buds. Once each stem was ≥ 1 cm long, it was tagged with a unique identifier and the respective date was recorded as the date of appearance. This provided the days from transplant to first visible flower bud (i.e., stems ≥ 1 cm), as well as insight into the rate of flower development (i.e., the time between visible flower bud appearance and harvest).

Flowers on 'Panama' plants were harvested twice weekly. Flowers were deemed harvestable once they developed one complete ring of matured anthers. Fresh mass, flower diameter (measured petal tip to petal tip on the widest part of the flower), and stem length (measured from heel to the base of the flower) were measured on each harvested flower. Flower quality was also classified subjectively as either marketable or unmarketable according to the severity of malformations and pest damage.

2.6. Statistical Analysis

The experiment was a block design with 5 treatments and 4 concurrent replications. All data sets were analyzed using JMP® (version 13; SAS Institute Inc., Cary, NC, USA, 1989–2017). Least squares analysis was used for light treatment uniformity; vegetative growth, rate of appearance of visible flower buds, and flower development metrics in 'Ultima'; and accumulated total and marketable flowers harvested per plant in 'Panama'. Flower yield metrics in 'Panama' were analyzed using the Mixed-Models add-in, which accounts for the different numbers of flower stems harvested from each plant. Data were evaluated using a significance level of $p \leq 0.05$ using Tukey's honestly significant difference (HSD) test. Days between the appearance of flower buds and harvest on 'Ultima' and accumulated total and marketable flowers harvested per plant on 'Panama' underwent regression analysis ($p \leq 0.05$), using total DLI (i.e., natural + supplemental) as the independent variable.

3. Results

Weekly average canopy-level natural DLI for the 17-week trial ranged from $\approx$ 1 to 6 mol m^{-2} d^{-1} with an overall average of 3.6 mol m^{-2} d^{-1} (Figure 1), which was consistent with previous years' light characterizations within the same experimental greenhouse (data not shown). Daytime (i.e., daily timeframe when supplemental lighting was on) and nighttime (i.e., daily timeframe when supplemental lighting was off) temperatures were (mean $\pm$ SD) 20.4 $\pm$ 2.0 °C and 16.6 $\pm$ 1.24 °C, respectively.

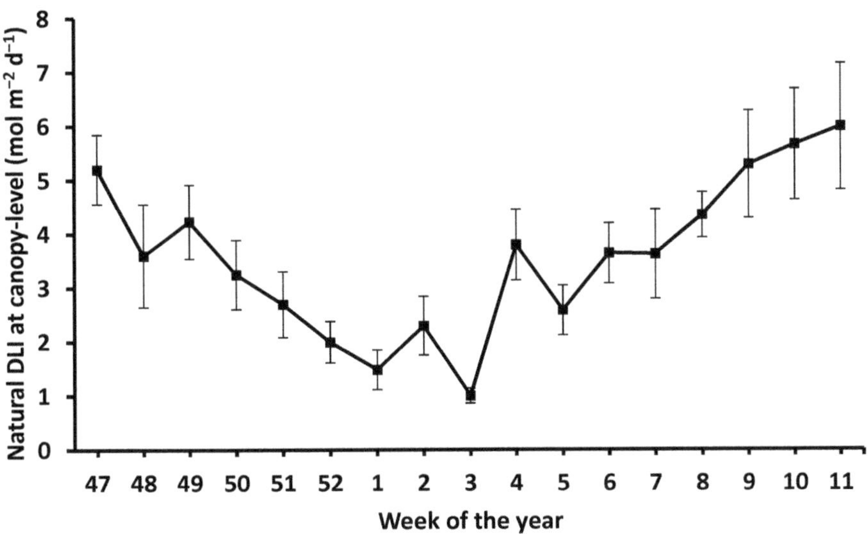

Figure 1. Weekly natural daily light integral (DLI) at canopy level (average $\pm$ SE, n = 7). The overall average natural DLI, during the 17-week trial, was 3.6 mol m^{-2} d^{-1}.

The supplemental PPFD treatments ranged from 40.7 to 179 μmol m^{-2} s^{-1}, corresponding to 1.8 to 7.7 mol m^{-2} d^{-1} of daily supplemental PAR with a 12-h photoperiod (Table 1).

Table 1. Canopy-level supplemental photosynthetic photon flux density (PPFD) of the five supplemental light-emitting diode (LED) treatments and their associated supplemental and total daily light integrals (DLI).

Treatment	PPFD (μmol m^{-2} s^{-1})			DLI (mol m^{-2} d^{-1})	
	Mean	Max	Min	LED [y]	Total [x]
T1	40.7 ± 1.3 a [z]	45.2	33.2	1.8	5.3
T2	76.1 ± 1.6 b	87.4	64.1	3.3	6.9
T3	133 ± 2.4 c	151	114	5.7	9.3
T4	167 ± 1.9 d	181	153	7.2	10.8
T5	179 ± 1.8 e	192	162	7.7	11.3

[z] There were no block or bench position effects on supplemental PPFD within each treatment, so data are pooled means for each treatment ± SE (n = 16). Values in the same column followed by the same letter are not different at $p < 0.05$, using Tukey's honestly significant difference (HSD). [y] DLI from supplemental LEDs were calculated using mean PPFD from each treatment and 12-h photoperiod. [x] Total DLI is the sum of supplemental DLI from LED treatments and experiment-wise mean natural DLI of 3.6 mol m^{-2} d^{-1}.

'Ultima' plants chosen for each treatment had uniform CCI and number of leaves at the start of the trial (9 November 2015). After one month of treatment (8 December 2015), plants in T5 had ≈ 4 more leaves than plants in T4. After two months of growth under the supplemental light treatments (6 January 2016), plants in T4 had higher CCI values than T1, T2, and T3, and plants in T5 had ≈ 6 more leaves than plants in T2 (Table 2).

Table 2. Chlorophyll content index (CCI) of the youngest fully-expanded leaf, and number of leaves per plant, measured at ≈ 4-week intervals post-transplant of 'Ultima' plants.

Date	Treatment [total DLI (mol m^{-2} d^{-1})]	CCI	No. of Leaves
	T1 (5.3)	43 ± 1.0 a [z]	9.6 ± 0.96 a
	T2 (6.9)	38 ± 1.6 a	9.5 ± 0.62 a
9 November 2015	T3 (9.3)	40 ± 2.2 a	9.3 ± 0.65 a
	T4 (10.8)	38 ± 1.8 a	7.8 ± 0.67 a
	T5 (11.3)	40 ± 1.9 a	8.1 ± 0.83 a
	T1 (5.3)	47 ± 1.5 a	11.8 ± 1.8 ab
	T2 (6.9)	47 ± 1.1 a	11.6 ± 1.2 ab
8 December 2015	T3 (9.3)	49 ± 1.5 a	12.9 ± 1.2 ab
	T4 (10.8)	48 ± 1.6 a	10.8 ± 1.2 a
	T5 (11.3)	49 ± 1.6 a	14.8 ± 1.3 b
	T1 (5.3)	47 ± 1.3 a	17.5 ± 2.2 ab
	T2 (6.9)	49 ± 1.0 a	16.9 ± 1.9 a
6 January 2016	T3 (9.3)	49 ± 1.2 a	21.5 ± 2.7 ab
	T4 (10.8)	50 ± 1.0 b	21.3 ± 1.8 ab
	T5 (11.3)	49 ± 1.0 ab	23.1 ± 2.0 b

[z] There were no block effects within each treatment at each measurement date, so data are pooled averages for each treatment ± SE (n = 8). Values in the same column with the same measuring day followed by the same letter are not different at $p < 0.05$, using Tukey's HSD.

Flowers in T5 matured (i.e., time between appearance of flower buds and harvest) ≈ 3.6 d faster than plants in T1, which represents ≈ 10% reduction in flower development time (Table 3).

There were only minor treatment effects in fresh flower harvest metrics on 'Panama' flowers (Table 4). Flowers grown in T5 had marginally shorter stems than flowers grown in T1, T2, and T3. Flowers grown in T3 were marginally larger and flowers grown in T1 were smaller than the other treatments (with < 0.2 cm difference in diameter). Flowers grown in T3 also had higher fresh mass than flowers grown in T1 and T2.

Table 3. Days between appearance of flower buds (i.e., stems ≥ 1 cm) and harvest for all 'Ultima' flowers harvested during the trial, for different total daily light integral (DLI) treatments.

Treatment (total DLI (mol m^{-2} d^{-1})	No. of Days Between Visual Appearance of Flower Bud and Harvest
T1 (5.3)	37.6 ± 0.90 a [z]
T2 (6.9)	35.8 ± 0.99 ab
T3 (9.3)	35.0 ± 0.86 ab
T4 (10.8)	34.5 ± 0.83 ab
T5 (11.3)	34.0 ± 0.71 b

[z] There were no block effects, so data are pooled averages for each treatment ± SE (n = 8). Values in the same column followed by the same letter are not different at $p < 0.05$, using Tukey's HSD.

Table 4. Stem length, flower diameter, and fresh mass of 'Panama' flowers harvested throughout the trial, for different total daily light integral (DLI) treatments.

Treatment (total DLI (mol m^{-2} d^{-1}))	Stem Length (cm)	Flower Diameter (cm)	Fresh Mass (g)
T1 (5.3)	46.5 ± 1.10 a [z]	9.9 ± 0.07 a	19.3 ± 0.57 a
T2 (6.9)	46.6 ± 1.05 a	10.1 ± 0.06 b	20.7 ± 0.52 ab
T3 (9.3)	46.7 ± 1.05 a	10.3 ± 0.06 c	22.9 ± 0.51 c
T4 (10.8)	45.9 ± 1.05 ab	10.1 ± 0.06 b	21.2 ± 0.51 bc
T5 (11.3)	44.4 ± 1.04 b	10.1 ± 0.06 b	21.1 ± 0.52 bc

[z] There were no block effects, so data are pooled means for each treatment ± SE (n = 8). Values in the same column followed by the same letter are not different at $p < 0.05$, using Tukey's HSD.

Regressing 'Panama' flower harvest numbers against total DLI indicated that every 1% increase in DLI increased cumulative flower yield by ≈ 1.5% (Figure 2). The trend was similar in terms of marketable flowers, where a 1% increase in DLI resulted in a concomitant ≈ 1% increase in the number of marketable flowers produced per plant.

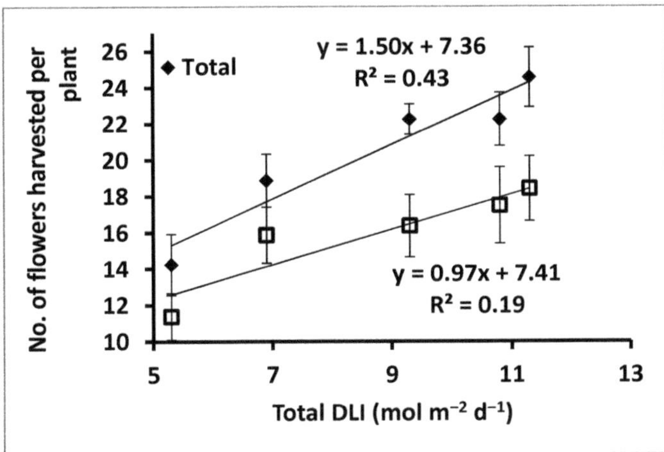

Figure 2. Cumulative total and marketable flowers harvested per plant, for 'Panama', in response to total daily light integral (DLI). Each point represents the treatment mean ± SE (n = 8); however, the equations are linear regressions of all of the harvest data on a per-plant basis.

4. Discussion

The range of supplemental PPFD levels used in this study raised the total canopy-level DLI to levels that approximately match the DLI range deemed necessary to produce minimum acceptable quality (6 mol m^{-2} d^{-1}) to high quality (12 mol m^{-2} d^{-1}) gerbera [23]. Vegetative growth and flower development indices were investigated using transplanted plugs of the 'Ultima' cultivar, while mature plants of the 'Panama' cultivar were used to assess the size, quality, and numbers of flowers produced.

There were no commercially-relevant LI treatment differences (or trends) in number of leaves or CCI of 'Ultima' plants. While there were also no LI treatment effects on the days from transplant to first visible flower (data not shown), flowers in T5 matured $\approx$ 10% faster than flowers in T1. Linear regression of the treatment means for days between appearance of first visible flower bud and harvest in 'Ultima' (in Table 3) against DLI indicates that each additional mol m^{-2} d^{-1} of DLI (e.g., $\approx$23 μmol m^{-2} s^{-1} of supplemental PAR over a 12-h photoperiod) shortened the time between flower bud appearance and harvest by 0.53 d. For example, adding $\approx$ 90 μmol m^{-2} s^{-1} of supplemental PAR with a 12-h photoperiod could shorten the flower production time by 2 d, during the darker months.

There were only minor (i.e., probably not commercially relevant) LI treatment effects on stem length, flower diameter, and fresh mass of marketable 'Panama' flowers. However, there were LI treatment effects on the total and marketable numbers of 'Panama' flowers harvested per plant, with plants in T5 producing $\approx$ 40% more flowers than plants in T1. Subjecting the cumulative flower production metrics to linear regression analysis showed that DLI could be used to predict the cumulative flowers produced per plant (Figure 2). Similarly, Bredmose [5,6] found linear relationships between supplemental light (HPS) intensity and numbers of flowers produced by mature plants of two rose cultivars, within the range of 0 to 174 μmol m^{-2} s^{-1}. Auito's [2] investigation on the effects of supplemental light intensity and photoperiod on cut gerbera production is the most comprehensive to date. However, insufficient information was provided about the natural lighting environment under which the crops were grown; making it difficult to assess the actual lighting conditions (e.g., total DLI) in these trials. Despite this drawback, the author concluded that cut gerbera plants utilize supplemental light for flower production most efficiently at shorter photoperiods (i.e., 12 h), which is in line with local production practices. Auito [2] noted some cultivar-specific responses to increased supplemental PAR, although total flowers per plant and total dry mass generally increased linearly with increasing supplemental DLI (between 3.2 and 13.0 mol m^{-2} d^{-1}, with a 12-h photoperiod)

In the present study, it was shown that doubling the total DLI from 6 to 12 mol m^{-2} d^{-1} by providing an additional 6 mol m^{-2} d^{-1} of supplemental PAR from LEDs could increase the number of flowers produced by nine flowers per plant (over 107 d). At typical commercial plant densities of 7 m^{-2}, this would result in monthly increases in flower production of $\approx$ 18 more flowers/m^2. In practical terms, if a grower provided 100 μmol m^{-2} s^{-1} of supplemental PAR, with a 12-h photoperiod, they could potentially increase the total number of flowers produced per plant during the darker months by $\approx$ 30%. To further contextualize in terms of energy cost, the efficacy factor of 1.29 μmol J^{-1} for the LumiGrow Pro 325 fixtures used in this study [24] can be used to estimate that $\approx$ 1.3 kWh m^{-2} d^{-1} would be needed to add 6 mol m^{-2} d^{-1} of supplemental PAR from LEDs, which would be $\approx$ 2 kWh per additional flower produced, in the above scenario. However, the efficacy of some horticultural LED fixtures has more than doubled versus the fixtures used in this study [25], which would reduce the energy input per flower to less than 1 kWh for modern LED fixtures.

Future research should include broadening the range of commodities investigated under supplemental LED lighting intensity regimens, as well as investigating applications of targeted spectrum treatments (especially at night, where applicable) for manipulating crop morphology. A promising example of spectrum-mediated change in morphology are the increases in stem extension rates without some of the negative "shade avoidance" effects of high far red (700–800 nm) treatments by using low fluence rates of monochromatic blue light, applied at nighttime [26].

5. Conclusions

This investigation examined the influence of different levels of supplemental PAR, supplied by red and blue LEDs, on the production of cut gerbera during the darker months at higher latitudes. While there were few commercially-relevant LI treatment effects in the vegetative growth and harvested flower quality indices, higher light was shown to proportionally increase the rate of flower development and cumulative numbers of flowers produced. These relationships can be used by growers to assess the economic viability of using supplemental LED lighting to produce cut gerbera within their own production environments.

Author Contributions: Conceptualization, D.L. and Y.Z.; methodology, D.L., K.S. and Y.Z.; validation, D.L. and Y.Z.; formal analysis, D.L.; investigation, K.S. and D.L.; writing—original draft, D.L. and Y.Z.; writing—review & editing, D.L. and Y.Z.; visualization, D.L.; supervision, Y.Z.; funding acquisition, Y.Z. All authors have read and agreed to the published version of the manuscript.

Funding: This research was funded by the International Cut Flower Growers Association and the Joseph H. Hill Memorial Foundation, Inc.

Acknowledgments: LumiGrow, Inc. supplied the LED lighting fixtures. Plant materials were donated by Van Geest Bros Ltd (Ontario, Canada) and Bayview Flowers Ltd. (Ontario, Canada). None of these contributors were involved in the conduction of the study or submission of this article for publication.

Conflicts of Interest: The authors declare no conflict of interest. The funders had no role in the design of the study; in the collection, analyses, or interpretation of data; in the writing of the manuscript, or in the decision to publish the results.

References

1. Marcelis, L.F.M.; Broekhuijsen, A.G.M.; Meinen, E.; Nijs, E.M.F.M.; Raaphorst, M.G.M. Quantification of the growth response to light quantity of greenhouse grown crops. *Acta Hortic.* **2006**, *711*, 97–103. [CrossRef]
2. Auito, J. Supplementary lighting regimes strongly affect the quantity of gerbera flower yield. *Acta Hortic.* **2000**, *515*, 91–98.
3. Spanomitsios, G.K.; Maloupa, E.M.; Grafiadellis, M.I. The effect of various environmental factors on yield of greenhouse gerbera plants. *Acta Hortic.* **1995**, *408*, 119–127. [CrossRef]
4. Nemali, K.S.; van Iersel, M.W. Light effects on wax begonia: Photosynthesis, growth respiration, maintenance respiration, and carbon use efficiency. *J. Am. Soc. Hortic. Sci.* **2004**, *129*, 416–424. [CrossRef]
5. Bredmose, N. Effects of year-round supplementary lighting on shoot development, flowering and quality of two glasshouse rose cultivars. *Sci. Hortic.* **1993**, *54*, 69–85. [CrossRef]
6. Bredmose, N. Year-round supplementary lighting at twelve photosynthetic photon flux densities for cut roses. *Acta Hortic.* **1997**, *418*, 59–64. [CrossRef]
7. Moe, R. Physiological aspects of supplemental lighting in horticulture. *Acta Hortic.* **1997**, *418*, 17–24. [CrossRef]
8. Pettersen, R.I.; Gislerød, H.R. Effects of lighting period and temperature on growth, yield and keeping quality of *Gerbera jamesonii* Bolus. *Eur. J. Hortic. Sci.* **2003**, *68*, 32–37.
9. Kong, Y.; Llewellyn, D.; Zheng, Y. Response of growth, yield and quality of pea shoots to supplemental LED lighting during winter greenhouse production. *Can. J. Plant Sci.* **2018**, *98*, 732–740. [CrossRef]
10. Mustapić-Karlić, J.; Teklić, T.; Paraðiković, N.; Vinković, T.; Lisjak, M.; Špoljarević, M. The effects of light regime and substrate on flower productivity and mineral composition in two gerbera cultivars. *J. Plant Nutr.* **2012**, *35*, 1671–1682. [CrossRef]
11. Gagnon, S.; Dansereau, B. Influence of light and photoperiod on growth and development of gerbera. *Acta Hortic.* **1990**, *272*, 145–151. [CrossRef]
12. Martineau, V.; Lefsrud, M.G.; Tahera-Naznin, M.; Kopsell, D.A. Comparison of supplemental greenhouse lighting from light emitting diode and high pressure sodium light treatments for hydroponic growth of Boston lettuce. *HortScience* **2012**, *47*, 477–482. [CrossRef]
13. Dueck, T.A.; Janse, J.; Eveleens, B.A.; Kempkes, F.L.K.; Marcelis, L.F.M. Growth of tomatoes under hybrid LED and HPS lighting. *Acta Hortic.* **2012**, *952*, 335–342. [CrossRef]

14. Gomez, C.; Morrow, R.C.; Bourget, C.M.; Massa, G.D.; Mitchell, C.A. Comparison of intracanopy light-emitting diode towers and overhead high-pressure sodium lamps for SL of greenhouse-grown tomatoes. *HortTechnology* **2013**, *23*, 93–98. [CrossRef]

15. Hernandez, R.; Kubota, C. Physiological, morphological, and energy-use efficiency comparisons of LED and HPS SL for cucumber transplant production. *HortScience* **2015**, *50*, 351–357. [CrossRef]

16. Currey, C.J.; Lopez, R.G. Cuttings of Impatiens, Pelargonium, and Petunia propagated under light-emitting diodes and high-pressure sodium lamps have comparable growth, morphology, gas exchange, and post-transplant performance. *HortScience* **2013**, *48*, 428–434. [CrossRef]

17. Poel, B.R.; Runkle, E.S. Seedling growth is similar under supplemental greenhouse lighting from high-pressure sodium lamps or light-emitting diodes. *HortScience* **2017**, *52*, 388–394. [CrossRef]

18. Randall, W.; Lopez, R.G. Comparison of SL from high-pressure sodium lamps and light-emitting diodes during bedding plant seedling production. *HortScience* **2014**, *49*, 589–595. [CrossRef]

19. Llewellyn, D.; Schiestel, K.; Zheng, Y. Light-emitting diodes can replace high-pressure sodium lighting for cut gerbera production. *HortScience* **2019**, *54*, 95–99. [CrossRef]

20. Bourget, C.M. An introduction to light-emitting diodes. *HortScience* **2008**, *43*, 1944–1946. [CrossRef]

21. Zheng, Y. Are LEDs the right choice for my operation? *Greenh. Can. Mag.* **2016**, *2*, 14–18.

22. Both, A.J. HPS or LED? Canadian Greenhouse Conference, Niagara Falls, Ontario, Canada, 2 November 2017. Available online: https://www.dropbox.com/sh/hyk410a3y3xvas1/AADIZ6GqTHycjx39Okx1VQTHa?dl=0&preview=2017+Both_LED+or+HPS.pdf (accessed on 31 August 2002).

23. Faust, J.E. Light. In *Ball Redbook: Crop Production*, 17th ed.; Hamrick, D., Ed.; Ball Publishing: Batavia, IL, USA, 2003; Volume 2, pp. 71–84.

24. Nelson, J.A.; Bugbee, B. Economic analysis of greenhouse lighting: Light emitting diodes vs. high intensity discharge fixtures. *PLoS ONE* **2014**, *9*, e99010. [CrossRef] [PubMed]

25. Design Lights Consortium, Qualified Product List. Available online: https://www.designlights.org/horticultural-lighting/search/ (accessed on 31 August 2020).

26. Ying, Q.; Kong, Y.; Zheng, Y. Overnight supplemental blue, rather than far-red, light improves microgreen yield and appearance quality without compromising nutritional quality during winter greenhouse production. *HortScience* **2020**, 1–7. [CrossRef]

Article

Interactive Effects of the CO_2 Enrichment and Nitrogen Supply on the Biomass Accumulation, Gas Exchange Properties, and Mineral Elements Concentrations in Cucumber Plants at Different Growth Stages

Xun Li [1], Jinlong Dong [1], Nazim S. Gruda [2], Wenying Chu [1,3,*] and Zengqiang Duan [1]

[1] State Key Laboratory of Soil and Sustainable Agriculture, Institute of Soil Science, Chinese Academy of Sciences, Nanjing 210008, China; xli@issas.ac.cn (X.L.); jldong@issas.ac.cn (J.D.); zqduan@issas.ac.cn (Z.D.)

[2] Department of Horticultural Sciences, Institute of Crop Science and Resource Conservation, University of Bonn, 53121 Bonn, Germany; ngruda@uni-bonn.de

[3] Department of Chemistry and Biochemistry, Old Dominion University, Norfolk, VA 23529, USA

* Correspondence: wchu@odu.edu; Tel.:+1-505-362-9619

Received: 24 December 2019; Accepted: 15 January 2020; Published: 17 January 2020

Abstract: The concentration changes of mineral elements in plants at different CO_2 concentrations ([CO_2]) and nitrogen (N) supplies and the mechanisms which control such changes are not clear. Hydroponic trials on cucumber plants with three [CO_2] (400, 625, and 1200 µmol mol^{-1}) and five N supply levels (2, 4, 7, 14, and 21 mmol L^{-1}) were conducted. When plants were in high N supply, the increase in total biomass by elevated [CO_2] was 51.7% and 70.1% at the seedling and initial fruiting stages, respectively. An increase in net photosynthetic rate (Pn) by more than 60%, a decrease in stomatal conductance (Gs) by 21.2–27.7%, and a decrease in transpiration rate (Tr) by 22.9–31.9% under elevated [CO_2] were also observed. High N supplies could further improve the Pn and offset the decrease of Gs and Tr by elevated [CO_2]. According to the mineral concentrations and the correlation results, we concluded the main factors affecting these changes. The dilution effect was the main factor driving the reduction of all mineral elements, whereas Tr also had a great impact on the decrease of [N], [K], [Ca], and [Mg] except [P]. In addition, the demand changes of N, Ca, and Mg influenced the corresponding element concentrations in cucumber plants.

Keywords: dry weight; N levels; elevated CO_2; open-top chamber; nutrient transportation; photosynthesis; transpiration; dilution effect

1. Introduction

In the cold of winter, most greenhouses in China are not opened or ventilated in order to keep a warmer air temperature for vegetable growth. Therefore, carbon dioxide (CO_2) is often depleted rapidly in these closed greenhouses, and this lack becomes one of the biggest adverse factors that depress the photosynthesis and growth of vegetables [1]. CO_2 enrichment has been widely implemented in greenhouses in Europe, North America, and Japan since the 1950s, and was introduced and used in greenhouses in China in the late 1980s [2–4]. CO_2 enrichment has been found to have a dramatic effect on faster growth, greater biomass, and higher yield [5,6] due to the increased photosynthesis and carbohydrate accumulation, particularly in C3 plants [7,8]. Nevertheless, the longer and further researches reported some drawbacks of CO_2 enrichment. These drawbacks included the decline of mineral concentrations in plant tissues [7,8], worsened taste due to the increased cellulose content [9,10], as well as the photosynthetic acclimation and the weak sustainability of its fertilization effects on yield improvement [11–13]. The main reason was the increased photosynthates and accumulated carbohydrates in plant tissues under elevated CO_2 concentrations ([CO_2]), which decreased the nitrogen to carbon ratio (N/C) and caused the imbalance between source and sink [7,13,14]. To deal with this problem, many researchers recommended higher N fertilization to minimize the reduction in N/C associated with high [CO_2] conditions [15–17]. However, when the dry matter accumulation outpaces N uptake, enriched CO_2 will still reduce N concentrations ([N]) in plants even if N uptake is enhanced by higher N supply [18].

A reduction in mineral concentrations has been frequently demonstrated in crops grown under elevated [CO_2] conditions [7–9,18]. Among these mineral elements, the possible mechanisms of the reduction of [N] in plants under high [CO_2] conditions have been extensively studied [9,10,19,20]. Four hypotheses are well-documented: (1) Dilution effect due to accumulation of non-structural carbohydrates [10,12]; (2) reduced mass flow and transpiration due to decreased stomatal conductance [21,22]; (3) decreased Rubisco protein concentrations and N demands due to increased plant N use efficiency [10,23,24]; (4) inhibited photorespiration-dependent nitrate assimilation under high [CO_2] [25,26]. However, the effects of elevated [CO_2] on other mineral elements in plants have received far less attention. In plants, N, phosphorus (P), and sulfur (S) are mainly bounded to C to form organic molecules, whereas potassium (K), calcium (Ca), and magnesium (Mg) tend to remain in ionic forms or be chelated with enzymes in plant tissues. Considering their different uptake pathways, existence forms, and physiological functions, concentration changes of different mineral elements in plants at different [CO_2] and N supply levels are possible with different mechanisms potentially contributing to each.

Cucumber (*Cucumis sativus* L.) is globally one of the most important vegetables that prefers to be cultivated in greenhouses with CO_2 enrichment [27]. While the growth, photosynthesis [28], nitrogen metabolism [29], yield [1], fruit quality [30], root morphology [31], root exudate [32], and water use efficiency [33] of cucumber grown in elevated [CO_2] conditions have been studied, the effects of different [CO_2] and N supply levels on mineral element concentrations in cucumber and the key factor leading to these changes have received far less attention. Moreover, the available information about the optimum N supply under different elevated [CO_2] is extremely limited. A better understanding of the concentration changes of mineral elements in cucumber plant responding to elevated [CO_2] and N supplies is necessary for optimizing [CO_2] and N fertilization in order to obtain high quality greenhouse products with higher C and N use efficiency [34,35] and to deal with future climate change scenarios with less CO_2 emission and fertilizer input [36,37].

In previous studies, we have found the optimum N supply was 7 and 14 mmol L^{-1} for the seeding and mature plants of cucumber, respectively, and the saturated and semi-saturated [CO_2] of cucumber plants under natural solar radiation (about 600 mol m^{-2} s^{-1}) was 1200 and 625 µmol mol^{-1}, respectively [28,30–32]. Therefore, in this study, we carried out the hydroponic trials on cucumber plants with three [CO_2] (400, 625, and 1200 µmol mol^{-1}) combined with five N supply levels (2, 4, 7, 14, and 21 mmol L^{-1}) during seedling and initial flowering stages. Then, we investigated the effects of [CO_2], N supply levels, and growth stages on cucumber growth, gas exchange, and macro-nutrient elements concentrations in different tissues. Pearson correlation coefficient is commonly used to quantify the degree of linear relationship between two factors and has been used to evaluate the relationships between root exudates and root morphological traits in our previous work [31]. In this study, we also used the Pearson correlation analysis to evaluate what were the key factors affecting the concentration changes of each mineral element in cucumber plants under different [CO_2] and N supply conditions.

2. Materials and Methods

2.1. Plant Culture and Growth Conditions

Three open-top chambers (OTCs) (2.3 m length × 0.8 m width × 1.4 m height) made of poly (methyl methacrylate) were established in the glasshouse at Institute of Soil Science, Chinese Academy of Sciences, Nanjing, P.R. China (32.0596° N, 118.8050° E). The OTCs were transparent for minimizing the shading effect and received solar radiation with natural day length. The OTCs also have a pair of opposite side doors, which can be opened like wings for inside operation and cooling. The experiments were carried out as a split-plot design where $[CO_2]$ was the main treatment, and N supplies were considered as the sub-plot treatment. Five N treatments were set at 2 (N1), 4 (N2), 7 (N3), 14 (N4), and 21 (N5) mmol L^{-1}, which were repeated six times in each chamber, and the thirty pots in each chamber were rotated within and among chambers every two weeks to minimize chamber effects. The $[CO_2]$ in three identical OTCs was set at 400 (ambient: C1), 625 (elevated: C2), and 1200 (super-elevated: C3) μmol mol^{-1} respectively and was reset to the corresponding treatment condition following plant rotation. The $[CO_2]$ in OTCs was controlled and monitored continuously through an infrared gas analyzer (Ultramat 6, Siemens, Munich, Germany) started on the day after transplanting (DAT). The $[CO_2]$ in OTCs was elevated from 0800 to 1700 h every sunny and cloudy day. The temperature and relative humidity within the OTCs were recorded by a L95-83 data logger (Hangzhou loggertech Co., Ltd., Hangzhou, China) every 15 min. The chambers were opened for cooling and CO_2 was not supplied when the temperature inside was above 35 °C. The accumulated CO_2 treating time was 338 h within the whole experiment period of 62 days. The average temperature in OTCs was 22.9–23.5 °C, and the average humidity was 61.2–63.2%, respectively.

Cucumber (*Cucumis sativus* L.) seeds of 'Jinyou 38' (Tianjin Lvfeng Co., Ltd., Tianjin, China) were germinated on moist filter paper in constant-temperature incubator at 28 °C and relative humidity of 70% for 48 h, and then seeds with radicles were sown into trays containing peat-vermiculite (2:1, *v/v*) substrate. When the third true leaf emerged, healthy seedlings were selected and transplanted to 5 L polyvinyl chloride polymer (PVC) pots with two plants per pot. Each pot was filled with 4 L modified Yamazaki nutrient solutions for cucumbers [38] with five nitrogen levels. To keep the same P, K, Ca, and Mg concentrations ([P], [K], [Ca], and [Mg]) in nutrient solutions with five N levels, anions or cations were balanced with SO_4^{2-}, NO_3^{-}, or NH_4^{+} respectively (Table 1). All the nutrient solutions contained the same concentration of micro-nutrients composed of (mg L^{-1}): Na$_2$Fe-EDTA (29.27), H_3BO_3 (2.86), $MnSO_4 \cdot 4H_2O$ (2.03), $ZnSO_4 \cdot 7H_2O$ (0.22), $CuSO_4 \cdot 5H_2O$ (0.08), and $(NH_4)_6Mo_7O_{24} \cdot 4H_2O$ (0.02). The pH of the nutrient solution was adjusted to 6.5 with dilute NaOH. All pots were aerated intermittently for 30 min in every hour and renewed every four days.

Table 1. Components of macro-nutrient solutions in different N levels.

N[1] Levels	Compounds (mmol L^{-1})									Elements (mmol L^{-1})							
	Ca(NO$_3$)$_2$ ·4H$_2$O	KNO$_3$	NH$_4$H$_2$PO$_4$	MgSO$_4$ ·7H$_2$O	NH$_4$NO$_3$	KH$_2$PO$_4$	K$_2$SO$_4$	Ca(H$_2$PO$_4$)$_2$ ·H$_2$O	CaSO$_4$ ·2H$_2$O	N			P	K	Ca	Mg	S
										Total-N	NH$_4^+$-N	NO$_3^-$-N					
N1	1	-[2]	-	2	-	-	3	0.5	2	2	0	2	1	6	3.5	2	7
N2	2	-	-	2	-	-	3	0.5	1	4	0	4	1	6	3.5	2	6
N3	3.5	-	-	2	-	1	2.5	-	-	7	0	7	1	6	3.5	2	4.5
N4	3.5	6	1	2	-	-	-	-	-	14	1	13	1	6	3.5	2	2
N5	3.5	6	1	2	3.5	-	-	-	-	21	4.5	16.5	1	6	3.5	2	2

[1] N level: N1, 2 mmol L^{-1}; N2, 4 mmol L^{-1}; N3, 7 mmol L^{-1}; N4, 14 mmol L^{-1}; N5, 21 mmol L^{-1}; [2] -, no chemicals.

2.2. Sampling and Measurements

2.2.1. Gas-Exchange Rate Measurements

The gas exchange properties of cucumber plants, including net photosynthetic rate (Pn), transpiration rate (Tr), and stomatal conductance (Gs) were measured using a portable photosynthesis system (Li-6400, Li-Cor Inc., Lincoln, OR, USA) with a standard leaf chamber (2 cm × 3 cm) (6400-02B) with a LED light source. The photosynthetic photon flux density, temperature, relative air humidity, and the air flow rate inside the leaf chamber were set at 1500 mol m^{-2} s^{-1}, 25 °C, 50%, and 500 µmol s^{-1}, respectively. The [CO_2] of the flow-in air was set the same as that in the corresponding OTC where the plant was grown. The measurements were conducted on 13, 38, and 60 DAT and six replicates of the third leaves from the top of cucumber plants in each treatment were used for measuring.

2.2.2. Plant Harvest and Biomass Determination

One plant in each pot was harvested at the seedling stage (T1) when the seedling had five to six true leaves (18 DAT) and the other was harvested at the initial fruiting stage (T2) when small fruits of 5–8 cm of length formed (62 DAT). After harvest, plants were separated to root, stem, and leaf samples and washed with tap water followed by distilled water. Dry weight (DW) of each tissue was determined by drying the fresh tissues at 105 °C for 30 min and then at 75 °C to a constant weight in an electro-thermostatic blast oven.

2.2.3. Mineral Element Concentration Determination

The dry samples were ground to pass through a 0.5-mm screen. Next, 0.2 g dry samples were soaked in 5 mL concentrated H_2SO_4 for 24 h then digested at 180 °C for 5 h, followed by intermittent addition of 0.5 mL H_2O_2 for 2 or 3 times. The extracted solution was diluted to 500 mL with deionized water and the [N] was analyzed using a discrete auto-analyzer (Smartchem200, Alliance, France) [28]. Another portion of 0.2 g dry samples was digested with 5 mL HNO_3-$HClO_4$ (85:15 *v/v*) at 190 °C, and [P], [K], [Ca], and [Mg] were determined by an inductively coupled plasma atomic emission spectrometer (IRIS Advantage, Thermo Elemental, Franklin, MA, USA) [39].

2.3. Statistical Analysis

Statistical analysis was performed using SPSS software (Version 22.0; IBM Corp., Armonk, NY, USA). All data were shown as mean ± standard error. The means of DW, gas exchange properties, and mineral concentrations with six replicates in each treatment were compared using Duncan's multiple range test at a significance level of $p = 0.05$ in one-way analysis of variance (ANOVA). The effects of N supply, [CO_2], growth stage, and their interaction on DW, gas exchange properties, and mineral concentrations were quantified using a general linear model. Correlation and significance tests between each mineral concentration in different tissues of cucumber and [CO_2], N supply, transpiration rate were calculated using the Pearson correlation coefficient with two-tailed test. All figures were generated by OriginPro (Version 8.0; OriginLab Corp., Northampton, MA, USA).

3. Results

3.1. Dry Weight and Root to Shoot Ratio

The effects of [CO_2] levels, N levels, growth stages, and their interactions on root, stem, leaf, and total DW as well as root to shoot ratios (R/S) of cucumber plants are shown in Table 2. The growth stage significantly affected the DW of roots, stems, leaves, and total biomass of cucumber plants. The average DW of total plants was increased from 1.01 g plant^{-1} at T1 stage to 6.14 g plant^{-1} at T2 stage. As the aerial parts of cucumbers grew faster than root, the R/S significantly decreased from 0.107 to 0.093 during a growth period of 44 days.

Table 2. The dry weight of roots, stems, leaves, whole plants and root/shoot ratios of cucumber grown under different [CO_2] and N levels at the seedling and initial fruiting stages ($n = 6$).

Stage [1]	CO_2 [2] Level	N [3] Level	Dry Weight (g plant^{-1})				
			Roots	Stems	Leaves	Total	Root/Shoot
T1	C1	N1	0.081 ± 0.005 Aa [4]	0.140 ± 0.012 Ab	0.505 ± 0.020 Ab	0.726 ± 0.022 Ab	0.127 ± 0.010 Aa
		N2	0.080 ± 0.008 Ba	0.162 ± 0.014 Bab	0.618 ± 0.045 Ba	0.860 ± 0.066 Bab	0.102 ± 0.005 Bab
		N3	0.076 ± 0.006 Aa	0.144 ± 0.009 Bb	0.589 ± 0.026 Bab	0.809 ± 0.036 Bab	0.104 ± 0.007 Aab
		N4	0.070 ± 0.005 Ba	0.158 ± 0.010 Bab	0.639 ± 0.027 Ba	0.867 ± 0.037 Ba	0.089 ± 0.007 Ab
		N5	0.070 ± 0.008 Ba	0.185 ± 0.013 Ba	0.674 ± 0.035 Ba	0.930 ± 0.045 Ba	0.083 ± 0.011 Ab
	C2	N1	0.104 ± 0.023 Aa	0.174 ± 0.014 Ab	0.550 ± 0.085 Ab	0.828 ± 0.119 Ab	0.138 ± 0.010 Aa
		N2	0.112 ± 0.012 Aa	0.209 ± 0.015 Aab	0.726 ± 0.050 ABa	1.047 ± 0.073 Aab	0.118 ± 0.007 Aab
		N3	0.091 ± 0.007 Aa	0.221 ± 0.017 Aab	0.744 ± 0.048 Aa	1.056 ± 0.069 Aab	0.095 ± 0.004 Ac
		N4	0.088 ± 0.005 Ba	0.227 ± 0.023 Aa	0.749 ± 0.033 Ba	1.064 ± 0.060 Bab	0.091 ± 0.003 Ac
		N5	0.104 ± 0.008 Aa	0.234 ± 0.012 ABa	0.797 ± 0.059 Ba	1.135 ± 0.070 Ba	0.102 ± 0.010 Abc
	C3	N1	0.111 ± 0.015 Aa	0.166 ± 0.009 Ab	0.560 ± 0.022 Ac	0.837 ± 0.044 Ac	0.151 ± 0.015 Aa
		N2	0.108 ± 0.006 Aa	0.218 ± 0.011 Ab	0.772 ± 0.021 Ab	1.098 ± 0.031 Ab	0.108 ± 0.005 Ab
		N3	0.089 ± 0.017 Aa	0.195 ± 0.020 Ab	0.763 ± 0.070 Ab	1.047 ± 0.102 Abc	0.091 ± 0.011 Ab
		N4	0.122 ± 0.015 Aa	0.281 ± 0.024 Aa	0.970 ± 0.058 Aa	1.373 ± 0.094 Aa	0.096 ± 0.006 Ab
		N5	0.134 ± 0.015 Aa	0.280 ± 0.028 Aa	0.995 ± 0.078 Aa	1.410 ± 0.112 Aa	0.105 ± 0.007 Ab
T2	C1	N1	0.298 ± 0.018 Ac	0.982 ± 0.058 Ad	1.546 ± 0.110 Ac	2.826 ± 0.157 Ac	0.119 ± 0.008 Aa
		N2	0.347 ± 0.031 Ac	1.294 ± 0.068 Acd	2.640 ± 0.175 Abc	4.281 ± 0.268 Abc	0.088 ± 0.004 Ac
		N3	0.401 ± 0.075 Bbc	1.684 ± 0.194 Abc	3.260 ± 0.507 Bab	5.345 ± 0.770 Bab	0.079 ± 0.004 Ac
		N4	0.551 ± 0.041 Aab	2.101 ± 0.138 Aab	4.076 ± 0.327 Ba	6.728 ± 0.483 Ba	0.091 ± 0.006 Abc
		N5	0.699 ± 0.102 Aa	2.375 ± 0.334 Ba	4.188 ± 0.611 Ba	7.262 ± 1.034 Ba	0.106 ± 0.006 Aab
	C2	N1	0.222 ± 0.028 Bc	0.700 ± 0.079 Bc	1.157 ± 0.166 Ac	2.079 ± 0.261 Bc	0.120 ± 0.005 Aa
		N2	0.310 ± 0.066 Abc	1.264 ± 0.235 Abc	2.256 ± 0.424 Abc	3.830 ± 0.719 Abc	0.087 ± 0.005 Ab
		N3	0.356 ± 0.059 Babc	1.532 ± 0.259 Aabc	3.405 ± 0.401 Bb	5.294 ± 0.614 Bb	0.073 ± 0.009 Ab
		N4	0.567 ± 0.151 Aab	2.097 ± 0.442 Aab	5.012 ± 0.730 Ba	7.677 ± 1.274 Ba	0.074 ± 0.011 Ab
		N5	0.635 ± 0.103 Aa	2.254 ± 0.367 Ba	5.117 ± 0.277 Ba	8.006 ± 0.674 Ba	0.084 ± 0.010 Ab
	C3	N1	0.223 ± 0.020 Bb	0.751 ± 0.068 Bc	1.194 ± 0.143 Ab	2.168 ± 0.220 Bc	0.116 ± 0.007 Aa
		N2	0.341 ± 0.038 Ab	1.287 ± 0.149 Ac	2.160 ± 0.265 Ab	3.789 ± 0.436 Ac	0.101 ± 0.007 Aab
		N3	0.729 ± 0.144 Aa	2.571 ± 0.471 Ab	5.728 ± 0.492 Aa	9.028 ± 0.965 Ab	0.084 ± 0.012 Ab
		N4	0.983 ± 0.195 Aa	3.370 ± 0.687 Aab	7.030 ± 0.729 Aa	11.38 ± 1.52 Aab	0.091 ± 0.011 Aab
		N5	0.902 ± 0.080 Aa	4.199 ± 0.404 Aa	7.248 ± 0.703 Aa	12.35 ± 1.00 Aa	0.082 ± 0.011 Ab

Table 2. *Cont.*

Stage [1]	CO_2 [2] Level	N [3] Level	Dry Weight (g plant^{-1})				
			Roots	Stems	Leaves	Total	Root/Shoot
	C		*** [6]	***	***	***	NS
	N		***	***	***	***	***
	T		***	***	***	***	***
	C × N		NS	*	***	***	NS
	C × T		**	***	***	***	NS
	N × T		***	***	***	***	NS
	C × N × T [5]		NS	*	**	**	NS

[1] Growth stage: T1, seedling stage (18 DAT); T2, initial fruiting stage (62 DAT); [2] CO_2 level: C1, C2, and C3: 400, 625, and 1200 µmol mol^{-1}; [3] N level: N1, N2, N3, N4, and N5: 2, 4, 7, 14, and 21 mmol L^{-1}; [4] Means within rows at the same stage not followed by the same lower case letters are significantly different among different N levels in the same CO_2 level, and not followed by the same upper case letters are significantly different among different CO_2 levels in the same N level, according to Duncan's test at $p < 0.05$; [5] C: [CO_2] level; N: N level; T: growth stage; [6] Asterisks (*) indicate significant differences (* $p < 0.05$; ** $p < 0.01$; *** $p < 0.001$). NS indicates non-significant differences ($p \geq 0.05$).

N supply levels also significantly affected the biomass accumulation of cucumber plants, especially at T2 stage. At T1 stage, the DW of stems, leaves as well as total biomass of cucumber plants was significantly increased with the increase of N supply, especially at higher [CO_2] treatments (C3), whereas the DW of roots was not influenced by N levels at all [CO_2] levels. Specifically, the increase of total biomass from N1 to N5 was 28.0%, 37.0%, and 68.5% in treatments C1, C2, and C3, respectively. At T2 stage, the positive effects of N supplies were more obvious than that at T1 stage, and the DW of all parts of cucumber plants were much greater in N5 treatment than those in N1 treatment. The total biomass was increased by 1.57, 2.85, and 4.70 folds from N1 to N5 in treatments C1, C2, and C3, respectively. Since the improvement of DW in the aerial parts was more noticeable than that in roots, the R/S at each [CO_2] treatment was significantly decreased with the increase of N supply at both growth stages.

With respect of [CO_2] levels, the increase of DW by super-elevated [CO_2] (C3) were more dramatic in moderate (N3) and high N supplies (N4 and N5), whereas there was little increase or even a decrease in low N supplies (N1 and N2). Generally, the increase of total biomass from C1 to C3 was 29.4%, 58.4%, and 51.7% at T1 stage, and was 68.9%, 69.2%, and 70.1% at T2 stage, in N3, N4, and N5 treatments, respectively. The R/S was not significantly affected by [CO_2] levels, irrespective of the N supply and growth stage.

The interactions of [CO_2] × N and [CO_2] × N × growth stage had significant effects on the DW of stems, leaves, and total cucumber plants but it was not significant on root DW. The interactions of [CO_2] × growth stage and N × growth stage had significant effects on the DW of all parts of and total cucumber plants. The interaction of neither two nor three of these factors had significant effects on R/S.

3.2. Gas Exchange

Generally, [CO_2] levels, N levels, growth stages and their interactions all had significant effects on the Pn of cucumber plants (Figure 1). On 13 DAT, super-elevated [CO_2] (C3) significantly increased the Pn in N3 and N4 treatments. On 38 DAT, the Pn under super-elevated [CO_2] was the highest in all N levels. On 60 DAT, the Pn under super-elevated [CO_2] was also the highest in all N levels except for N2. Compared with ambient [CO_2] (C1), the increase of the Pn by super-elevated [CO_2] (C3) was 60.1%, 115.5%, and 77.7% in N3, N4, and N5 treatment, respectively. However, elevated [CO_2] (C2) did not significantly increase the Pn compared with C1 in all N levels at three growth stages. The Pn was usually increased with the N supply increasing at the same [CO_2] and growth stage. The increase of the Pn from N1 to N5 in C3 treatment was 2.35, 0.89, and 2.24 folds on 13, 38, and 60 DAT, respectively. As the plant grew, the Pn was gradually increased in high N supplies (N4 and N5) under super-elevated [CO_2]. Whereas in low N supplies (N1 and N2), the Pn reached its highest value earlier on 38 DAT.

The Gs of cucumber plants was significantly influenced by [CO_2] levels, N levels, growth stages, and the interactions of [CO_2] × growth stage and N × growth stage (Figure 1). The Gs was much lower in N1 treatment on 13 DAT and in N1 and N2 treatments on 60 DAT than that in other higher N levels in the same [CO_2] level. The Gs was also depressed by higher [CO_2] (C2 and C3) in N5 treatment on 13 DAT and in N1, N2, and N5 treatments on 60 DAT. In other N treatments and growth stages, there was only a decreasing trend not a significant decrease in the Gs by higher [CO_2] treatments. Averaged across all N treatments, the decrease of Gs from C1 to C3 was 27.4%, 27.7%, and 21.2%, on 13, 38, and 60 DAT, respectively.

The changes of Tr of cucumbers grown under different N and [CO_2] levels at three growth stages were similar to those of Gs (Figure 1). [CO_2] levels, N levels, growth stages, and the interactions of [CO_2] × growth stage and N × growth stage had significant effects on the Tr of cucumber plants. The Tr was inhibited by super-elevated [CO_2] (C3) in N2, N4, and N5 treatments on 13 DAT, in N4 on 38 DAT, and in N1, N2, and N3 on 60 DAT. In other treatments, a decreasing trend in C3 was usually observed compared with C1. The average decrease of Tr from C1 to C3 among all N treatments was 31.9%, 26.4%, and 22.3%, on 13, 38, and 60 DAT, respectively. Increasing N supply improved the Tr in C2 and C3 treatments on 60 DAT, and the increase from N1 to N5 was 72.6% and 130.2% in C2 and C3 treatment, respectively. The Tr differences among the three growth stages were not significant.

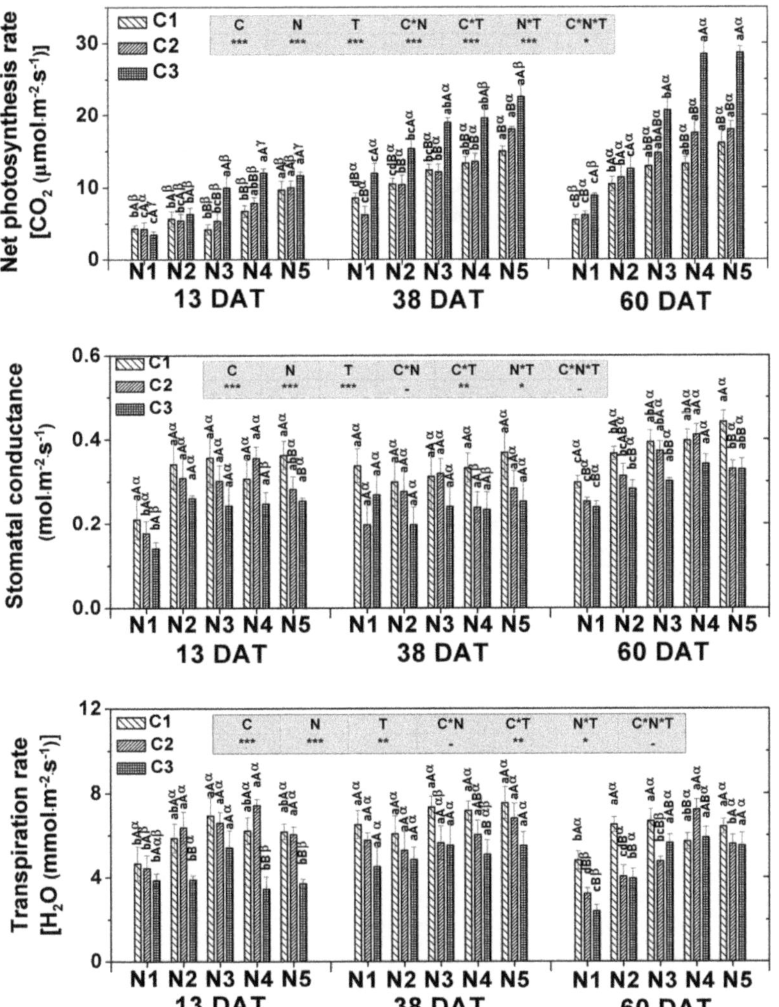

Figure 1. Net photosynthesis rate, stomatal conductance, and transpiration rate of cucumbers grown under different N and $[CO_2]$ levels at three growth stages ($n = 6$). Bars represent standard errors. CO_2 levels: C1, C2, and C3: 400, 625, and 1200 μmol mol^{-1}. N levels: N1, N2, N3, N4, and N5: 2, 4, 7, 14, and 21 mmol L^{-1}. Means not followed by the same lower case letters are significantly different among different N levels in the same CO_2 level and growth stage, and not followed by the same upper case letters are significantly different among different CO_2 levels in the same N level and growth stage, and not followed by the same Greek letters are significantly different among different growth stages in the same CO_2 and N level, according to Duncan's test at $p < 0.05$. In the internal table, C: $[CO_2]$ level; N: N level; T: growth stage. Asterisks (*) indicate significant differences (* $p < 0.05$; ** $p < 0.01$; *** $p < 0.001$); - indicates non-significant differences ($p \geq 0.05$).

3.3. Mineral Nutrient Concentration

As shown in Figure 2, $[CO_2]$ levels, N levels, growth stages, and the interaction of $[CO_2] \times$ N had significant effects on the [N] in all three parts of cucumber plants, and the interaction of $[CO_2] \times$ growth stage, N $\times$ growth stage, and $[CO_2] \times$ N $\times$ growth stage also had significant effects on the [N] in leaves of cucumbers. [N] in roots, stems, and leaves of cucumbers was significantly increased with the increasing N supply. At T1 stage (18 DAT), the average increase of [N] from N1 to N5 among all [CO_2] treatments was 71.4%, 85.9%, and 61.2% in roots, stems, and leaves respectively, whereas the value was 50.0%, 71.7%, and 27.7% at T2 stage (62 DAT). [N] in leaves was significantly decreased by super-elevated [CO_2] (C3) compared with ambient [CO_2] (C1) in all N treatments at both growth stages except for N1 and N2 at T2 stage. Averaged across all N treatments, the decrease of [N] in leaves from C1 to C3 was 6.1% and 9.3%, at T1 and T2 stage, respectively. However, [CO_2] levels did not affect the [N] in roots and stems in most treatments. As the plants grew, the [N] in stems in N2, N3, and N4 treatments in all [CO_2] levels, in leaves in N1, N4, and N5 treatments in higher [CO_2] levels (C2 and C3) were decreased.

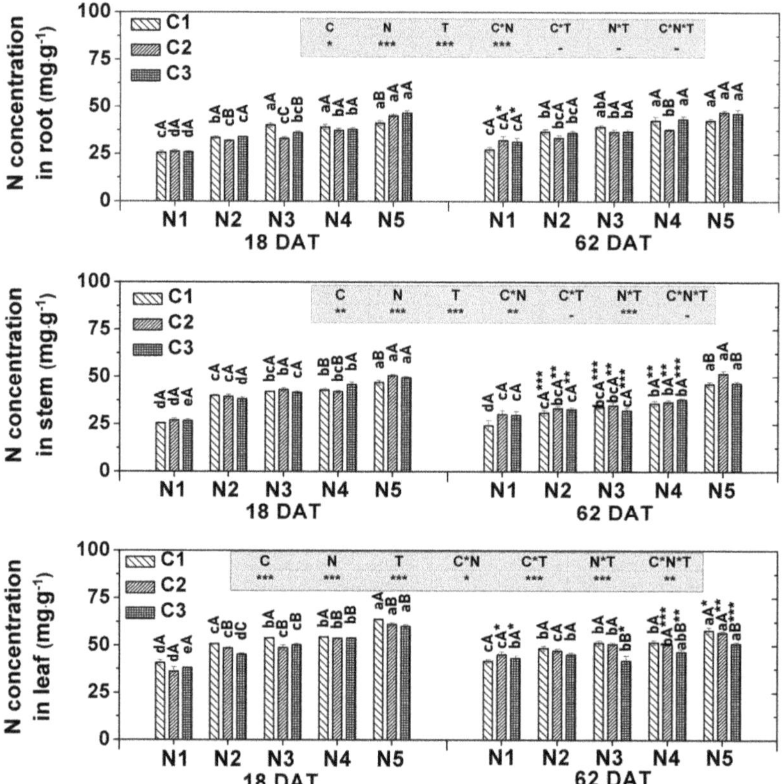

Figure 2. Nitrogen concentrations in different tissues of cucumber plants grown under different N and [CO_2] levels at two growth stages (*n* = 6). Bars represent standard errors. CO_2 levels: C1, C2, and C3: 400, 625, and 1200 μmol mol^{-1}. N levels: N1, N2, N3, N4, and N5: 2, 4, 7, 14, and 21 mmol L^{-1}. Means not followed by the same lower case letters are significantly different among different N levels in the same CO_2 level and growth stage, and not followed by the same upper case letters are significantly different among different CO_2 levels in the same N level and growth stage, according to Duncan's test at *p* < 0.05. Means with asterisks are significantly different between two growth stages in the same CO_2 and N level (**p* < 0.05; ** *p* < 0.01; *** *p* < 0.001). In the internal table, C: [CO_2] level; N: N level; T: growth stage. Asterisks (*) indicate significant differences (* *p* < 0.05; ** *p* < 0.01; *** *p* < 0.001); - indicates non-significant differences (*p* ≥ 0.05).

The [P] in different tissues of cucumber plants grown under different N and [CO_2] levels at two growth stages was shown in Figure 3. [CO_2] levels, N levels, growth stages, and the interaction of N × growth stage had significant effects on the [P] in all three parts of cucumber plants. At T1 stage, [P] in stems was gradually decreased with the N supply increased, and the average decrease from N1 to N5 among all [CO_2] treatments was 17.3%. At T2 stage, [P] in roots and stems was significantly decreased as the N supply increasing, and the corresponding average decrease from N1 to N5 was 44.2% and 38.3%, respectively. At T1 stage, [P] in moderate (N3) and high (N4 and N5) N supplies were usually lower in C3 than that in C1 treatment, specifically in roots of the N3 and N4 treatments, stems of the N4 and N5 treatments, and leaves of N3 and N5 treatments. At T2 stage, the differences of [P] between C1 and C3 were not significant except for N1 and N4 in stems and N2 and N4 in leaves.

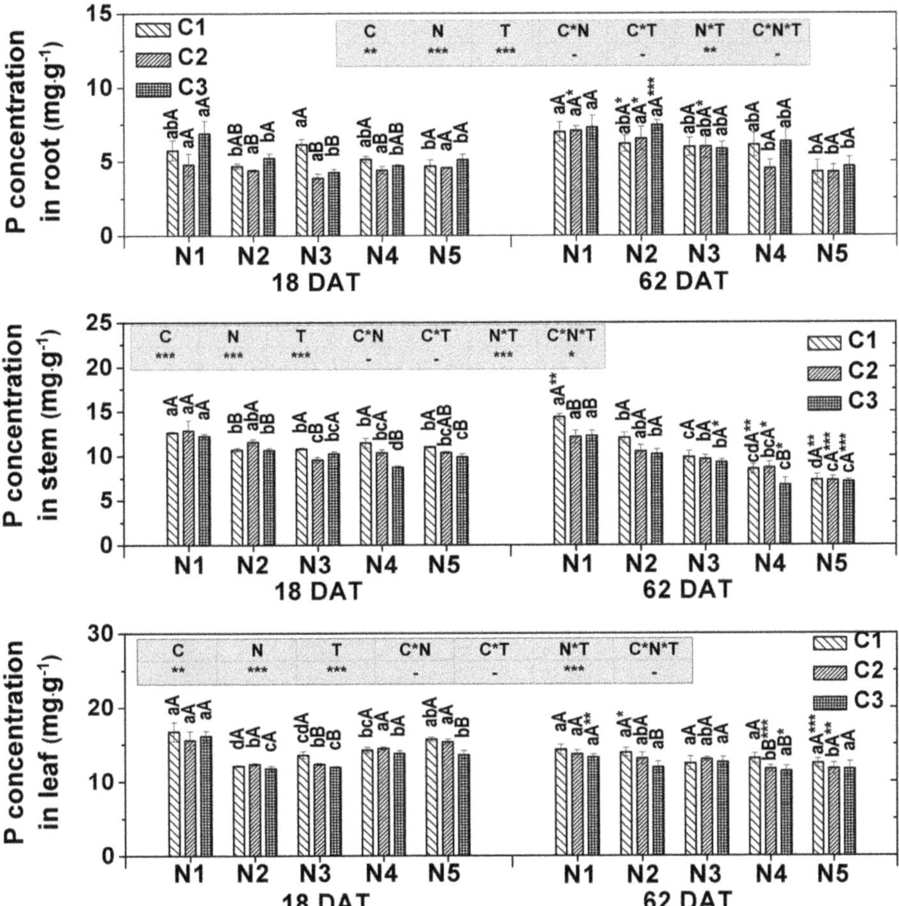

Figure 3. Phosphorus concentrations in different tissues of cucumber plants grown under different N and [CO_2] levels at two growth stages (n = 6). Bars represent standard errors. CO_2 levels: C1, C2, and C3: 400, 625, and 1200 μmol mol^{-1}. N levels: N1, N2, N3, N4, and N5: 2, 4, 7, 14, and 21 mmol L^{-1}. Means not followed by the same lower case letters are significantly different among different N levels in the same CO_2 level and growth stage, and not followed by the same upper case letters are significantly different among different CO_2 levels in the same N level and growth stage, according to Duncan's test at $p < 0.05$. Means with asterisks are significantly different between two growth stages in the same CO_2 and N level (* $p < 0.05$; ** $p < 0.01$; *** $p < 0.001$). In the internal table, C: [CO_2] level; N: N level; T: growth stage. Asterisks (*) indicate significant differences (* $p < 0.05$; ** $p < 0.01$; *** $p < 0.001$); - indicates non-significant differences ($p \geq 0.05$).

Figure 4 showed the [K] in different tissues of cucumber plants grown under different treatments. Similar to the [P], [CO$_2$] levels, N levels, growth stages, and the interaction of N × growth stage had significant effects on the [K] in all three parts of cucumber plants. At T1 stage, [K] in roots was gradually decreased as the N level elevating, and the average decrease from N1 to N5 among all [CO$_2$] treatments was 40.5%, whereas they were gradually increased in leaves as the N level elevated, and the average increase from N1 to N5 was 13.8%. Compared with ambient [CO$_2$] (C1), [K] was significantly decreased by super-elevated [CO$_2$] (C3) in N3 and N4 in roots, N3, N4, and N5 in stems and N2, N4, and N5 in leaves. At T2 stage, a decrease of [K] by elevated [CO$_2$] was observed in N4 and N5 in roots and N4 in leaves. In terms of growth stages, [K] was decreased from T1 stage to T2 stage in stems of cucumbers in all treatments.

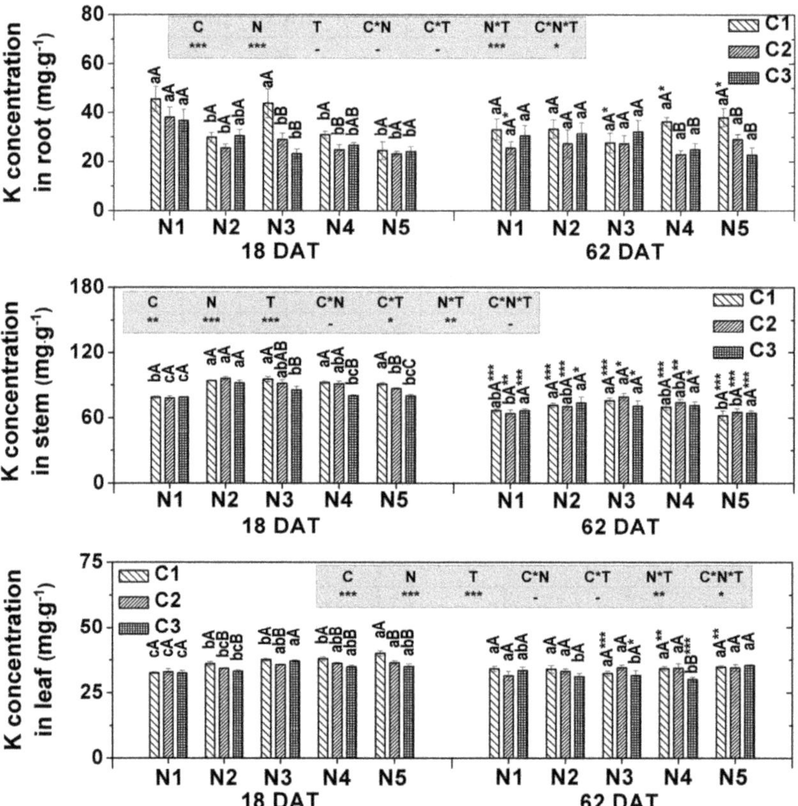

Figure 4. Potassium concentrations in different tissues of cucumber plants grown under different N and [CO$_2$] levels at two growth stages (n = 6). Bars represent standard errors. CO$_2$ levels: C1, C2, and C3: 400, 625, and 1200 µmol mol^{-1}. N levels: N1, N2, N3, N4, and N5: 2, 4, 7, 14, and 21 mmol L^{-1}. Means not followed by the same lower case letters are significantly different among different N levels in the same CO$_2$ level and growth stage, and not followed by the same upper case letters are significantly different among different CO$_2$ levels in the same N level and growth stage, according to Duncan's test at $p < 0.05$. Means with asterisks are significantly different between two growth stages in the same CO$_2$ and N level (* $p < 0.05$; ** $p < 0.01$; *** $p < 0.001$). In the internal table, C: [CO$_2$] level; N: N level; T: growth stage. Asterisks (*) indicate significant differences (* $p < 0.05$; ** $p < 0.01$; *** $p < 0.001$); - indicates non-significant differences ($p \geq 0.05$).

[CO$_2$] levels, N levels, growth stages, and the interaction of [CO$_2$] × growth stages had significant effects on the [Ca] in all three parts of cucumber plants (Figure 5). [Ca] was always the lowest in highest N treatment (N5) at two growth stages in all three parts of cucumbers except for that in roots at T1 stage. The average decrease from N1 to N5 among all [CO$_2$] treatments was 16.9% and 15.6% in stems and leaves respectively at T1 stage, and was 32.4%, 33.9%, and 15.3% in roots, stems, and leaves respectively at T2 stage. [Ca] was also decreased by elevated [CO$_2$] in all three tissues regardless of N supply at T1 stage. The average decrease from C1 to C3 among all N treatments was 16.4%, 14.3%, and 10.4% in roots, stems, and leaves, respectively. Growing caused a significant decrease of [Ca] in 10 of 15 treatments in root and 11 of 15 treatments in leaves.

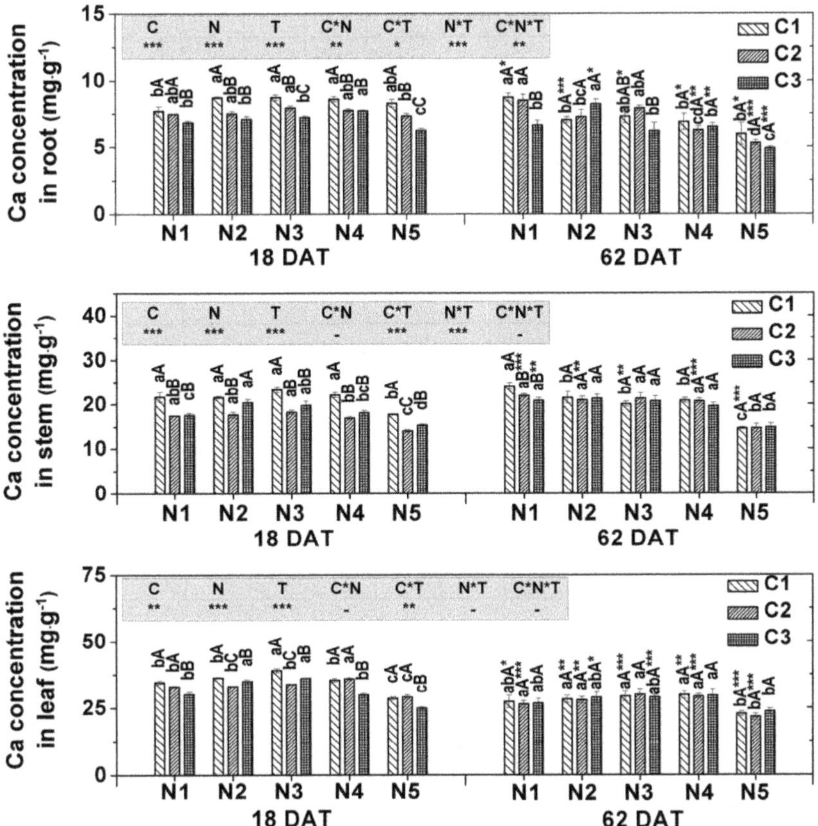

Figure 5. Calcium concentrations in different tissues of cucumber plants grown under different N and [CO$_2$] levels at two growth stages ($n = 6$). Bars represent standard errors. CO$_2$ levels: C1, C2, and C3: 400, 625, and 1200 µmol mol^{-1}. N levels: N1, N2, N3, N4, and N5: 2, 4, 7, 14, and 21 mmol L^{-1}. Means not followed by the same lower case letters are significantly different among different N levels in the same CO$_2$ level and growth stage, and not followed by the same upper case letters are significantly different among different CO$_2$ levels in the same N level and growth stage, according to Duncan's test at $p < 0.05$. Means with asterisks are significantly different between two growth stages in the same CO$_2$ and N level (* $p < 0.05$; ** $p < 0.01$; *** $p < 0.001$). In the internal table, C: [CO$_2$] level; N: N level; T: growth stage. Asterisks (*) indicate significant differences (* $p < 0.05$; ** $p < 0.01$; *** $p < 0.001$); - indicates non-significant differences ($p \geq 0.05$).

The [Mg] in different tissues of cucumber plants growing under different N and [CO$_2$] levels at two growth stages was shown in Figure 6. Generally, [CO$_2$] levels, N levels, and growth stages had significant effects on the [Mg] in all three parts of cucumber plants. [Mg] was almost the highest in lowest N treatments (N1) and the lowest in highest N treatment (N5) in stems and leaves at T1 stage and in roots and leaves at T2 stage. The corresponding average decrease from N1 to N5 among all [CO_2] treatments was 19.1%, 27.1%, 24.5%, and 24.8%. At T1 stage, [Mg] was significantly decreased by higher [CO_2] (C2 and C3) compared with ambient [CO_2] (C1) at all N levels in roots except for N1, and in leaves except for N2. The average decrease among all N treatments was 26.4%, 0.2% (not significant), and 6.6% in roots, stems, and leaves respectively from C1 to C2, and the decrease from C1 to C3 was 16.9%, 10.0%, and 7.8%. Similar to [Ca], growing also caused a significant decrease of [Mg] in 10 of 15 treatments in root and 11 of 15 treatments in leaves.

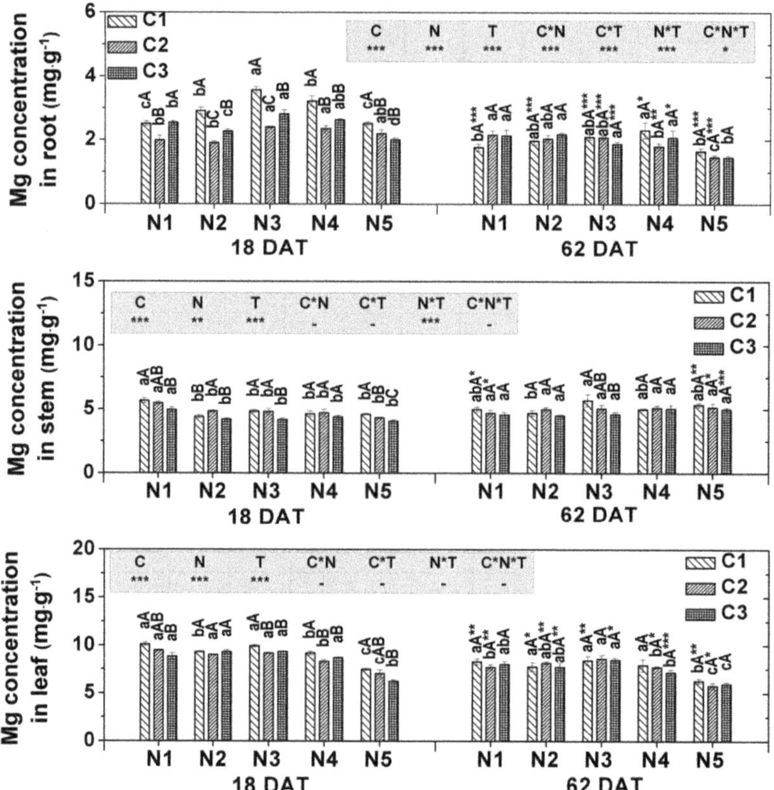

Figure 6. Magnesium concentrations in different tissues of cucumber plants grown under different N and [CO_2] levels at two growth stages ($n = 6$). Bars represent standard errors. CO_2 levels: C1, C2, and C3: 400, 625, and 1200 µmol mol^{-1}. N levels: N1, N2, N3, N4, and N5: 2, 4, 7, 14, and 21 mmol L^{-1}. Means not followed by the same lower case letters are significantly different among different N levels in the same CO_2 level and growth stage, and not followed by the same upper case letters are significantly different among different CO_2 levels in the same N level and growth stage, according to Duncan's test at $p < 0.05$. Means with asterisks are significantly different between two growth stages in the same CO_2 and N level (* $p < 0.05$; ** $p < 0.01$; *** $p < 0.001$). In the internal table, C: [CO_2] level; N: N level; T: growth stage. Asterisks (*) indicate significant differences (* $p < 0.05$; ** $p < 0.01$; *** $p < 0.001$); - indicates non-significant differences ($p \geq 0.05$).

3.4. Correlations between [CO$_2$], N Supply, Transpiration Rate, and Mineral Nutrient Concentration

To further evaluate the relationship between mineral nutrient concentration, [CO$_2$], N supply levels, and Tr, the Pearson correlation coefficient was calculated (Figure 7). Tr was negatively correlated to [CO$_2$] at both growth stages and was positively correlated to N supply levels at T2 stage. [N] in all parts of cucumber plants was significantly positively correlated to N supply levels and Tr at both growth stages except for the insignificant positive correlation between [N] in root and Tr at T1 stage. [N] was negatively correlated to [CO$_2$] only in leaves at both growth stages. [P] in all parts of cucumber plants was significantly negatively correlated to N supply except for levels at T2 stage. [P] was only negatively correlated to [CO$_2$] in stems at T1 stage and in leaves at T2 stage, and was not significantly correlated to Tr. [K] in all parts of cucumber plants was significantly negatively correlated to [CO$_2$] at T1 stage and in leaves at T2 stage. [K] in all parts of cucumber plants was also significantly positively correlated to Tr at both growth stages except for that in roots at T1 stage. [K] in leaves was also significantly positively correlated to N levels, whereas it was negatively correlated to N levels in root at T1 stage. [Ca] in all parts of cucumber plants was significantly negatively correlated to [CO$_2$] at T1 stage, whereas this correlation was not significant at T2 stage. [Ca] was also significantly negatively correlated to N levels in all parts of cucumber plants at both growth stages except for that in roots at T1 stage. [Ca] in all parts of plants at T1 stage and in leaves at T2 stage also had a strong positive correlation to Tr. [Mg] in all parts of cucumber plants at T1 stage and in stems at T2 stage was significantly negatively correlated to [CO$_2$]. [Mg] in stems and leaves at T1 stage and in roots and leaves at T2 stage was also significantly negatively correlated to N levels. [Mg] in aerial parts of cucumber was all significantly positively correlated to Tr at both growth stages.

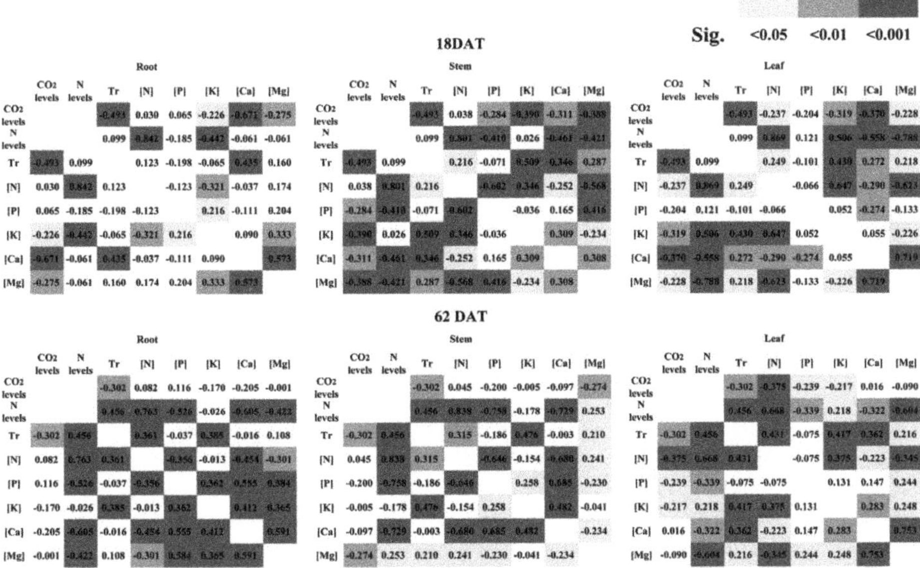

Figure 7. Pearson correlations between [CO$_2$], N supply, transpiration rate and mineral nutrient concentration of cucumber grown under different N and [CO$_2$] levels at two growth stages. Tr: transpiration rate.

4. Discussion

4.1. Impacts of [CO₂] and N Supply on the Growth of Cucumbers

The beneficial effects of CO_2 enrichment on stimulating plant growth and biomass accumulation of crops have been extensively reported [2,5,6,40]. An increase in the DW ranging from 13.5% to 34.4% has been reported for cucumbers, when [CO_2] was elevated to 500–760 µmol mol^{-1} [1,31,33,41,42], and ranged from 43.9% to 128.0% when [CO_2] was elevated to 1000–1200 µmol mol^{-1} [28–31,43]. In the present work, in moderate N (N3), the increase in the DW was 30.5% and not significant when [CO_2] was elevated to 625 µmol mol^{-1}, and 29.4% and 68.9% when [CO_2] was elevated to 1200 µmol mol^{-1}, at T1 and T2 stage, respectively (Table 2). This increased DW caused by more fixation of CO_2 and accumulation of biomass under elevated [CO_2] combined with moderate and high N supplies [5,6]. Our results showed that the increase in Pn was greater than 60% by super-elevated [CO_2] compared with ambient [CO_2] in moderate and high N supplies (Figure 1), which were in close conformity with this explanation.

Moreover, the present work also clearly demonstrated that the DW accumulation by CO_2 enrichment depended on the N supply, in which there was significant increase in DW in moderate and high N supplies but no change or decrease in low N supplies (Table 2). This is consistent with the previous findings that limited N will inhibit the synthesis of photosynthetic related proteins, lower the photosynthetic capacity, and reduce the photo-assimilate accumulation [12,15,44,45]. The significant decrease of [N] in leaves at T1 stage, as well as no change or decrease in Pn by elevated [CO_2] observed in low N supply treatments in this work, also gave a clear indication that the stimulation of Pn at high [CO_2] was only partial or counteracted when N supply was limited (Figures 1 and 2). It is worth mentioning that the improvement in Pn by CO_2 enrichment maintained at 60 DAT even when the [N] in leaves was also deceased in high N supplies (Figures 1 and 2). A possible reason is that the [N] in leaves treated with high N supplies was still enough for guaranteeing the RuBP regeneration and Rubisco activity to match the increased C-fixation [10,46]. These results also confirmed that increasing N supply could alleviate or prevent the photosynthetic acclimation under elevated [CO_2] condition and ensure the sustainability of the [CO_2] enrichment fertilization effects on crop growth [15–17,33].

Decreased Gs and Tr are the most obvious and universal changes observed in C3 plants including cucumbers grown in elevated [CO_2] condition [41,47–49]. Elevated [CO_2] causes partial stomatal closure and decreases the Gs by 8–44% for C3 plants, consequently with a reduction of Tr by 20–40% [2,50]. In the present work, the decrease in Gs and Tr of cucumbers in super-elevated [CO_2] compared with the ambient [CO_2] was 12.2–27.7% and 22.9–31.9%, respectively (Figure 1), which was in good agreement with previous reports. Additionally, we found a greater reduction in Gs and Tr in low N supply treatments (Figure 1). Low N causes reductions in Rubisco concentration and activity hereby forces the reduction in Gs and Tr in order to maintain a constant ratio of internal leaf [CO_2] to that of outside air [50]. The improvement in Gs and Tr under higher N treatments might also result from the synthesis of more photosynthates and increased cell wall rigidity [51].

[CO_2] and N Supplies also affect the biomass allocation of the plants. In the present work, the R/S was significantly decreased with N supplies increasing, but was not affected by [CO_2] levels (Table 2). A similar phenomenon was observed in other reports and our previous works [23,28,30,31,52]. Incorporated with current views in the literature, two possible reasons were usually proposed. Firstly, the accumulation of nitrate in shoots under high N supply down-regulates the growth of roots relative to shoots, resulting in lower R/S [53]. Secondly, plants always allocate more biomass to the apparatus in nutrient-limiting conditions, so more photosynthates will be invested in roots for exploring and acquiring more nutrients in N-deficient conditions [54].

4.2. Key Factors Affecting the Mineral Nutrient Concentrations

The changes of [N] under different [CO_2] and N supply conditions have been extensively studied. The results that [N] was decreased as [CO_2] increasing and increased as N supply increasing have been frequently observed for crops and vegetables [8,9,19]. The reasons for the increase in [N] with increasing N supply are transparent, whereas the reasons for the decrease in [N] with increasing $[CO_2]$ are more complicated. The most common reason is that the [N] is diluted by more accumulated carbohydrates in high $[CO_2]$ conditions [9,10,12]. However, tissues respond differently to elevated $[CO_2]$ conditions, and it has been reviewed that the average decrease in [N] in leaves was 16%, which was larger than that in stems (9%) and roots (9%) [7,9]. In the present work, only the [N] in leaves was significantly decreased by 6.1–9.3% in super-elevated $[CO_2]$ treatment, whereas [N] in stems and roots were not influenced by $[CO_2]$ levels (Figures 2 and 7). Therefore, the dilution effect is not the only cause of [N] decrease in elevated $[CO_2]$. Previous elevated $[CO_2]$ studies have found that the NO_3^- assimilation was enhanced in roots by increased photosynthate translocation to roots, whereas it was inhibited in leaves caused by the competition for reductants between the carbon fixation and NO_3^- reduction [17,25,26]. Hence, the dilution effect on [N] could be counteracted by the enhanced NO_3^- assimilation in roots while aggravated by the inhibited NO_3^- assimilation in leaves. Additionally, the decrease in Tr caused by the closed stomata could result in a reduced flow of NO_3^- from roots to leaves, which also leads to more decrease in [N] in leaves than that in roots [17,19,21,22,55]. In the present work, a significant positive correlation between [N] and Tr, especially in leaves, also gave convincing evidence for this reason (Figure 7).

In this study, [P] in moderate and high N supplies was decreased in higher $[CO_2]$ levels at T1 stage, and was significantly decreased in roots and stems with the increasing N supply (Figure 3). The similar results were also found in tomatoes [17], beans [56], and soybeans [57]. This decrease of [P] in elevated $[CO_2]$ especially associated with higher N supplies has been considered as the result of the dilution effect [10,17]. When the larger biomass was accumulated in higher N supplies and elevated $[CO_2]$ conditions, [P] in plants will decease if the P supply was not changed [20]. Besides, we also found there was little correlation between [P] and Tr (Figure 7). This lack of response of [P] to Tr may be explained by the free transportation pathway of phosphate in the xylem [58,59]. When phosphate is delivered from root to shoot in xylem sap, it can also be redistributed between different tissues according to their own demands, and excess phosphate will be stored in the vacuoles to maintain the cellular phosphate homeostasis, which is less affected by Tr.

In moderate and high N supplies, a decrease of [K] by elevated $[CO_2]$ was found in the whole plant at T1 stage, and in roots and leaves at T2 stage (Figure 4). There was also a significant negative correlation between [K] and $[CO_2]$ level in the whole plant at T1 stage, and in leaves at T2 stage (Figure 7). These results are consistent with the average decrease of 10% [K] in plants by elevated $[CO_2]$ in previous reviews [7,22]. The dilution effect and reducing Tr are considered as two key factors driving this decrease [17,22,56,60,61]. The significant positive correlation between [K] and Tr at both growth stages also confirmed the importance of Tr on the transportation of K in cucumber plants (Figure 7). Interestingly, at T1 stage, [K] in leaves was significantly increased with the increasing N supply, whereas [K] in roots was decreased. A possible reason is that K^+ is always transported accompanied with NO_3^- from roots to shoots to maintain the balance between K^+ and NO_3^- in xylem sap [62,63]. So, transportation of more NO_3^- in high N supply will cause a synchronous increase of K^+ in xylem sap, and result in an increase in [K] in leaves and a decrease in [K] in roots in higher N treatments.

Although an average decrease of 8% in leaf [Ca] has been reviewed [7,22], the change of [Ca] in different tissues and species under elevated $[CO_2]$ were different [60,61,64]. On one hand, being different from N, P, and K, the largest [Ca] are always found in cell walls in plants, where Ca^{2+} are stably fixed not only by electrostatic interactions with carboxylic groups of pectin, but also by coordination linkage with hydroxylic groups of polysaccharides [65]. So, when plant growth is improved under elevated $[CO_2]$, the demands of Ca^{2+} are also increased, which could partially offset the dilution effect [64]. On the other hand, Ca in xylem sap is mainly in the form of ions or chelate, so its transportation from roots to the aerial parts largely depends on the Tr [22]. However, due to the inhibited transportation of Ca in the phloem, Ca could hardly move and be reused from old tissues to young tissues [65].

Therefore, the accumulation of Ca could happen in old tissues and counterbalance the negativity of the dilution effect and reduced Tr on [Ca]. In the present work, an average decrease of [Ca] by 16.4%, 14.3%, and 10.4% in roots, stems, and leaves was observed under [CO_2] enrichment at T1 stage, respectively (Figure 5). The significant negative correlation between [Ca] and [CO_2], and the positive correlation between [Ca] and Tr, implied that the dilution effect and reducing Tr were two key factors driving [Ca] decrease in young plants (Figure 7) [7,63]. At T2 stage, elevated [CO_2] had little effects on [Ca], and this might be due to the accumulation of Ca in older tissues that offsets the dilution effect and reduced Tr (Figure 5). Since the leaves we analyzed did not include the old leaves that had fallen, the [Ca] in leaves was still influenced by Tr, and the positive correlation between [Ca] and Tr was observed only in leaves. Besides, the decrease of [Ca] with the increasing N supply may also be the result of the dilution effect by the increased biomass accumulation in higher N levels [18].

A decrease in [Mg] in elevated [CO_2] conditions has been frequently reported, and the average decrease value was 10% in leaves [7,18,60,64]. McGrath and Lobell found a 20% decrease in [Mg] but only a 10% decrease of other mineral elements in leaves of wheat [22]. Based on the chlorophyll concentration analysis and mass flow experiment, they calculated that the dilution effect accounted for a 10% reduction of [Mg], reduced Tr accounted for 3–10%, and reduced chlorophyll content accounted for 1–5% (represents the reduced demands). In the present work, we found an average decrease in [Mg] by 16.9%, 10.0%, and 7.8% in roots, stems, and leaves in elevated [CO_2] conditions, respectively, and an average decrease of [Mg] by 19.1–27.1% under higher N supply (Figure 5). The significant negative correlation between [Mg] and [CO_2] as well as [Mg] and N supply indicated the dilution effect had a strong impact on [Mg] (Figure 7). Meanwhile, the significant positive correlation between [Mg] in the aerial parts and Tr implied that reduced Tr also had detrimental effects on [Mg] (Figure 7).

5. Conclusions

According to our results, the cucumber biomass accumulation could be significantly increased by elevated [CO_2] accompanied by high N supplies. High N supplies could further improve the Pn and offset the decrease of the Gs and Tr by elevated [CO_2]. Thus, increasing N supply could alleviate or prevent the photosynthetic acclimation under elevated [CO_2] conditions. Based on the mineral nutrient concentrations in different [CO_2] and N supply treatments and the correlation analysis, we proposed the key factors affecting the concentration changes of each mineral element. The dilution effect was the main factor that reduced all mineral elements, whereas Tr had a large impact on the decrease of [N], [K], [Ca], and [Mg] except [P]. The decreased demands of N and Mg and the increased demands of Ca also influenced the concentrations of the corresponding elements in cucumber plants. However, this study was just a qualitative analysis. A quantitative analysis of the effect of each factor on the concentration changes is urgently needed. When we have better understanding of the mechanisms controlling the mineral concentration changes in cucumber plant responding to elevated [CO_2], we could optimize the mineral fertilization in order to improve the growth of cucumber plant under elevated [CO_2] conditions. Thus, a sustainable vegetable production with higher C and N use efficiency and less CO_2 emission and fertilizer input will be achieved.

Author Contributions: Conceptualization, X.L., J.D., W.C., and Z.D.; methodology, X.L., J.D., and W.C.; investigation, X.L., J.D., and W.C.; data curation, X.L. and W.C.; writing—original draft preparation, X.L. and W.C.; writing—review and editing, J.D. and N.S.G.; funding acquisition, W.C., X.L., and Z.D. All authors have read and agreed to the published version of the manuscript.

Funding: This research was funded by the National Natural Science Foundation of China (41877103), the Strategic Priority Research Program of the Chinese Academy of Science (XDA23020401), and the Frontier Project of Knowledge Innovation Program of Institute of Soil Science, Chinese Academy of Sciences (ISSASIP1635).

Acknowledgments: The authors are grateful to Hua Gong for assistance with the ICP analyses and to two anonymous reviewers for their constructive comments and suggestions.

Conflicts of Interest: The authors declare no conflict of interest.

References

1. Kläring, H.P.; Hauschild, C.; Heißner, A.; Bar-Yosef, B. Model-based control of CO_2 concentration in greenhouses at ambient levels increases cucumber yield. *Agric. For. Meteorol.* **2007**, *143*, 208–216. [CrossRef]
2. Mortensen, L.M. Review: CO_2 enrichment in greenhouses crop responses. *Sci. Hortic.* **1987**, *33*, 1–25. [CrossRef]
3. Gruda, N. Impact of Environmental Variables on Product Quality of Greenhouse Vegetables for Fresh Consumption. *Crit. Rev. Plant Sci.* **2005**, *24*, 227–247. [CrossRef]
4. Gruda, N.; Tanny, J. Protected crops. In *Horticulture: Plants for People and Places*; Dixon, G.R., Aldous, D.E., Eds.; Springer: Dordrecht, The Netherlands, 2014; Volume 1, pp. 327–405.
5. Kimball, B.A. Carbon Dioxide and Agricultural Yield: An Assemblage and Analysis of 430 Prior Observations. *Agron. J.* **1983**, *75*, 779–788. [CrossRef]
6. Van Der Kooi, C.J.; Reich, M.; Löw, M.; De Kok, L.J.; Tausz, M. Growth and yield stimulation under elevated CO_2 and drought: A meta-analysis on crops. *Environ. Exp. Bot.* **2016**, *122*, 150–157. [CrossRef]
7. Loladze, I. Rising atmospheric CO_2 and human nutrition: Toward globally imbalanced plant stoichiometry? *Trends Ecol. Evol.* **2002**, *17*, 457–461. [CrossRef]
8. Dong, J.L.; Gruda, N.; Lam, S.K.; Li, X.; Duan, Z.Q. Effects of Elevated CO_2 on Nutritional Quality of Vegetables: A Review. *Front. Plant Sci.* **2018**, *9*, 924. [CrossRef]
9. Cotrufo, M.F.; Ineson, P.; Scott, A. Elevated CO_2 reduces the nitrogen concentration of plant tissues. *Glob. Chang. Biol.* **1998**, *4*, 43–54. [CrossRef]
10. Gifford, R.M.; Barrett, D.J.; Lutze, J.L. The effects of elevated [CO_2] on the C:N and C:P mass ratios of plant tissues. *Plant Soil* **2000**, *224*, 1–14. [CrossRef]
11. LaNoue, J.; Leonardos, E.D.; Khosla, S.; Hao, X.; Grodzinski, B. Effect of elevated CO_2 and spectral quality on whole plant gas exchange patterns in tomatoes. *PLoS ONE* **2018**, *13*, e0205861. [CrossRef]
12. Reich, P.B.; Hobbie, S.E.; Lee, T.; Ellsworth, D.S.; West, J.B.; Tilman, D.; Knops, J.M.H.; Naeem, S.; Trost, J. Nitrogen limitation constrains sustainability of ecosystem response to CO_2. *Nature* **2006**, *440*, 922–925. [CrossRef] [PubMed]
13. Arp, W.J. Effects of source-sink relations on photosynthetic acclimation to elevated CO_2. *Plant Cell Environ.* **1991**, *14*, 869–875. [CrossRef]
14. Niinemets, Ü.; Tenhunen, J.D.; Canta, N.R.; Chaves, M.M.; Faria, T.; Pereira, J.S.; Reynolds, J.F. Interactive effects of nitrogen and phosphorus on the acclimation potential of foliage photosynthetic properties of cork oak, Quercus suber, to elevated atmospheric CO_2 concentrations. *Glob. Chang. Biol.* **1999**, *5*, 455–470. [CrossRef]
15. Reich, P.B.; Hungate, B.A.; Luo, Y. Carbon-Nitrogen Interactions in Terrestrial Ecosystems in Response to Rising Atmospheric Carbon Dioxide. *Annu. Rev. Ecol. Evol. Syst.* **2006**, *37*, 611–636. [CrossRef]
16. Sanz-Sáez, Á.; Erice, G.; Aranjuelo, I.; Nogués, S.; Irigoyen, J.J.; Sánchez-Díaz, M. Photosynthetic down-regulation under elevated CO_2 exposure can be prevented by nitrogen supply in nodulated alfalfa. *J. Plant Physiol.* **2010**, *167*, 1558–1565. [CrossRef] [PubMed]
17. Halpern, M.; Bar-Tal, A.; Lugassi, N.; Egbaria, A.; Granot, D.; Yermiyahu, U. The Role of Nitrogen in Photosynthetic Acclimation to Elevated [CO_2] in Tomatoes. *Plant Soil* **2019**, *434*, 397–411. [CrossRef]
18. Duval, B.D.; Blankinship, J.C.; Dijkstra, P.; Hungate, B.A. CO_2 Effects on Plant Nutrient Concentration Depend on Plant Functional Group and Available Nitrogen: A Meta-analysis. *Plant Ecol.* **2012**, *213*, 505–521. [CrossRef]
19. Taub, D.R.; Wang, X. Why are Nitrogen Concentrations in Plant Tissues Lower under Elevated CO_2? A Critical Examination of the Hypotheses. *J. Integr. Plant Biol.* **2008**, *50*, 1365–1374. [CrossRef]
20. Sardans, J.; Grau, O.; Chen, H.Y.H.; Janssens, I.A.; Ciais, P.; Piao, S.; Peñuelas, J. Changes in nutrient concentrations of leaves and roots in response to global change factors. *Glob. Chang. Biol.* **2017**, *23*, 3849–3856. [CrossRef]
21. McDonald, E.P.; Erickson, J.E.; Kruger, E.L. Can Decreased Transpiration Limit Plant Nitrogen Acquisition in Elevated CO_2? *Funct. Plant Biol.* **2002**, *29*, 1115–1120. [CrossRef]
22. Mcgrath, J.M.; Lobell, D.B. Reduction of Transpiration and Altered Nutrient Allocation Contribute to the Nutrient Decline of Crops Grown in Elevated CO_2 Concentrations. *Plant Cell Environ.* **2013**, *36*, 697–705. [CrossRef] [PubMed]

23. Stitt, M.; Krapp, A. The interaction between elevated carbon dioxide and nitrogen nutrition: The physiological and molecular background. *Plant Cell Environ.* **1999**, *22*, 583–621. [CrossRef]

24. Davey, P.A.; Wadge, K.; Parsons, A.; Atkinson, L.; Long, S.P. Does photosynthetic acclimation to elevated CO_2 increase photosynthetic nitrogen-use efficiency? A study of three native UK grassland species in open-top chambers. *Funct. Ecol.* **1999**, *13*, 21–28. [CrossRef]

25. Bloom, A.J.; Burger, M.; Asensio, J.S.R.; Cousins, A.B. Carbon Dioxide Enrichment Inhibits Nitrate Assimilation in Wheat and Arabidopsis. *Science* **2010**, *328*, 899–903. [CrossRef] [PubMed]

26. Bloom, A.J.; Burger, M.; Kimball, B.A.; Pinter, J.P.J. Nitrate assimilation is inhibited by elevated CO_2 in field-grown wheat. *Nat. Clim. Chang.* **2014**, *4*, 477–480. [CrossRef]

27. Gruda, N.; Balliu, A.; Sallaku, G. Crop Technologies: Cucumber. In *Good Agricultural Practices for Greenhouse Vegetable Production in the South East European Countries—Principles for Sustainable Intensification of Smallholder Farms*; Baudoin, W., Nersisyan, A., Shamilov, A., Hodder, A., Gutierrez, D., de Pascale, S., Nicola, S., Gruda, N., Urban, L., Tanny, J., Eds.; Food and Agriculture Organization of the United Nations: Roma, Italy, 2017; Chapter 2 of Part III; pp. 287–300.

28. Dong, J.; Li, X.; Chu, W.; Duan, Z. High nitrate supply promotes nitrate assimilation and alleviates photosynthetic acclimation of cucumber plants under elevated CO_2. *Sci. Hortic.* **2017**, *218*, 275–283. [CrossRef]

29. Agüera, E.; Ruano, D.; Cabello, P.; De La Haba, P. Impact of atmospheric CO_2 on growth, photosynthesis and nitrogen metabolism in cucumber (*Cucumis sativus* L.) plants. *J. Plant Physiol.* **2006**, *163*, 809–817. [CrossRef]

30. Dong, J.L.; Xu, Q.; Gruda, N.; Chu, W.Y.; Li, X.; Duan, Z.Q. Elevated and Super-elevated CO_2 Differ in Their Interactive Effect with Nitrogen Availability on Fruit Yield and Quality of Cucumber. *J. Sci. Food Agric.* **2018**, *98*, 4509–4516. [CrossRef]

31. Li, X.; Dong, J.L.; Chu, W.Y.; Chen, Y.J.; Duan, Z.Q. The relationship between root exudation properties and root morphological traits of cucumber grown under different nitrogen supplies and atmospheric CO_2 concentrations. *Plant Soil* **2018**, *425*, 415–432. [CrossRef]

32. Li, X.; Chu, W.Y.; Dong, J.L.; Duan, Z.Q. An Improved High-performance Liquid Chromatographic Method for the Determination of Soluble Sugars in Root Exudates of Greenhouse Cucumber Grown under CO_2 Enrichment. *J. Am. Soc. Hortic. Sci.* **2014**, *139*, 356–363. [CrossRef]

33. Sánchez-Guerrero, M.; Lorenzo, P.; Medrano, E.; Baille, A.; Castilla, N. Effects of EC-based irrigation scheduling and CO_2 enrichment on water use efficiency of a greenhouse cucumber crop. *Agric. Water Manag.* **2009**, *96*, 429–436. [CrossRef]

34. Gruda, N.; Bisbis, M.; Tanny, J. Impacts of protected vegetable cultivation on climate change and adaptation strategies for cleaner production—A review. *J. Clean. Prod.* **2019**, *225*, 324–339. [CrossRef]

35. Gruda, N.; Bisbis, M.; Tanny, J. Influence of climate change on protected cultivation: Impacts and sustainable adaptation strategies—A review. *J. Clean. Prod.* **2019**, *225*, 481–495. [CrossRef]

36. Bisbis, M.B.; Gruda, N.S.; Blanke, M.M. Securing Horticulture in a Changing Climate—A Mini Review. *Horticulturae* **2019**, *5*, 56. [CrossRef]

37. Gruda, N.S. Increasing Sustainability of Growing Media Constituents and Stand-Alone Substrates in Soilless Culture Systems. *Agronomy* **2019**, *9*, 298. [CrossRef]

38. Yamazaki, K. *Nutrient Solution Culture*; Pak-Kyo Co.: Tokyo, Japan, 1982; p. 251.

39. Zhao, F.; McGrath, S.P.; Crosland, A.R. Comparison of three wet digestion methods for the determination of plant sulphur by inductively coupled plasma atomic emission spectroscopy (ICP-AES). *Commun. Soil Sci. Plant Anal.* **1994**, *25*, 407–418. [CrossRef]

40. Broberg, M.C.; Högy, P.; Feng, Z.; Pleijel, H. Effects of Elevated CO_2 on Wheat Yield: Non-Linear Response and Relation to Site Productivity. *Agronomy* **2019**, *9*, 243. [CrossRef]

41. Jiang, Y.P.; Cheng, F.; Zhou, Y.H.; Xia, X.J.; Shi, K.; Yu, J.Q. Interactive effects of CO_2 enrichment and brassinosteroid on CO_2 assimilation and photosynthetic electron transport in Cucumis sativus. *Environ. Exp. Bot.* **2012**, *75*, 98–106. [CrossRef]

42. Hartz, T.; Baameur, A.; Holt, D. Carbon Dioxide Enrichment of High-value Crops under Tunnel Culture. *J. Am. Soc. Hortic. Sci.* **1991**, *116*, 970–973. [CrossRef]

43. Slack, G.; Hand, D.W. The effect of winter and summer CO_2 enrichment on the growth and fruit yield of glasshouse cucumber. *J. Hortic. Sci.* **1985**, *60*, 507–516. [CrossRef]

44. Peterson, A.G.; Ball, J.T.; Luo, Y.; Field, C.B.; Reich, P.B.; Curtis, P.S.; Griffin, K.L.; Gunderson, C.A.; Norby, R.J.; Tissue, D.T.; et al. The photosynthesis—Leaf nitrogen relationship at ambient and elevated atmospheric carbon dioxide: A meta-analysis. *Glob. Chang. Biol.* **1999**, *5*, 331–346. [CrossRef]

45. Lee, T.D.; Tjoelker, M.G.; Ellsworth, D.S.; Reich, P.B. Leaf Gas Exchange Responses of 13 Prairie Grassland Species in the Field Under Elevated Carbon Dioxide and Increased Nitrogen Supply. *New Phytol.* **2001**, *150*, 405–418. [CrossRef]

46. Sage, R.F.; Sharkey, T.D.; Seemann, J.R. Acclimation of Photosynthesis to Elevated CO_2 in Five C3 Species. *Plant Physiol.* **1989**, *89*, 590–596. [CrossRef] [PubMed]

47. Cure, J.D.; Acock, B. Crop responses to carbon dioxide doubling: A literature survey. *Agric. For. Meteorol.* **1986**, *38*, 127–145. [CrossRef]

48. Saralabai, V.; Vivekanandan, M.; Babu, R.S. Plant Responses to High CO_2 Concentration in the Atmosphere. *Photosynthetica* **1997**, *33*, 7–37. [CrossRef]

49. Nederhoff, E.M.; De Graaf, R. Effects of CO_2 on leaf conductance and canopy transpiration of greenhouse grown cucumber and tomato. *J. Hortic. Sci.* **1993**, *68*, 925–937. [CrossRef]

50. Kimball, B.; Kobayashi, K.; Bindi, M. Responses of Agricultural Crops to Free-Air CO_2 Enrichment. *Adv. Agron.* **2002**, *77*, 293–368.

51. Song, X.D.; Zhou, G.S.; Ma, B.L.; Wu, W.; Ahmad, I.; Zhu, G.L.; Yan, W.K.; Jiao, X.R. Nitrogen Application Improved Photosynthetic Productivity, Chlorophyll Fluorescence, Yield and Yield Components of Two Oat Genotypes under Saline Conditions. *Agronomy* **2019**, *9*, 115. [CrossRef]

52. Rogers, H.H.; Runion, G.; Krupa, S.V. Plant responses to atmospheric CO_2 enrichment with emphasis on roots and the rhizosphere. *Environ. Pollut.* **1994**, *83*, 155–189. [CrossRef]

53. Stitt, M. Nitrate Regulation of Metabolism and Growth. *Curr. Opin. Plant Biol.* **1999**, *2*, 178–186. [CrossRef]

54. Stulen, I.; den Hertog, J. Root Growth and Functioning under Atmospheric CO_2 Enrichment. *Vegetatio* **1993**, *104*, 99–115. [CrossRef]

55. Lotfiomran, N.; Köhl, M.; Fromm, J. Interaction Effect between Elevated CO_2 and Fertilization on Biomass, Gas Exchange and C/N Ratio of European Beech (*Fagus sylvatica* L.). *Plants* **2016**, *5*, 38. [CrossRef] [PubMed]

56. Soares, J.; Deuchande, T.; Valente, L.M.; Pintado, M.; Vasconcelos, M.W. Growth and Nutritional Responses of Bean and Soybean Genotypes to Elevated CO_2 in a Controlled Environment. *Plants* **2019**, *8*, 465. [CrossRef] [PubMed]

57. Lenka, N.K.; Lenka, S.; Singh, K.K.; Kumar, A.; Aher, S.B.; Yashona, D.S.; Dey, P.; Agrawal, P.K.; Biswas, A.K.; Patra, A.K. Effect of elevated carbon dioxide on growth, nutrient partitioning, and uptake of major nutrients by soybean under varied nitrogen application levels. *J. Plant Nutr. Soil Sci.* **2019**, *182*, 509–514. [CrossRef]

58. Bieleski, R.L. Phosphate Pools, Phosphate Transport, and Phosphate Availability. *Annu. Rev. Plant Physiol.* **1973**, *24*, 225–252. [CrossRef]

59. Hamburger, D.; Rezzonico, E.; Petétot, J.M.C.; Somerville, C.; Poirier, Y. Identification and Characterization of the Arabidopsis PHO1 Gene Involved in Phosphate Loading to the Xylem. *Plant Cell* **2002**, *14*, 889–902. [CrossRef]

60. Peet, M.M.; Huber, S.C.; Patterson, D.T. Acclimation to High CO_2 in Monoecious Cucumbers: II. Carbon Exchange Rates, Enzyme Activities, and Starch and Nutrient Concentrations. *Plant Physiol.* **1986**, *80*, 63–67. [CrossRef]

61. Jauregui, I.; Aparicio-Tejo, P.M.; Avila, C.; Cañas, R.; Sakalauskiene, S.; Aranjuelo, I. Root-shoot Interactions Explain the Reduction of Leaf Mineral Content in Arabidopsis Plants Grown under Elevated [CO_2] Conditions. *Physiol. Plant* **2016**, *158*, 65–79. [CrossRef]

62. Blevins, D.G.; Barnett, N.M.; Frost, W.B. Role of Potassium and Malate in Nitrate Uptake and Translocation by Wheat Seedlings. *Plant Physiol.* **1978**, *62*, 784–788. [CrossRef]

63. Triplett, E.W.; Barnett, N.M.; Blevins, D.G.; Reed, A.J.; Below, F.E.; Hageman, R.H. Organic Acids and Ionic Balance in Xylem Exudate of Wheat during Nitrate or Sulfate Absorption. *Plant Physiol.* **1980**, *65*, 610–613. [CrossRef]

64. Jauhiainen, J.; Vasander, H.; Silvola, J. Nutrient concentration in shape Sphagna at increased N-deposition rates and raised atmospheric CO_2 concentrations. *Plant Ecol.* **1998**, *138*, 149–160. [CrossRef]

65. Demarty, M.; Morvan, C.; Thellier, M. Calcium and the cell wall. *Plant Cell Environ.* **1984**, *7*, 441–448. [CrossRef]

 *agronomy*

Article

Antioxidant Seasonal Changes in Soilless Greenhouse Sweet Peppers

Damianos Neocleous [1] and Georgios Nikolaou [2,*]

[1] Department of Natural Resources and Environment, Agricultural Research Institute, 1516 Nicosia, Cyprus; d.neocleous@ari.gov.cy

[2] Department of Agriculture Crop Production and Rural Environment, School of Agricultural Sciences, University of Thessaly, 38446 Volos, Greece

* Correspondence: gnicolaounic@gmail.com; Tel.: +30-24-2109-3249

Received: 10 October 2019; Accepted: 6 November 2019; Published: 8 November 2019

Abstract: This study was commissioned to study the effect of the growing season on the antioxidant components of greenhouse sweet pepper crops, which is of scientific interest because of their possible beneficial health effects. The total antioxidant activity (estimated by ferric reducing antioxidant power-FRAP assay) major antioxidants (ascorbic acid, phenolics and carotenoids) and taste fruit quality characteristics (soluble solids, titratable acidity, dry matter and sugars) were recorded in soilless-grown sweet pepper cultivars of red, orange, yellow and green color at four harvesting season months, i.e., February (winter), May (spring), July (summer) and October (autumn). The results showed seasonal variations in antioxidant components and activity of pepper fruits. In most cases measured parameters showed higher values in spring (May) and summer (July) compared with winter (February) and autumn (October) growing seasons. This study indicates that during late autumn and winter, lower levels of solar irradiance, ultraviolet radiation and temperature in Mediterranean greenhouses can be insufficient to stimulate phytochemicals production in peppers; thus, plant–light interception must be more actively managed.

Keywords: *Capsicum annuum* L.; colored sweet peppers; antioxidant activity; phenolics; ascorbic acid; carotenoids; solar and ultraviolet radiation; soilless culture

1. Introduction

Peppers are among the most consumed vegetables worldwide. In the United States consumption of fresh bell peppers (*Capsicum annuum* L.) increased up to 20% the last decade and averaged 11.2 pounds per person in 2018 [1]. Bell peppers have become one of the most important cultivated fruiting vegetable in Mediterranean greenhouses. Particularly, solanaceous crops (tomato, pepper and eggplant) constitute about 60% of greenhouse-cultivated areas, which are often cultivated in soilless culture to enhance yield, product quality, water use efficiency and sustainability [2,3]. To date, peat-based substrates are most widely used in fruiting vegetable production systems, although rockwool is still the dominant soilless culture system in Europe [4]. Furthermore, alternative ecofriendly substrates (e.g., biowaste materials) with a lower carbon footprint coupled with soilless culture systems may be considered as a useful tool in sustainable greenhouse horticulture [5].

Recently, Mediterranean sweet peppers gained a growing interest to produce brightly colored (e.g., red, yellow, orange) fruits throughout the year based on the favorable climatic conditions (high radiation and mild autumn and winter temperatures) of the region [6], modern soilless culture technologies and the good marketability of the product. In this context, EU [7] has established specific marketing standards for sweet peppers. However, many consumers have additional quality requirements, which can go beyond legislation and standards [8]. Thus, the interest in foods of plant origin as a source of phytochemicals, increases throughout the years [9]. Specifically, the consumption of natural antioxidant compounds coming from fruits and vegetables such as phenolics, ascorbic acid and carotenoids have been associated with prevention of chronic diseases (e.g., cardiovascular disease and different forms of cancer) due to their ability to neutralize free radicals in the human body [10,11]. Peppers are among vegetable crops that are considered as naturally abundant in plant phytochemicals and their composition is of major importance for the beneficial health effects of the product [12,13]. As already mentioned peppers are widely grown in Mediterranean greenhouses, however, the lack of information regarding seasonal influences on the accumulation of dietary antioxidants in colored pepper fruits represents a drawback (e.g., in diets based on antioxidant intake). It is relevant that in other crops such as tomatoes and lettuce, seasonality can affect their antioxidant composition [14,15].

It is well known that genetic and environmental factors may directly affect the antioxidant composition of plant parts [16,17]. Particularly, under Mediterranean climatic conditions, increased solar irradiation and mild temperatures during winter months can affect plant antioxidant content and eventually fruit quality [18,19]. Although different species may have different responses [20], there is a general notion that solar ultraviolet radiation (UV total radiation 280–400 nm) has relevant biological effects on agroecosystems and induce the accumulation of phenolic compounds in the plant [14,19,20]. However, global UV fraction (i.e., ratio of the UV to global solar radiation) is highly dependent on variations in the concentration of clouds, water vapor and aerosols in the atmosphere and may vary from 2.0% to 9.5% [21,22]. Such environmental factors impact the quality of greenhouse vegetables. For example, a positive correlation between light levels and levels of secondary metabolites such as ascorbic acid in sweet peppers has been reported [8]. In accordance, the increase levels of Ultraviolet B (UVB) radiation (280–315 nm) enhanced several defense compounds such as carotenoids and flavonoids in bell peppers [23]. Ultraviolet A (UVA) radiation (315–400 nm) also enhanced the amounts of secondary metabolites, soluble carbohydrates, free amino acids and proteins in greenhouse peppers [24]. Consequently, it is assumed that changes in global solar and UV radiation and air temperature levels at different harvesting seasons, may affect the antioxidant components of greenhouse soilless-grown peppers, which is of scientific interest because of their possible beneficial effects on human health.

To document this response, this study was designed to evaluate seasonal effects in total antioxidant activity (estimated by ferric reducing antioxidant power-FRAP assay) and important antioxidant compounds (i.e., ascorbic acid, phenolics, carotenoids) so as in other fruit quality characteristics (soluble solids, titratable acidity, pH, dry matter, reducing sugars) in soilless sweet peppers of red, orange, yellow, and green color, in Mediterranean greenhouses.

2. Materials and Methods

2.1. Plant Material and Agronomic Features

Data was collected from the same plants of a year round sweet pepper crop, giving harvests in 2017, at greenhouse facilities of Agricultural Research Institute of Cyprus (34°94′ N, 33°19′ E, altitude 40 m). Three colored pepper (*Capsicum annuum* L.) cultivars (Agroglobal, Hungary), namely red (*cv. Castello*), orange, (*cv. Donat*), yellow (*cv. Solero*) and one green local variety (*cv. Glikes*) were grown on rockwool substrates (Grodan Company, Denmark; slabs dimensions 100 cm × 20 cm × 7.5 cm). Substrates were placed into Polygal-gutters (Mapal Plastics, Israel) 12-m long, which were supported by metal frames in 12 single rows. Each experimental unit consisted of one Polygal-gutter planted with 24 plants. Three replications (one Polygal-gutter per replication) for each cultivar were randomly arranged in three blocks. The plants were vertically supported ('V' system) giving a planting density of 2.0 plants m^{-2} [25]. Colored fruits (red, orange, yellow) ripened on the vine following the mature green stage. The crop was transplanted on rockwool slabs 3 months prior harvesting, which started in February and terminated in October 2017. Harvested fruits of each experimental unit were weighted and counted to determine fresh yield and average weight of the fruit. Total marketable fruit yield was the combined total of Extra Class and Class I according to EU marketing standards [7]. For quality analysis, fruits were sampled at four harvesting times, particularly in February (winter), May (spring), July (summer) and October (autumn).

The irrigation schedule was controlled by Fertimix hydroponic head unit (Galgon, Kfar Blum, Israel) and adjusted to light conditions [26]. The start of irrigation was depended from light sums according to the growth stage (1500–2800 kJ/m^2) targeting a leaching fraction of about 20%. Drip emitters delivered the nutrient solution directly to the root zone of pepper plants. Electrical conductivity (EC) and pH values were monitored in both irrigation and drainage water. The target EC levels of the irrigation nutrient solution were adjusted in response to radiation differences ($\pm$ 0.3 dS/m; higher EC at low radiation and lower EC at high radiation). The hydroponic fertigation head prepared a nutrient solution (NS) for growing soilless peppers in Mediterranean greenhouses with NS composition originating from the literature [2]. The following NS was delivered to the plants at the vegetative stage: 5.4 mM K$^+$, 4.65 mM Ca^{2+}, 1.6 mM Mg^{2+}, 1.2 mM NH$_4^+$, 13.7 mM NO$_3^-$, 1.2 mM H$_2$PO$_4^-$, 1.85 mM SO$_4^{2-}$, 15 µM Fe as Fe-EDDHA, 10 µM Mn, 5 µM Zn, 0.8 µM Cu, 30 µM B, and 0.5 µM Mo. Corresponding EC and pH values were 2.20 dS/m and 5.6, respectively. At the reproductive stage the plants were fed with the following nutrient solution: 5.8 mM K$^+$, 4.5 mM Ca^{2+}, 1.40 mM Mg^{2+}, 0.6 mM NH$_4^+$, 13.0 mM NO$_3^-$, 1.2 mM H$_2$PO$_4^-$, 1.75 mM SO$_4^{2-}$, 15 µM Fe as Fe-EDDHA, 10 µM Mn, 5 µM Zn, 0.8 µM Cu, 30 µM B, and 0.5 µM Mo. The EC and pH values of this NS were 2.10 dS/m and 5.6, respectively.

2.2. Greenhouse Facilities and Climatic Data

The experiment was conducted in a North–South oriented greenhouse with a total ground area of 216 m^2, with cutter height 3.50 m, ridge height 5.26 m, spans width 6 m and total length 18 m. The gable end and side walls were covered with double-walled polycarbonate and the roof was covered with a common polyethylene film (88% light global transmission, 55% light diffused transmission and 88% thermal efficiency). In each greenhouse span there was a single continuous rooftop window for natural ventilation. In addition, evaporative cooling was performed by a fan-pad system consisted of four fans, two at each span and a wetted pad. The greenhouse floor was completely covered by a white, water permeable polypropylene sheet.

External climatic parameters measured were air relative humidity (RHo, %) and temperature (To, °C) (Sensor type PT 100; Galcon, Kfar Blum, Israel) and net solar radiation (Gh, kJ/m^2) with a pyranometer at 3 m above the greenhouse (Sensor type TIR-4P; Bio Instruments Company, Chisinau, Moldova). The same types of sensors were used for monitoring relative humidity and air temperature within the greenhouse. All measurements were recorded every 30 s on a data logger system (Galileo controller; Galcon, Kfar Blum, Israel) and a ten-minute average was estimated. Vapor-pressure deficit (VPD) was estimated based on greenhouse air temperature and relative humidity. The mean daily value of ultraviolet radiation over a month was calculated based on global solar radiation, following Equation (1). According to this formula [27], the hourly and daily values of both radian fluxes are highly correlated with a general linear relationship of the following form providing coefficients of determination of R^2 always greater than 0.91 for hourly and 0.88 for daily fittings in the case of Cyprus.

$$Guv = aGh \tag{1}$$

where Guv is the solar global ultraviolet radiation (kJ/m^2); Gh is the solar global radiation (kJ/m^2); a is the slope corresponding to measurements.

2.3. Fruit Quality Measurements

The fruit quality characteristics (i.e., ascorbic acid, sugars, total soluble solids, pH, titratable acidity and dry matter) were determined at commercial maturity stage (Figure 1) in randomly selected samples excluding outliers, from each experimental unit. The edible part of the fruit was homogenized and soluble solids (°Brix; Atago PR-1, Tokyo, Japan), pH (Mettler Toledo, Switzerland), titratable acidity (titration with sodium hydroxide solution to pH 8.2, % citric acid), and fruit dry matter (g/100 g FW, drying at 70 °C) were recorded. The ration between total soluble solids and titratable acidity was calculated (TSS/TA). The content of fruits in ascorbic acid (mg AA/100 g FW) and reducing sugars (mg Glucose+Fructose/g FW) were determined by a Merck RQflex reflectometer. Briefly, ascorbic acid reduces yellow molybdophosphoric acid to phosphormolybdenum blue that is determined reflectometrically as reducing sugars after enzymatic conversion with glucose-6-phosphate dehydrogenase and diaphorase according to the company protocols (Merck, Darmstadt, Germany). For the determination of the total phenolic content and antioxidant activity, subsamples of fruits were kept-frozen at −30 °C until the date of analysis. Quantitative determination of phenolic substances was performed in fruits samples (10 g) homogenized with 25 mL acidified acetone (acetone: water: acetic acid 70:29.5:0.5, v:v:v), following the Folin–Ciocalteu procedure [28]. The absorbance of the reaction mixtures (0.25 mL extract, 2.5 mL FolinCiocalteu's reagent (previously diluted 1:10 with deionized water) and 2 mL 7.5% Na_2CO_3) after 5 min at 50 °C was measured at 760 nm (UV-Vis spectrophotometer Helios Zita, Thermo Fisher Scientific, USA). The results were expressed in gallic acid equivalents (mg GAE) per g of fresh weight, using a calibration curve (GAE/g FW) [29]. For the determination of the antioxidant capacity of pepper fruits by the ferric reducing antioxidant power method (FRAP; [30,31], sample extracts (100 µL) were mixed with 3 mL FRAP reagent (1:1:10 mixture of 20 mM $FeCl_3$, 10 mM TPTZ and 0.3 M acetate buffer at pH 3.6) and after 4 min at 37 °C the absorbance at 593 nm was recorded. Ascorbic acid (AA) was used as standard and the results were expressed per g of fresh weight (µmol AA/g FW) as previously described [29]. Chlorophyll content was determined in green fruit samples blended with 80% acetone measuring the absorbance of the supernatant at 648 and 664 nm [29]. Total carotenoids content in colored fruits extracts (hexane: acetone: ethanol 50:25:25, v:v:v) was determined at 450 nm following concentration calculations as previously reported [32]. The results were expressed as mg β-carotene per g of FW.

(a)   **(b)**   **(c)**   **(d)**

Figure 1. Fruit maturity at the time of harvest in (**a**) red; (**b**) orange; (**c**) yellow; (**d**) green sweet pepper (*Capsicum annuum* L.) cultivars grown in greenhouse soilless culture.

2.4. Statistical Analysis

Experimental layout in the greenhouse consisted of three replicates for each cultivar arranged in a randomized complete block design. SAS software system (ver. 9.2, Cary, NC, USA) was used for analysis of variance (ANOVA) for all traits studied and means were separated using DMRT at 5% level of significance. Pearson correlation coefficients between antioxidant variables studied were calculated.

3. Results

3.1. Greenhouse Microclimate and External Climatic Data

The monthly mean values, of outdoor climate data (i.e., air temperature and relative humidity) and inside greenhouse microclimate; global solar radiation and calculated ultraviolet radiation are presented in Table 1. The monthly variability of both radiant fluxes, Gh and Guv, is shown in Figure 2.

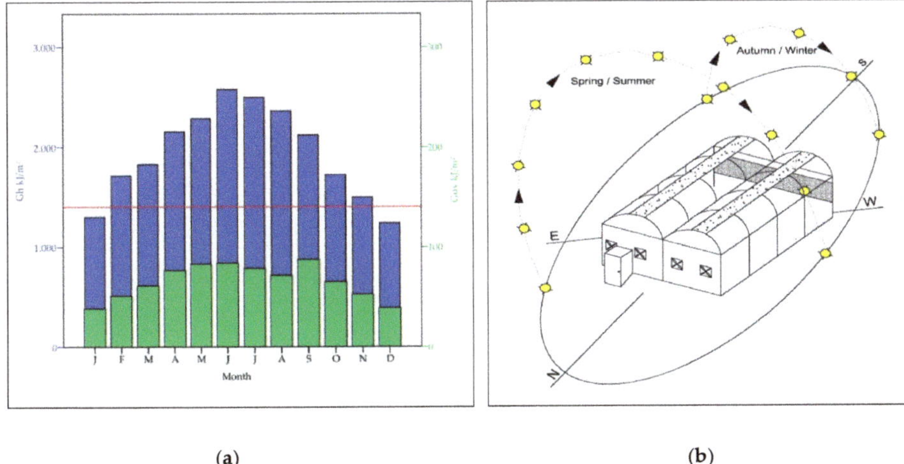

(a)   (b)

Figure 2. (**a**) Monthly means of global solar and ultraviolet radiation (kJ/m^2; the bars are in relation but not proportional to the data they encode); (**b**) Sun orientation during the experiment; Straight-line embedded in the graph (**a**) represents minimum radiation requirements for cultivation of thermophilic vegetable species in Mediterranean greenhouses [33].

The mean estimated values kJ/m^2 (±standard deviation) of Gh and Guv were respectively 1378 (799.99) and 42 (24.08) in winter (D-J-F), 2087 (1216.19) and 74 (43.48) in spring (M-A-M), 2451 (1222.93) and 77 (38.76) in summer (J-J-A) and 1788 (976.99) and 69 (39.42) in autumn (S-O-N). Seasonal variations of Guv value followed seasonal variations of Gh. Particularly, higher Guv values observed during summer and lower values at winter; as affected by yearly length of a day and the solar zenith angle. However, from Table 1 we can observe that, despite the decrease of Gh from August to September by 15%, Guv values increased by 17%. The line in Figure 2a represents minimum radiation requirements for cultivation of thermophilic vegetable species in Mediterranean greenhouses during N-D-J according to the literature [33].

Table 1. Monthly mean values (±standard deviation) of outdoor climate data and inside greenhouse microclimate for daylight hours.

	J	F	M	A	M	J	J	A	S	O	N	D
Gh	1075 (810)	1509 (1022)	1801 (1096)	2269 (1234)	2528 (1211)	2838 (1135)	2746 (1144)	2515 (1148)	2151 (1095)	1681 (987)	1251 (857)	972 (751)
Guv	32 (24)	45 (30)	60 (37)	81 (44)	91 (44)	92 (37)	86 (36)	76 (34)	89 (45)	63 (37)	44 (30)	31 (24)
To	14.8 (4.2)	20.5 (1.7)	22.6 (4.2)	26.0 (2.2)	28.6 (2.5)	31.7 (2.1)	33.5 (3.1)	30.0 (2.9)	28.5 (3.0)	22.7 (3.4)	20.5 (2.6)	12 (3.2)
RHo	55.7 (14.7)	49.3 (12.5)	46.8 (8.4)	48.4 (5.9)	53.5 (9.5)	65.8 (8.5)	63.9 (14.8)	72.0 (11.5)	55.4 (11.0)	47.2 (15.0)	49.2 (11.9)	53.0 (9.8)
Ti	21.0 (4.9)	23.7 (2.4)	24.5 (2.9)	25.4 (3.5)	26.2 (2.9)	27.9 (2.0)	28.0 (2.0)	27.9 (2.0)	24.2 (2.2)	23.8 (4.7)	23.0 (3.6)	21.8 (2.3)
RHi	69.8 (11.8)	59.2 (12.5)	60.2 (11.3)	61.5 (7.8)	67.2 (6.6)	69.8 (5.3)	65.3 (9.1)	75.1 (14.3)	60.3 (12.8)	62.9 (14.7)	54.9 (11.7)	61.5 (6.3)
VPD	0.9 (0.5)	1.2 (0.4)	1.1 (0.2)	1.1 (0.3)	1.1 (0.3)	1.3 (0.4)	1.4 (0.2)	1.1 (0.3)	1.2 (0.3)	1.2 (0.2)	1.3 (0.2)	0.9 (0.3)

Gh, global solar radiation (kJ/m^2); Guv, ultraviolet radiation (kJ/m^2); To, outside greenhouse air temperature (°C); RHo, outside greenhouse air relative humidity (%); Ti, inside greenhouse air temperature (°C); RHi, inside greenhouse air relative humidity (%); VPD, inside greenhouse air vapor pressure deficit (kPa).

3.2. Antioxidants and Other Fruit Quality and Yield Parameters

Season and cultivar were in most cases significant sources of variation (Table 2). Because of some interactions observed between season and cultivar, data were graphically presented within each cultivar (Figure 3). The antioxidant activity (FRAP values; μmol AA/g FW) of the pepper cultivars tested showed higher values in spring (May) and summer (July) compared with winter (February) and autumn (October) (Figure 3). An increase was also observed for total phenolics (GAE/g FW) during May compared with February in orange and yellow cultivars, however, in red and green cultivars the increase was not significant ($p < 0.05$; Figure 3). Accordingly, ascorbic acid content (mg AA/100 g FW) showed higher value in May and July and lower in February and October in all cases (Figure 3). Similarly, sugars (mg Glucose + Fructose/g FW) were accumulated at higher levels during May and July compared with the other two months in red, orange and yellow pepper fruits, whereas in green fruits the values observed were not differentiated with harvest time (Figure 3). Changes in total soluble solids (°Brix) with harvesting time-followed alterations of the sugar content as may be expected. However, in some cases (orange cultivar) differences were not consistent (Figure 3). The titratable acidity (% citric acid) was higher during July compared with February for red, orange and yellow cultivars, whereas no variation was observed among harvest times for the green cultivar. Yet importantly, carotenoids content (mg β-carotene/g FW) at harvesting times May and July was enhanced in colored pepper fruits in relation with the other two months depending on the cultivar (Figure 3). On the contrary, total chlorophyll (a + b) content in the green cultivar remained unaffected by the growing season (Figure 3), so as the dry matter content in most of the cases. Similarly, the estimated ratio total soluble solids to titratable acidity (TSS/TA) was not differentiated among harvest months in the cultivars tested. Overall, total marketable fruit yield (kg/m^2) and mean fruit weight (g/fruit) was greater in colored peppers in relation to the green cultivar (Figure 4). On the contrary, more fruits per m^2 were produced by the green than the rest of the colored cultivars (Figure 4). Last but not least, FRAP values were highly correlated ($p < 0.001$) with phenolics ($r = 0.81$) and ascorbic acid ($r = 0.84$), whereas pigment phytochemicals had a lower influence to the reducing potential. In addition, phenolics were highly correlated with ascorbic acid ($r = 0.77$) so as both with reducing sugars ($r = 0.60$ and $r = 0.83$, respectively).

Table 2. Analysis of variance table and levels of significance ($p < 0.05$, $p < 0.01$, $p < 0.001$).

	F Probability								
Source	**FRAP**	**Ph**	**AA**	**Sug**	**TSS**	**TA**	**Car**	**Chl**	**DM**
Season	0.0010	0.6085	<0.0001	<0.0001	0.0012	<0.0001	<0.0001	0.1736	0.1724
Cultivar	<0.0001	<0.0001	<0.0001	<0.0001	<0.0001	<0.0001	<0.0001	<0.0001	<0.0001
S × C	0.0305	0.0042	0.0534	0.1100	0.6260	0.9321	0.0035	<0.0001	0.0016

Antioxidant activity (FRAP), phenolics (Ph), ascorbic acid (AA), sugars (Sug), total soluble solids (TSS), titratable acidity (TA), carotenoids (Car), total chlorophyll (Chl) and dry matter (DM).

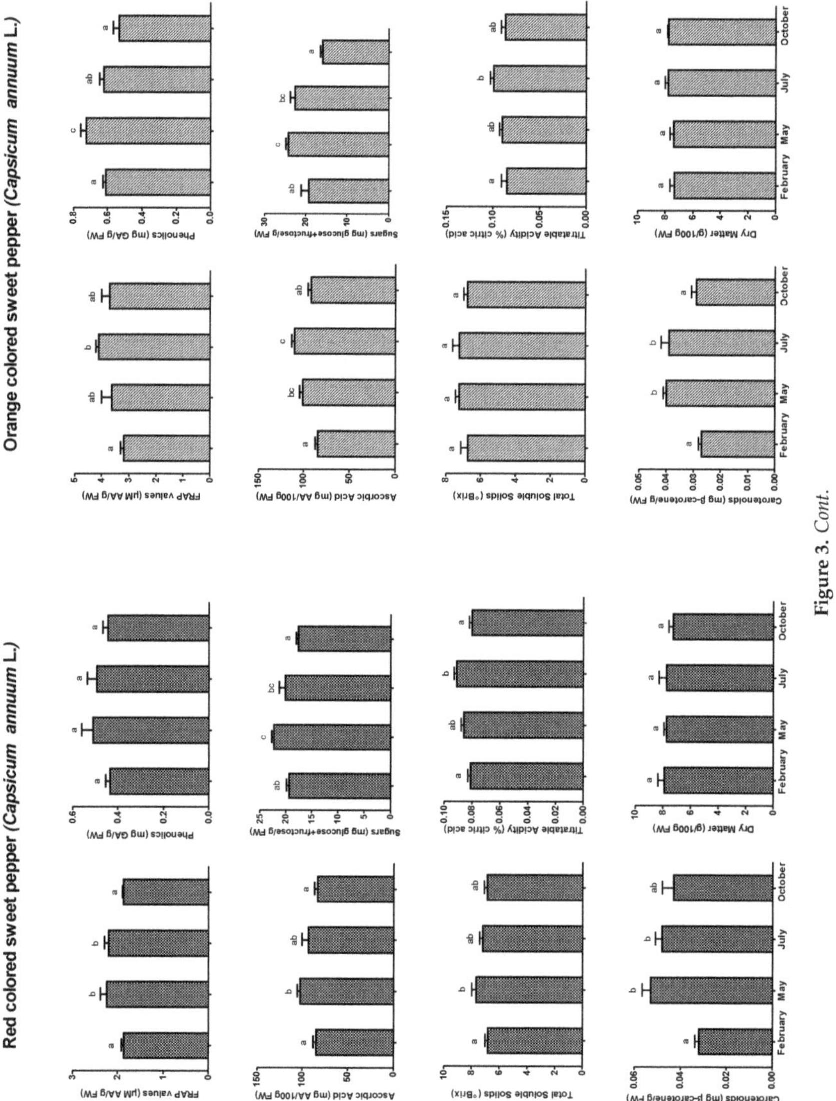

Figure 3. Cont.

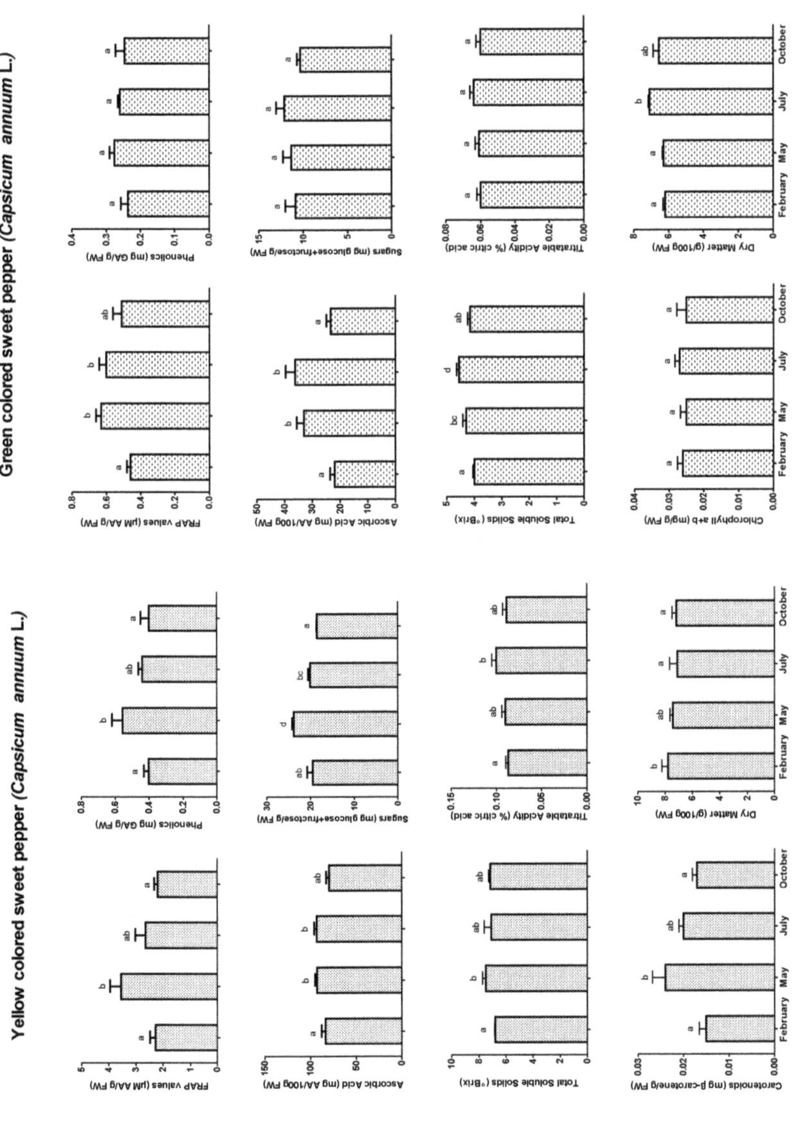

Figure 3. Antioxidant activity (ferric reducing antioxidant power (FRAP) values), phenolics, ascorbic acid, sugars, total soluble solids (°Brix), titratable acidity, carotenoids, total chlorophyll (a + b) and dry matter changes in red, orange, yellow and green sweet pepper cultivars at four harvest months, February, May, July and October 2017. Different lower-case letters above the bars indicate significant differences between mean values at *p* < 0.05 according to Duncan's test. Error bars indicate ± standard errors of the mean.

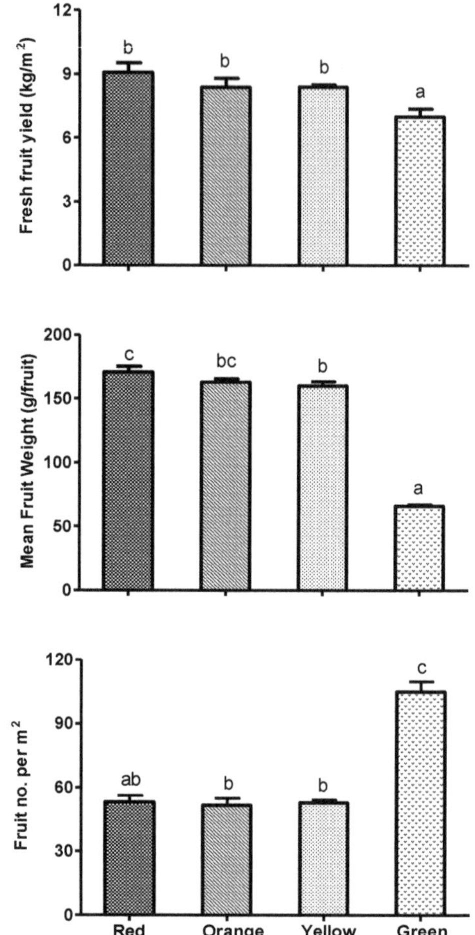

Figure 4. Total marketable fruit yield (kg/m^2), average fruit weight (g/fruit) and fruit number per m^2, in red, orange, yellow and green sweet pepper cultivars from February 2017 until October 2017. Different lower-case letters above the bars indicate significant differences between mean values at $p < 0.05$ according to Duncan's test. Error bars indicate ± standard errors of the mean.

4. Discussion

Pepper plants were grown in a plastic greenhouse under soilless conditions giving harvests from February until October 2017, to study seasonal variations in fruit antioxidants and antioxidant activity in cultivars of red, orange, yellow and green color. Experimental results on other crops such as spinach and tomatoes demonstrated that antioxidant activity and phytochemicals including phenolics and ascorbic acid were greatly affected by the growing season [14]. In this study, total antioxidant activity and major antioxidant components including, phenolics, ascorbic acid, and carotenoids were higher in pepper fruits during harvesting on May and July in relation to the other two harvesting periods (i.e., February and October) depending on the cultivar (Figure 3). This increase was associated with an increase (avg. 30%) of solar and ultraviolet (UV) radiation and elevated temperature conditions inside the greenhouse from the autumn–winter to spring–summer period (Table 1). Ultraviolet radiation (i.e., Guv values) followed seasonal variations of global solar radiation (i.e., Gh values) as affected by the yearly length of a day and the solar zenith angle Figure 2; [27]. In this context, literature suggests that light intensity is closely related with the biosynthesis of bioactive (i.e., biologically active) compounds of secondary metabolism in plants such as phenolics, because light increases the activities of key enzymes in the phenolic synthesis such as phenylalanine ammonia lyase (PAL) [14]. It is also known that the synthesis of secondary metabolites in plants is involved in the defense mechanism against several stresses such as UV (280–400 nm) radiation [11]. In accordance, other authors [9] reported phenolics accumulation in different fruits and vegetables in response to UVB (280–315 nm) exposure due to increase expression of phenylpropanoid pathway genes. For example, UV-treated sweet pepper plants contained higher amounts of bioactive compounds such as phenolics so as soluble carbohydrates and photosynthetic pigments at earlier reports [24]. In addition, biosynthesis of phenolic constituents with well-known antioxidant properties in *Capsicum annuum* and other *Capsicum* species including flavonoids, quercetin and luteolin [34], were connected to UVB radiation [11]. These irradiance effects on metabolic functions may also be used to explain phenolics changes with growing season in the current study. In absolute values, even though phenolic and ascorbic acid accumulation is greatly affected by pre- and post-harvest factors [16], the values observed in this study were in the same range with those reported in other experiments for hydroponic sweet peppers grown in a Mediterranean type climate [35].

Moreover, light and temperature in the optimal range stimulate photosynthesis, which leads to the accumulation of reducing sugars and soluble solids in the fruits [36]. Indeed, the increase of ascorbic acid in colored pepper fruits of *C. annuum* L. during May and July was accompanied by an increase in reducing sugars and soluble solids, which supports previous findings for other *Capsicum* species, correlating ascorbic acid biosynthesis with light intensity [8] and reducing sugars (ascorbic acid is synthesized from D-glucose) [37]. Overall, ascorbic acid in fruits of red, orange and yellow colors varied in a range of 80–110 mg/100 g FW, which seems to fall close to that reported previously for greenhouse-grown colored sweet peppers in Spain [38]. Noteworthy, the values observed during spring–summer months were higher than 90 mg/100 g FW and the values during autumn-winter months were lower than 90 mg/100 g FW (Figure 3). Taking into consideration that 90 mg/day is the threshold of ascorbic acid recommended daily allowances for adult men set by the Food and Nutrition Board of the Institute of Medicine in the United States as cited in [37], this study clearly shows that the challenge to eliminate nutritional variations all year round of the selected crops is fundamental. Accordingly, growing season affected carotenoids formations in colored pepper fruits, with higher values observed during spring and summer at elevated light and temperature conditions. Earlier studies have clearly demonstrated that greater exposure to sunlight and higher temperature enhances carotenoid biosynthesis (isoprenoid pathway) in fruits [32]. Thus, the physiological mechanism implicated in the differences between seasons in the current is presumably based on biosynthesis of secondary metabolites from carbon skeletons derived from photosynthetic process [36]. Thus, it is reasonable to conclude that pepper fruit biochemistry was upregulated in response to prevailing environmental conditions (light and temperature) as previously suggested [8]. Considering the total antioxidant activity, the higher activity in spring and summer months was in accordance with the elevated concentrations of phenolics, ascorbic acid and carotenoids in pepper fruits, which confirms the close relationship between these antioxidant components with antioxidant activity [35,39] and their synergistic effect [40]. In general, antioxidant activity reflects the cumulative antioxidant function of a food product [13] and may serve as a tool in epidemiological studies [12]. Particularly, peppers had the second highest total antioxidant capacity among 34 vegetables as reported previously [13]. Summarizing, these results let us suggest that in Mediterranean greenhouses during late autumn and winter light conditions, they need to be more carefully managed (Figure 2) to stimulate brightly colored peppers with higher content of phytochemicals. On the other hand, there is a growing interest among vegetable producers to better control pest and diseases using UV-absorbing films as greenhouse material [24,41], however, UV exclusion may lead to lower concentrations of secondary metabolites in plants and deterioration of product nutritional quality [8]. Therefore, it can be hypothesized that much stricter selection of the greenhouse covers UV blocking or transmitting properties in conjugation with the cultivated crop and production practices (e.g., crop orientation, harvesting time, planting density), would be beneficial to the growers to reduce pesticide use without a negative effect on phytochemical composition of selected crops in Mediterranean greenhouses. Improvement of product quality in soilless cultivations by manipulating nutrient solution composition has also been stated in several cases [42]. Moreover, the use of artificial light sources (e.g., UV light-emitting diodes) in the greenhouse could not be ruled out, however at the moment is of low usability in horticulture due to operating costs and law restrictions [43]. Furthermore, the data set of this study indicated that although variations in total soluble solids and titratable acidity of pepper fruits may exist at different times of the year, the sensory TSS/acid ratio remained unaffected with time. This may suggest that taste quality of peppers would probably not greatly vary among harvest months in any of the cultivars tested, which is of importance for the market value of the product but may not always coincide with the micronutritional quality of the fruits [44]. Yield results also revealed that there is always a need to validate the results of the earlier studies with the new high yielding cultivars, modern growing systems and prevailing environmental conditions. Indicatively, red cultivar showed greater yield and average fruit weight, followed by orange and yellow, with the lowest values observed in the green one. Total marketable fruit yield and mean fruit weight varied from 8.4 to 9.1 (kg/m^2) and 162 to 171 (g/fruit), respectively, for colored fruited peppers. In this content, greenhouse pepper production in Spain yields about 7 kg/m^2 yearly, whereas colored peppers grown in Florida yielded 6.9 to 11.3 kg/m^2 in a harvesting period from October to March with an average fruit weight from 161 to 212 g/fruit [45].

5. Conclusions

This study clearly shows the challenge to eliminate fruit antioxidant phytochemical variations in yearly grown greenhouse colored pepper crops. It was clearly shown that the total antioxidant activity and major antioxidant components including phenolics, ascorbic acid, and carotenoids tend to accumulate in higher amounts in sweet pepper fruits at harvesting times with higher solar and ultraviolet radiation and elevated temperature (i.e., spring and summer). Collectively, these results indicate that in Mediterranean greenhouses during late autumn and winter, light conditions can be insufficient to stimulate brightly colored peppers with elevated content of antioxidants, thus the antioxidant activity. This further suggests that a proper selection of greenhouse type and cover material in response to plant–light interception in conjugation with the selected crop and cultivation system may be a prerequisite to optimize environmental conditions for plant growth and elevated antioxidant phytochemicals in yearly grown sweet colored peppers in Mediterranean greenhouses.

Author Contributions: Formal analysis, D.N. and G.N.; Investigation, D.N.; Methodology, D.N.; Writing— review & editing, D.N. and G.N.

Funding: This work was supported by the Agricultural Research Institute of Cyprus and authors did not receive any specific grant from funding agencies in the public, commercial, or not-for-profit sectors.

Conflicts of Interest: The authors declare no conflict of interest.

References

1. USDA, United States Department of Agriculture, Economic Research Service. Vegetables and Pulses Yearbook Tables. Available online: https://www.ers.usda.gov/data-products/vegetables-and-pulses-data/ (accessed on 10 June 2019).
2. Savvas, D.; Gianquinto, G.P.; Tuzel, Y.; Gruda, N. Soilless Culture. In *Good Agricultural Practices for Greenhouse Vegetable Crops. Principles for Mediterranean Climate Areas*; FAO, Ed.; Food and Agriculture Organization of the United Nations: Rome, Italy, 2013; pp. 303–354.
3. Gruda, N. Do soilless culture systems have an influence on product quality of vegetables? *J. Appl. Bot. Food* **2009**, *82*, 511–519.

4.   Gruda, N.; Qaryouti, M.M.; Leonardi, C. Growing Media. In *Good Agricultural Practices for Greenhouse Vegetable Crops. Principles for Mediterranean Climate Areas*; FAO, Ed.; Food and Agriculture Organization of the United Nations: Rome, Italy, 2013; pp. 301–371.

5.   Gruda, N. Increasing Sustainability of Growing Media Constituents and Stand-Alone Subsstrates in Soilless Culture Systems. *Agronomy* **2019**, *9*, 298. [CrossRef]

6.   Gázquez, J.C.; López, J.C.; Pérez-Parra, J.J.; Baeza, E.J.; Saéz, M.; Parra, A. Greenhouse Cooling Strategies for Mediterranean Climate Areas. *Acta Hortic.* **2008**, *801*, 425–432. [CrossRef]

7.   EUR-Lex. Access to European Union Law, Commission Implementing Regulation (EU) No 543/2011. Available online: https://eur-lex.europa.eu/legal-content/GA/TXT/?uri=CELEX:32011R0543 (accessed on 10 June 2019).

8.   Gruda, N. Impact of Environmental Factors on Product Quality of Greenhouse Vegetables for Fresh Consumption. *Crit. Rev. Plant Sci.* **2015**, *24*, 227–247. [CrossRef]

9.   Santin, M.; Lucini, L.; Castagna, A.; Rocchetti, G.; Hauser, M.-T.; Ranieri, A. Comparative "Phenol-Omics" and Gene Expression Analyses in Peach (Prunus Persica) Skin in Response to Different Postharvest UV-B Treatments. *Plant Physiol. Biochem.* **2019**, *135*, 511–519. [CrossRef] [PubMed]

10. Prakash, D.; Gupta, K.R. The Antioxidant Phytochemicals of Nutraceutical Importance. *Open Nutraceuticals* **2009**, *2*, 20–35. [CrossRef]

11. Pandey, K.B.; Rizvi, S.I. Plant Polyphenols as Dietary Antioxidants in Human Health and Disease. *Oxid. Med. Cell. Longev.* **2009**, *2*, 270–278. [CrossRef] [PubMed]

12. O'Sullivan, A.M.; O'Callaghan, Y.C.; O'Grady, M.N.; Queguineur, B.; Hanniffy, D.; Troy, D.J.; Kerry, J.P.; O'Brien, N.M. In Vitro and Cellular Antioxidant Activities of Seaweed Extracts Prepared from Five Brown Seaweeds Harvested in Spring from the West Coast of Ireland. *Food Chem.* **2011**, *126*, 1064–1070. [CrossRef]

13. Pellegrini, N.; Serafini, M.; Colombi, B.; Del Rio, D.; Salvatore, S.; Bianchi, M.; Brighenti, F. Total Antioxidant Capacity of Plant Foods, Beverages and Oils Consumed in Italy Assessed by Three Different In Vitro Assays. *J. Nutr.* **2003**, *133*, 2812–2819. [CrossRef] [PubMed]

14. Toor, R.K.; Savage, G.P.; Lister, C.E. Seasonal Variations in the Antioxidant Composition of Greenhouse Grown Tomatoes. *J. Food Compos. Anal.* **2006**, *19*, 1–10. [CrossRef]

15. Li, Q.; Kubota, C. Effects of Supplemental Light Quality on Growth and Phytochemicals of Baby Leaf Lettuce. *Environ. Exp. Bot.* **2009**, *67*, 59–64. [CrossRef]

16. Blokhina, O. Antioxidants, Oxidative Damage and Oxygen Deprivation Stress: A Review. *Ann. Bot.* **2003**, *91*, 179–194. [CrossRef] [PubMed]

17. Shan, B.; Cai, Y.Z.; Sun, M.; Corke, H. Antioxidant Capacity of 26 Spice Extracts and Characterization of Their Phenolic Constituents. *J. Agric. Food Chem.* **2005**, *53*, 7749–7759. [CrossRef] [PubMed]

18. Conte, A.; Conversa, G.; Scrocco, C.; Brescia, I.; Laverse, J.; Elia, A.; Del Nobile, M.A. Influence of Growing Periods on the Quality of Baby Spinach Leaves at Harvest and during Storage as Minimally Processed Produce. *Postharvest Biol. Technol.* **2008**, *50*, 190–196. [CrossRef]

19. Sun, J.; Chu, Y.-F.; Wu, X.; Liu, R.H. Antioxidant and Antiproliferative Activities of Common Fruits. *J. Agric. Food Chem.* **2002**, *50*, 7449–7454. [CrossRef] [PubMed]

20. Grifoni, D.; Agati, G.; Bussotti, F.; Michelozzi, M.; Pollastrini, M.; Zipoli, G. Different Responses of Arbutus Unedo and Vitis Vinifera Leaves to UV Filtration and Subsequent Exposure to Solar Radiation. *Environ. Exp. Bot.* **2016**, *128*, 1–10. [CrossRef]

21. Del-Castillo-Alonso, M.-Á.; Diago, M.P.; Tomás-Las-Heras, R.; Monforte, L.; Soriano, G.; Martínez-Abaigar, J.; Núñez-Olivera, E. Effects of Ambient Solar UV Radiation on Grapevine Leaf Physiology and Berry Phenolic Composition along One Entire Season under Mediterranean Field Conditions. *Plant Physiol. Biochem.* **2016**, *109*, 374–386. [CrossRef] [PubMed]

22. Escobedo, J.F.; Gomes, E.N.; Oliveira, A.P.; Soares, J. Modeling Hourly and Daily Fractions of UV, PAR and NIR to Global Solar Radiation under Various Sky Conditions at Botucatu, Brazil. *Appl. Energy* **2009**, *86*, 299–309. [CrossRef]

23. León-Chan, R.G.; López-Meyer, M.; Osuna-Enciso, T.; Sañudo-Barajas, J.A.; Heredia, J.B.; León-Félix, J. Low Temperature and Ultraviolet-B Radiation Affect Chlorophyll Content and Induce the Accumulation of UV-B-Absorbing and Antioxidant Compounds in Bell Pepper (*Capsicum Annuum*) Plants. *Environ. Exp. Bot.* **2017**, *139*, 143–151. [CrossRef]

24. Dáder, B.; Gwynn-Jones, D.; Moreno, A.; Winters, A.; Fereres, A. Impact of UV-A Radiation on the Performance of Aphids and Whiteflies and on the Leaf Chemistry of Their Host Plants. *J. Photochem. Photobiol. B Biol.* **2014**, *138*, 307–316. [CrossRef] [PubMed]

25. Olympios, C.M. *The Cultivation Technique of Vegetables in Greenhouses (In Greek)*; Stamoulis Publications: Athens, Greece, 2001.

26. Katsoulas, N.; Kittas, C.; Dimokas, G.; Lykas, C.H. Effect of Irrigation Frequency on Rose Flower Production and Quality. *Biosyst. Eng.* **2006**, *93*, 237–244. [CrossRef]

27. Jacovides, C.; Assimakopoulos, V.; Tymvios, F.; Theophilou, K.; Asimakopoulos, D. Solar Global UV (280–380 nm) Radiation and Its Relationship with Solar Global Radiation Measured on the Island of Cyprus. *Energy J.* **2006**, *31*, 2728–2738. [CrossRef]

28. Scalbert, A.; Monties, B.; Janin, G. Tannins in Wood: Comparison of Different Estimation Methods. *J. Agric. Food Chem.* **1989**, *37*, 1324–1329. [CrossRef]

29. Neocleous, D.; Ntatsi, G. Seasonal Variations of Antioxidants and Other Agronomic Features in Soilless Production of Selected Fresh Aromatic Herbs. *Sci. Hortic.* **2018**, *234*, 290–299. [CrossRef]

30. Benzie, F.F.I.; Strain, J.J. Ferric reducing/antioxidant power assay: Direct measure of total antioxidant activity of biological fluids and modified version for simultaneous measurements of total antioxidant power and ascorbic acid concentration. *Methods Enzymol.* **1999**, *299*, 15–27. [PubMed]

31. Benzie, F.F.I.; Strain, J.J. The ferric reducing ability of plasma (FRAP) as a measure of "Antioxidant Power". The FRAP assay. *Anal. Biochem.* **1996**, *239*, 70–76. [CrossRef] [PubMed]

32. Rodriguez-Amaya, D.B. *A Guide to Carotenoid Analysis*; ILDI Press: Washington, DC, USA, 2001.

33. Gastilla, N.; Baeza, E. Greenhouse site selection. In *Good Agricultural Practices for Greenhouse Vegetable Crops. Principles for Mediterranean Climate Areas*; FAO, Ed.; Food and Agriculture Organization of the United Nations: Rome, Italy, 2013; pp. 21–33.

34. Zimmer, A.R.; Leonardi, B.; Miron, D.; Schapoval, E.; de Oliveira, J.R.; Gosmann, G. Antioxidant and Anti-Inflammatory Properties of Capsicum Baccatum: From Traditional Use to Scientific Approach. *J. Ethnopharmacol.* **2012**, *139*, 228–233. [CrossRef] [PubMed]

35. Ghasemnezhad, M.; Sherafati, M.; Payvast, G.A. Variation in Phenolic Compounds, Ascorbic Acid and Antioxidant Activity of Five Coloured Bell Pepper (Capsicum Annum) Fruits at Two Different Harvest Times. *J. Funct. Foods* **2011**, *3*, 44–49. [CrossRef]

36. Taiz, L.; Zeiger, E. *Plant Physiology*, 3rd ed.; Sinauer Associates Inc.: Sunderland, MA, USA, 2002.

37. Perla, V.; Nimmakayala, P.; Nadimi, M.; Alaparthi, S.; Hankins, G.R.; Ebert, A.W.; Reddy, U.K. Vitamin C and Reducing Sugars in the World Collection of Capsicum Baccatum, L. Genotypes. *Food Chem.* **2016**, *202*, 189–198. [CrossRef] [PubMed]

38. Mateos, R.M.; Jiménez, A.; Román, P.; Romojaro, F.; Bacarizo, S.; Leterrier, M.; Gómez, M.; Sevilla, F.; del Río, L.A.; Corpas, F.J.; et al. Antioxidant Systems from Pepper (*Capsicum annuum* L.): Involvement in the Response to Temperature Changes in Ripe Fruits. *Int. J. Mol. Sci.* **2013**, *14*, 9556–9580. [CrossRef] [PubMed]

39. Loizzo, M.R.; Pugliese, A.; Bonesi, M.; Menichini, F.; Tundis, R. Evaluation of chemical profile and antioxidant activity of twenty cultivars from *Capsicum annuum*, *Capsicum baccatum*, *Capsicum chacoense* and *Capsicum chinense*: A comparison between fresh and processed peppers. *LWT—Food Sci. Technol.* **2015**, *64*, 623–631. [CrossRef]

40. Prior, R.L.; Wu, X.; Schaich, K. Standardized Methods for the Determination of Antioxidant Capacity and Phenolics in Foods and Dietary Supplements. *J. Agric. Food Chem.* **2005**, *53*, 4290–4302. [CrossRef] [PubMed]

41. Kittas, C.; Tchamitchian, M.; Katsoulas, N.; Karaiskou, P.; Papaioannou, C.H. Effect of Two UV-Absorbing Greenhouse-Covering Films on Growth and Yield of an Eggplant Soilless Crop. *Sci. Hortic.* **2006**, *110*, 30–37. [CrossRef]

42. Schnitzler, W.; Gruda, N. Hydroponics and Product Quality. In *Hydroponic Production of Vegetables and Ornamentalss*; Savvas, D., Passam, H., Eds.; Embryo Publications: Athens, Greece, 2002; pp. 373–411. ISBN 960-8002-12-5.

43. Neugart, S.; Schreiner, M. UVB and UVA as eustressors in horticultural and agricultural crops. *Sci. Hortic.* **2018**, *234*, 370–381. [CrossRef]

44. Gruda, N.; Savvas, D.; Colla, G.; Rouphael, Y. Impacts of Genetic Material and Current Technologies on Product Quality of Selected Greenhouse Vegetables—A Review. *Eur. J. Hortic. Sci.* **2018**, *83*, 319–328. [CrossRef]

45. Shaw, N.L.; Cantliffe, D.J. Brightly colored pepper cultivars for greenhouse production in Florida. *Proc. Fla. State Hort. Soc.* **2002**, *115*, 236–241.

 *agronomy*

Article

Effect of UV Radiation and Salt Stress on the Accumulation of Economically Relevant Secondary Metabolites in Bell Pepper Plants

Jan Ellenberger *, Nils Siefen, Priska Krefting, Jan-Bernd Schulze Lutum, Daniel Pfarr, Maja Remmel, Lukas Schröder and Simone Röhlen-Schmittgen

INRES Horticultural Sciences, University of Bonn, Auf dem Huegel 6, 53121 Bonn, Germany; nils.siefen@googlemail.com (N.S.); priska-krefting@gmx.de (P.K.); jbsl@uni-bonn.de (J.-B.S.L.); daniel-pfarr@gmx.de (D.P.); s7maremm@uni-bonn.de (M.R.); lukas-schroeder@onlinehom.de (L.S.); s.schmittgen@uni-bonn.de (S.R.-S.)
* Correspondence: ellenberger@uni-bonn.de

Received: 19 December 2019; Accepted: 14 January 2020; Published: 18 January 2020

Abstract: The green biomass of horticultural plants contains valuable secondary metabolites (SM), which can potentially be extracted and sold. When exposed to stress, plants accumulate higher amounts of these SMs, making the extraction and commercialization even more attractive. We evaluated the potential for accumulating the flavones cynaroside and graveobioside A in leaves of two bell pepper cultivars (Mavras and Stayer) when exposed to salt stress (100 mM NaCl), UVA/B excitation (UVA 4–5 W/m^2; UVB 10–14 W/m^2 for 3 h per day), or a combination of both stressors. Plant age during the trials was 32–48 days. HPLC analyses proved the enhanced accumulation of both metabolites under stress conditions. Cynaroside accumulation is effectively triggered by high-UV stress, whereas graveobioside A contents increase under salt stress. Highest contents of secondary metabolites were observed in plants exposed to combined stress. Effects of stress on overall plant performance differed significantly between treatments, with least negative impact on above ground biomass found for high-UV stressed plants. The usage of two non-destructive instruments (Dualex and Multiplex) allowed us to gain insights into the ontogenetical effects at the leaf level and temporal development of SM contents. Indices provided by those devices correlate fairly with amounts detected via HPLC (Cynaroside: $r^2 = 0.46$–0.66; Graveobioside A: $r^2 = 0.51$–0.71). The concentrations of both metabolites tend to decrease at leaf level during the ontogenetical development even under stress conditions. High-UV stress should be considered as a tool for enriching plant leaves with valuable SM. Effects on the performance of plants throughout a complete production cycle should be evaluated in future trials. All data is available online.

Keywords: *Capsicum annuum*; flavonoids; fluorescence monitoring; bio-waste utilization

1. Introduction

1.1. Green Biomass as a Source of Valuable Chemicals

Commercial vegetable production is accompanied by large quantities of so far under-utilized green biomass in all stages of production and especially after harvest [1]. While the use of biomass for the purpose of energy production is becoming a standard procedure in northern Europe in recent years [2], the extraction and the use of high-value secondary metabolites (SMs) from vegetable plant leaves are just being developed. Research strategies in Europe are heading toward a cascade use of agricultural byproducts and pave the way for extracting and using "valuable substances or molecules before ultimately discarding the left-overs" [3]. The pharmaceutical industry—as an example—is highly dependent on plant SMs, since approximately 60% of anticancer compounds and 75% of drugs for infectious diseases are derived from plants [4]. In this frame, research on targeted enrichment of valuable substances in plant biomass is gaining importance [5].

1.2. Plant Stress as a Measure to Increase Leaf Secondary Metabolite Content

The biochemical background of enhanced accumulation of SMs in plant leaves as a measure to cope with stress is a well-described phenomenon [2,6,7]. In short, the cultivation of plants under suboptimal conditions leads to an increased amount of reactive oxygen species (ROS) in plant tissues. Accumulation of SMs is a plant strategy to avoid oxidative damage caused by reactive oxygen species [8]. In theory, both biotic and abiotic stressors could lead to higher amounts of valuable SMs in plants. While biotic stressors such as fungi and insects are hard to control and may cause major phytosanitary problems, abiotic stressors are easier to manage and applicable by practitioners. The results of several studies in recent years indicate that abiotic stressors are a useful tool for SM accumulation in leaves of horticultural plants. Secondary metabolites in *Centella asiatica* leaves increase under enhanced UV-B light, especially in the epidermis [9]. In bell pepper, increased flavonoid contents can be found in leaves exposed to elevated UV [10]. The promoting effect of UV-B radiation on flavonoid accumulation in plant leaves has recently been reviewed [11]. The effects of salt stress on the antioxidant machinery may be adverse and depend on the plant's tolerance [12] and salt concentration in the rootzone [13]. Another extensive study on leaf metabolism in bell pepper under different levels of salt stress revealed an increasing reduction in growth with increasing NaCl contents in the rootzone [14]. While tolerant plants increase leaf SM contents to cope with salt stress, sensitive plants do not have this mechanism and senesce, finally dying off if the stressor is persistent [12]. Studies directly comparing effects of salt and UV stress on leaf SMs are rare. One study shows both stressors to similarly affect leaf contents of the flavonoids quercetin and luteolin in *Ligustrum vulgare* [15]. Abiotic stressors such as drought and salt stress are easily applicable in commercial greenhouse production in soilless systems, which are the predominant systems in many parts of the world, including Europe [16].

1.3. Non-Invasive Monitoring of Secondary Metabolites in Plant Leaves

Quantification of secondary metabolites including flavonoids with portable optical devices is well established in plant sciences [17]. The use of non-invasive optical sensors to investigate plant leaf components has several advantages over laboratory analyses: data acquisition is faster and more cost effective than laboratory analyses [18]. Moreover, considerate handling of leaves allows for several measurements of the same leaf, enabling to gain insights in temporal developments. Several studies demonstrated the viability of optical devices to access secondary metabolites in plant leaves: a multiparametric fluorescence sensor was used to evaluate the influence of nutrient deficiency on the chemical properties of tomato leaves and to quantify the content of the flavonoids rutin and solanesol [19,20]. In bell pepper, a fluorescence sensor was used to evaluate the impact of priming plants with high light conditions on leaf flavonoid content [10].

1.4. Cynaroside and Graveobioside A

The vast diversity and chemical complexity of plant SMs often prohibit an economically feasible chemical synthesis. Therefore, extraction either from wild or cultivated plants often represents the best source of supply [1].

Cynaroside (Luteolin-7-glucoside) potentially has a range of medicinal applications: it has the capability to prevent ROS-induced apoptosis in heart cells [21]. Cynaroside furthermore diminishes kidney injury as a side effect of cancer treatments with the chemotherapeutic drug cisplatin. A potential medicinal use of graveobioside A (Luteolin-7-apiosyl-glucoside) is proven by a patent on its application in preparation of drugs for preventing hyperuricemia and gout [22]. Graveobioside A was shown to be contained in several plants, such as celery seeds, parsley, and bell pepper [23,24].

Several SMs in Solanaceae leaves have the potential to biologically control insects [25]. Graveobioside A is such a potential natural insecticide, since oviposition of the American serpentine leafminer fly (*Liriomyza trifolii*) was shown to drop in kidney bean leaves treated with a graveobioside A containing solution [24]. It is expected that the demand for natural insecticides will increase across the EU due to more rigid legislation [26].

We hypothesize that cynaroside and graveobioside A contents in bell pepper leaves can be enhanced by abiotic stressors that are potentially applicable by practitioners in the future. Another aim is to check whether non-invasive devices can be used for assessments of cynaroside and graveobioside A in bell pepper leaves. Furthermore, we attempt to get insights in interactions between different stressors and differences in stress response between two bell pepper cultivars.

2. Material and Methods

2.1. Plant Material and Growth Conditions

Seeds of sweet pepper plants (*Capsicum annuum*) cultivar 'Stayer' (Rijk Zwaan, De Lier, The Netherands) and 'Mavras' (Enza Zaden, Enkhuizen, The Netherlands) were sown in soil under greenhouse conditions. Fourteen-days old pepper plants were transplanted into small rockwool cubes (3 × 3 × 5 cm) and one further week later into larger cubes (10 × 10 × 7.5 cm) (Grotop Master, Grodan, The Netherlands). On day 32 after seeding, plants were transferred to a grow chamber to ensure a stable environment. From that day on, stress was applied for 16 days, resulting in a plant age of 48 days at the end of the trial. A longer trial was not feasible due to limitations of the chosen facility. All plants received all nutrients mandatory for optimal growth prepared from two stock solutions (17.2 mM nitrogen, 5.4 mM calcium, 4.7 mM potassium, 0.4 mM phosphorous, 5.4 mM sulfur, 2.4 mM magnesium, 0.01 mM iron and micronutrients; electrical conductivity 2.5 mS cm^{-1}; pH 5.5). Plants were cultivated at the greenhouse facility in Bonn-Endenich (University of Bonn, Bonn, Germany) at day/night temperatures of 24.5 °C ± 5.4 and supplemental light intensity of 203–540 μm m^{-2}s^{-1} provided by sodium vapor lamps (Philips Lighting GmbH, Hamburg, Germany).

To apply salt stress, a salt concentration of 100 mM NaCl for a period of 16 days was added to the standard nutrient solution, since that concentration was shown to trigger a higher total phenolic content in leaves of bell pepper seedlings in a previous study [14]. To apply UV stress, plants were exposed to UV light (UVA 4–5 W m^{-2}; UVB 10–14 W m^{-2}) for 3 h per day (Philips Lighting GmbH, Hamburg, Germany) over a 16-day period. In addition, some plants were exposed to combined salt and UV stress. Plant age at stress onset was 32 days. A total of 5 plants per treatment (control, salt stress, UV stress, combined stress) were randomized in the growth chamber.

2.2. Non-Destructive Recordings

Non-destructive measurements were performed on all leaves per plant, from mature to young. Measurements were conducted using two well-established devices in stress physiology monitoring. First device is the multiparametric fluorescence excitation system Multiplex® (Multiplex®, Force-A, Orsay, France), described in previous studies [27]. All recordings with this device were done at a constant distance of 0.10 m to the leaf surface and a frontal cover plate with an aperture of 4 cm in diameter opening to assess the index of epidermal flavonols (FLAV index): $\log \frac{FRF\_R}{FRF\_UV}$.

Secondly, the transmittance-based fluorescence measurements were conducted with the Dualex sensor (Force-A, Orsay, France). The Dualex is a device with a leaf-clip; measurements were taken with virtually no distance to the leaf surface. The device is extensively described in the literature [28,29].

2.3. Plant Harvest

Plants were harvested 16 days after treatment inception (DATI) at a plant age of 48 days. The total fresh weight of shoots was determined immediately. Leaves were dried for 7 days at 50 °C (drying oven) to collect dry weights.

2.4. Leaf Sample Preparation and Laboratory Analysis

Samples were taken at the harvesting at 16 DATI, of the mature leaf 4 and the young leaves 10 and 12, to assess the impact of stress application on the amount of the two luteolins, graveobioside A and cynaroside. All leaf numbers are given as the number of true leaves, counted from the base of the plant. The samples were freeze-dried and then stored at −20 °C until further processing. Ground leaf samples were prepared for HPLC determination (Agilent 1260 Infinity HPLC System Agilent Technology Deutschland GmbH, Ratingen, Germany). An amount of 0.3 g was extracted with water-diluted methanol (60:40, *v/v*) for 10 min in an ultrasonic bath, centrifuged for 10 min at 4 °C with 13,000 rpm (Centrifuge 5415R, Eppendorf AG, Hamburg, Deutschland) repeated four times. The supernatants were collected and stored at −20 °C until HPLC analysis. The samples were filtrated through a membrane filter (Phenomenex, Aschaffenburg, Germany) prior to injection. The HPLC system consisted of an autosampler, a diode array UV–Vis detector and was coupled with a quaternary solvent delivery system. The column (Nocleodur C18, 3 × 150 mm, 3 μm, Macherey-Nagel, GmbH & Co. KG, Düren, Germany) was isocratically eluted with a binary mixture of water and methanol (60:40) adjusted to pH 2.8 with phosphoric acid. The flow rate was 0.3 mL min^{-1}; 10 μL samples were injected onto the column equilibrated at 25 °C (detection at 355 nm). Graveobioside A peak was detected at 14.1 min, and cynaroside at 15.6 min. Both calibration curves were obtained from diluted series of standards provided by PhytoLab (Vestenbergsgreuth, Germany).

2.5. Data Analysis and Statistics

All data is available online [30]. Data analysis was performed with R (R Core Team, Vienna, Austria) [31] in RStudio (R Studio Team, Boston, USA) [32]. According to the data structure, e.g., balanced or imbalanced, type I or type III ANOVA were used to compare group means. Applied post-hoc test was Tukey's HSD. Figures were created in RStudio, with the package ggplot2 [33].

3. Results

3.1. Stress-Related Effect Varies Among Secondary Metabolites and Cultivars

A treatment effect was observed on contents of both cynaroside and graveobioside A, while no significant effect of the variable cultivar on either metabolite content was found. There was a strong tendency for higher graveobioside A in 'Mavras' as compared to Stayer ($p = 0.055$). No interactions between cultivar and treatment were observed (Table 1). Both combined-stressed plants and plants under UV-exposure accumulated significantly higher amounts of cynaroside in their leaves than control and salt-stressed plants (Figure 1, A + B). Plants of the cultivar 'Mavras' accumulated significantly higher graveobioside A amounts in salt-stressed and combined-stressed plants than in control and UV-stressed plants (Figure 1C). No significant treatment effect on graveobioside A content in plants of the cultivar Stayer was found (Figure 1D). Levels of SM in leaves of different ontogenetical stages are shown as an illustration of uneven distribution within the plants. SM contents decrease with leaf ontogenetical stage (Figure 1).

Table 1. Interaction and main effect for treatments (control, salt-stress, combined-stress, UV-stress) and cultivars (Mavras and Stayer), calculated with a type I two-way ANOVA. Grayed area indicates significant effect ($p \leq 0.001$).

Factor	Cultivar	Treatment	Cultivar × Treatment
Cynaroside	0.179	$<2 \times 10^{-16}$	0.917
Graveobioside A	0.055	1.25×10^{-5}	0.141
Dry Weight	0.00082	3.8×10^{-12}	0.426
Fresh Weight	0.00017	1.15×10^{-15}	0.146

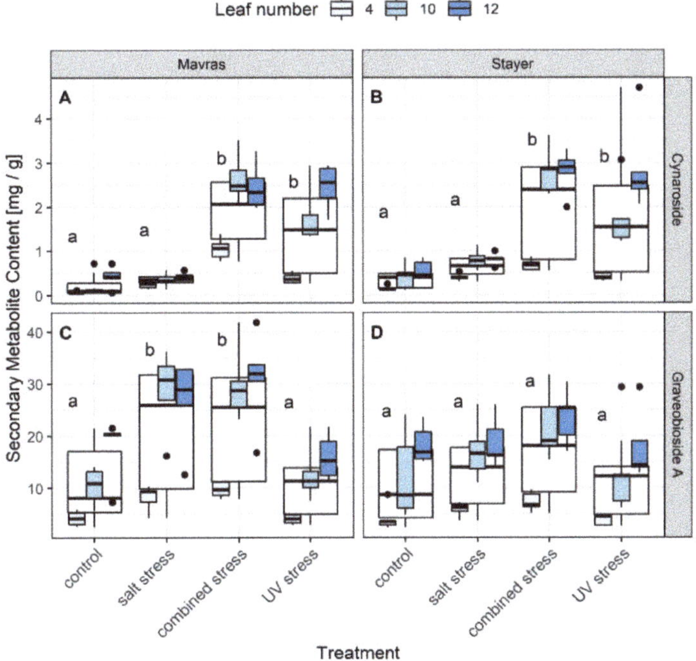

Figure 1. HPLC-determined leaf cynaroside (**A,B**) and graveobioside A (**C,D**) contents, for bell pepper cultivars 'Mavras' (**A,C**) and 'Stayer' (**B,D**) under different growth conditions, 15 days after treatment inception ($n = 5$). Transparent boxplots show pooled data from all leaves ($n = 15$). Colored boxplots represent leaf age—subgroups (Leaf 4, 10, and 12 as counted from the base, with darkest colors for youngest leaves). Letters (a,b) indicate differences within each cultivar × secondary metabolite—combination (Tukey HSD, $p < 0.05$).

Both fresh and dry weight of bell pepper plants differed significantly depending on the cultivar, with Stayer attaining higher weights than Mavras. Treatment had a significant effect on both fresh and dry weight. There was no interaction between the treatment and cultivar regarding plant's fresh or dry weight. Dry weight of plants of the cultivar Mavras was significantly higher in control plants than in any other treatment (Table 1). UV-stressed plants of both tested cultivars exhibited higher fresh and dry weights than plants under salt-stress and combined-stress conditions (Figure 2). Observed mean fresh weight decreased in salt-stress and combined-stress plants compared to control and UV stress, which were in the magnitude of 50% (Figure 2C,D). The mean dry weight tended to be higher for salt-stressed plants as compared to plants under combined stress, but lower than the dry weights of plants experiencing UV stress or control conditions (Figure 2A,B).

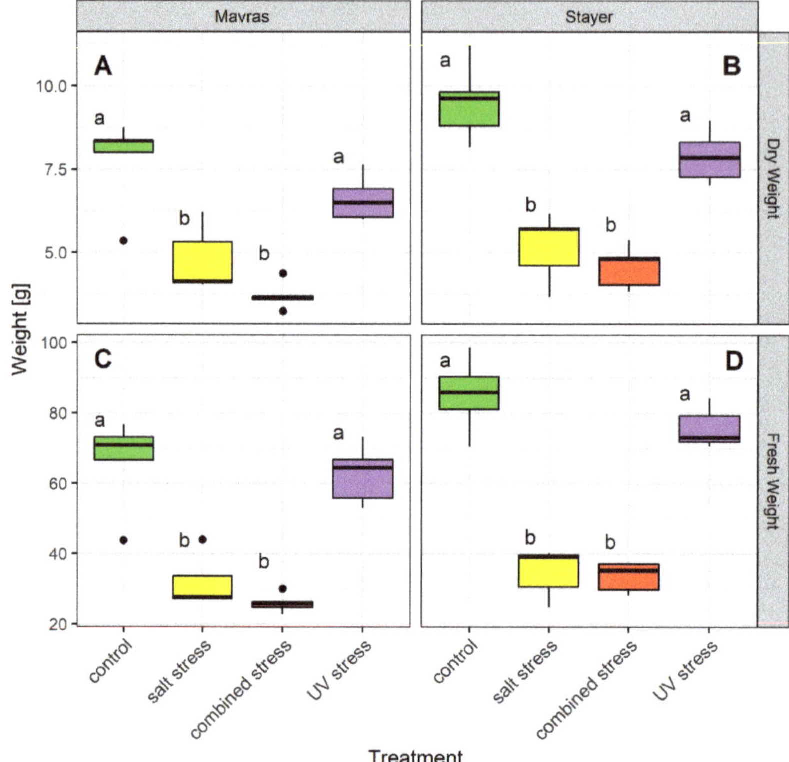

Figure 2. Aboveground biomass (dry weight: (**A**), (**B**); fresh weight: (**C**), (**D**)) of bell pepper cultivars "Mavras" (**A**), (**C**) and "Stayer" (**B**), (**D**) under different growth conditions, 15 days after treatment induction ($n = 5$). Letters (a,b) indicate differences within each cultivar × dry/fresh weight—combination (Tukey HSD, $p < 0.05$).

3.2. Non-Invasive Monitoring of Secondary Metabolites

Figure 3 shows exponential regressions between three indices (Multiplex indices FLAV and NBI_R; Figure 3A–D and Dualex index Flav; Figure 3E,F) and leaf contents of the SMs cynaroside (Figure 3A,C,E) and graveobioside A (Figure 3B,D,F), respectively. Predictions of graveobioside A contents based on the indices are better than predictions of cynaroside contents. Multiplex indices are more accurate predictors than the Dualex index, as outlined by correlation coefficients (r^2). Index values level off at cynaroside contents above 1.5 mg g^{-1}. The connection between graveobioside A and the indices is more linear, but still leveling off at graveobioside A contents above approximately 25 mg g^{-1}.

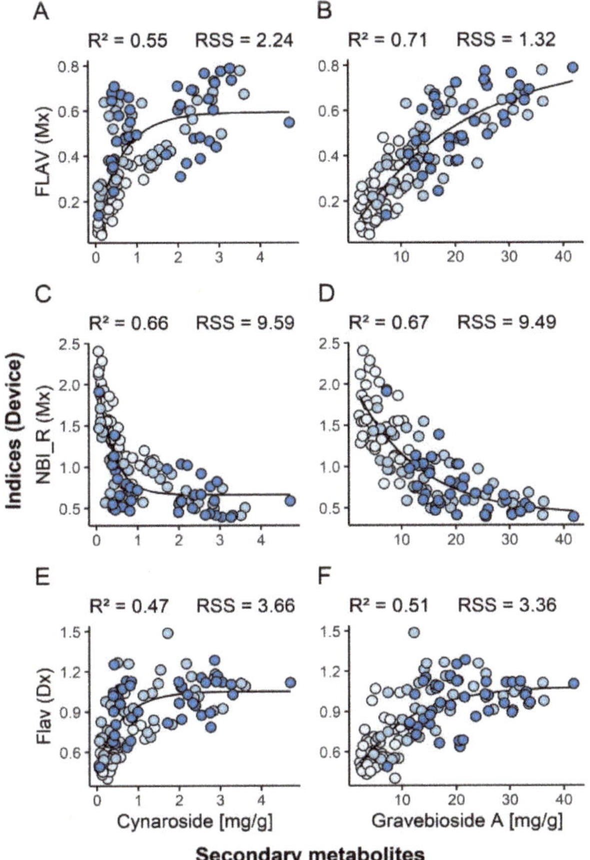

Figure 3. Exponential regression between indices of non-invasive devices and leaf secondary metabolite concentrations in bell pepper leaves, determined via HPLC. Contents of cynaroside and graveobioside A correlated with FLAV (Mx) (**A**), (**B**), NBI_R (Mx) (**C**), (**D**), and Flav (Dx) (**E**), (**F**). Color of points represents leaf age (Leaf 4, 10, and 12 as counted from the base, with darkest colors for youngest leaves). Lines indicate exponential regressions ($n = 60$). RSS, residual sum of squares.

3.3. Spatial and Temporal Development of Secondary Metabolite Contents

The only significant changes in FLAV values within cultivar × treatment groups were seen among the fourth leaves of combined-stressed Mavras plants at days 0 versus 9 and 0 versus 15, respectively (Figure 4C). A clear trend was observed for the fourth leaves of combined-stressed Stayer plants at days 0 versus 15 (TukeyHSD, $p = 0.053$) (Figure 4D). Generally, FLAV values for stressed plants tend to increase, while the values for control leaves tend to decrease. A comprehensive overview of associated main effects is given in Table 2.

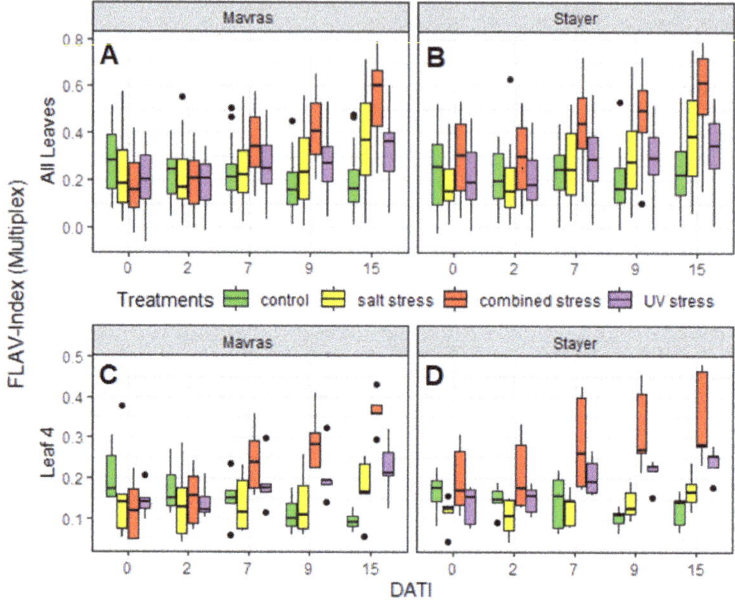

Figure 4. Temporal development of secondary metabolites in leaves of bell pepper cultivars "Mavras" and "Stayer", expressed with the FLAV-index (Multiplex). (**C**), (**D**), n = 5; (**A**), (**B**), n = 5–50; DATI, day after treatment initiation.

Table 2. Interaction and main effect for treatments (control, salt-stress, combined-stress, UV-stress) and DATI (0, 2, 7, 9, 15). To account for the unbalanced design (e.g., unequal numbers of observations within each level of DATI), type III ANOVA was selected to compare differences between factor means for FLAV values of "All leaves". Grayed area indicates significant effect at $p \leq 0.05$ (light), $p \leq 0.01$ (medium), and $p \leq 0.001$ (dark).

Leaf	Cultivar	Treatment	DATI	Treatment × DATI
All leaves	Mavras	0.085	0.027	$<2 \times 10^{-16}$
	Stayer	0.079	0.509	2.17×10^{-6}
Leaf 4	Mavras	0.00011	0.055	0.00027
	Stayer	8.37×10^{-12}	0.00484	0.081

4. Discussion

We are among the first groups accessing the amount of graveobioside A in pepper leaves [4]. For cynaroside, the range of values detected corresponds to the results of other studies [34,35].

4.1. Stress-Related Effect Varies According to Secondary Metabolites and Cultivars

Since cynaroside contents under single UV-stress and combined UV- and salt-stress are not significantly different (Figure 1A,B), cynaroside accumulation appears to be triggered mainly by high radiation conditions. Interestingly, and in contrast to cynaroside, graveobioside A accumulation is triggered more effectively by salt stress than by UV-stress, especially in the cultivar Mavras (Figure 1C). This is a surprising result, since biosynthesis of flavonoids is said to be enhanced similarly by UV radiation and salinity [15,36]. On the other hand, some authors report that the regulation of SM production in response to salt stress differs between salt-sensitive (upregulation) and salt-tolerant (downregulation) plants [12]. However, differences in salt-stress tolerance between the cultivars used in this study are not supported by differing plant biomasses (Figure 2). The chemical group of flavonoids is highly diverse, and metabolic pathways are not entirely understood to date. At this point, it remains unclear how exactly upregulation of cynaroside synthesis under UV stress and upregulation of graveobioside A synthesis under salt stress occurs.

Our results indicate—as expected—that salt-stressed plants acquire a significantly lower biomass than both control plants and UV-stressed plants. Stunted growth is a well-described symptom of severe salt stress in plants [12,37]. If the applied salt concentration would have been lower, negative effects could probably have been avoided to a certain extent, as recently discussed in a review on the potential of seawater use in soilless culture [13]. Reaction of plants to UV-B exposure varies from growth reduction to enhancement, depending on species, cultivar, and stress level [11,38]. Since the overall aim of the stress application is the accumulation of higher amounts of secondary metabolites in the plant's green biomass, it is necessary to consider not only the share of desired metabolite in the plant's biomass, but also the biomass reduction caused by the treatment. Considering this background, we can state that stressors with minor negative effects on plant biomass accumulation, but major positive effects on contents of desired metabolites in the plant tissues, are necessary to achieve these aims. Finding the perfect trade-off between biomass and fruit yield loss, on the one hand, and SM increase, on the other hand, will be crucial to improve the production system. In our specific setup with two single stressors and one combined stress, with respective levels of stress described above, the single UV stress is most promising, whereas salt stress (100 mM NaCl), although promoting the accumulation of graveobioside A, is less promising as a tool to enhance whole plant SM amounts, due to the decrease in total biomass. Effects on plants grown over a whole season are a matter of ongoing research.

4.2. Non-Invasive Monitoring

The indices provided by both optical devices deliver better estimates for leaf graveobioside A contents than for leaf cynaroside contents. That is an expected result, since the amount of graveobioside A as determined via HPLC is up to ten-fold higher than the amount of cynaroside (0–4 versus 2–40 mg g^{-1}) and both secondary metabolites share similar optical properties. Any estimate of concentrations based on non-invasive, optical devices will be best for the predominant fraction of a group of metabolites with similar optical properties. By the same token, signals of metabolites that occur in small quantities are more likely to be superimposed by other signals and therefore difficult to quantify. Additional factors known to influence non-invasive assessment of leaf compounds include the concentration of other pigments potentially influencing the measurement [39], leaf thickness [40], and the device used [41].

In our study, the FLAV-index of the Multiplex shows an almost linear response to changes in leaf graveobioside A content (Figure 2B). The same applies for the NBI_R index, which correlates negatively with the actual graveobioside A content. Both indices use the far-red fluorescence of leaves excited with UV-light and normalize that signal for the red fluorescence emitted after excitation with red light [29]. As an enhanced graveobioside A content leads to a stronger absorption of UV light in the leaf epidermis, less radiation penetrates into the mesophyll, which in turn leads to a lower chlorophyll fluorescence. We have to highlight the broad distribution of fluorescence values, though, which prohibits a precise prediction of actual graveobioside A levels on the individual leaf level. The Flav-index of the Dualex is almost indifferent to changes at graveobioside A levels above 25 mg g^{-1}.

None of the indices is strongly related to the leaf cynaroside contents quantified by HPLC. Neither the Dualex nor the Multiplex provide any indices that allow to quantify cynaroside contents higher than approximately 1 mg g^{-1} dry weight. An exact evaluation of high levels of this specific SM in bell pepper leaves is therefore not possible with the tested devices. However, the correlations we have identified between the FLAV index and HPLC measurements still allow us to analyze the gradual changes in SM contents as they occur during the prolonged period of stress.

4.3. Insights in Spatial and Temporal Accumulation of Secondary Metabolites

The usage of non-invasive phenotyping tools such as the Multiplex and Dualex devices allows to analyze leaf constituents during ontogenesis. The observed drop of the flavonol content in leaves of unstressed plants during ontogenesis (Figure 4C,D) is in line with the theories that (a) the production of phenolics, such as flavonols, is mainly caused by photodamage [42] and (b) that ontogenetically young leaves are, in general, more prone to be affected by high light stress than older leaves, since their photosynthetic apparatus is not yet well developed [43] and the photoprotective cuticula is thinner compared with older leaves [44]. Therefore, young leaves show stress-related reactions in conditions that are neither stressful for older leaves nor for the entire plant. However, the described ontogenetic effects tend to be overcompensated by stress-related effects in all three stress treatments (Figure 4C,D). Thus, flavonol contents of the fourth leaf as measured with the FLAV (Mx) index slightly increased in plants experiencing single stresses, while plants exposed to combined stress showed major increases in leaf flavonol contents (Figure 4C,D).

4.4. Implications and Future Challenges

The present study proves that abiotic stresses, in particular, salt stress and UV stress, can enhance the amount of economically valuable SMs, namely cynaroside and graveobioside A, in bell pepper leaves. The main objective of growing bell pepper plants, however, is the production of fruits of adequate quantity and quality for human nutrition. Considering the decline in plant biomass in response to stress conditions, it is very likely that the stressors applied would also lead to a reduction in fruit production. Severe salt stress, in particular, is known to be an important factor limiting crop productivity [45]. We have shown that the type of stressor has magnificent effects on both plant biomass and leaf secondary metabolite content. Other studies have proven that this also applies for different levels of abiotic stress [14,46]. The search for the best stressors and stress levels for the accumulation of secondary metabolites in plant leaves with negligible effects on fruit yield is a major future challenge for research in stress physiology. Several authors reported neutral or positive responses of product quality to mild stress [46]. For salt stress, several studies in the model-crop tomato reveal positive impacts of mild stress on fruit quality (e.g., antioxidant capacity and nutritional value) [47,48]. Low UV radiation reduces the antioxidative capacity and, therefore, the fruit quality of bell pepper fruits [49]. Additional UV radiation may help to overcome this problem and, at the same time, induce the production of valuable SM in the leaves. Cultivation of plants under mild water stress conditions can also enhance water use efficiency. To avoid any competition with food production, post-harvest treatment of leaves could be an appropriate measure to achieve high contents of promising metabolites [50,51]. These effects should also be taken into account when evaluating the value of production systems that are based on commercialization of both fruits and SMs in leaves of horticultural plants.

To enhance precision of non-invasive estimation of SMs in pepper leaves, future studies should consider hyperspectral sensors as well as chlorophyll fluorescence-based sensors, ideally a combination of both. Sensors covering the UV range are just entering the market and appear as a promising tool to access SMs in plants, as they cover absorption bands of flavones and other phenolic leaf compounds [52].

5. Conclusions

Both additional UV light and salt stress can enhance concentrations of the two SMs graveobioside A and cynaroside in bell pepper leaves. Highest concentrations were reached by combining both treatments. Stressed bell pepper leaves contain up to 30 mg graveobioside A and about 2 mg cynaroside per gram dry weight. While salt stress (100 mM NaCl) has a major negative impact on plant vegetative growth, UV stress (UVA 4–5 W m^{-2}; UVB 10–14 W m^{-2}; 3 h per day) has no significant impact on the fresh mass of the plants. The tendency of decreasing SM contents in leaves during ontogenesis is outweighed by the stress treatments. Graveobioside A contents can be assessed with the multiparametric fluorescence sensor Multiplex. Reliable quantification of cynaroside is not possible with the non-invasive sensors used. If future experiments exclude major negative impacts on fruit quality, UV stress can be recommended as one tool to enhance valuable SMs in bell pepper leaves and potentially in vegetable leaves in general. A less-intense salt stress should also be considered in future experiments.

Author Contributions: Conceptualization, S.R.-S.; data analysis, J.E., N.S., P.K., J.-B.S.L., D.P., M.R., and L.S.; methodology, S.R.-S.; writing—original draft preparation, J.E.; writing—review and editing, J.E., S.R.-S., N.S., P.K., J.-B.S.L., D.P., M.R., and L.S. All authors have read and agreed to the published version of the manuscript.

Funding: This research was supported by the German Federal Ministry of Education and Research (grant number: 031B0361C).

Acknowledgments: The authors are grateful to Libeth Schwager for her support in laboratory analysis, and for plant cultivation by the staff members of the plant service team "Dienstleistungsplattform" of the University of Bonn. We thank Eduardo Fernandez for discussions and support in data visualization. We appreciate the support of Katharina Krah, Simone Klein, Miriam Brink, and Mark Schmutzler during the measurements. We are thankful for the great support by our project partners Manuel Lück, Anika Wiese-Klinkenberg, Julia J. Reimer, and Alexandra Wormit.

Conflicts of Interest: The authors declare no conflict of interest.

Abbreviations

DATI	Days after treatment inception
HPLC	High performance liquid chromatography
ROS	Reactive oxygen species
SM	Secondary metabolite

References

1. Wormit, A.; Bröring, S.; Carraresi, L.; Junker, L.; Jupke, A.; Lück, M.; Noga, G.; Reimer, J.J.; Schmittgen, S.; Thiele, B.; et al. TaReCa–Kaskadennutzung gartenbaulicher Biomasse für eine ressourceneffiziente Produktion von wertvollen bioaktiven Substanzen. *Julius-Kühn-Arch.* **2018**, *460*, 10–13.
2. Junker-Frohn, L.V.; Lück, M.; Schmittgen, S.; Wensing, J.; Carraresi, L.; Thiele, B.; Groher, T.; Reimer, J.J.; Bröring, S.; Noga, G.; et al. Tomato's Green Gold: Bioeconomy Potential of Residual Tomato Leaf Biomass as a Novel Source for the Secondary Metabolite Rutin. *ACS Omega* **2019**, *4*, 19071–19080. [CrossRef]
3. *European Commission A sustainable bioeconomy for Europe strengthening the connection between economy, society and the environment: updated bioeconomy strategy*; European Commission: Brussels, Belgium, 2018; ISBN 978-92-79-94144-3.
4. Carvalho Lemos, V.; Reimer, J.J.; Wormit, A. Color for Life: Biosynthesis and Distribution of Phenolic Compounds in Pepper (Capsicum annuum). *Agriculture* **2019**, *9*, 81. [CrossRef]
5. Taylor, M.A.; Fraser, P.D. Solanesol: Added value from Solanaceous waste. *Phytochemistry* **2011**, *72*, 1323–1327. [CrossRef] [PubMed]
6. Junker, L.V. Induktion des Sekundärmetabolismus in Tomatenpflanzen zur alternativen Verwendung der Blattbiomasse. *DGG-Proc. Ger. Soc. Hortic. Sci. DGG* **2017**, *7*, 1–5.
7. Groher, T.; Schmittgen, S.; Noga, G.; Hunsche, M. Limitation of mineral supply as tool for the induction of secondary metabolites accumulation in tomato leaves. *Plant Physiol. Biochem.* **2018**, *130*, 105–111. [CrossRef] [PubMed]
8. Fini, A.; Brunetti, C.; Di Ferdinando, M.; Ferrini, F.; Tattini, M. Stress-induced flavonoid biosynthesis and the antioxidant machinery of plants. *Plant Signal. Behav.* **2011**, *6*, 709–711. [CrossRef]
9. Müller, V.; Lankes, C.; Schmitz-Eiberger, M.; Noga, G.; Hunsche, M. Estimation of flavonoid and centelloside accumulation in leaves of Centella asiatica L. Urban by multiparametric fluorescence measurements. *Environ. Exp. Bot.* **2013**, *93*, 27–34. [CrossRef]

10. Hoffmann, A.M.; Noga, G.; Hunsche, M. High blue light improves acclimation and photosynthetic recovery of pepper plants exposed to UV stress. *Environ. Exp. Bot.* **2015**, *109*, 254–263. [CrossRef]

11. Escobar-Bravo, R.; Klinkhamer, P.G.L.; Leiss, K.A. Interactive Effects of UV-B Light with Abiotic Factors on Plant Growth and Chemistry, and Their Consequences for Defense against Arthropod Herbivores. *Front. Plant Sci.* **2017**, *8*, 278. [CrossRef]

12. Acosta-Motos, J.; Ortuño, M.; Bernal-Vicente, A.; Diaz-Vivancos, P.; Sanchez-Blanco, M.; Hernandez, J. Plant Responses to Salt Stress: Adaptive Mechanisms. *Agronomy* **2017**, *7*, 18. [CrossRef]

13. Atzori, G.; Mancuso, S.; Masi, E. Seawater potential use in soilless culture: A review. *Sci. Hortic.* **2019**, *249*, 199–207. [CrossRef]

14. Hand, M.J.; Taffouo, V.D.; Nouck, A.E.; Nyemene, K.P.J.; Tonfack, B.; Meguekam, T.L.; Youmbi, E. Effects of Salt Stress on Plant Growth, Nutrient Partitioning, Chlorophyll Content, Leaf Relative Water Content, Accumulation of Osmolytes and Antioxidant Compounds in Pepper (*Capsicum annuum* L.) Cultivars. *Not. Bot. Horti Agrobot. Cluj-Napoca* **2017**, *45*, 481–490. [CrossRef]

15. Agati, G.; Biricolti, S.; Guidi, L.; Ferrini, F.; Fini, A.; Tattini, M. The biosynthesis of flavonoids is enhanced similarly by UV radiation and root zone salinity in L. vulgare leaves. *J. Plant Physiol.* **2011**, *168*, 204–212. [CrossRef] [PubMed]

16. Gruda, N.S. Increasing Sustainability of Growing Media Constituents and Stand-Alone Substrates in Soilless Culture Systems. *Agronomy* **2019**, *9*, 298. [CrossRef]

17. Cerovic, Z.G.; Samson, G.; Morales, F.; Tremblay, N.; Moya, I. Ultraviolet-induced fluorescence for plant monitoring: present state and prospects. *Agronomie* **1999**, *19*, 543–578. [CrossRef]

18. Gitelson, A.A.; Keydan, G.P.; Merzlyak, M.N. Three-band model for noninvasive estimation of chlorophyll, carotenoids, and anthocyanin contents in higher plant leaves. *Geophys. Res. Lett.* **2006**, *33*, L11402. [CrossRef]

19. Groher, T.; Schmittgen, S.; Fiebig, A.; Noga, G.; Hunsche, M. Suitability of fluorescence indices for the estimation of fruit maturity compounds in tomato fruits. *J. Sci. Food Agric.* **2018**, *98*, 5656–5665. [CrossRef]

20. Groher, T.; Röhlen-Schmittgen, S.; Fiebig, A.; Noga, G.; Hunsche, M. Influence of supplementary LED lighting on physiological and biochemical parameters of tomato (*Solanum lycopersicum* L.) leaves. *Sci. Hortic.* **2019**, *250*, 154–158. [CrossRef]

21. Sun, X.; Sun, G.; Wang, M.; Xiao, J.; Sun, X. Protective effects of cynaroside against H2O2-induced apoptosis in H9c2 cardiomyoblasts. *J. Cell. Biochem.* **2011**, *112*, 2019–2029. [CrossRef]

22. Chen, J.; Yang, T.; Huang, X.; Ma, Y.; Zhang, X.; Shen, Y.; Geng, C.; Lu, C. Application of Graveobioside A in the preparation of medicines or health foods against hyperuricemia and gout 2016. Patent CN105560262A, 11 May 2016.

23. Lin, L.-Z.; Lu, S.; Harnly, J.M. Detection and Quantification of Glycosylated Flavonoid Malonates in Celery, Chinese Celery, and Celery Seed by LC-DAD-ESI/MS. *J. Agric. Food Chem.* **2007**, *55*, 1321–1326. [CrossRef] [PubMed]

24. Kashiwagi, T.; Horibata, Y.; Mekuria, D.B.; Tebayashi, S.; Kim, C.-S. Ovipositional Deterrent in the Sweet Pepper, Capsicum annuum, at the Mature Stage against Liriomyza trifolii (Burgess). *Biosci. Biotechnol. Biochem.* **2005**, *69*, 1831–1835. [CrossRef] [PubMed]

25. Castillo-Sánchez, L.E.; Jiménez-Osornio, J.J.; América, M. Secondary Metabolites of the Annonaceae, Solanaceae and Meliaceae Families used as Biological Control of Insects. *Trop. Subtrop. Agroecosyst.* **2010**, *12*, 445–462.

26. Villaverde, J.J.; Sandín-España, P.; Sevilla-Morán, B.; López-Goti, C.; Alonso-Prados, J.L. Biopesticides from Natural Products: Current Development, Legislative Framework, and Future Trends. *BioResources* **2016**, *11*, 5618–5640. [CrossRef]

27. Leufen, G.; Noga, G.; Hunsche, M. Physiological response of sugar beet (Betavulgaris) genotypes to a temporary water deficit, as evaluated with a multiparameter fluorescence sensor. *Acta Physiol. Plant.* **2013**, *35*, 1763–1774. [CrossRef]

28. Goulas, Y.; Cerovic, Z.G.; Cartelat, A.; Moya, I. Dualex: A new instrument for field measurements of epidermal ultraviolet absorbance by chlorophyll fluorescence. *Appl. Opt.* **2004**, *43*, 4488. [CrossRef] [PubMed]

29. Cerovic, Z.G.; Moise, N.; Agati, G.; Latouche, G.; Ben Ghozlen, N.; Meyer, S. New portable optical sensors for the assessment of winegrape phenolic maturity based on berry fluorescence. *J. Food Compos. Anal.* **2008**, *21*, 650–654. [CrossRef]

30. Schmittgen, S.; Ellenberger, J. A dataset on secondary metabolites in bell pepper leaves. **2019**. [CrossRef]

31. R Core Team. R: A Language and Environment for Statistical Computing. Available online: https://www.R-project.org/ (accessed on 16 January 2020).

32. RStudio Team. *RStudio: Integrated Development Environment for R*; Boston, MA, USA, 2016.

33. Wickham, H. *Ggplot2: Elegant Graphics for Data Analysis*; Springer International Publishing A&G: Cham, Switzerland, 2016.

34. Mudrić, S.Ž.; Gašić, U.M.; Dramićanin, A.M.; Ćirić, I.Ž.; Milojković-Opsenica, D.M.; Popović-Đorđević, J.B.; Momirović, N.M.; Tešić, Ž.L. The polyphenolics and carbohydrates as indicators of botanical and geographical origin of Serbian autochthonous clones of red spice paprika. *Food Chem.* **2017**, *217*, 705–715. [CrossRef]

35. Kim, W.-R.; Kim, E.O.; Kang, K.; Oidovsambuu, S.; Jung, S.H.; Kim, B.S.; Nho, C.W.; Um, B.-H. Antioxidant Activity of Phenolics in Leaves of Three Red Pepper (*Capsicum annuum*) Cultivars. *J. Agric. Food Chem.* **2014**, *62*, 850–859. [CrossRef]

36. Ramakrishna, A.; Ravishankar, G.A. Influence of abiotic stress signals on secondary metabolites in plants. *Plant Signal. Behav.* **2011**, *6*, 1720–1731. [PubMed]

37. Hernandez, J.A.; Almansa, M.S. Short-term effects of salt stress on antioxidant systems and leaf water relations of pea leaves. *Physiol. Plant.* **2002**, *115*, 251–257. [CrossRef] [PubMed]

38. Jansen, M.A.K.; Hectors, K.; O'Brien, N.M.; Guisez, Y.; Potters, G. Plant stress and human health: Do human consumers benefit from UV-B acclimated crops? *Plant Sci.* **2008**, *175*, 449–458. [CrossRef]

39. Barnes, P.W.; Searles, P.S.; Ballaré, C.L.; Ryel, R.J.; Caldwell, M.M. Non-invasive measurements of leaf epidermal transmittance of UV radiation using chlorophyll fluorescence: field and laboratory studies. *Physiol. Plant.* **2000**, *109*, 274–283. [CrossRef]

40. Marenco, R.A.; Antezana-Vera, S.A.; Nascimento, H.C.S. Relationship between specific leaf area, leaf thickness, leaf water content and SPAD-502 readings in six Amazonian tree species. *Photosynthetica* **2009**, *47*, 184–190. [CrossRef]

41. Padilla, F.M.; de Souza, R.; Peña-Fleitas, M.T.; Gallardo, M.; Giménez, C.; Thompson, R.B. Different Responses of Various Chlorophyll Meters to Increasing Nitrogen Supply in Sweet Pepper. *Front. Plant Sci.* **2018**, *9*. [CrossRef]

42. Close, D.C.; McArthur, C. Rethinking the role of many plant phenolics–protection from photodamage not herbivores? *Oikos* **2002**, *99*, 166–172. [CrossRef]

43. Close, D.C.; Beadle, C.L. The Ecophysiology of Foliar Anthocyanin. *Bot. Rev.* **2003**, *69*, 149–161. [CrossRef]

44. Riederer, M.; Muller, C. *Annual Plant Reviews, Biology of the Plant Cuticle*; John Wiley & Sons: Hoboken, NJ, USA, 2008; ISBN 978-1-4051-7157-1.

45. Parihar, P.; Singh, S.; Singh, R.; Singh, V.P.; Prasad, S.M. Effect of salinity stress on plants and its tolerance strategies: A review. *Environ. Sci. Pollut. Res.* **2015**, *22*, 4056–4075. [CrossRef]

46. Costa, J.M.; Ortuño, M.F.; Chaves, M.M. Deficit Irrigation as a Strategy to Save Water: Physiology and Potential Application to Horticulture. *J. Integr. Plant Biol.* **2007**, *49*, 1421–1434. [CrossRef]

47. Sgherri, C.; Navari-Izzo, F.; Pardossi, A.; Soressi, G.P.; Izzo, R. The Influence of Diluted Seawater and Ripening Stage on the Content of Antioxidants in Fruits of Different Tomato Genotypes. *J. Agric. Food Chem.* **2007**, *55*, 2452–2458. [CrossRef] [PubMed]

48. Sgherri, C.; Kadlecová, Z.; Pardossi, A.; Navari-Izzo, F.; Izzo, R. Irrigation with Diluted Seawater Improves the Nutritional Value of Cherry Tomatoes. *J. Agric. Food Chem.* **2008**, *56*, 3391–3397. [CrossRef] [PubMed]

49. Neocleous, D.; Nikolaou, G. Antioxidant Seasonal Changes in Soilless Greenhouse Sweet Peppers. *Agronomy* **2019**, *9*, 730. [CrossRef]

50. Sun, M.; Gu, X.; Fu, H.; Zhang, L.; Chen, R.; Cui, L.; Zheng, L.; Zhang, D.; Tian, J. Change of secondary metabolites in leaves of Ginkgo biloba L. in response to UV-B induction. *Innov. Food Sci. Emerg. Technol.* **2010**, *11*, 672–676. [CrossRef]

51. Gogo, E.O.; Förster, N.; Dannehl, D.; Frommherz, L.; Trierweiler, B.; Opiyo, A.M.; Ulrichs, C.; Huyskens-Keil, S. Postharvest UV-C application to improve health promoting secondary plant compound pattern in vegetable amaranth. *Innov. Food Sci. Emerg. Technol.* **2017**, *45*, 426–437. [CrossRef]
52. Brugger, A.; Behmann, J.; Paulus, S.; Luigs, H.-G.; Kuska, M.T.; Schramowski, P.; Kersting, K.; Steiner, U.; Mahlein, A.-K. Extending Hyperspectral Imaging for Plant Phenotyping to the UV-Range. *Remote Sens.* **2019**, *11*, 1401. [CrossRef]

Article

Promising Composts as Growing Media for the Production of Baby Leaf Lettuce in a Floating System

Almudena Giménez [1], Juan A. Fernández [1,2,*], José A. Pascual [3], Margarita Ros [3], José Saez-Tovar [4], Encarnación Martinez-Sabater [4], Nazim S. Gruda [5] and Catalina Egea-Gilabert [1,2]

[1] Departamento de Ingeniería Agronómica, Universidad Politécnica de Cartagena, Paseo Alfonso XIII, 48, 30203 Cartagena, Spain; almudena.gimenez@upct.es (A.G.); catalina.egea@upct.es (C.E.-G.)
[2] Instituto de Biotecnología Vegetal, Edificio I+D+i, Campus Muralla del Mar, 30202 Cartagena, Spain
[3] Conservación de Suelos y Agua y Manejo de Residuos Orgánicos, CEBAS-CSIC, Campus de Espinardo, E-30100 Murcia, Spain; jpascual@cebas.csic.es (J.A.P.); margaros@cebas.csic.es (M.R.)
[4] Departamento de Agroquímica y Medioambiente, Universidad Miguel Hernández, UMH-EPSO, ctra. Beniel Km 3.2, 03312 Orihuela, Spain; jose.saezt@umh.es (J.S.-T.); e.martinezs@umh.es (E.M.-S.)
[5] Department of Horticultural Sciences, University of Bonn Germany, INRES, 53121 Bonn, Germany; ngruda@uni-bonn.de
* Correspondence: juan.fernandez@upct.es

Received: 9 September 2020; Accepted: 5 October 2020; Published: 10 October 2020

Abstract: The floating system is a successful strategy for producing baby leaf vegetables. Moreover, compost from agricultural and agri-food industry wastes is an alternative to peat that can be used as a component of growing media in this cultivation system. In this study, we experimented with three composts containing tomato (*Solanum lycopersicum* L.), leek (*Allium porrum* L.), grape (*Vitis vinifera* L.), and/or olive (*Olea europaea* L.) mill cake residues, which were used as the main component (75/25 volume/volume) of three growing media (GM1, GM2 and GM3) to evaluate their effect on the growth and quality of red baby leaf lettuce (*Lactuca sativa* L.). We used a commercial peat substrate as a control treatment (100% volume) and in mixtures (25% volume) with the composts. The plants were cultivated over two growing cycles, in spring and summer, and harvested twice in each cycle when the plants had four to five leaves. We found that the percentage of seed germination was significantly higher in plants grown in peat than in those grown in compost growing media. The yield was affected by the growing media in the summer cycle, and we obtained the highest value with GM1. Furthermore, the second cut was more productive than the first one for all the growing media in both cycles. The lettuce quality was also affected by the growing media. In general, the total phenolic content and antioxidant capacity in the leaves was higher in plants grown in the compost growing media, particularly in the second cut, but the nitrate content in the leaves was greater in some of the compost treatments compared with the peat treatment. In addition, an in vitro suppressive activity study demonstrated that the interaction between different fungi and bacteria observed through metagenomics analysis could contribute to the effectiveness of the compost in controlling *Pythium irregulare*. The use of compost as a component of the growing media in the production of baby leaf vegetables in a floating system does not only favor the crop yield and product quality, but also shows suppressive effects against *P. irregulare*.

Keywords: germination; nitrate content; phenolic content; antioxidant capacity; microbial community

1. Introduction

Nowadays, there is a high demand among consumers for ready-to-eat vegetables due to a growing interest in healthy, fresh convenience foods. Demand for baby leaf vegetables has especially increased [1]. Baby leaf vegetables come in a wide variety of textures, colors and flavors, which makes them very attractive for consumption. Lettuce is considered to be a health-beneficial food due to the high concentrations of vitamins, minerals, dietary fiber, and antioxidant compounds different lettuces contain [2]. A wide range of varieties can be used for baby leaf production.

Among the hydroponic methods used to produce baby-leaf vegetables, the floating system is a successful strategy for producing baby leaf vegetables, which consists of trays floating on a waterbed or hydroponic nutrient solution, which can be operated as a closed system [1], resulting in a more environmentally friendly crop production strategy (Nicola et al., 2016) [3]. Among other reasons for their use, floating systems make it possible to obtain clean and safe products for the processing industry and to reduce crop cycle duration with respect to soil culture [4]. In addition, some baby leaf crops can be harvested more than once, if regrowth is allowed. With this latter approach, the time to harvest is shortened, resulting in a lesser environmental impact and a reduction in the economic cost [5].

Peat is the usual substrate used to fill the holes in the trays used for growing baby leaf vegetables in floating systems [6]. Nevertheless, peat increases susceptibility to some diseases, such as damping off, which is caused by fungi or oomycetes like *Pythium* spp., which can lead to significant production losses [7]. Moreover, peat comes from peatland ecosystems, and harvesting peat despoils ecologically important peat bog areas [8]; degraded peatlands negatively and disproportionally contribute to released stored carbon and an increase in global greenhouse gas emissions, affecting the environment and CO_2 balance [9,10]. The search for organic materials that can be used as peat alternatives has become increasingly important.

Compost from agricultural and from agri-food industry waste can be an alternative to peat in soilless culture systems. Furthermore, compost can control different plant pathogens, like *Fusarium* sp. [11] and *Pythium irregulare*, and improve the yield and quality of the final product [12]. In addition, compost use is an environmentally friendly practice in light of the circular economy [13]. Depending on its composition, compost made with so-called green materials, such as pruning waste, can in exceptional cases be used directly as a standalone substrate, but it is usually used as a growing media constituent [8,14]. The main limitations to the use of composts in growing media are their physical properties, salinity, high pH, and rate of residual degradation over time [15].

Compost can also be considered an important resource for the biofertilization and bio-stimulation of crops. During composting, organic matter is decomposed and transformed by microorganisms after a polymerization process to form humic substances [16], which have a very important effect on improving soil fertility, because they are rich in mature organic matter. Besides humic substances, the hormone-like molecules secreted by microbes and nutritional elements are compost components that may play a crucial role in the bio-stimulation of plants [17]. The compost microbiome composition plays an important role in the complex relationships that occur in the rhizosphere [18]. High-throughput sequencing technologies have provided an important way to determine compost microbiome information [19], rather than the isolation and identification of microorganism species.

The objective of this study was to characterize three composts from agro-industrial wastes and evaluate their impact as a growing media component on the yield and quality of a red baby leaf lettuce crop growing in a floating system. Our hypothesis was that composts could provide a biostimulant and biofertilizing effect on baby leaf lettuce in addition to its suppressive activity against *P. irregulare*.

2. Materials and Methods

2.1. Compost Characterisation

Three types of compost produced at the University Miguel Hernandez composting site were used for the experiment. In the compost feedstocks, the following raw materials were used: vineyard pruning, tomato and leek processing by-products, and olive mill cake. Their proportion in the composts is described in Table 1. The composting process for the three composts lasted 210 days and consisted of a composting phase with a mesophilic and thermophilic phase of 166 days and a maturation phase of 44 days. The temperatures reached were >60 °C. The physical properties of the compost and peat (bulk density, total pore space, and total water holding capacity) were measured as described by Bustamante et al. [20] pH and electrical conductivity (EC) were measured in a water-soluble extract 1:10 (*w/v*) using a conductivity/pH meter (Crison). The total organic carbon (TOC) and total nitrogen (N) were measured using a LECO TruSpec C/N Elemental Analyzer. P, K, Ca, Mg, B, Fe, Mn, Mo, and Zn were determined by inductively coupled plasma-mass spectrophotometry (ICP-MS PQExCell, VG-Thermo Elemental, Winsford, Cheshire, UK), after $HNO_3/HClO_4$ high pressure digestion. Organic N, nitrate, and ammonium N were determined following the McKenzie and Young [21] method. Available phosphorus was extracted with ammonium citrate pH 7 and it colorimetrically determined on the extracts according to Watanabe and Olsen [22]. Available K was extracted with ammonium acetate pH 7 and later filtered through whatman 0.22 mm^2; it was determined by inductively coupled plasma-mass spectrophotometry, as the rest of the above-measured elements. All the analyses were performed in triplicate. Available humic acids were measured according to Sanchez-Monedero et al. [23]. For the biological characteristics, the bacterial and fungal colony forming units (CFUs) were counted after plating different tenfold serial dilutions of water extract from the composts/peat in Trypto-casein Soy Agar (TSA) plus cycloheximide (100 mg mL^{-1}) and potato dextrose agar (PDA) plus streptomycin (50 mg mL^{-1}), respectively. The Petri plates were incubated at 28 °C, and a standard plate count (SPC) was performed to determine the number of colonies of bacteria and fungi grown on the respective media after 7 and 5 days, respectively. The CFUs were counted and the values were multiplied by the dilution factor and expressed in log CFU g^{-1} of dry compost. Finally, dehydrogenase activity (DHA) was determined according to García et al. [24]

Table 1. Composition of composts in percentage of dry matter.

Composts.		C1	C2	C3
	Vineyard wastes	54	42	41
Feedstocks	Tomato wastes	46	25	21
(% dry matter)	Leek wastes	-	-	8
	Olive mill cake	-	33	30

C1, C2, and C3 represent the composts used.

The main physical, chemical, and biological characteristics of the composts are shown in Table 2.

Table 2. Main physical, chemical, and biological characteristics of the composts used.

	Peat	C1	C2	C3	
		Physical characteristics			
BD (g cm^{-3})	0.38 ± 0.01 b	0.19 ± 0.1 a	0.20 ± 0.1 a	0.20 ± 0.1 a	***
TPS (Vol %)	75.12 ± 0.10	88.4 ± 0.1	87.6 ± 0.2	87.9 ± 0.1	n.s.
AC (Vol %)	20.45 ± 1.23 a	20.6 ± 0.4 a	32.7 ± 0.3 b	34.6 ± 0.7 b	***
WHC (Vol %)	547 ± 11.25 a	678 ± 4.5 b	548 ± 4.9 a	533 ± 6.5 a	***
		Chemical characteristics			
pH	5.6 ± 0.03 a	8.41 ± 0.02 b	8.59 ± 0.03 b	8.84 ± 0.08 b	***
EC (dS m^{-1})	1.2 ± 0.01 a	5.44 ± 0.01 d	3.65 ± 0.04 b	4.75 ± 0.01 c	***
C/N	49.6 ± 0.3	10.9 ± 0.1	13.2 ± 0.5	12.2 ± 0.3	
TOC (g kg^{-1})	466 ± 0.2 b	378 ± 1 a	433 ± 1 b	404 ± 3 b	***
HA (g kg^{-1})	252.1 ± 9.0 c	44.2 ± 2.0 a	70.9 ± 1.0 b	70.5 ± 0.3 b	***
Total N (g kg^{-1})	9.4 ± 0.3 a	36.5 ± 0.5 b	35.2 ± 0.6 b	31.3 ± 0.4 b	***
Organic N (g kg^{-1})	9.3 ± 0.1 a	35.1 ± 0.5 b	34.1 ± 0.5 b	29.8 ± 0.5 b	***
Nitric N (g kg^{-1})	0.10 ± 0.03 a	0.81 ± 0.07 b	0.51 ± 0.09 b	1.1 ± 0.09 b	***
Ammonium N (g kg^{-1})	<0.01 a	0.61 ± 0.02 b	0.6 ± 0.03 b	0.5 ± 0.02 b	***

Table 2. *Cont.*

	Peat	C1	C2	C3	
	Chemical characteristics				
Total P (P_2O_5, g kg^{-1})	4.5 ± 0.2 a	21.4 ± 7.8 b	12.9 ± 5.6 b	15.1 ± 7.7 b	***
Available P (P_2O_5, g kg^{-1})	4.1 ± 0.1 a	19.6 ± 0.7 b	11.7 ± 0.8 b	14.4 ± 0.6 b	***
Total K (K_2O, g kg^{-1})	3.2 ± 0.3 a	28.4 ± 1.8 b	27.4 ± 3.8 b	32.3 ± 2.8 b	***
Available K (K_2O, g kg^{-1})	2.8 ± 0.3 a	25.4 ± 3.7 b	23.9 ± 2.8 b	29.6 ± 3.0 b	***
Ca (g kg^{-1})	18.1 ± 3.0 a	40.0 ± 3.0 b	22.5 ± 2.2 a	28.3 ± 0.41 ab	***
Mg (g kg^{-1})	1.8 ± 0.8 a	6.4 ± 0.1 b	4.0 ± 0.3 b	4.4 ± 0.8 b	***
B (mg kg^{-1})	0.3 ± 0.1	0.6 ± 0.1	0.4 ± 0.1	0.4 ± 0.1	n.s.
Cu (mg kg^{-1})	5.5 ± 1.1 a	21.6 ± 1.2 b	19.3 ± 1.5 b	20.1 ± 1.1 b	***
Fe (g kg^{-1})	1.2 ± 0.1	1.2 ± 0.1	0.9 ± 0.1	1.2 ± 0.1	n.s.
Mn (mg kg^{-1})	70.8 ± 5.7 a	120.1 ± 2.7 b	82.4 ± 5.7 ab	98.6 ± 5.6 ab	***
Mo (mg kg^{-1})	1.5 ± 0.1	1.2 ± 0.1	0.7 ± 0.1	1.0 ± 0.3	n.s.
Zn (mg kg^{-1})	14.3 ± 1.1 a	46.4 ± 1.3 b	28.9 ± 1.8 b	28.0 ± 1.9 b	***
	Biological characteristics				
Total fungi (log(10) CFUs g^{-1})	4.88 ± 0.26 a	4.51 ± 0.02 a	5.40 ± 0.29 b	5.18 ± 0.04 b	***
Total Bacteria (log(10) CFUs g^{-1})	7.23 ± 0.08 a	8.60 ± 0.02 b	9.19 ± 0.04 c	9.41 ± 0.01 c	***
DHA (μmol INTF g^{1-} h^{-1})	3.7 ± 0.1 a	16.43 ± 0.59 b	20.49 ± 0.89 b	17.35 ± 0.15 b	***

Values are the mean $\pm$ SD ($n = 3$). Asterisk indicates significances at *** $p < 0.001$; n.s: non-significant. Different letters indicate significant differences. C1, C2, and C3 represent the compost used. BD: Bulk density; TPS: total pore space; AC: air capacity; WHC: water holding capacity; EC: electrical conductivity; TOC: total organic carbon; HA: humic acids; CFUs: colony formed units; DHA: dehydrogenase activity; INTF: p-iodonitrotetrazolium formazan.

2.2. Compost Microbial Community

Total DNA was extracted from 500-mg compost samples using the DNeasy PowerSoil Kit (Qiagen, Hilden, Germany), following the modification described by Taskin et al. [25] For bacteria, the V4 region of bacterial 16S rDNA was amplified using the barcoded primers 515F and 806R [26]. For fungi, the ITS2 region was amplified with the primer pair gITS7/ITS4 [27]. Each sample was amplified in triplicate as described previously by Žifčáková et al. [28] Amplicons were purified using the QIAquick PCR Purification Kit (Qiagen, Hilden, Germany), and the DNA concentration was measured by Qubit (Thermo Fisher, Waltham, MA, USA). A TruSeq PCR-Free kit was used for library preparation. Sequencing of the bacterial and fungal communities was performed on Illumina MiSeq, and the sequences were generated with the MiSeq Reagent Kit v2 on a paired-end mode with sizes of 251 base pairs (Institute of Microbiology of the CAS, Prague, Czech Republic).

2.3. In Vitro Suppressiveness against Pythium Irregulare

Pytium irregulare isolate, originally recovered from over-used floating trays where baby-leaf lettuces were grown, was selected from the pathogen culture collection of CEBAS-CSIC. The pathogenicity of the isolate was tested every three months, by passing it through baby-leaf plants and re-isolating it again to assure that pathogenicity was not lost due over culture in petri dishes. The mycelial growth of *P. irregulare* was estimated on potato dextrose agar plates (PDA). An 8-mm agar disk of *P. irregulare* was placed on the edge of one side of the plate, and 0.5 mL of dilution 10^{-4} of each compost and peat water extract was spread over the PDA surface on the other side of the plate. As a control, 0.5 mL of sterile water was spread on a PDA surface. Three replicates were performed per treatment. The plates were incubated in the dark at 28 °C, and the radial growth of the pathogen was measured every 24 h for 7 days. Growth inhibition was expressed by Mycelia Growth Inhibition, MGI (%) [29].

MGI% = ((RGcontrol − RGcompost)/RGcontrol) × 100;
RGcontrol = radial growth of pathogen in control plates;
RGcompost = radial growth of pathogen in plates with compost.

2.4. Experimental Conditions

The experiments were conducted at the 'Tomás Ferro' Experimental Agro Food Station of the Technical University of Cartagena (UPCT; lat. 37_410 N; long. 0_570 W). A cultivar of red baby leaf lettuce (*Lactuca sativa* L., cv. 'Ligier') from Rijk Zwaan, De Lier, the Netherlands, was cultivated in a floating system in an unheated greenhouse covered with thermal polyethylene. In the greenhouse, the light conditions during the experiments were an average daily light integral (DLI) of 14.07 mol m^{-2} d^{-1}; the minimum, maximum, and average air temperatures, in the spring cycle were 8.10 °C, 39.10 °C, and 19.41 °C, respectively; in the summer cycle, the average DLI was 15.48 mol m^{-2} d^{-1} and the minimum, maximum, and average air temperatures were 14.02 °C, 44.15 °C and 28.69 °C, respectively. Two crop cycles were carried out with sowings on 29 March 2019 (spring) and 14 June 2019 (summer).

Seeds were sown in 60 × 40-cm styrofloat trays [30] filled with the three compost-growing media (GM1, GM2, GM3) composed using each compost (C1, C2, C3) mixed with commercial peat (75:25, *v/v*). A commercial peat 315 (Blond/black 60/40 Turbas y Coco Mar Menor S.L.) was used as a control. The main chemical characteristics of the peat were as follows: pH 5.6; EC 1 dS m^{-1}; total C 466.8 g kg^{-1}; total N 9.4 g kg^{-1}; total P 0.3 g kg^{-1}; and total K 0.9 g kg^{-1}. After sowing, the trays were placed in a climatic chamber at 18 °C and 90% relative humidity and left in the dark for 48 h to improve germination. After seedling emergence, the trays were transferred to flotation beds (1.35 × 1.25 × 0.2 m). Each level of treatment (peat, GM1, GM2, and GM3) was carried out in beds randomly located at three places inside the greenhouse described above, in both growing seasons; each bed had three floating trays of 60 cm × 41 cm. The trays were floating on tap water with an EC of 1.1 dS m^{-1} and pH 7.8. Aeration was provided using a blow pump connected to a perforated pipe trellis positioned at the bottom of each flotation bed.

A week after sowing, the lettuce plants were thinned, and 10 plants were left per cell (2000 plants m^{-2}). At the same time, the tap water in the beds was replaced with the nutrient solution [31]. The nutrient solution was adjusted to EC 2.5 dS m^{-1} and pH 5.8. The EC and temperature of the nutrient solution were monitored throughout the growing cycles using Campbell CS547 sensors (Campbell Scientific In. Logan, UT) with an average of 2.76 dS m^{-1} and 19.53 °C in spring and 2.76 dS m^{-1} and 29.66 °C in summer, respectively. The oxygen concentrations were monitored using Campbell CS512 sensors located in each flotation bed with an average of 7.09 mg L^{-1} and 6.97 mg L^{-1} in the spring and summer cycles, respectively.

Harvesting was carried out twice per cycle at the same phenological stage for both cycles, when the plants had four to five leaves. The plants were harvested on April 25 (1st cut) and May 6 (2nd cut) in the spring cycle and on July 5 (1st cut) and July 12 (2nd cut) in the summer cycle. For each growing media, 90 plants from three cells fissure were randomly chosen from each tray for harvest; they were then stored at −80 °C for analysis.

2.5. Germination

To calculate the germination percentage, we used nine trays, i.e., three trays per growing media. Twenty baby leaf lettuce seeds were sown in each fissure on the tray, with 154 fissures per replication. After two days in a germination chamber at 18 °C and 90% relative humidity, the trays were transferred to flotation beds randomly placed on three stainless steel beds located in the greenhouse described above with tap water for five days (7 days after sowing (das)), with temperature conditions of 10.17 °C, 31.39 °C, and 17.63 °C as minimum, maximum, and average air temperature, respectively. Then, the percentage of seed germination with respect to the total seeds sown was calculated.

2.6. Plant Analysis at Harvesting

At harvesting time, the following parameters were analyzed in both cycles: biomass production (yield), calculated as g of fresh mass plant^{-1}; nitrate content in leaves and in the nutrient solution; and the total phenolic content and antioxidant capacity in the leaves.

The nitrate content was determined by ion chromatography following Lara et al. [32]

The total phenolic content (TPC) was determined by the Folin-Ciocalteu colorimetric method, as previously described by Singleton and Rossi [33], with modifications previously reported by Martínez-Hernández et al. [34]

The total antioxidant capacity (TAC) was determined as described by Klug et al. [35], using three different approaches: via the free radical scavenging capacity with 2,2–diphenyl–1–picrylhydrazil (DPPH) [36]; the ferric-reducing antioxidant power (FRAP) [37]; and 2 2'–azino–bis (3–ethylbenzothiazoline– 6–sulfonic acid) (ABTS) [38]. The DPPH method was conducted by measuring the decrease in absorbance at 515 nm for 30 min. The TAC extract (21 µL) was mixed with a volume (194 µL) of DPPH solution ($\approx$ 0.8 mM and adjusted to Abs_{515} = 1.1 ± 0.02) and allowed to react for 30 min. The ABTS method was conducted by measuring the absorbance increase at 734 nm for 30 min. A volume (200 µL) of ABTS solution (14 mM $ABTS^+$ and 4.9 mM $K_2S_2O_8$ by 1:1 (v/v)) was added to each extract sample (11 µL) and allowed to react for 30 min. The FRAP method was conducted by measuring the increase in absorbance at 593 nm for 60 min. The freshly made FRAP solution (prepared at a ratio of 10:1:1 ($v/v/v$) using sodium acetate buffer, pH 3.6; 10 mM TPTZ solution in 40 mM HCl; and 20 mM $FeCl_3$, respectively, and preincubated at 37 °C for 2 h) was added (198 µL) to the extract (6 µL) and allowed to react for 60 min. All TAC reactions were conducted at room temperature in darkness, and absorbance was measured using the same microplate reader that was used for TPC. TAC was expressed as mg of Trolox equivalent per 100 g DW of lettuce leaves, as the mean of three replicates per each treatment and cut.

2.7. Statistical Analysis

Data were analyzed using Statgraphics Plus. To determine the compost characteristics, we performed an analysis of variance of measured parameters (one-way ANOVA). For the greenhouse experiment, we performed an analysis of variance of the measured parameters (two-way ANOVA), in which the growing media (peat, GM1, GM2, GM3) and time of cutting (1st cut and 2nd cut) were included for each crop cycle. When the interaction between factors was significant, ANOVA was carried out for each factor independently.

The amplicon sequencing data were processed using the SEED 2 program [39,40]. Pair-end reads were merged using fastq-join [41], and whole amplicons were processed. Chimeric sequences were detected using Usearch 7.0.1090 [42] and removed. Non-chimeric sequences were clustered to 97% similarity using UPARSE implemented within Usearch [43]. Consensus sequences were constructed for each cluster, and the closest hits at the genus level were identified using BLASTn against the GenBank databases for both bacteria and fungi [44]. Sequences identified as non-bacterial or non-fungal were excluded from subsequent analyses.

3. Results

3.1. Compost Characterisation

The three composts showed a similar BD (ca. 0.20 g cm^{-3}), but it was significantly lower than in peat (Table 2). There were no significant differences between treatments for TPS. Composts C2 and C3 showed a significantly higher AC (more than 30 vol %) than C1 and peat (ca. 20 vol %). However, the same two composts and peat showed a significantly lower WHC than C1 678 mL L^{-1}. The three composts showed a basic pH higher than 8.0, significantly higher to peat pH (5.6). The EC values of the three composts were significantly higher than peat, with compost C1 showing the highest EC values, followed by C3 and C2 (Table 2). Composts C2 and C3 showed significantly higher TOC than compost C1 and peat (Table 2). HA was significantly higher in peat with respect to the composts, C2 and C3 also being significantly higher than C1. Composts showed significantly higher values in total and available N, P, and K. In general, peat had lower values with respect to composts for Ca, Mg, Cu, Mn, and Zn. Finally, there were no significant differences between treatments for Mo, Fe, and B.

The fungal and bacterial content (CFUs) of composts C2 and C3 was significantly higher than in C1 and peat (Table 2). The total bacteria content (CFUs) and DHA activity of composts was significantly higher than in peat, CFUs values in C2 and C3 being significantly higher than in C1. Nevertheless, no significant differences were observed between the three composts in terms of DHA activity (Table 2).

3.2. Compost Microbial Community

The Shannon and Simpson diversity index for bacteria did not show significant differences among composts, while for fungi, the compost C2 showed the highest diversity indexes followed by C3 and C1 (Table 3). The coverage value was estimated to be >99% and did not significantly differ between composts.

Table 3. The Shannon and Simpson diversity index for compost bacteria and fungi.

Compost		C1	C2	C3	
	Diversity Index				
Bacteria	Shannon (H)	7.88 ± 0.14	7.80 ± 0.11	7.92 ± 0.04	n.s.
	Simpson (Ds)	0.99 ± 0.01	0.98 ± 0.01	0.99 ± 0.01	n.s.
Fungi	Shannon (H)	1.92 ± 0.07 a	2.60 ± 0.10 b	2.09 ± 0.14 a	***
	Simpson (Ds)	0.49 ± 0.03 a	0.73 ± 0.01 c	0.60 ± 0.03 b	***

Values are the mean ± SD ($n = 3$). Asterisk indicates significances at *** $p < 0.001$; n.s: non-significant. Different letters indicate significant differences.

The dominant bacteria and fungi genera are shown in Figure 1. We identified different bacterial genera belonging to phyla Proteobateria (*Pseudomonas, Pseudoxanthomonas, Pseudofulvimonas, Luteimonas* or *Acinetobacter*); Bacteroidete (*Sphingobacterium, Prevotella* or *Chryseolina*); Firmicutes (*Weisella, Lactobacillus, Megasphaera, Clostridium* or *Brevibacillus*); and Thermus (*Truepera*). As for fungi, we recognized different genera belonging to Ascomycota (*Aspergillus, Thermomyces, Myceliophthora, Mycothermus, Madurella* and *Scedeosporium*); Basicomycota (*Coprinellus, Coprinopsis* and *Coprinus*); and Mucoromycota (*Mortierella*).

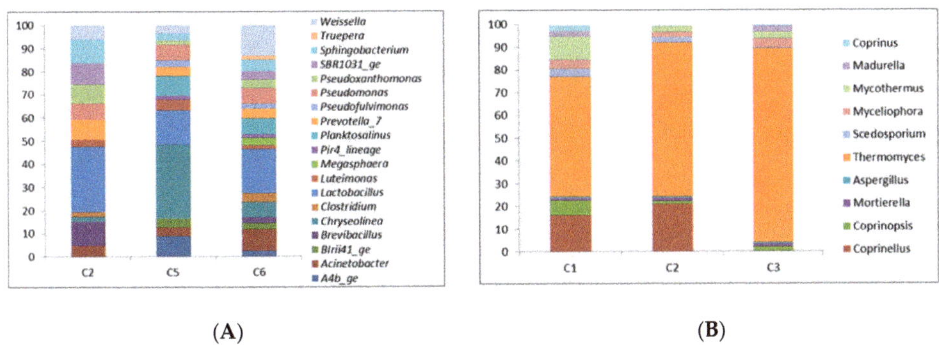

(A) **(B)**

Figure 1. Relative abundance of different genera of bacteria (**A**) and fungi (**B**) in the three composts.

3.3. Suppressiveness: Composts with Added Value

In vitro, the three composts used in this study showed a higher percentage of mycelium growth inhibition (MGI) of *P. irregular* in comparison with peat, where no inhibition was observed. Compost C1 showed the highest MGI (100%), followed by C2 (73%), C3 (65%), and peat (49%) ($F = 22.44$, $p = 0.001$).

3.4. Compost as a Component of Growing Media in Floating Systems

3.4.1. Percentage of Germination and Yield

The percentage of seed germination was significantly higher ($F = 10.37$, $p = 0.0001$) in plants grown in peat, at 94%, compared with those grown in compost growing media (GM) (82–84%).

The yield, on the other hand, was only affected by the growing media in the summer cycle (Table 4). The highest yield was observed in GM1, reaching more than 3 g/plant of the total yield (adding the two cuts), which means an increase of about 23% with respect to that obtained in peat (Table 4). Comparing the different cuts, the second cut was more productive than the first in both cycles, independent of the growing media.

Table 4. Yield and nitrate content in baby leaf red lettuce grown on different growing media (peat, GM1, GM2, GM3) in a floating system.

	Spring		Summer	
	Yield (g Plant^{-1})	Nitrate Content (mg kg^{-1} FW)	Yield (g Plant^{-1})	Nitrate Content (mg kg^{-1} FW)
Substrate (A)				
Peat	1.95 ± 0.15	1618.6 ± 24.6 a	1.28 ± 0.08 a	1968.2 ± 39.4 a
GM1	1.91 ± 0.17	1806.7 ± 27.3 b	1.55 ± 0.08 b	1955.9 ± 58.6 a
GM2	1.94 ± 0.18	1848.6 ± 47.3 b	1.31 ± 0.06 a	2035.1 ± 32.0 b
GM3	2.14 ± 0.22	1833.6 ± 80.6 b	1.40 ± 0.09 a, b	2000.4 ± 22.2 a, b
Cut (B)				
1st cut	1.34 ± 0.04 a	1654.5 ± 21.0 a	1.26 ± 0.04 a	1855.8 ± 18.4 a
2nd cut	2.63 ± 0.09 b	1899.2 ± 40.3 b	1.51 ± 0.07 b	2124.0 ± 16.4 b
A × B				
Peat × 1st cut	1.48 ± 0.05	1584.0 ± 46.7 a	1.06 ± 0.04	1835.6 ± 34.7 b
Peat × 2nd cut	2.42 ± 0.19	1653.3 ± 9.4 a, b	1.49 ± 0.12	2100.7 ± 31.8 d, e
GM1 × 1st cut	1.24 ± 0.07	1757.0 ± 20.2 b, c	1.37 ± 0.03	1722.3 ± 25.0 a
GM1 × 2nd cut	2.57 ± 0.09	1856.3 ± 46.2 c	1.76 ± 0.13	2189.6 ± 18.5 f
GM2 × 1st cut	1.27 ± 0.09	1699.8 ± 34.5 a, b	1.32 ± 0.08	1906.8 ± 12.2 c
GM2 × 2nd cut	2.61 ± 0.13	1997.5 ± 52.6 d	1.35 ± 0.05	2163.4 ± 9.0 e, f
GM3 × 1st cut	1.36 ± 0.10	1577.3 ± 31.9 a	1.33 ± 0.07	1958.6 ± 5.5 c
GM3 × 2nd cut	2.91 ± 0.24	2090.0 ± 99.7 d	1.46 ± 0.16	2042.1 ± 40.4 d
Analysis of variance				
Substrate (A)	n.s.	***	*	*
Cut (B)	***	***	**	***
A × B	n.s.	***	n.s.	***

Asterisks indicate significances at * $p < 0.05$; ** $p < 0.01$; *** $p < 0.001$; n.s.: non-significant. Different letters indicate significant differences. FW: fresh weight.

3.4.2. Nitrate Content in Leaves and in the Nutrient Solution

Regarding the nitrate content in leaves, the two-way ANOVA indicated a significant interaction between the growing media and cut in both cycles (Table 4). In the spring cycle, the highest nitrate content values were obtained in the 2nd cut for GM2 and GM3. In the summer, the nitrate content was greater in the 2nd cut than in the 1st cut in all growing media, and the highest values were obtained with GM1 and GM2 in the 2nd cut.

During the spring cycle, in the 1st cut, the nitrate content measured in the nutrient solution was higher in plants grown in compost growing media than in plants grown in peat (Figure 2). In the 2nd cut, the nitrate concentrations decreased slightly for every compost growing media but not for peat, which remained constant, thus equalizing the values of all treatments by the end of the cycle. In the summer cycle, the lowest concentrations of nitrate in the nutrient solution in the 1st cut were found for GM1. In the 2nd cut, the nitrate concentration increased for every growing media, while GM1 maintained the lowest value.

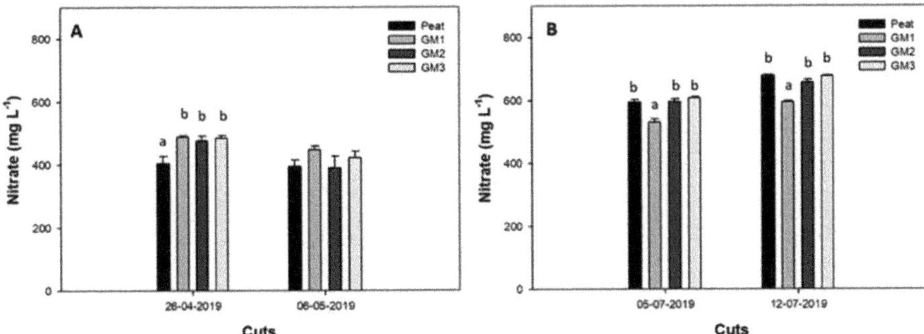

Figure 2. Nitrate content in the nutrient solution for both cuts in the spring (**A**) and summer (**B**) cycles using different growing media (peat, GM1, GM2, GM3). Values are the mean ± SD ($n = 9$).

3.4.3. Total Phenolic and Antioxidant Capacity in Red Baby Leaf Lettuce Leaves and Roots

Regarding the total phenolic content, there was an interaction between growing media and cuts in both cycles (Table 5). The plants showed a similar phenolic content pattern in leaves when they were cultivated in the different growing media, although those with compost showed higher values, particularly in the 2nd cut. In general, the lowest values were found in the 1st cut in plants grown with peat. GM3 stood out for the high phenolic content values found in leaves in the 2nd cut in both cycles. The total phenolic content was always higher in summer than in spring.

Table 5. Total phenolic content in the leaves of baby leaf red lettuce grown on different growing media (peat, GM1, GM2, GM3) and harvested twice (1st and 2nd cut), cultivated in spring and summer cycles in a floating system.

Total Phenolic Content (mg GA 100 g^{-1} DW)		
	Spring	**Summer**
Substrate (A)		
Peat	2467.8 ± 225.6 a	2638.8 ± 177.5 a
GM1	2741.8 ± 226.9 a, b	2932.7 ± 156.2 a, b
GM2	2838.3 ± 94.4 a, b	3046.8 ± 216.5 b, c
GM3	2976.6 ± 192.2 b	3268.4 ± 24.1 c
Cut (B)		
1st cut	2461.0 ± 99.2 a	2577.0 ± 68.0 a
2nd cut	3051.2 ± 119.7 b	3366.4 ± 114.4 b
A × B		
Peat × 1st cut	2023.5 ± 133.4 a	2304.9 ± 121.4 a
Peat × 2nd cut	2912.2 ± 198.3 b, c	2972.8 ± 176.4 b, c
GM1 × 1st cut	2488.5 ± 144.5 a, b	2624.2 ± 128.1 a, b
GM1 × 2nd cut	2995.1 ± 415.1 b, c	3241.1 ± 102.0 c, d
GM2 × 1st cut	2641.7 ± 75.6 a, b, c	2615.0 ± 72.5 a, b
GM2 × 2nd cut	3035.0 ± 12.2 b, c	3478.6 ± 206.8 d, e
GM3 × 1st cut	2690.5 ± 182.8 b, c	2763.9 ± 94.8 b
GM3 × 2nd cut	3262.7 ± 263.5 c	3772.9 ± 171.9 e
Analysis of variance		
Substrate (A)	*	**
Cut (B)	***	***
A × B	*	***

Asterisks indicate significances at * $p < 0.05$; ** $p < 0.01$; *** $p < 0.001$; n.s.: non-significant. Different letters indicate significant differences. GA: gallic acid. DW: dry weight.

The antioxidant capacity, measured by the ABTS and FRAP methods, highlighted a similar pattern to that found in the total phenolic content for both cycles, with higher values in the compost growing media, particularly in the 2nd cut and in the leaves of plants grown with GM3 (Table 6). However, there were some exceptions: we did not find significant differences between the growing media in terms of antioxidant capacity measured in leaves using either the ABTS method in the spring cycle or the FRAP method in the summer.

Table 6. Total antioxidant capacity (FRAP and ABTS methods) in leaves of baby leaf red lettuce grown on different substrates (peat, GM1, GM2, GM3) and harvested twice (1st and 2nd cut), cultivated in spring and summer cycles in a floating system.

	Total Antioxidant Capacity (mg Trolox Equivalents 100 g^{-1} DW)			
	Spring		Summer	
	ABTS	FRAP	ABTS	FRAP
Substrate (A)				
Peat	2552.4 ± 76.0	6444.5 ± 183.4 a	2593.0 ± 114.6 a	7223.6 ± 98.7
GM1	2477.0 ± 93.9	6717.3 ± 196.9 a, b	2654.5 ± 61.4 a	7807.5 ± 258.6
GM2	2755.9 ± 76.9	70004.7 ± 224.9 b, c	2990.6 ± 226.9 b	7518.7 ± 281.2
GM3	2574.5 ± 133.3	7290.6 ± 197.3 c	2906.4 ± 246.5 b	7661.5 ± 392.2
Cut (B)				
1st cut	2461.4 ± 49.4 a	6557.0 ± 108.6 a	2456.8 ± 49.9 a	7158.3 ± 129.8 a
2nd cut	2717.5 ± 71.8 b	7171.5 ± 161.2 b	3115.4 ± 110.4 b	7947.4 ± 183.5 b
A × B				
Peat × 1st cut	2448.2 ± 103.9 a	6221.2 ± 97.0 a	2423.4 ± 133.9 a	7224.9 ± 202.2 a, b
Peat × 2nd cut	2656.5 ± 85.2 a, b, c	6667.7 ± 330.1 a, b	2762.7 ± 137.7 b	7222.8 ± 88.6 a, b
GM1 × 1st cut	2432.9 ± 107.9 a	6449.6 ± 245.8 a, b	2545.9 ± 82.5 a, b	7361.2 ± 335.2 a, b
GM1 × 2nd cut	2521.1 ± 247.0 a, b	6985.0 ± 248.5 b, c	2763.0 ± 15.7 b	8253.7 ± 151.2 c
GM2 × 1st cut	2635.12 ± 82.72 a, b, c	6518.8 ± 102.4 a, b	2493.3 ± 114.5 a, b	6999.64 ± 36.10 a
GM2 × 2nd cut	2872.74 ± 92.97 c	7490.6 ± 79.2 c	3484.8 ± 12.7 c	8037.8 ± 353.1 b, c
GM3 × 1st cut	2329.41 ± 6.17 a	7038.5 ± 20.3 b, c	2361.6 ± 84.1 a	7047.2 ± 417.5 a
GM3 × 2nd cut	2819.57 ± 169.57 b, c	7542.7 ± 367.6 c	3451.1 ± 10.8 c	8275.8 ± 466.0 c
Analysis of variance				
Substrate (A)	n.s.	**	**	n.s.
Cut (B)	**	**	***	**
A × B	*	*	***	*

Asterisks indicate significances at * $p < 0.05$; ** $p < 0.01$; *** $p < 0.001$; n.s.: non-significant. Different letters indicate significant differences. DW: dry weight.

4. Discussion

To date, there have been few studies that have investigated the ideal growing media for a floating system. Among them, Cros et al. [4] demonstrated that a peat-based floating cultivation system can be considered the most suitable growing medium to grow purslane, because of its ideal physical and chemical characteristics. Nicola et al. [43] also recommended a peat-based horticultural medium for baby leaf vegetables grown in floating system. However, in recent years, there have been increasing environmental and ecological concerns about the use of peat as a growing medium because its harvest is jeopardizing endangered wetland ecosystems worldwide [45]. Furthermore, increasing demand and rising costs for peat as growing media in horticulture have led to a search for high-quality and low-cost substrates as an alternative. Compost may have physical, chemical, and biological properties that can contribute to partial peat reduction in growing media formulations [46]. In the case of a floating system, the three assayed composts showed physical properties similar to peat [47], although C1 showed a significantly higher WHC than peat, which could bring higher moisture and cause some negative aspects on plant growth [40]. Yet this issue was easily overcome given the type of trays used in this study, which contain a low volume of substrate per hole, and the system of cultivation (floating trays), where substrates obtain the water that they need and the roots mostly grow into the nutrient solution [30]. This makes the compost physical properties in this cultivation system a non-limiting factor for plant growth.

The C/N ratio of the three composts was less than 25, which indicates that the composts can be considered matured [48]. By using unmatured composts and/or unstable organic materials with a high C/N ratio, such as wood fibers, as the plant material degrades, N-immobilization occurs. This is accompanied by a decrease in soil volume (shrinkage), pore space, and air content [49–51]. Both pH and EC have an essential influence on seedling quality and plant growth. The three composts used in this study showed a basic pH ranging from 8.4 to 8.8; these values are higher than those recommended for growing media [52]. By mixing the composts with 25% peat, however, the pH was reduced for all growing media used (to 7.7, 7.8 and 7.9, for GM1, GM2 and GM3, respectively). As a result, we obtained a good germination rate and good seedling growth. These results are in accordance with those of Morales et al. [53]. With respect to EC, an EC $\leq$ 3.5 dS m^{-1} is considered to be the limit for seedling growth in a growing medium [54]. Moreover, an EC > 4 dS m^{-1} has been reported to inhibit seed germination [55]. In our study, the growing media presented a percentage of seed germination: ca. >82%, a level of germination that can be considered standard within the range normally found for this species.

Even if C1 and C3 exceeded the abovementioned EC limit, the peat used in the mixtures served as a thinner and reduced the salt concentration of the growing media to EC 3.3, 1.8, and 2.6 dS m^{-1} for GM1, GM2, and GM3, respectively. Moreover, the high water holding capacity of C1 may also have positively influenced seed germination in GM1, because water retention is a decisive factor to this process [56]. Nevertheless, other factors could influence seed germination, due to the complexity of the mechanisms involved in it.

Furthermore, the growing media did not have any adverse effects on plant growth. In fact, the compost growing media promoted plant growth to a greater extent than peat, reaching higher yields. This beneficial effect of compost on yield could be due to the availability of nutrients and the production of auxin-like components from humic substances (Table 2). According to Trevisan et al. [57], compost acts as a reservoir for nutrients, ensuring their slow release to plant roots [58]. Moreover, some microorganisms found in our compost have been described as plant growth promoters (PGP). According to Castellano-Hinojosa et al. [59], strains of *Pseudoxanthomonas* promote plant growth via the production of ACC deaminase and siderophore and the solubilization of phosphate. In addition, Kuan et al. [60] found that inoculating maize with N2-fixing PGP strains belonging to genera *Acinetobacter* significantly increased the total N content and dry biomass of plants.

The time that the plants needed to reach the adequate phenological stage for the first cut was longer (27 and 21 days in spring and summer cycle, respectively) than the time from the first to the second cut (11 and 7 days in spring and summer cycle, respectively). As Jasper et al. [61] demonstrated recently in rocket plants, the growth rate prior to the first cut is slower than the subsequent regrowth rate due to the initial plant establishment. In addition, the second cut was more productive than the first in every growing media. Awan and Ahmad [62] and Suzuki et al. [63] also found that spinach foliage weight at the second harvest was greater than the weight at the first harvest, suggesting that new roots, which developed vigorously during the regrowth period, had a positive effect on the absorption of water and nutrients by the plants [63].

The three composts used in our study all showed suppressive activity against *P. irregulare*, which was not observed in the peat [64,65]. Hoitink et al. [66] pointed out a direct relationship between compost microbial activity and the suppression of *Pythium* and *Phytophthora* root rots. In our study, compost C1 showed the highest suppressiveness, which may be related to the fact that the primed plants displayed a faster and stronger activation of the various cellular defenses [64], it could also be due to microbial antagonism, nutrient competition, parasitism, and antibiosis [65]. Dehydrogenase activity, a potential indicator of general microbial activity [11], cannot be considered as an indicator for determining compost suppressiveness in this study, since no significant differences for this parameter were observed among the composts tested. Several studies have shown that compost microbial composition primarily depends on the microbial competition for nutrients in different types of feedstock, which deeply influence compost recolonization during curing time [67,68]. The composts investigated here were recolonized by different microbial communities. Among them, we found microorganisms belonging to specific beneficial groups, such as *Aspergillus*, *Pseudomonas*, and *Morteriella* sp. These microorganisms are effective against different pathogens [69,70] and can induce systemic resistance in plants [71]. *Brevibacillus*, which has also been found to produce bioactive compounds against pathogens [72], showed higher relative abundance than other beneficial microorganisms in C1. Interaction between different beneficial microorganisms in the composts studied could contribute to their effectiveness in controlling *P. irregulare*.

According to Nicola et al. [1], baby leaf vegetables are a significant source of nitrates, so the nitrate content is an important quality characteristic to consider. Our data reveal that the nitrate concentrations did not exceed the maximum level allowed by the EU for this type of lettuce and cultivation system, although the use of compost did increase the amount of nitrates in comparison with peat. In both cycles, the nitrate content in the 2nd cut was higher than in the first. This fact could be linked to changes in the nutrient solution, due to both a gradual release of nitrate from the compost growing media and the evaporation of water from the floating beds, particularly in summer. Moreover, the nitrogen mineralization rate from organic substrates is higher in summer than in spring, due to the higher temperatures [73]. The influence of enriched nitrates in the nutrient solution could overcome the effect produced by the higher LDI in the summer cycle on nitrate reductase activity, which would increase the conversion rate of nitrate to amino acids, reducing nitrate levels in the leaves [74].

Plants grown in compost growing media showed a higher total phenolic content levels than plants grown in peat. The compost feedstocks used (tomato, leek, vineyard, and olive mill cake residues) are rich in compounds with the capacity to activate an oxidative process in plants [75]. As suggested by Santos et al. [76], these kinds of compounds induce the stimulatory effects of secondary metabolites on different parts of lettuce grown in different agro-industrial composts. The season also influenced the accumulation of phenolic compounds in the lettuce leaves: the total phenolics were higher in summer. This agrees with the results of Marín et al. [77], who found a positive correlation between the total phenolic content and temperatures. Besides, the root zone temperature has been found to influence the production of plant metabolites in several plants [78–80]. Temperature increases in the root zone in leafy vegetables can lead to alterations in the production of some secondary metabolites in greenhouse cultivation, phenolic compounds being the most pronounced secondary metabolites [81,82]. Furthermore, the higher phenolic compound levels after the 2nd cut could be linked to the increase in some phenolic metabolism enzymes due to the signals that spread from the injured tissue to the adjacent non-injured tissue after wounding, as observed by Salveit [83], who reported a 6 to 12-fold increase in PAL activity within 24 h after cutting in Batavia lettuce.

In our study, antioxidant capacity, measured by ABTS or FRAP methods, had a positive correlation with the total phenolic content in the leaves. The correlation coefficients were $r = 0.67$ and $r = 0.86$ for the first and second cuts in spring and $r = 0.91$ and $r = 0.85$ in summer, respectively. Santos et al. [76] found similar results in lettuce using agro-industrial composts as substrates. Moreover, we found a higher antioxidant capacity after the second cut. This agrees with the results of Kang and Saltveit [84], who demonstrated that the antioxidant capacity of lettuce leaf tissue increases after wounding. The baby leaf lettuce in our study showed the highest antioxidant capacity and total phenolic content in GM2 and GM3 growing media. This is most probably due to the original presence of olive mill cake, given that Chrysargyris et al. [85] reported an increase in antioxidant enzyme metabolism in marigold and petunia grown in soilless media, using up to 30% olive mill cake in place of peat. In general, these findings suggest that compost amendments can help add value to lettuce by increasing its antioxidant activity to a greater extent than other organic resources such as peat [76].

5. Conclusions

Composts from different raw materials like vineyard wastes, tomato wastes, leek wastes, and olive mill cake can be an alternative to peat as a central component of the growing media in the production of baby leaf vegetables in a floating system. They not only increase crop yields due to their biofertilizer activity, but also boost the final quality of the product with a higher total phenolic content and antioxidant capacity. Moreover, composts were able to control the effect of *P. irregulare* due to their suppressive effect of bacterial–fungal interactions. However, the percentage of seed germination was higher in plants grown in peat than in those grown in compost growing media and the nitrate content in the leaves was greater in some of the compost treatments than in the peat treatment. Further studies are needed on the standardization of feedstocks origin, composting and stabilization processes in order to obtain standard growing media for this cultivation system.

Author Contributions: Conceptualization, M.R., J.A.P., J.A.F., C.E.G.; data curation, A.G., J.S.-T., E.M.-S.; formal analysis, A.G., M.R.; funding acquisition, J.A.F., J.A.P., C.E.-G., M.R.; investigation, A.G., J.S.-T., E.M.-S.; project administration J.A.F., J.A.P.; supervision, M.R.; J.A.F., C.E.G.; visualization, C.E.-G.; writing—original draft J.A.F., C.E.G., M.R.; writing—review and editing, J.A.F., C.E.-G., J.A.P., N.S.G. All authors have read and agreed to the published version of the manuscript.

Funding: This research was funded by the Spanish Ministry of Economy and Competitiveness: Reference project: AGL2017-84085-C3-1-R, AGL2017-84085-C3-2-R, AGL2017-84085-C3-3-R.

Acknowledgments: This work was supported by projects AGL2017-84085-C3-1-R, AGL2017-84085-C3-2-R, AGL2017-84085-C3-3-R from the Ministry of Economy and Competitiveness of Spain.

Conflicts of Interest: The authors declare no conflict of interest.

References

1. Nicola, S.; Egea-Gilabert, C.; Niñirola, D.; Conesa, E.; Pignata, G.; Fontana, E.; Fernández, J.A. Nitrogen and aeration levels of the nutrient solution in soilless cultivation systems as important growing conditions affecting inherent quality of baby leaf vegetables: A review. *Acta Hortic.* **2015**, *1099*, 167–177. [CrossRef]

2. Mampholo, B.M.; Maboko, M.M.; Soundy, P.; Svakumar, D. Phytochemicals and overall quality of baby leaf lettuce (*Lactuca sativa* L.) varieties grown in closed hydroponic system. *J. Food Qual.* **2016**, *39*, 805–815. [CrossRef]

3. Nicola, S.; Pignata, G.; Casale, M.; Lo Turco, P.E.; Gaino, W. Overview of a Lab-scale Pilot Plant for Studying Baby leaf Vegatables Grown in Soiless Culure. *Hortic. J.* **2016**, *85*, 97–104. [CrossRef]

4. Cros, V.; Martínez-Sánchez, J.J.; Franco, J.A. Good yields of common purslane with a high fatty acid content can be obtained in a peat-based floating system. *HortTechnology* **2007**, *17*, 14–20. [CrossRef]

5. Pennisi, G.; Sanyé-Mengual, E.; Orsini, F.; Crepaldi, A.; Nicola, S.; Ochoa, J.; Fernandez, J.A.; Gianquinto, G. Modelling environmental burdens of indoor-grown vegetables and herbs as affected by red and blue LED lighting. *Sustainability* **2019**, *11*, 4063. [CrossRef]

6. Pane, C.; Spaccini, R.; Piccolo, A.; Scala, F.; Bonanomi, G. Compost amendments enhance peat suppressiveness to *Pythium ultimum*, *Rhizoctonia solani* and *Sclerotinia minor*. *Biol. Control* **2011**, *56*, 115–124. [CrossRef]

7. Limpens, J.; Berendse, F.; Blodau, C.; Canadell, J.G.; Freeman, C.; Holden, J.; Roulet, N.; Rydin, H.; Schaepman-Strub, G. Peatlands and the carbon cycle: From local processes to global implications—A synthesis. *Biogeosciences* **2008**, *5*, 1475–1491. [CrossRef]

8. Gruda, N.S. Increasing sustainability of growing media constituents and stand-alone substrates in soilless culture systems. *Agronomy* **2019**, *9*, 298. [CrossRef]

9. Bonn, A.; Reed, M.S.; Evans, C.D.; Joosten, H.; Bain, C.; Farmer, J.; Emmer, I.; Couwenberg, J.; Moxey, A.; Artz, R.; et al. Investing in nature: Developing ecosystem service markets for peatland restoration. *Ecosyst. Serv.* **2014**, *9*, 54–65. [CrossRef]

10. Gruda, N.; Bisbis, M.B.; Tanny, J. Impacts of protected vegetable cultivation on climate change and adaptation strategies for cleaner production—A review. *J. Clean. Prod.* **2019**, *225*, 324–339. [CrossRef]

11. Blaya, J.; Lloret, E.; Ros, M.; Pascual, J.A. Identification of predictor parameters to determine agro-industrial compost suppressiveness against *Fusarium oxysporum* and *Phytophthora capsici* disease in muskmelon and pepper seedlings. *J. Sci. Food Agric.* **2015**, *95*, 1482–1490. [CrossRef] [PubMed]

12. Giménez, A.; Fernández, J.A.; Pascual, J.A.; Ros, M.; López-Serrano, M.; Egea-Gilabert, C. An agroindustrial compost as alternative to peat for production of baby leaf red lettuce in a floating system. *Sci. Hortic.* **2019**, *246*, 907–915. [CrossRef]

13. Razza, F.; D'Avino, L.; L'Abate, G.; Lazzeri, L. The role of compost in bio-waste management and circular economy. In *Designing Sustainable Technologies, Products and Policies*; Benetto, E., Gericke, K., Guiton, M., Eds.; Springer: Cham, Switzerland, 2018. [CrossRef]

14. Prasad, M.; Maher, M.J. The use of composted green waste (cgw) as a growing medium component. *Acta Hortic.* **2001**, *549*, 107–114. [CrossRef]

15. Raviv, M. Composts in growing media: What's new and what's next? *Acta Hortic.* **2013**, *982*, 39–52. [CrossRef]

16. Tan, K.H. *Humic Matter in Soil and the Environment*, 2nd ed.; CRC Press: Boca Raton, FL, USA, 2014. [CrossRef]

17. Zaccardelli, M.; Pane, C.; Scotti, R.; Maria, A.; Colturali, S.; Basilicata, U. Impiego di compost-tea come bioagrofarmaci e biostimolanti in orto- frutticoltura. *Italus Horto* **2012**, *19*, 17–28.

18. Hadar, Y.; Papadopoulou, K.K. Suppressive composts: Microbial ecology links between abiotic environments and healthy plants. *Annu. Rev. Phytopathol.* **2012**, *50*, 133–153. [CrossRef]

19. Ros, M.; Blaya, J.; Baldrian, P.; Bastida, F.; Richnow, H.; Jehmlich, N.; Pascual, J. In vitro elucidation of suppression effects of composts to soil-borne pathogen *Phytophthora nicotianae* on pepper plants using 16S amplicon sequencing and metaproteomics. *Renew. Agric. Food Syst.* **2020**, *35*, 206–214. [CrossRef]

20. Bustamante, M.A.; Paredes, C.; Moral, R.; Agullo, E.; Perez-Murcia, M.D.; Abad, M. Compost from distillery wastes as peat subsitutes for transplant production. *Resour. Conserv. Recycl.* **2008**, *52*, 792–799. [CrossRef]

21. McKenzie, J.R.; Young, P.N. Determination of ammonia-, nitrate-, and organic nitrogen in water and wastewater with an ammonia gas-sensing electrode. *Analyst* **1975**, *100*, 620–628. [CrossRef]

22. Watanabe, F.S.; Olsen, S.R. Test of an ascorbic acid method for determining phosphorus in water and $NaHCO_3$ extracts from soil. *Soil Sci. Soc. Am. Proc.* **1965**, *29*, 677–678. [CrossRef]

23. Sánchez-Monedero, M.A.; Roig, A.; Martínez-Pardo, C.; Cegarra, J.; Paredes, C. A microanalysis method for determining total organic carbón in extracts of humic substances. Relationships between total organic carbón and oxidable carbón. *Bioresour. Technol.* **1996**, *57*, 291–295. [CrossRef]

24. Garcia, C.; Hernandez, T.; Costa, F. Potential use of dehydrogenase activity as an index of microbial activity in degraded soil. *Commun. Soil Plant* **2008**, *28*, 123–134. [CrossRef]

25. Taskin, B.; Gozen, A.G.; Duran, M. Selective quantification of viable *Escherichia coli* bacteria in biosolids by quantitative PCR with propidium monoazide modification. *Appl. Environ. Microbiol.* **2011**, *77*, 4329–4335. [CrossRef] [PubMed]

26. Caporaso, J.G.; Lauber, C.L.; Walters, W.A.; Berg-Lyons, D.; Huntley, J.; Fierer, N.; Owens, S.M.; Betley, J.; Fraser, L.; Bauer, M.; et al. Ultra-high-throughput microbial community analysis on the Illimina HiSeq and MiSeq platforms. *ISME J.* **2012**, *6*, 1621–1624. [CrossRef]

27. Ihrmark, K.; Bödeker, I.T.M.; Cruz-Martinez, K.; Friberg, H.; Kubartova, A.; Schenck, J.; Strid, Y.; Stenlid, J.; Brandström-Durling, M.; Clemmensen, K.E.; et al. New primers to amplify the fungal ITS2 region–evaluation by454-sequencing of artificial and natural communities. *FEMS Microbiol. Ecol.* **2012**, *82*, 666–677. [CrossRef]

28. Žifčáková, L.; Vetrovsky, T.; Howe, A.; Baldrian, P. Microbial activity in forest soil reflects the changes in ecosystem properties between summer and winter. *Environ. Microbiol.* **2016**, *18*, 288–301. [CrossRef]

29. De Corato, U.; Viola, E.; Arcieri, G.; Valerio, V.; Zimbardi, F. Use of composted agro-energy co-products and agricultural residues against soil-borne pathogens in horticultural soil-less system. *Sci. Hortic.* **2016**, *210*, 166–179. [CrossRef]

30. Niñirola, D.; Fernández, J.A.; Conesa, E.; Martínez, J.A.; Egea-Gilabert, C. Combined effects of growth cycle and different levels of aeration in nutrient solution on productivity, quality and shelf-life of watercress (*Nasturtium officinale* R. Br.) plants. *HortScience* **2014**, *49*, 567–573. [CrossRef]

31. Egea-Gilabert, C.; Fernández, J.A.; Migliaro, D.; Martínez-Sánchez, J.J.; Vicente, M.J. Genetic variability in wild vs. cultivated *Eruca vesicaria* populations as assessed by morphological, agronomical and molecular analyses. *Sci. Hortic.* **2009**, *121*, 260–266. [CrossRef]

32. Lara, L.J.; Egea-Gilabert, C.; Niñirola, D.; Conesa, E.; Fernández, J.A. Effect of aeration of the nutrient solution on the growth and quality of purslane (*Portulaca oleracea*). *J. Hortic. Sci. Biotechnol.* **2011**, *86*, 603–610. [CrossRef]

33. Singleton, V.L.; Orthofer, R.; Lamuela-Raventós, R.M. Analysis of total phenols and other oxidation substrates and antioxidants by means of Folin–Ciocalteu reagent. *Methods Enzymol.* **1999**, *299*, 152–178. [CrossRef]

34. Martínez-Hernández, G.B.; Gómez, P.A.; Pradas, I.; Artés, F.; Artés-Hernández, F. Moderate UV-C pretreatment as a quality enhancement tool in fresh-cut Bimi®broccoli. *Postharvest Biol. Technol.* **2011**, *62*, 327–337. [CrossRef]

35. Klug, T.V.; Martínez-Hernández, G.B.; Collado, E.; Artés, F.; Artés-Hernández, F. Effect of microwave and high-pressure processing on quality of an innovative broccoli hummus. *Food Bioprocess. Technol.* **2018**, *11*, 1464–1477. [CrossRef]

36. Brand-Williams, W.; Cuvelier, M.E.; Berset, C. Use of free radical method to evaluate antioxidant activity. *LWT—Food Sci. Technol.* **1995**, *28*, 25–30. [CrossRef]

37. Benzie, L.F.; Strain, J.J. Ferric reducing/antioxidant power assay: Direct measure of total antioxidant activity of biological fluids and modified version for simultaneous measurement of total antioxidant power and ascorbic acid concentration. *Methods Enzymol.* **1999**, *299*, 15–27. [CrossRef] [PubMed]

38. Cano, A.; Hernández-Ruíz, J.; García-Cánovas, F.; Acosta, M.; Arnao, M.B. An end-point method for estimation of the total antioxidant activity in plant material. *Phytochem. Anal.* **1998**, *9*, 196–202. [CrossRef]

39. Větrovský, T.; Baldrian, P.; Morais, D. SEED 2: A user-friendly platform for amplicon high-throughput sequencing data analyses. *Bioinformatics* **2018**, *34*, 2292–2294. [CrossRef]

40. Větrovský, T.; Baldrian, P. The variability of the 16S rRNA gene in bacterial genomes and its consequences for bacterial community analyses. *PLoS ONE* **2013**, *8*, e57923. [CrossRef]

41. Aronesty, E. Comparison of sequencing utility programs. *Open Bioinform. J.* **2013**, *7*, 1–8. [CrossRef]

42. Edgar, R.C. Search and clustering orders of magnitude faster than BLAST. *Bioinformatics* **2010**, *1*, 2460–2461. [CrossRef]

43. Edgar, R.C. UPARSE: Highly accurate OTU sequences from microbial amplicon reads. *Nat. Methods* **2013**, *10*, 996–998. [CrossRef] [PubMed]

44. Kõljalg, U.; Nilsson, R.H.; Abarenkov, K.; Tedersoo, L.; Taylor, A.F.; Bahram, M.; Bates, S.T.; Bruns, T.D.; Bengtsson-Palme, J.; Callaghan, T.M.; et al. Towards a unified paradigm for sequence-based identification of fungi. *Mol. Ecol.* **2013**, *22*, 5271–5277. [CrossRef] [PubMed]

45. Zaller, J.G. Vermicompost as a substitute for peat in potting media: Effects on germination, biomass allocation, yields and fruit quality of three tomato varieties. *Sci. Hortic.* **2007**, *112*, 191–199. [CrossRef]

46. Clemmensen, A.W. Physical characteristics of Miscanthus composts compared to peat and wood fiber growth substrates. *Compost Sci. Util.* **2004**, *12*, 219–224. [CrossRef]

47. Pascual, J.A.; Ceglie, F.; Tuzel, Y.; Koller, M.; Koren, A.; Hitchings, R.; Tittarelli, F. Organic substrate for transplant production in organic nurseries. A review. *Agron. Sustain. Dev.* **2018**, *38*, 35. [CrossRef]

48. Chen, J.; McConnell, D.B.; Robinson, C.A.; Caldwell, R.D.; Huang, Y. Rooting foliage plant cuttings in compost formulated substrates. *HorTechnology* **2003**, *13*, 110–114. [CrossRef]

49. Gruda, N.; Tucher, S.V.; Schnitzler, W.H. N-Immobilization of wood fiber substrates in the production of tomato transplants (*Lycopersicon lycopersicum* (L.) Karst. Ex Farw.). *J. Appl. Bot.* **2000**, *74*, 32–37. (In German)

50. Gruda, N.; Schnitzler, W.H. Suitability of wood fiber substrates for production of vegetable transplants. I. Physical properties of wood fiber substrates. *Sci. Hortic.* **2004**, *100*, 309–322. [CrossRef]

51. Gruda, N.; Schnitzler, W.H. Suitability of wood fiber substrates for production of vegeta-ble transplants. II. The effect of wood fiber substrates and their volume weights on the growth of tomato transplants. *Sci. Hortic.* **2004**, *100*, 333–340. [CrossRef]

52. Bunt, A.C. *Media and Mixes for Container-Grown Plants (Second Edition of Modern Potting Composts): A Manual on the Preparation and Use of Growing Media for Pot Plants*; Academic Division of Unwin Hyman Ltd.: Broadwick Street, London, UK, 1988.

53. Morales, A.B.; Bustamante, M.A.; Marhuenda-Egea, F.C.; Moral, R.; Ros, M.; Pascual, J.A. Agri-food sludge management using different co-composting strategies: Study of the added value of the composts obtained. *J. Clean. Prod.* **2016**, *121*, 186–197. [CrossRef]

54. Lemaire, F.; Dartigues, A.; Rivière, L.M. Properties of substrate made with spent mushroom compost. *Acta Hortic.* **1985**, *172*, 13–30. [CrossRef]

55. Singh, S.; Nain, L. Microorganisms in the conversion of agricultural wastes to compost. *Proc. Indian Natl. Sci. Acad.* **2014**, *80*, 473–481. [CrossRef] [PubMed]

56. Paradelo, R.; Basanta, R.; Barral, M.T. Water-holding capacity and plant growth in compost-based substrates modified with polyacrylamide, guar gum or bentonite. *Sci. Hortic.* **2019**, *243*, 344–349. [CrossRef]

57. Trevisan, S.; Francioso, O.; Quaggiotti, S.; Nardi, S. Humic substances biological activity at the plant-soil interface: From environmental aspects to molecular factors. *Plant Signal. Behav.* **2010**, *5*, 635–643. [CrossRef] [PubMed]

58. Rady, M.M.; Semida, W.M.; Hemida, K.A.; Abdelhamid, M.T. The effect of compost on growth and yield of *Phaseolus vulgaris* plants grown under saline soil. *Int. J. Recycl. Organ. Waste Agric.* **2016**, *5*, 311–321. [CrossRef]

59. Castellano-Hinojosa, A.; Correa-Galeote, D.; Palau, J.; Bedmar, E.J. Isolation of N2-fixing rhizobacteria from *Lolium perenne* and evaluating their plant growth promoting traits. *J. Basic Microbiol.* **2016**, *56*, 85–91. [CrossRef]

60. Kuan, K.B.; Othman, R.; Abdul Rahim, K.; Shamsuddin, Z.H. Plant growth-promoting rhizobacteria inoculation to enhance vegetative growth, nitrogen fixation and nitrogen remobilisation of maize under greenhouse conditions. *PLoS ONE* **2016**, *11*, e0152478. [CrossRef]

61. Jasper, J.; Wagstaff, C.; Bell, L. Growth temperature influences postharvest glucosinolate concentrations and hydrolysis product formation in first and second cuts of rocket salad. *Postharvest Biol. Technol.* **2020**, *163*, 111157. [CrossRef]

62. Awan, D.A.; Ahmad, F.; Imdad, M. Influence of row spacing and frequency of cuttings on spinach (*Spinacia oleracea*) production. *J. Bioresour. Manag.* **2016**, *31*, 9. [CrossRef]

63. Suzuki, T.; Kamada, E.; Ishii, T. Effects of ratoon harvesting on the root systems of processing spinach. *Plant Root* **2019**, *13*, 23–28. [CrossRef]

64. Conrath, U.; Beckers, G.J.M.; Flors, V.; García-Agustín, P.; Jakab, G.; Mauch, F.; Newman, M.A.; Pieterse, C.M.J.; Poinssot, B.; Pozo, M.J. Priming: Getting ready for battle. *Mol. Plant Microbe Interact.* **2006**, *19*, 1062–1071. [CrossRef] [PubMed]

65. Sang, M.K.; Kim, K.D. Biocontrol activity and primed systemic resistance by compost water extracts against anthracnoses of pepper and cucumber. *Phytopathology* **2011**, *101*, 732–740. [CrossRef] [PubMed]

66. Hoitink, H.A.J.; Grebus, M.E. Status of biological control of plant diseases with composts. *Compost Sci. Util.* **1994**, *2*, 6–12. [CrossRef]

67. Castaño, R.; Borrero, C.; Avilés, M. Organic matter fractions by SP-MAS 13C NMR and microbial communities involved in the suppression of Fusarium wilt in organic growth media. *Biol. Control* **2011**, *58*, 286–293. [CrossRef]

68. Ting, A.S.Y.; Hermanto, A.; Peh, K.L. Indigenous actinomycetes from empty fruit bunch compost of oil palm: Evaluation on enzymatic and antagonistic properties. *Biocatal. Agric. Biotechnol.* **2014**, *3*, 310–315. [CrossRef]

69. González-Sánchez, M.A.; Pérez-Jiménez, R.M.; Pliego, C.; Ramos, C.; De Vicente, A.; Cazorla, F.M. Biocontrol bacteria selected by a direct plant protection strategy against avocado white root rot show antagonism as a prevalent trait. *J. Appl. Microbiol.* **2009**, *109*, 65–78. [PubMed]

70. Földes, T.; Bánhegyi, I.; Herpai, Z.; Varga, L.; Szigeti, J. Isolation of Bacillus strains from the rhizosphere of cereals and in vitro screening for antagonism against phytopathogenic, food-borne pathogenic and spoilage micro-organisms. *J. Appl. Microbiol.* **2000**, *89*, 840–846. [CrossRef]

71. Tian, B.; Yang, J.; Zhang, K.Q. Bacteria used in the biological control of plant-parasitic nematodes: Populations, mechanisms of action, and future prospects. *FEMS Microbiol. Ecol.* **2007**, *61*, 197–213. [CrossRef]

72. Ahmed, H.H. Isolation and identification of biphenyl and polychlorinated biphenyl degrading microbes. *Appl. Biochem. Biotechnol.* **2017**, *170*, 391–398.

73. Kirschbaum, M.U. The temperature dependence of soil organic matter decomposition, and the effect of global warming on soil organic C storage. *Soil Biol. Biochem.* **1995**, *27*, 753–760. [CrossRef]

74. Burns, I.G.; Zhang, K.; Turner, M.K.; Edmondson, R. Iso-osmotic regulation of nitrate accumulation in lettuce. *J. Plant Nutr.* **2010**, *34*, 283–313. [CrossRef]

75. Lakhdar, A.; Falleh, H.; Ouni, Y.; Oueslati, S.; Debez, A.; Ksouri, R.; Abdelly, C. Municipal solid waste compost application improves productivity, polyphenol content, and antioxidant capacity of *Mesembryanthemum edule*. *J. Hazard. Mater.* **2011**, *191*, 373–379. [CrossRef] [PubMed]

76. Santos, F.T.; Goufo, P.; Santos, C.; Botelho, D.; Fonseca, J.; Queirós, A.; Costa, M.S.S.M.; Trindade, H. Comparison of five agro-industrial waste-based composts as growing media for lettuce: Effect on yield, phenolic compounds and vitamin C. *Food Chem.* **2016**, *209*, 293–301. [CrossRef] [PubMed]

77. Marín, A.; Ferreres, F.; Barberá, G.G.; Gil, M.I. Weather variability influences color and phenolic content of pigmented baby leaf lettuces throughout the season. *J. Agric. Food Chem.* **2015**, *63*, 1673–1681. [CrossRef]

78. Adebooye, O.C.; Schmitz-Eiberger, M.; Lankes, C.; Noga, G.J. Inhibitory effects of sub-optimal root zone temperature on leaf bioactive components, photosystem II (PS II) and minerals uptake in *Trichosanthes cucumerina* L. Cucurbitaceae. *Acta Physiol. Plant.* **2010**, *32*, 67–73. [CrossRef]

79. Sakamoto, M.; Suzuki, T. Effect of root-zone temperature on growth and quality of hydroponically grown red leaf lettuce (*Lactuca sativa* L. cv. Red Wave). *Am. J. Plant Sci.* **2015**, *10*, 2350. [CrossRef]

80. Yan, Q.-Y.; Duan, Z.-Q.; Mao, J.-D.; Li, X.; Dong, F. Low root zone temperature limits nutrient effects on cucumber seedling growth and induces adversity physiological response. *J. Integr. Agric.* **2013**, *12*, 1450–1460. [CrossRef]

81. Chadirin, Y.; Hidaka, K.; Sago, Y.; Wajima, T.; Kitano, M. Aplication of temperature stress to root zone II. Effect of the high temperature pre-treatment on quality. *Environ. Control Biol.* **2011**, *49*, 157–164. [CrossRef]

82. Costa, L.D.; Tomasi, N.; Gottardi, S.; Iacuzzo, F.; Cortella, G.; Manzocco, L.; Pinton, R.; Mimmo, T.; Cesco, S. The effect of growth medium temperature on corn salad [*Valerianella locusta* (L.) laterr] baby leaf yield and quality. *HortScience* **2011**, *46*, 1619–1625. [CrossRef]

83. Salveit, M.E. Wound induced changes in phenolic metabolism and tissue browning are altered by heat shock. *Postharvest Biol. Technol.* **2000**, *21*, 61–69. [CrossRef]

84. Kang, H.M.; Saltveit, M.E. Antioxidant capacity of lettuce leaf tissue increases after wounding. *J. Agric. Food Chem.* **2002**, *26*, 7536–7541. [CrossRef] [PubMed]

85. Chrysargyris, A.; Antoniou, O.; Tzionis, A.; Prasad, M.; Tzortzakis, N. Alternative soilless media using olive-mill and paper waste for growing ornamental plants. *Environ. Sci. Pollut. Res.* **2018**, *25*, 35915–35927. [CrossRef] [PubMed]

Article

Characterization of Physicochemical and Hydraulic Properties of Organic and Mineral Soilless Culture Substrates and Mixtures

Mohammad R. Gohardoust [1], Asher Bar-Tal [2], Mohaddese Effati [1] and Markus Tuller [1,*]

[1] Department of Environmental Science, The University of Arizona, Tucson, AZ 85721, USA; gohardoust@email.arizona.edu (M.R.G.); effati@email.arizona.edu (M.E.)

[2] Institute of Soil, Water and Environmental Sciences, The Volcani Center, Agricultural Research Organization (ARO), Bet Dagan 50250, Israel; abartal@volcani.agri.gov.il

* Correspondence: mtuller@email.arizona.edu; Tel.: +1-520-621-7225

Received: 20 August 2020; Accepted: 12 September 2020; Published: 16 September 2020

Abstract: Many arid and semiarid regions of the world face serious water shortages that are projected to have significant adverse impacts on irrigated agriculture and create unprecedented challenges for providing food and water security for the rapidly growing human population in a changing global climate. Consequently, there is a momentous incentive to shift to more resource-efficient soilless greenhouse production systems. Though there is considerable empirical and theoretical research devoted to specific issues related to control and management of soilless culture systems, a comprehensive approach that quantitatively considers relevant physicochemical processes within containerized soilless growth modules is missing. An important first step towards development of advanced soilless culture management strategies is a comprehensive characterization of hydraulic and physicochemical substrate properties. In this study we applied state-of-the-art measurement techniques to characterize six soilless substrates and substrate mixtures [i.e., coconut coir, perlite, volcanic tuff, perlite/coconut coir (50/50 vol.-%), tuff/coconut coir (70/30 vol.-%), and Growstone®/coconut coir (50/50 vol.-%)] that are used in commercial production in Israel and the United States. The measured substrate properties include water retention characteristics, saturated hydraulic conductivity, packing and particle densities, as well as phosphorus and ammonium adsorption isotherms. In addition, integral water availability and integral energy parameters were calculated to compare investigated substrates and provide valuable information for irrigation and fertigation management.

Keywords: soilless culture; organic and mineral substrates and mixtures; laboratory characterization; hydraulic properties; physicochemical properties

1. Introduction

The projected growth of the world population to around 9.7 billion by 2050 [1] poses unprecedented challenges for providing and sustaining food and water security and mitigating associated economic inequalities and social tensions that threaten global security [2,3]. This is further exacerbated by climate change via alterations of precipitation patterns, more likely occurrence of climate extremes (e.g., prolonged droughts), and modification of diurnal and seasonal temperature regimes [4] and soil degradation that leads to an alarming reduction of arable land. Because of these imminent challenges as well as a strong demand for high-quality, out-of-season vegetables, fruits, and ornamentals in many industrial countries and the ban of methyl bromide fumigation of horticultural field soils, there is an increasing incentive to shift from soil to more resource-efficient soilless culture [5].

Substrates used in soilless culture systems exhibit major advantages over soils. Besides the alleviated risk for spreading soil-borne pathogens, physicochemical properties of growth substrates can be controlled within narrow margins, which commonly leads to healthier plants and higher yields when compared to soil-based production [6,7].

Organic substrates that are extensively used in soilless culture include peat moss, compost, coconut coir, bark and other wood-based materials, and biochar, all of which are commonly mixed with inorganic substrates such as perlite, volcanic tuff, expanded clay granules, pumice, zeolite, and sand, in order to improve their physicochemical properties [8,9].

Though the same physical principles apply to both soilless substrates and soils, their physical and hydraulic properties are vastly different, which is of significant importance for management and control of soilless growth systems. In addition, there are fundamental differences with regard to dynamic water, air, and nutrient distribution processes, and root growth and development between spatially confined containerized production systems and unconfined field soils. While water flow and nutrient transport in growth containers is restricted by an impermeable container bottom with drainage holes, water drains and redistributes to much deeper layers in agricultural soils unless natural impediments exist. This leads to vastly different infiltration and redistribution dynamics requiring more intensive management of soilless systems. The smaller the root zone the more intensive the production system needs to be managed to provide a stress-free rhizosphere environment for optimum plant growth [5].

While soils are well-researched, and many discovered soil physical principles are readily available for application to soilless substrates, their adaptation and translation to substrates appear to lag behind. The physical properties of essence for the design and management of containerized soilless production systems include bulk density (*BD*), particle density (i.e., specific gravity), the water characteristic (*WC*), and hydraulic conductivity (*K*) [10,11]. The substrate *WC* [12] that relates the water content to the matric potential (i.e., capillary and adsorptive surface forces that hold water under subatmospheric pressure within the substrate matrix) and *K* are the most important physical properties that govern water flow and distribution processes [13] and aeration [14] in containerized soilless systems. The matric potential (*h*) [15] determines the "ease" for plant roots to extract water from the substrate and is commonly expressed as a negative (subatmospheric) pressure. In general, all irrigation practices that explicitly attempt to avoid water stress in soilless production are confined to a matric potential range from 0 to −8 kPa. In some substrates, such as rockwool, the range is even narrower, with the onset of water stress occurring if *h* is allowed to attain values < −5 kPa. In contrast, matric potentials encountered in field soils may well go as low as −75 kPa; in such systems matric potentials rarely rise above −10 kPa, except during or immediately after irrigation [16].

Several concepts related to the *WC* and *K* have been introduced to determine plant water availability. These include plant available water capacity [17], easily available water and water buffering capacity [18], container capacity [19–21], limiting and least limiting water range [22,23], and the integral water capacity [24], the later accounting for aeration and root penetration. The more recently introduced integral energy concept calculates the energy required to extract water from the growth medium [14,25]. The water flux in soilless substrates that may significantly vary due to large changes of hydraulic conductivity within a narrow *h* range is another important parameter for irrigation management to avoid plant water stress [26,27]. Accurate measurements of hydraulic substrate properties (i.e., *WC* and *K*) are also essential for the parameterization of numerical computer codes for the simulation of water and nutrient dynamics in containerized soilless systems. Such simulations aid with the optimization of substrate mixtures for specific plants as well as with the design and management of soilless systems (i.e., container geometry and irrigation amount and frequency), which may reduce costly and time consuming trial and error greenhouse experiments [14].

Nutrient supply in conjunction with irrigation (i.e., fertigation) is another important aspect of soilless culture management that requires insights about the adsorption of nutrients on substrate surfaces. For example, phosphorus and nitrogen need to be continuously supplied due to limited container volumes and associated restricted nutrient buffering capacities [28–30]. Rapid depletion of phosphorus after fertigation is a well-documented phenomenon caused by electrostatic adsorption onto substrate surfaces and slow formation of new solid metal-phosphorus compounds [31,32].

To increase phosphorus uptake by plant roots, high frequency fertigation is commonly applied to induce nonequilibrium conditions [33–35]. Ammonium promotes optimum plant development and growth when the NH_4-N/total-N ratio does not exceed plant specific thresholds that depend on species, rooting medium, root zone temperature, and pH [36–41]. For proper nutrient management, adsorption isotherms need to be determined to not only assure optimal growth conditions, but also to minimize nutrient loss in open-loop soilless culture systems.

The presented collaborative project that involves research teams from the U.S. and Israel was motivated by the rapidly growing demand for soilless growth media due to an ongoing momentous shift to more resource-efficient containerized soilless greenhouse production systems. It should be noted that the choice of soilless substrates and the selection of measured substrate properties was guided by ongoing production-scale greenhouse trials and the goal to utilize the obtained properties to parameterize a three-dimensional numerical code for simulation of water and nutrient dynamics in containerized growth modules to aid with their design and management. In the following we first discuss the selected substrates, then present a solid procedure for preparation of substrate mixtures, which is followed by an introduction of the applied state-of-the-art characterization techniques for the WC, saturated hydraulic conductivity (K_{sat}), and particle density as well as for the measurement of phosphorus and ammonium adsorption isotherms. We conclude the paper with a thorough discussion of obtained results.

2. Materials and Methods

2.1. Investigated Soilless Substrates

Six soilless substrates and substrate mixtures, including perlite (Figure 1a), volcanic tuff (Figure 1b), coconut coir (Figure 1c), a 50/50 vol.-% perlite/coconut coir mixture (Figure 1d), a 70/30 vol.-% volcanic tuff/coconut coir mixture (Figure 1e), and a 50/50 vol.-% foamed glass aggregate (i.e., Growstone®)/coconut coir mixture (Figure 1f), were investigated.

Horticultural perlite (Figure 1a) is a naturally occurring amorphous volcanic glass with high water holding capacity, typically formed through hydration of obsidian [42,43]. Perlite is usually sieved and then heated to 1000 °C. At high temperature water evaporates, and when rehydrated perlite expands to 4 to 20 times of its original volume [9], which yields a lightweight substrate with high porosity. Perlite aggregates are chemically inert and pathogen free [44], two desired attributes when plants remain in the same substrate for prolonged time periods [9,45]. However, if perlite is applied in high amounts, a negative impact on plant growth due to nutrient leaching may occur [46,47].

Tuff is a common name for pyroclastic volcanic material, exhibiting high porosity and surface area (Figure 1b). The physicochemical properties of tuff are mainly dependent on mineral composition and the weathering stage [48,49]. In addition, grinding and sieving processes may alter these properties. Tuff commonly exhibits a *BD* between 0.8 and 1.5 g cm^{-3} and a total porosity between 60 and 80%. Tuff possesses a high buffering capacity and may adsorb or release nutrients, especially phosphorus, throughout the plant growth period [49,50].

Coconut coir (Figure 1c) is the mesocarp of Cocos nucifera L., containing short and medium length fibers left from industrial applications. Depending on origin and industrial source, there is a difference in physical and chemical characteristics [51,52]. The coconut coir dust is commonly sieved to desired sizes and washed to leach excess salts. Coconut coir exhibits remarkable physical and chemical properties such as high water holding capacity, good drainage and aeration properties, and high cation exchange capacity. It is also commonly used as a surrogate for peat and mixed with mineral substrates [53–55].

Foamed glass aggregates (Growstone®, Growstone, LLC, Santa Fe, NM, USA) are made of recycled glass bottles and windows. The production process starts with crushing and grinding glass into a fine powder of vitreous soda lime glass, which is mixed with calcium carbonate (2% on weight basis) that acts as a foaming agent. When the mixture is heated it expands, thereby creating a network of fine pores [9,56]. After the cooling process, the solid block of foamed glass is crushed, tumbled, and sieved to various aggregate sizes. The aggregates are commonly mixed with organic substrates.

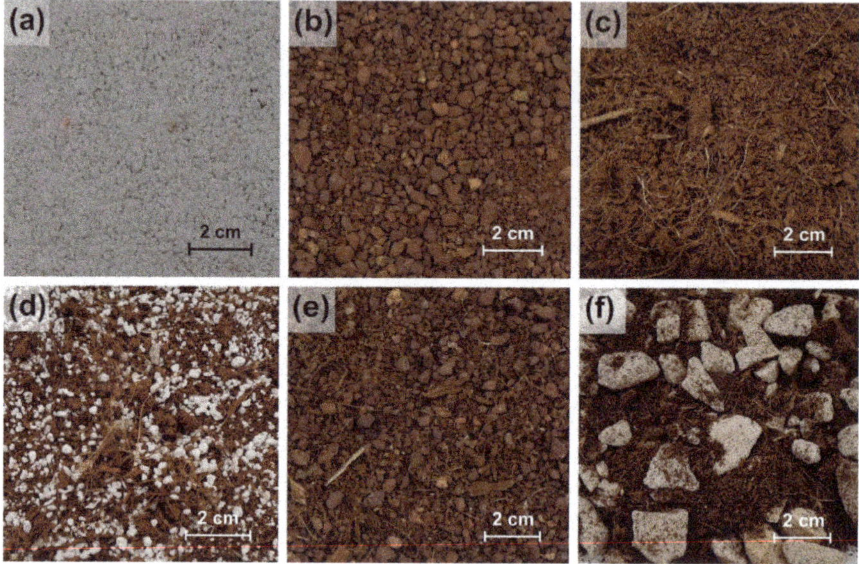

Figure 1. Investigated soilless substrates and substrate mixtures. (**a**) perlite, (**b**) tuff, (**c**) coconut coir, (**d**) 50/50 vol.-% perlite/coconut coir mixture, (**e**) 70/30 vol.-% volcanic tuff/coconut coir mixture, and (**f**) 50/50 vol.-% foamed glass aggregate/coconut coir mixture.

2.2. Sample Preparation

To obtain uniform and reproducible substrate samples for hydraulic characterization we first performed comprehensive compaction trials to determine the lowest and highest achievable dry bulk densities for the considered soilless substrates. The average dry bulk densities were then used as initial target bulk densities for preparation of samples for substrate WC and K_{sat} measurements. Because of particle segregation during transport, the 50/50 vol.-% Growstone®/coconut coir mixture (Figure 1f) supplied by Growstone, LLC was separated, remixed, and homogenized. All tests were performed in sextuplicate for each substrate and substrate mixture. We used air-dry samples as this is the most realistic scenario for large-scale greenhouse applications and also to avoid potential problems with hydrophobicity of coconut coir that may be induced during oven drying.

Subsamples of perlite, tuff, and coconut coir were first oven-dried to determine the air-dry gravimetric water content. Then, the thoroughly homogenized air-dried substrates were compacted into cylinders with known volume (V_C) in multiple layers to achieve a uniform packing density. To achieve the lowest potential packing density, the substrates were poured into and carefully manually distributed within the cylinders without imposing a significant compaction force. Only on the very top the substrate particles were gently pushed inside the cylinder to obtain a smooth surface. To achieve the highest potential packing density, the substrates were compacted layer by layer with a rubber stopper mounted on a push rod. At the end, the lowest and highest dry bulk densities were determined, and the average values were used as the target density for sample preparation for WC and K_{sat} measurements.

Compaction trials were also performed for the 50/50 vol.-% perlite/coconut coir mixture, the 70/30 vol.-% tuff/coconut coir mixture, and the 50/50 vol.-% Growstone®/coconut coir mixture. First, several subsamples of the individual substrates to be mixed were collected and oven-dried to determine their air-dried gravimetric water content. Once the gravimetric water content of the individual mixture components was known, the substrates were poured into two separate cylinders of known volumes and compacted in the same fashion as described above for the lowest packing density. The air-dried mass of the substrates occupying a specific volume was then measured and the oven-dried masses per volume were calculated. The dry mass ratio (ϑ) may then be defined as:

$$\vartheta = \frac{M_{ODv1}}{M_{ODv2}} \times R_V \tag{1}$$

with M_{ODv} as the oven-dried mass of substrates 1 or 2 occupying a specific volume and R_V the volumetric substrate mixing ratio (i.e., 50/50 vol.-% for perlite/coconut coir; 70/30 vol.-% for tuff/coconut coir; and 50/50 vol.-% for Growstone®/coconut coir).

For the compaction trials the air-dried substrate components were then mixed at the desired volumetric substrate mixing ratio and the resulting mixture was meticulously homogenized. The homogenized air-dried mixture was then compacted into cylinders in the same fashion as the individual substrates to obtain the lowest and highest achievable potential packing densities. After compaction, the mass of the air-dried mixture occupying the cylinder, M_{ADmix}, was determined and the oven-dried masses of the individual components composing the sample were calculated as:

$$M_{OD1} = \left[M_{ADmix} - \frac{M_{ADmix}}{\vartheta\left(\frac{1 + \theta_{m1}}{1 + \theta_{m2}}\right) + 1} \right] \cdot \frac{1}{1 + \theta_{m1}} \tag{2}$$

$$M_{OD2} = \frac{M_{ADmix}}{\vartheta\left(\frac{1 + \theta_{m1}}{1 + \theta_{m2}}\right) + 1} \cdot \frac{1}{1 + \theta_{m2}} \tag{3}$$

where M_{OD1} and M_{OD2} are the oven-dry masses of substrate 1 and 2, respectively, and θ_m is the gravimetric water content. The dry bulk density of the mixture $(\rho_{b\text{-}mix})$, which is used as target value for further measurements, was derived as:

$$\rho_{b\text{-}mix} = \frac{M_{OD1} + M_{OD2}}{V_C} \tag{4}$$

with V_C as the cylinder volume. The mass of air-dry substrate required to fill a distinct volume (V) at target bulk density was calculated as:

$$M_{ADmix} = \left[\frac{1 + \theta_{m1}}{1 + \vartheta^{-1}} + \frac{1 + \theta_{m2}}{1 + \vartheta} \right] \cdot \rho_{b\text{-}mix} \cdot V_C \tag{5}$$

2.3. Substrate Water Characteristic and Integral Energy and Water Storage

Tempe cells (Soilmoisture Equipment Corp., Santa Barbara, CA, USA) were used to measure the substrate WC curve. The Tempe cells were connected to a pressure manifold (Figure 2) with a high-resolution pressure/vacuum regulator and initially saturated samples were sequentially desaturated by applying increasing pressures. Each pressure step was maintained until the sample was in equilibrium with the applied pressure and the outflow ceased. A detailed description of the pressure desaturation method is provided in [12]. All measurements were performed in quintuplicate and averaged values are reported.

Figure 2. Setup of the Tempe cell experiment.

The van Genuchten model [57] (Equation (6)) was fitted to *WC* measurements for substrates exhibiting unimodal behavior and the Durner model [58] (Equation (7)) was fitted to *WC* measurements exhibiting bimodal behavior.

$$\theta(h) = \theta_r + (\theta_s - \theta_r)\left[\frac{1}{1 + |\alpha h|^n}\right]^m \tag{6}$$

$$\theta(h) = \theta_r + (\theta_s - \theta_r)\left[(1 - w)\left(\frac{1}{1 + |\alpha_1 h|^{n_1}}\right)^{m_1} + w\left(\frac{1}{1 + |\alpha_2 h|^{n_2}}\right)^{m_2}\right] \tag{7}$$

where θ is the volumetric water content, θ_r and θ_s are the residual and saturated water contents respectively, h is the matric potential, and α, n, and m are shape parameters with $m = 1 - 1/n$. w is a weighting factor that varies between 0 and 1 and the indices 1 and 2 refer to the first and second substrate in the mixture, respectively.

To capture effects of the *WC* curve shape on plant water availability, rather than relying on two matric potential thresholds such as proposed in [17,22,23], we calculated the integral water (W_I) and energy (E_I) storage following [24,25] as:

$$W_I\big[h_i, h_f\big] = \frac{1}{|h_i - h_f|}\int_{h_f}^{h_i}\theta(h)dh \tag{8}$$

where the indices *i* and *f* are the wet and dry matric potential cut-offs, respectively. W_I has units of volumetric water content and represents the weighted average of water contents between h_i and h_f. The integral energy (i.e., the energy required to extract water from θ_i to θ_f) was calculated as:

$$E_I\big[\theta_i, \theta_f\big] = \frac{1}{\theta_i - \theta_f}\int_{\theta_f}^{\theta_i} h(h\theta)d\theta \tag{9}$$

The integrals in Equations (8) and (9) were numerically solved with the MATLAB®—Version R2019a software package (MathWorks, Natick, MA, USA). Based on obtained *WC* parameters and selection of plant specific cut-off values for the wet- and dry-end matric potentials and water contents, substrate water availability can be determined as:

$$R = \frac{W_I}{E_I} \tag{10}$$

R indicates the amount of water that can be extracted via exertion of a unit amount of energy by plant roots within the range of the wet- and dry-end thresholds. The higher R, the easier it is for plants to extract water.

2.4. Saturated Hydraulic Conductivity

For the K_{sat} measurements we designed and fabricated an automated constant head device that was placed on a load cell attached to a laboratory jack and connected to a flow cell filled with substrate (Figure 3). The load cell was connected to a datalogger to record and monitor the weight change of the constant head container (i.e., Marriot tank) while water was flowing through the sample. In addition, the water temperature was continuously measured with a thermocouple and used to convert mass to volume change. The setup was initially thoroughly tested and adjusted to minimize flow resistance in the tubing and connectors. Each substrate was compacted into a flow cell at average bulk density (see Section 2.2). Before slowly saturating samples with water from the Marriot tank, they were flushed with CO_2 for about 10 min at very low flow rate to enhance the saturation process. After sample saturation the constant head was adjusted with the lab jack and the experiment initiated. The experiment was terminated after several hours of steady state flow. Each substrate was measured in duplicate at 20, 15, 10, and 5 cm hydraulic heads. Darcy's law was applied to calculate K_{sat} from the measured water flux density and set hydraulic heads [59].

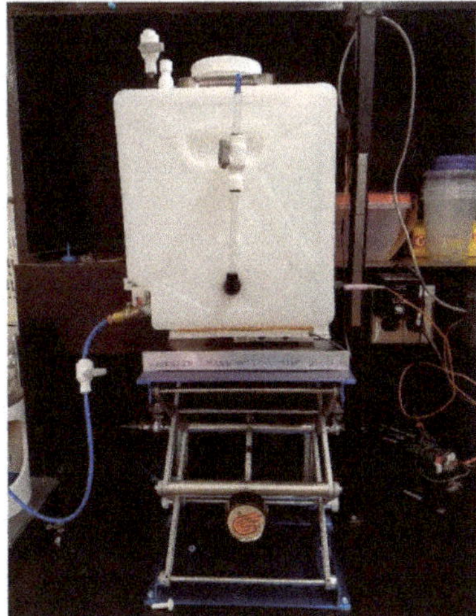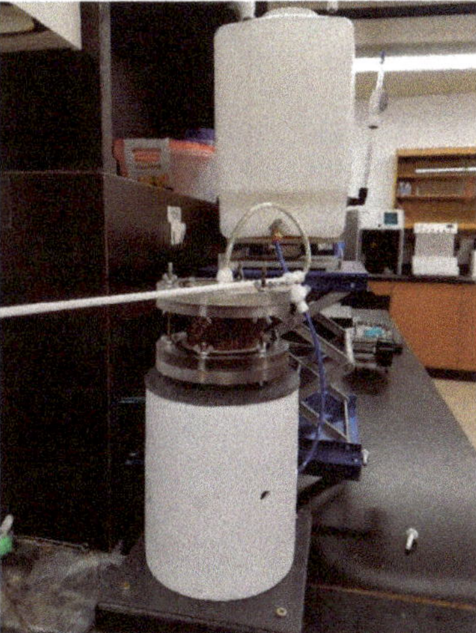

Figure 3. Automated constant head setup for K_{sat} measurements.

2.5. Particle Density

While a standard water pycnometer [60] was used to measure the particle densities of tuff and coconut coir, nitrogen gas pycnometry was applied for the lighter perlite and Growstone® substrates. A Multipycnometer (Quantachrome Corp., Boynton Beach, FL, USA) with nitrogen as probing gas was used for the latter measurements. Gas pycnometry is based on Archimedes' fluid displacement principle and Boyle's gas expansion law. The volume of a solid or powder sample is determined via measuring the pressure drop that occurs when a known amount of pressurized gas initially contained in a reference cell with known volume (V_R) is allowed to expand into a cell of known volume (V_C) that contains the sample. The sample volume vs. is calculated as:

$$V_S = V_C - V_R\left[\frac{P_1}{P_2} - 1\right]$$ (11)

where P_1 and P_2 are the pressures before and after gas expansion into the sample cell. All measurements were performed in quintuplicate. From the known oven-dry mass of the sample and its determined volume V_S, the particle density can be calculated. The particle densities of the mixtures were calculated based on their mixing ratios.

2.6. Phosphorus Adsorption Isotherms

Phosphorus adsorption isotherms were measured in triplicate with adsorption batch experiments. The substrates were air dried and a 1-g subsample was added to a 50 mL equilibration tube. Then 20 mL of KH_2PO_4 solution with concentrations of 0, 1, 5, 10, 50, and 100 mg KH_2PO_4-P l^{-1} in the background of 0.01 M $CaCl_2$ was added to the tubes to obtain a soil/solution ratio of 1:20. The pH of the solution was adjusted with 1M sodium hydroxide to fall between 6.5 and 7.0. The samples were left to equilibrate for 24 h in an end-over-end shaker. The supernatant was extracted after centrifuging for 15 min at 12,000 rpm and filtering with 0.2 μm paper filters.

The analysis of the filtrate soluble reactive phosphorus was carried out with the ascorbic acid colorimetric method for perlite, tuff, and tuff-coir substrates. Required reagents were prepared as follows: Molybdate Reagent: 12.0 g of ammonium molybdate was dissolved in 250 mL of deionized water and 0.1455 g of antimony potassium tartrate was also dissolved in 500 mL of 5N H_2SO_4. Then 125 mL of ammonium molybdate solution was thoroughly mixed with the 500 mL H_2SO_4-antimony potassium tartrate solution and diluted to one liter with deionized water using a volumetric flask. Color developing reagent was prepared as follows: in a 1L volumetric flask, 0.739 g of ascorbic acid was dissolved in deionized water and 70 mL of the molybdate reagent added and brought to volume. A series of standard PO_4-P solutions with concentrations of 0, 1, 2, 3, and 4 ppm, were prepared for calibration of the spectrophotometer each time a measurement was made. 1 mL of sample solution was mixed with 9 mL of color developing reagent in a small tube and its P concentration was measured after about 1 h with a spectrophotometer at 880 nm wavelength.

Because colorimetric determination of the phosphorus concentration requires a clear solution, which was not the case for samples containing considerable amounts of coconut coir (i.e., coconut coir, perlite/coconut core mixture, and Growstone®/coconut coir mixture), the total phosphorus concentrations for these substrates were measured with Inductively Coupled Plasma Mass Spectrometry (ICP-MS) at the Arizona Laboratory for Emerging Contaminants (ALEC).

The linearized Langmuir adsorption equation was fitted to the measured data to obtain the substrate sorption parameters:

$$\frac{C}{S} = \frac{1}{kS_{max}} + \frac{C}{S_{max}}$$ (12)

where S is the total amount of P retained (mg kg^{-1}), C is concentration of P after 24 h equilibration (mg l^{-1}), S_{max} is the maximum P sorption capacity (mg kg^{-1}), and k is a constant related to the bonding energy, l (mg P)$^{-1}$. Additional details are provided in [61].

2.7. Ammonium Adsorption Isotherms

Ammonium adsorption isotherms were calorimetrically determined in triplicate in batch experiments. Ammonium solutions were prepared in concentrations of 0, 1, 5, 10, 50, and 100 mg NH_4Cl-N l^{-1}. One gram of each substrate was agitated with 20 mL of the ammonium solutions in a centrifuge tube for 3 h after adjusting the pH with 1M sodium hydroxide to fall between 6.5 and 7.0. Samples were then centrifuged and filtered with 0.2 μm filter paper. The concentration of ammonium was measured with the salicylate method following [62] with the following reagents:

1. Sodium salicylate-sodium nitroprusside solution (reagent 1): 33.0 g of $NaC_7H_5O_3$ and 20.0 mg of $Na_2Fe(CN)_5NO.5H_2O$ was dissolved in deionized water and diluted to 100 mL.
2. Buffer solution (reagent 2): 9.33 g of sodium citrate dihydrate and 4.0 g of NaOH were dissolved in deionized water and diluted to 100 mL.
3. *Hypochlorite solution (reagent 3)*: 5 mL of hypochlorite (10% active chlorine) was dissolved in 25 mL deionized water.

Four mL of extracted ammonium solution was mixed in a glass tube with 0.9 mL combined reagent (i.e., one part of reagent 1 mixed with two parts of reagent 2). Then within one minute 0.1 mL of reagent 3 was added to the tube, which was then placed in a dark room for 120 min to allow the establishment of the emerald blue color. The absorbance of the chromophore was measured with a spectrophotometer at 647 nm wavelength and the Langmuir adsorption model (Equation (12)) was fitted to the measured data.

3. Results and Discussion

3.1. Bulk and Particle Densities

The lowest and highest dry bulk densities achieved with the packing procedures described in Section 2.2 are listed in Table 1. The average values were used as the target bulk densities for the samples used for the *WC* and K_{sat} measurements. Perlite was the lightest of the investigated substrates with an average bulk density of 0.076 g cm^{-3}, followed by the perlite/coconut coir mixture. Tuff exhibited the highest bulk density with an average value of 1.15 g cm^{-3}. From transportation and handling point of view low bulk densities are desirable [63]. The determined dry mass ratios for the substrate mixtures are also displayed in Table 1.

Table 1. Dry bulk densities and oven-dried mass ratios of mixtures determined with compaction experiments.

Substrates	Mixing Ratio (vol.-%)	Dry Mass Ratio (ϑ)	Dry Bulk Density (g cm^{-3})		
			Lowest	Highest	Average
Perlite	-	-	0.072	0.080	0.076
Tuff	-	-	1.100	1.200	1.150
Coconut coir	-	-	0.100	0.120	0.110
Perlite/coir	50/50	0.73	0.082	0.094	0.088
Tuff/coir	70/30	25.82	0.875	0.975	0.925
Growstone®/coir	50/50	2.89	0.180	0.190	0.185

The average particle densities and associated standard errors (SE) are listed in Table 2. As discussed in Section 2.5, a standard water pycnometer was employed for tuff and coconut coir and a gas pycnometer was used for the perlite and Growstone® substrates. Perlite exhibits the lowest particle density. The obtained value of 0.739 g cm^{-3} falls within the reported range of 0.28–0.98 g cm^{-3} [64,65]—the variations are attributable to differences in the production process. The highest particle density of 2.653 g cm^{-3} was determined for tuff, which is due to the presence of significant amounts of metal oxides such as aluminum, iron, and magnesium [48].

Table 2. Measured particle densities.

Substrates	Particle Density (g cm^{-3})	Standard Error
Perlite	0.739	0.004
Tuff	2.653	0.015
Coconut coir	1.717	0.069
Growstone®	1.621	0.014

3.2. Substrate Water Characteristic and Saturated Hydraulic Conductivity

The continuous parametric *WC* models of van Genuchten (VG) [57] and Durner [58] were fitted to the measured matric potential and volumetric water content pairs (Figure 4). For calculation of integral water storage (W_I) and integral energy (E_I) the threshold matric potential at the wet-end (h_i) of the *WC* was set at −2 cm H$_2$O below the substrate's air-entry potential (i.e., the potential at which the largest pore in the system starts draining and water is displaced by air—the transition from fully water-saturated to partially saturated) and the potential at the dry-end (h_f) at −440 cm H$_2$O. The latter was adapted from [66] for spring tomatoes, which is the major crop of our greenhouse trials (Figure 4). It should be noted that while the matric potential has a negative subatmospheric pressure (lower potential means larger negative number—−440 cm is lower than −2 cm), out of convenience it is commonly plotted on a positive scale with a minus sign in front of the units. It is also common to use units of lengths of H$_2$O column (e.g., m) for the matric potential, which can be converted to pressure units (e.g., kPa) via multiplication with the density of water (kg m^{-3}) and the acceleration due to gravity (m s^{-1}). More details are provided in [12,15].

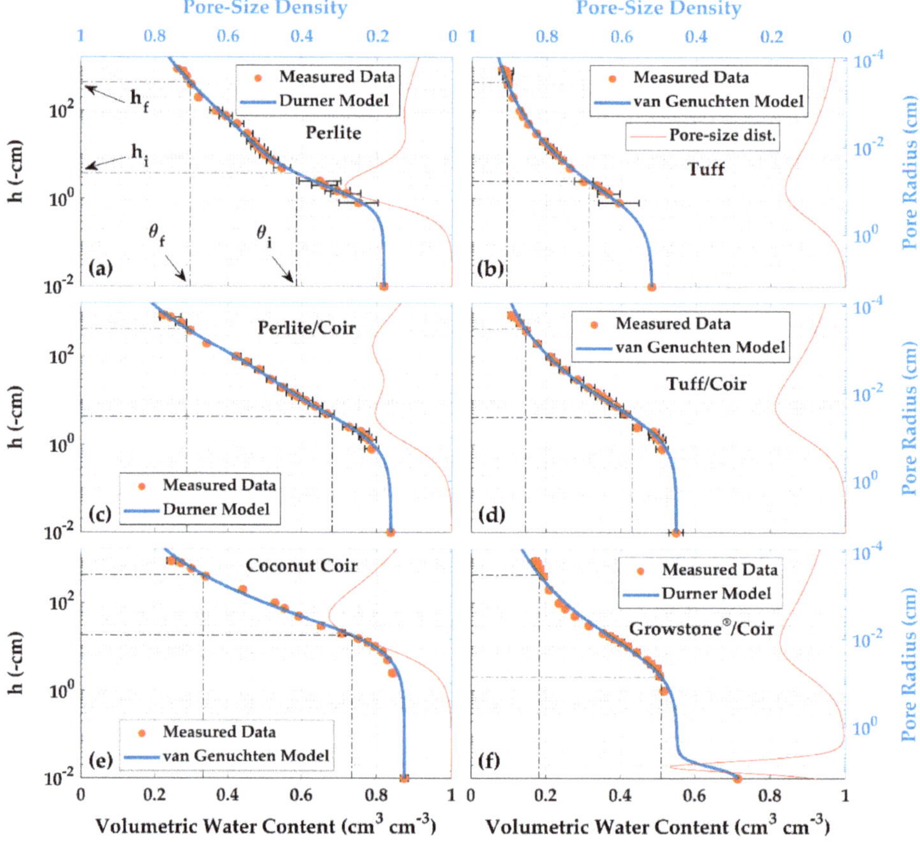

Figure 4. Measured *WC* data displayed with the fitted unimodal van Genuchten (Equation (6)) or bimodal Durner (Equation (7)) *WC* models for: (**a**) perlite, (**b**) tuff, (**c**) perlite/coconut coir, (**d**) tuff/coconut coir, (**e**) coconut coir, and (**f**) Growstone®/coconut coir. The error bars represent the standard deviation of the measured volumetric water contents. The wet- and dry-end matric potential thresholds, h_i and h_f, and their corresponding water contents, θ_i and θ_f, are marked with dash-dotted lines. The pore size distributions associated with the *WC* curves are plotted on the right side.

The hydraulic substrate properties consisting of the *WC* model parameters and the saturated hydraulic conductivity (K_{sat}) are listed in Table 3.

Table 3. Substrate WC parameters and measured K_{sat} (for parameter definitions see Equations (6) and (7) in Section 2.3).

Substrate	θ_r (cm^3 cm^{-3})	θ_s (cm^3 cm^{-3})	α_1 (cm^{-1})	n_1	α_2 (cm^{-1})	n_2	w	K_{sat} (cm h^{-1})	SE * (cm h^{-1})
Perlite	0.001	0.818	0.822	1.820	0.032	1.164	0.558	305.1	16.7
Tuff	0.014	0.483	6.970	1.249	-	-	-	304.2	12.1
Coconut Coir	0.010	0.874	0.062	1.296	-	-	-	56.2	5.1
Perlite/Coir	0.005	0.837	0.599	1.331	0.011	1.318	0.416	165.1	8.4
Tuff/Coir	0.014	0.549	0.458	1.267	-	-	-	110.7	11.6
Growstone®/Coir	0.004	0.722	56.290	5.146	0.295	1.232	0.765	172.4	9.9

* Standard error of K_{sat} measurements.

The integral water storage and energy values calculated for each substrate are displayed in Table 4 together with their wet- and dry-end threshold water contents and R values (Equation (10)).

Table 4. Integral water storage and energy values and associated threshold water contents (for parameter definitions see Equations (8) to (10) in Section 2.3).

Substrate	θ_i (cm^3 cm^{-3})	θ_f (cm^3 cm^{-3})	W_I (cm^3 cm^{-3})	E_I (cm)	R (cm^3 cm^{-3} cm^{-1}) $\times 10^{-3}$
Perlite	0.586	0.298	0.348	80.03	4.35
Tuff	0.314	0.093	0.120	54.69	2.19
Coconut Coir	0.734	0.333	0.431	120.65	3.57
Perlite/Coir	0.681	0.288	0.372	96.94	3.83
Tuff/Coir	0.431	0.144	0.187	69.02	2.71
Growstone®/Coir	0.511	0.182	0.231	67.78	3.41

Because of its aggregated structure, perlite exhibits a bimodal pore size distribution (Figure 4a) with distinct contributions of inter- and intra-aggregate pores [67]. The bimodal pore structure and WC that was well approximated with the Durner model (Figure 4a) is consistent with observations by [68], who applied mercury intrusion porosimetry to measure the pore size distribution of uncrushed expanded perlite. It should be noted that a distinct bimodal pore structure of perlite was not reported in [69–71]. The saturated hydraulic conductivity of perlite that was slightly above that of tuff (Table 3) falls within the range provided in [70]. Based on the W_I, E_I, and R values listed in Table 4, it is obvious that the plant water availability (accessibility) of perlite is the highest of all investigated substrates. In other words, perlite yields the highest water amount between the respective threshold water contents (θ_i and θ_f) per unit energy exerted by plant roots.

In contrast to perlite, tuff provides the lowest water yield of the investigated substrates between θ_i and θ_f (Table 4). This in conjunction with its high K_{sat} (Table 3) indicates rapid drainage of the fertigation solution from the substrate, which provides valuable insights for irrigation and fertigation management to avoid problems with water and nutrient deficiencies. For example, an increase in irrigation frequency to keep the matric potential above −200 cm would double the plant water yield for the same applied energy. Wallach et al. [17] measured hydraulic characteristics of two red tuff varieties and reported K_{sat} values of 130 and 439 cm h^{-1} and associated dry bulk densities of 1.227 and 1.091 g cm^{-3}, respectively. They also evaluated the capability of the Mualem hydraulic conductivity model [72] to estimate unsaturated hydraulic conductivity from van Genuchten WC model parameters and found good agreement with data measured for tuff.

Coconut coir exhibits the lowest K_{sat} of the investigated substrates—about one-sixth of that of perlite and tuff (Table 3). Because of differences in industrial source and pretreatment of coconut coir, considerable variations in physicochemical and hydraulic properties can be expected [51]. The measured

K_{sat} of 56.2 cm h^{-1} is about half of that measured by [73], who compacted the samples to a similar bulk density as used in this study. The K_{sat} value reported in [74] is more than one order of magnitude higher than our measurements, but due to the lack of information about the associated bulk density a direct comparison is not feasible. Their extremely high K_{sat} is most likely due to a much lower bulk density of the coconut coir in the narrow glass columns that were used in their experiments, which is also evident from the van Genuchten *WC* model α-parameter reported in [74]. In general, horticultural coconut coir does not contain a significant number of large pores. This is why it is commonly mixed with aggregated mineral substrates to enhance aeration properties [53]. In terms of plant water availability (i.e., *R*-value), pure coconut coir yields more water per unit of energy exerted by plant roots within the θ_i–θ_f range than the tuff/coconut coir and Growstone$^{®}$/coconut coir mixtures (Table 4).

The perlite/coconut coir and tuff/coconut coir mixtures exhibit hydraulic properties that fall in between the properties of their constituents (Figure 4c,d; Tables 3 and 4). This includes their air-entry potentials, which enhances aeration relative to sole coconut coir. The addition of coconut coir to perlite and tuff also lowers the K_{sat} of the mixtures, slowing down drainage and increasing the water yield (availability) within the θ_i–θ_f range. For example, the 70/30 vol.-% tuff/coconut coir mixture has a 19% higher *R* value than the sole tuff substrate (Table 4). Such information may be utilized to optimize (engineer) substrate mixtures via varying mixing ratios to achieve optimum plant specific growth environments in terms of total porosity, air-filled porosity, available water, aeration, and bulk density [10,14,75] as well as to provide guidance for selection of container geometry and irrigation and fertigation management [76–78].

The Growstone$^{®}$/coconut coir mixture has the highest (i.e., least negative) air-entry potential of the investigated substrates (Figure 4), which is also evident from the high α_1 Durner *WC* model parameter (Table 3). From Figure 4, it is evident that a −1 cm change in matric potential will cause an almost instantaneous drainage of water from about 25% of the entire pore space. As shown in [14], where both the *WC* and aeration properties of four soilless substrates were measured, caution is required when assessing aeration properties of mixtures containing large aggregates as water blockage and pore discontinuities might occur.

3.3. Phosphorus and Ammonium Adsorption

The Langmuir adsorption isotherm parameters for both phosphorus and ammonium are summarized in Table 5. The maximum amount of phosphorus adsorbed onto perlite, coconut coir, and the perlite/coconut coir mixture of about 20 mg per kilogram of solid is negligibly small. The low phosphorus absorptivity of perlite was previously reported by [63,79], who evaluated perlite as a potential filtration medium for urban runoff. Low phosphorus adsorption onto coconut coir was indicated in [80,81]. Tuff and the Growstone$^{®}$/coconut coir mixture exhibited the highest phosphorus adsorption per unit substrate mass, about 15 and 12 times that of perlite and coconut coir (Table 5), respectively. It should be noted that while tuff and the Growstone$^{®}$/coconut coir mixture show about the same capacity for phosphorus adsorption per unit substrate mass, the adsorption onto tuff within the same container volume will be more than six times higher than that onto the Growstone$^{®}$/coconut coir mixture because of the significantly higher dry bulk density of tuff (Table 1).

Figure 5 depicts the measured equilibrium concentrations for both ammonium and phosphorus together with the fitted Langmuir isotherms. The low coefficients of determination (R^2) for perlite, coconut coir, and Growstone$^{®}$/coconut coir are attributable to low adsorption values (perlite and coconut coir) and the uncertainty inherent to the measurement procedure.

Table 5. Langmuir adsorption isotherm parameters for phosphorus and ammonium.

Substrate	Phosphorus		Ammonium	
	S_{max} (mg kg^{-1})	k (l mg^{-1} KH$_2$PO$_4$-P)	S_{max} (mg kg^{-1})	k (l mg^{-1} NH$_4$-N)
Perlite	18.0	0.984	43.6	3.376
Tuff	270.6	0.066	432.8	0.135
Coconut Coir	23.0	0.548	1419.5	0.036
Perlite/Coir	24.2	0.327	809.0	0.056
Tuff/Coir	241.8	0.102	517.3	0.083
Growstone®/Coir	265.5	0.036	473.6	0.054

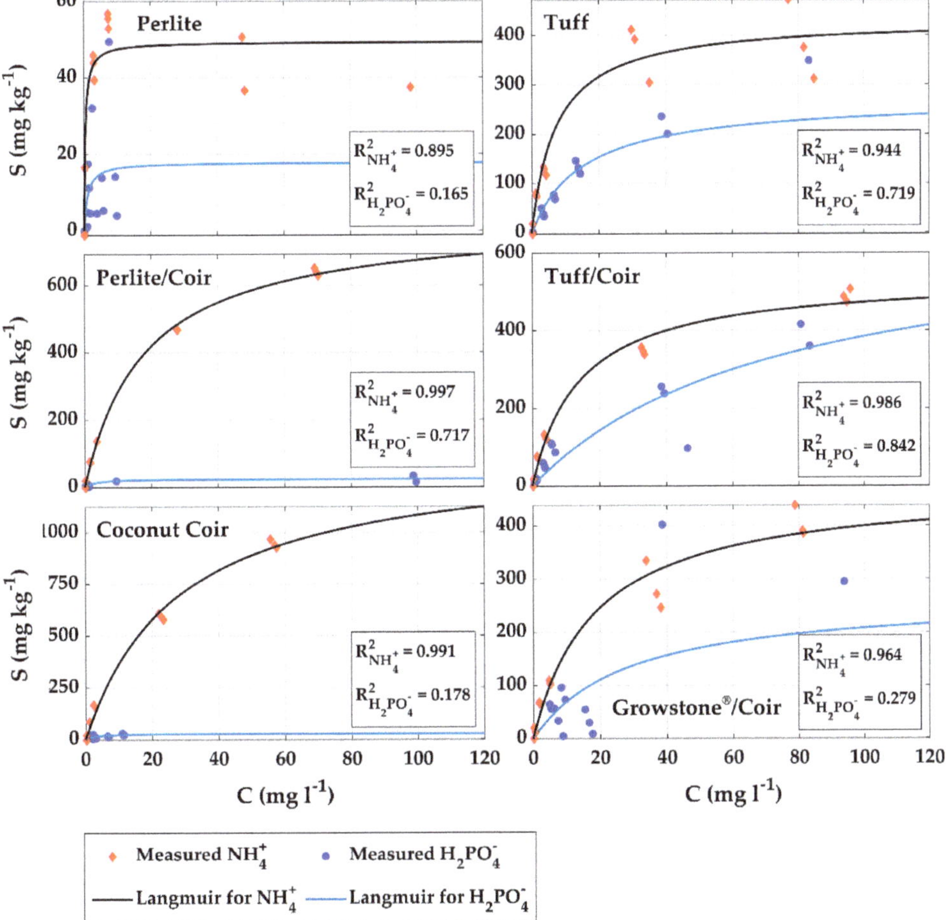

Figure 5. Measured NH_4^+ and $H_2PO_4^-$ equilibrium concentrations displayed with the fitted Langmuir adsorption isotherms and the associated coefficients of determination.

The k coefficient in the Langmuir equation represents the affinity of the adsorbed species to the adsorbent (i.e., the higher k, the stronger the affinity). When the affinity is stronger, maximal adsorption is attained at lower adsorbate concentrations and there is a sharp increase in the adsorbed amount at low concentrations. The k values in Table 5 indicate that the order of affinities of phosphorus to the substrates is perlite > coconut coir > perlite/coconut coir mixture > tuff/coconut coir mixture > tuff > Growstone®/coconut coir mixture. Despite the high affinity of phosphorus to perlite, the importance of phosphorus adsorption is small due to the combination of low S_{max} and low bulk density. It was expected that the k values of mixtures of two components fall between the values of the pure components. However, note that the k of the perlite/coconut coir mixture is smaller than that of coconut coir, most likely due to chemical interactions between the coconut coir and perlite surfaces. The k value of the tuff/coconut coir mixture is much closer to that of tuff, which may be attributed to the much higher bulk density of tuff.

Because of its high cation exchange capacity (CEC) [51,82], which is the most important factor for ammonium adsorption [83], the maximum amount (S_{max}) of ammonium was adsorbed onto coconut coir. This translates to the mixtures containing coconut coir (Table 5). Perlite exhibited the lowest S_{max} value of the investigated substrates, which together with its low bulk density indicates that ammonium adsorption onto perlite is rather negligible. It should be noted that because the substrates were mixed on a volume basis, the dry mass ratio parameter (ϑ) in Equation (1) is crucial for estimation of adsorption properties of the substrate mixtures. For example, the 30 vol.-% coconut coir contained in the tuff/coconut coir mixture does not significantly increase ammonium adsorption. In contrast, the 50 vol.-% coconut coir contained in the perlite/coconut coir mixture significantly impacts ammonium adsorption due to an almost 30 times lower ϑ than that of the tuff/coconut coir mixture. The k values in Table 5 indicate that the order of affinities of ammonium to the substrates is perlite > tuff > tuff/coconut coir > perlite/coconut coir > coconut coir > Growstone®/coconut coir. Similar to phosphorus, the importance of ammonium adsorption to perlite is small due to the combination of low S_{max} and low bulk density. As expected, the k values for ammonium in the two component mixtures fall in between the values of the pure components.

4. Conclusions

A thorough physicochemical and hydraulic characterization of six soilless substrates and substrate mixtures that were selected based on ongoing greenhouse trials was presented. The investigated substrates included perlite, volcanic tuff, coconut coir, a 50/50 vol.-% perlite/coconut coir mixture, a 70/30 vol.-% volcanic tuff/coconut coir mixture, and a 50/50 vol.-% foamed glass aggregate (i.e., Growstones®)/coconut coir mixture. After developing a precise sample preparation procedure to assure high repeatability, the substrate WC, K_{sat}, particle densities, average bulk densities, as well as phosphorus and ammonium adsorption isotherms were measured with state-of-the-art techniques. The WC measurements were used to parameterize the unimodal van Genuchten [57] and bimodal Durner [58] WC models to derive integral water and energy storage parameters to estimate the amount of water that can be extracted from a specific volumetric water content range per unit energy exerted by plant roots. From integral energy calculations, it is evident that plant water availability (accessibility) of perlite is the highest of all investigated substrates, followed by the perlite/coconut coir mixture. Perlite also exhibits favorable nutrient adsorption characteristics. Despite the high affinity of phosphorus to perlite the importance of P adsorption is small due to a low maximum adsorption capacity and the low bulk density of perlite. In addition, ammonium adsorption to perlite is rather negligible. The obtained soilless substrate parameters can not only be applied for optimization (engineering) of soilless substrates via mixing of organic and inorganic constituents at different ratios to meet specific plant physiological demands, but also used for the parameterization of three-dimensional numerical computer codes for simulation of water and nutrient dynamics in containerized growth modules to aid with their design and management as well as to provide scientifically sound data for the design of greenhouse trials to avoid costly trial and error experiments, which motivated this study and is part of our ongoing research.

Author Contributions: Conceptualization, A.B.-T., M.R.G., and M.T.; methodology, M.R.G and M.T.; investigation, M.E. and M.R.G.; data analysis, M.E., M.R.G., and M.T.; writing—original draft preparation, M.E., M.R.G., and M.T.; writing—review and editing, A.B.-T., M.E., M.R.G., and M.T.; funding acquisition, A.B.-T. and M.T. All authors have read and agreed to the published version of the manuscript.

Funding: This research was funded by The United States-Israel Binational Agricultural Research and Development Fund (BARD), grant number US-4764-14 R, and by the United States Department of Agriculture (USDA) National Institute of Food and Agriculture (NIFA), Hatch/Multi-State project number ARZT-1370600-R21-189.

Conflicts of Interest: The authors declare no conflict of interest.

References

1. United Nations, Department of Economic and Social Affairs, Population Division. *World Population Prospects 2019: Highlights (ST/ESA/SER.A/423)*; United Nations: New York, NY, USA, 2019.
2. Artuso, F.; Guijt, I. *Global Megatrends—Mapping the Forces that Affect Us All*; Oxfam GB: Oxford, UK, 2020; ISBN 978-1-78748-564-8.
3. Newman, E. Hungry, or Hungry for Change? Food Riots and Political Conflict, 2005–2015. *Stud. Confl. Terror.* **2020**, *43*, 300–324. [CrossRef]
4. Wheeler, T.; von Braun, J. Climate change impacts on global food security. *Science* **2013**, *341*, 508–513. [CrossRef] [PubMed]
5. Raviv, M.; Lieth, J.H.; Bar-Tal, A. Significance of Soilless Culture in Agriculture. In *Soilless Culture: Theory and Practice*, 2nd ed.; Raviv, M., Lieth, J.H., Bar-Tal, A., Eds.; Academic Press: London, UK, 2019; pp. 3–14.
6. Gruda, N.S. Increasing sustainability of growing media constituents and stand-alone substrates in soilless culture systems. *Agronomy* **2019**, *9*, 298. [CrossRef]
7. Savvas, D. Hydroponics: A modern technology supporting the application of integrated crop management in greenhouse. *J. Food Agric. Environ.* **2003**, *1*, 80–86.
8. Raviv, M. Can compost improve sustainability of plant production in growing media? In Proceedings of the International Symposium on Growing Media, Composting and Substrate Analysis-SusGro2015, Vienna, Austria, 7–11 September 2015; pp. 119–134.
9. Bar-Tal, A.; Saha, U.K.; Raviv, M.; Tuller, M. Inorganic and Synthetic Organic Components of Soilless Culture and Potting Mixtures. In *Soilless Culture: Theory and Practice*, 2nd ed.; Raviv, M., Lieth, J.H., Bar-Tal, A., Eds.; Academic Press: London, UK, 2019; pp. 259–301.
10. Yeager, T.; Gilliam, C.; Bilderback, T.E.; Fare, D.; Niemiera, A.; Tilt, K. *Best Management Practices: Guide for Producing Container-Grown Plants*; Southern Nursery Association: Acworth, GA, USA, 2000.
11. Caron, J.; Pepin, S.; Periard, Y. Physics of growing media in a green future. In Proceedings of the International Symposium on Growing Media and Soilless Cultivation, Leiden, Netherlands, 17–21 June 2013; pp. 309–317.
12. Tuller, M.; Or, D. Water Retention and Characteristic Curve. In *Encyclopedia of Soils in the Environment*; Hillel, D., Ed.; Elsevier: London, UK, 2005; pp. 278–289.
13. Wallach, R.; Da Silva, F.F.; Chen, Y. Hydraulic characteristics of tuff (scoria) used as a container medium. *J. Am. Soc. Hortic. Sci.* **1992**, *117*, 415–421. [CrossRef]
14. Deepagoda, T.K.K.C.; Lopez, J.C.C.; Møldrup, P.; de Jonge, L.W.; Tuller, M. Integral parameters for characterizing water, energy, and aeration properties of soilless plant growth media. *J. Hydrol.* **2013**, *502*, 120–127. [CrossRef]
15. Or, D.; Tuller, M.; Or, D.; Wraith, J.M. Water Potential. In *Encyclopedia of Soils in the Environment*; Hillel, D., Ed.; Elsevier: London, UK, 2005; pp. 270–277.
16. Lieth, J.H.; Oki, L.R. Irrigation in soilless production. In *Soilless Culture: Theory and Practice*, 1st ed.; Raviv, M., Lieth, J.H., Eds.; Elsevier Science: London, UK, 2007; pp. 117–156.
17. Veihmeyer, F.J.; Hendrickson, A.H. Soil moisture in relation to plant growth. *Annu. Rev. Plant Physiol.* **1950**, *1*, 285–304. [CrossRef]
18. De Boodt, M.; Verdonck, O. The physical properties of the substrates in horticulture. In Proceedings of the III Symposium on Peat in Horticulture, Dublin, Ireland, 28 June–3 July 1971; pp. 37–44.
19. White, J.W.; Mastalerz, J.W. Soil moisture as related to container capacity. In Proceedings of the American Society for Horticultural Science; Am. Soc. Horticultural Science: Alexandria, VA, USA, 1966; Volume 89, p. 758.
20. Bilderback, T.E.; Fonteno, W.C. Effects of container geometry and media physical properties on air and water volumes in containers. *J. Environ. Hortic.* **1987**, *5*, 180–182. [CrossRef]

21. Fonteno, W.C. An approach to modeling air and water status of horticultural substrates. In Proceedings of the Symposium on Substrates in Horticulture other than Soils in Situ, Dublin, Ireland, 12–16 September 1988; pp. 67–74.

22. Letey, J. Relationship between soil physical properties and crop production. In *Advances in Soil Science*; Springer: New York, NY, USA, 1985; pp. 277–294.

23. Da Silva, A.P.; Kay, B.D.; Perfect, E. Characterization of the least limiting water range of soils. *Soil Sci. Soc. Am. J.* **1994**, *58*, 1775–1781. [CrossRef]

24. Groenevelt, P.H.; Grant, C.D.; Semetsa, S. A new procedure to determine soil water availability. *Soil Res.* **2001**, *39*, 577–598. [CrossRef]

25. Minasny, B.; McBratney, A.B. Integral energy as a measure of soil-water availability. *Plant Soil* **2003**, *249*, 253–262. [CrossRef]

26. Da Silva, F.F.; Wallach, R.; Chen, Y. Hydraulic properties of sphagnum peat moss and tuff (scoria) and their potential effects on water availability. In *Optimization of Plant Nutrition*; Springer: Berlin/Heidelberg, Germany, 1993; pp. 569–576.

27. Raviv, M.; Wallach, R.; Silber, A.; Medina, S.; Krasnovsky, A. The effect of hydraulic characteristics of volcanic materials on yield of roses grown in soilless culture. *J. Am. Soc. Hortic. Sci.* **1999**, *124*, 205–209. [CrossRef]

28. Loneragan, J.F. Plant nutrition in the 20th and perspectives for the 21st century. In *Plant Nutrition for Sustainable Food Production and Environment*; Springer: Berlin/Heidelberg, Germany, 1997; pp. 3–14.

29. Savvas, D. Nutritional management of vegetables and ornamental plants in hydroponics. *Crop Manag. Postharvest Handl. Hortic. Prod.* **2001**, *1*, 37–87.

30. Silber, A.; Bar-Tal, A. Nutrition of Substrate-Grown Plants. In *Soilless Culture: Theory and Practice*, 2nd ed.; Raviv, M., Lieth, J.H., Bar-Tal, A., Eds.; Academic Press: London, UK, 2019; pp. 197–257.

31. Silber, A. Chemical characteristics of soilless media. In *Soilless Culture: Theory and Practice*, 2nd ed.; Raviv, M., Lieth, J.H., Bar-Tal, A., Eds.; Academic Press: London, UK, 2019; pp. 113–148.

32. Willard, L.L. *Chemical Equilibria in Soils*; John Wiley & Sons: Chichester, UK, 1979.

33. Silber, A.; Xu, G.; Levkovitch, I.; Soriano, S.; Bilu, A.; Wallach, R. High fertigation frequency: The effects on uptake of nutrients, water and plant growth. *Plant Soil* **2003**, *253*, 467–477. [CrossRef]

34. Silber, A.; Bruner, M.; Kenig, E.; Reshef, G.; Zohar, H.; Posalski, I.; Yehezkel, H.; Shmuel, D.; Cohen, S.; Dinar, M. High fertigation frequency and phosphorus level: Effects on summer-grown bell pepper growth and blossom-end rot incidence. *Plant Soil* **2005**, *270*, 135–146. [CrossRef]

35. Xu, S.; Lam, J. Positive real control for uncertain singular time-delay systems via output feedback controllers. *Eur. J. Control.* **2004**, *10*, 293–302. [CrossRef]

36. Gerendás, J.; Zhu, Z.; Bendixen, R.; Ratcliffe, R.G.; Sattelmacher, B. Physiological and biochemical processes related to ammonium toxicity in higher plants. *Z. Pflanz. Bodenkd.* **1997**, *160*, 239–251.

37. Savvas, D.; Karagianni, V.; Kotsiras, A.; Demopoulos, V.; Karkamisi, I.; Pakou, P. Interactions between ammonium and pH of the nutrient solution supplied to gerbera (*Gerbera jamesonii*) grown in pumice. *Plant Soil* **2003**, *254*, 393–402. [CrossRef]

38. Bar-Tal, A.; Aloni, B.; Karni, L.; Rosenberg, R. Nitrogen nutrition of greenhouse pepper. II. Effects of nitrogen concentration and NO_3:NH_4 ratio on growth, transpiration, and nutrient uptake. *HortScience* **2001**, *36*, 1252–1259. [CrossRef]

39. Claussen, W.; Lenz, F. Effect of ammonium or nitrate nutrition on net photosynthesis, growth, and activity of the enzymes nitrate reductase and glutamine synthetase in blueberry, raspberry and strawberry. *Plant Soil* **1999**, *208*, 95–102. [CrossRef]

40. Magalhaes, J.R.; Wilcox, G.E. Growth, free amino acids, and mineral composition of tomato plants in relation to nitrogen form and growing media. *J. Am. Soc. Hortic. Sci.* **1984**, *109*, 406–411.

41. Ganmore-Neumann, R.; Kafkafi, U. Root Temperature and Percentage $NO3-/NH4$ Effect on Tomato Plant Development I. Morphology and Growth. *Agron. J.* **1980**, *72*, 758–761. [CrossRef]

42. Alkan, M.; Do, M. Surface Titrations of Perlite Suspensions. *J. Colloid Interface Sci.* **1998**, *207*, 90–96. [CrossRef] [PubMed]

43. Kaufhold, S.; Reese, A.; Schwiebacher, W.; Dohrmann, R.; Grathoff, G.H.; Warr, L.N.; Halisch, M.; Muller, C.; Schwarz-Schampera, U.; Ufer, K. Porosity and distribution of water in perlite from the island of Milos, Greece. *Springerplus* **2014**, *3*, 598. [CrossRef] [PubMed]

44. Noland, D.A.; Spomer, L.A.; Williams, D.J. Evaluation of pumice as a perlite substitute for container soil physical amendment. *Commun. Soil Sci. Plant Anal.* **1992**, *23*, 1533–1547. [CrossRef]

45. Kingston, P.H.; Scagel, C.F.; Bryla, D.R.; Strik, B.C. Influence of Perlite in Peat-and Coir-based Media on Vegetative Growth and Mineral Nutrition of Highbush Blueberry. *HortScience* **2020**, *55*, 658–663. [CrossRef]

46. Muñoz, C.; Soto, R.; Valenzuela, J. Effect of chemical and physical potting media characteristics on growth of container-grown rabbiteye blueberries. In Proceedings of the V International Symposium on Vaccinium Culture, Melbourne, Australia, 14–16 January 1993; pp. 162–172.

47. Bugbee, G.J.; Frink, C.R. Aeration of potting media and plant growth. *Soil Sci.* **1986**, *141*, 438–441. [CrossRef]

48. Silber, A.; Bar-Yosef, B.; Singer, A.; Chen, Y. Mineralogical and chemical composition of three tuffs from northern Israel. *Geoderma* **1994**, *63*, 123–144. [CrossRef]

49. Silber, A.; Bar-Yosef, B.; Chen, Y. pH-dependent kinetics of tuff dissolution. *Geoderma* **1999**, *93*, 125–140. [CrossRef]

50. Silber, A.; Raviv, M. Effects on chemical surface properties of tuff by growing rose plants. *Plant Soil* **1996**, *186*, 353–360. [CrossRef]

51. Evans, M.R.; Konduru, S.; Stamps, R.H. Source variation in physical and chemical properties of coconut coir dust. *HortScience* **1996**, *31*, 965–967. [CrossRef]

52. Abad, M.; Fornes, F.; Carrión, C.; Noguera, V.; Noguera, P.; Maquieira, Á.; Puchades, R. Physical properties of various coconut coir dusts compared to peat. *HortScience* **2005**, *40*, 2138–2144. [CrossRef]

53. Londra, P.A.; Paraskevopoulou, A.T.; Psychoyou, M. Evaluation of water–Air balance of various substrates on Begonia growth. *HortScience* **2012**, *47*, 1153–1158. [CrossRef]

54. Mak, A.T.Y.; Yeh, D.M. Nitrogen nutrition of Spathiphyllum 'Sensation' grown in Sphagnum peat-and coir-based media with two irrigation methods. *HortScience* **2001**, *36*, 645–649. [CrossRef]

55. Meerow, A.W. The Potential of Coir (Coconut Mesocarp Pith) as a Peat Substitute in Container Media. *HortScience* **1994**, *29*, 452. [CrossRef]

56. Evans, M.R. Physical properties of and plant growth in peat-based root substrates containing glass-based aggregate, perlite, and parboiled fresh rice hulls. *HortTechnology* **2011**, *21*, 30–34. [CrossRef]

57. Van Genuchten, M.T. A closed-form equation for predicting the hydraulic conductivity of unsaturated soils. *Soil Sci. Soc. Am. J.* **1980**, *44*, 892–898. [CrossRef]

58. Durner, W. Hydraulic conductivity estimation for soils with heterogeneous pore structure. *Water Resour. Res.* **1994**, *30*, 211–223. [CrossRef]

59. Reynolds, W.D.; Elrick, D.E.; Youngs, E.G.; Booltink, H.W.G.; Bouma, J. Saturated and Field-Saturated Water Flow Parameters: Laboratory Methods. In *Methods of Soil Analysis: Part 4—Physical Methods, 5.4*; Dane, J.H., Topp, G.C., Eds.; Soil Science Society of America, Inc.: Madison, WI, USA, 2002; pp. 802–816.

60. Flint, A.L.; Flint, L.E. Particle Density. In *Methods of Soil Analysis: Part 4—Physical Methods, 5.4*; Dane, J.H., Topp, G.C., Eds.; Soil Science Society of America, Inc.: Madison, WI, USA, 2002; pp. 229–240.

61. Kovar, J.L.; Pierzynski, G.M. Methods of phosphorus analysis for soils, sediments, residuals, and waters, 2nd edition. *South. Coop. Ser. Bull.* **2009**, *408*, 33–37.

62. Kempers, A.J.; Zweers, A. Ammonium determination in soil extracts by the salicylate method. *Commun. Soil Sci. Plant Anal.* **1986**, *17*, 715–723. [CrossRef]

63. Ma, J.; James Lenhart, P.E.; WRE, D. Phosphorus Removal in Urban Runoff Using Adsorptive Filtration Media. In Proceedings of the StormCon, Anaheim, CA, USA, 16–20 August 2009; pp. 16–20.

64. Ghazvini, R.F.; Payvast, G.; Azarian, H. Effect of clinoptiloliticzeolite and perlite mixtures on the yield and quality of strawberry in soil-less culture. *Int. J. Agric. Biol.* **2007**, *9*, 885–888.

65. Demirboğa, R.; Gül, R. The effects of expanded perlite aggregate, silica fume and fly ash on the thermal conductivity of lightweight concrete. *Cem. Concr. Res.* **2003**, *33*, 723–727. [CrossRef]

66. Thompson, R.B.; Gallardo, M.; Valdez, L.C.; Fernández, M.D. Using plant water status to define threshold values for irrigation management of vegetable crops using soil moisture sensors. *Agric. Water Manag.* **2007**, *88*, 147–158. [CrossRef]

67. Smettem, K.R.J.; Kirkby, C. Measuring the hydraulic properties of a stable aggregated soil. *J. Hydrol.* **1990**, *117*, 1–13. [CrossRef]

68. Jamei, M.; Guiras, H.; Chtourou, Y.; Kallel, A.; Romero, E.; Georgopoulos, I. Water retention properties of perlite as a material with crushable soft particles. *Eng. Geol.* **2011**, *122*, 261–2718. [CrossRef]

69. Londra, P.A. Simultaneous Determination of Water Retention Curve and Unsaturated Hydraulic Conductivity of Substrates Using a Steady-state Laboratory Method. *HortScience* **2010**, *45*, 1106–1112. [CrossRef]

70. Giuffrida, F.; Consoli, S. Reusing perlite substrates in soilless cultivation: Analysis of particle size, hydraulic properties, and solarization effects. *J. Irrig. Drain. Eng.* **2016**, *142*, 4015047. [CrossRef]

71. Al Naddaf, O.; Livieratos, I.; Stamatakis, A.; Tsirogiannis, I.; Gizas, G.; Savvas, D. Hydraulic characteristics of composted pig manure, perlite, and mixtures of them, and their impact on cucumber grown on bags. *Sci. Hortic.* **2011**, *129*, 135–141. [CrossRef]

72. Mualem, Y. A new model for predicting the hydraulic conductivity of unsaturated porous media. *Water Resour. Res.* **1976**, *12*, 513–522. [CrossRef]

73. Quintero, M.F.; González, C.A.; Florez-Roncancio, V.J. Physical and hydraulic properties of four substrates used in the cut-flower industry in Colombia. In Proceedings of the III International Symposium on Models for Plant Growth, Environmental Control and Farm Management in Protected Cultivation, Wageningen, Netherlands, 29 October–2 November 2006; pp. 499–506.

74. Raviv, M.; Lieth, J.H.; Burger, D.W.; Wallach, R. Optimization of Transpiration and Potential Growth Rates of Kardinal Rose with Respect to Root-zone Physical Properties. *J. Am. Soc. Hortic. Sci.* **2001**, *126*, 638–643. [CrossRef]

75. Chamindu Deepagoda, T.K.K.; Moldrup, P.; Tuller, M.; Pedersen, M.; Lopez, C.; Choc, J.; Wollesen de Jonge, L.; Kawamoto, K.; Komatsu, T. Gas diffusivity-based design and characterization of greenhouse growth substrates. *Vadose Zone J.* **2013**, *12*, 1–13. [CrossRef]

76. Riga, P.; Álava, S.; Usón, A.; Blanco, F.; Garbisu, C.; Aizpurua, A.; Tejero, T.; Larrea, A. Evaluation of recycled rockwool as a component of peat-based mixtures for geranium (*Pelargonium peltatum* L.) production. *J. Hortic. Sci. Biotechnol.* **2003**, *78*, 213–218. [CrossRef]

77. Caron, J.; Elrick, D.E.; Beeson, R.; Boudreau, J. Defining critical capillary rise properties for growing media in nurseries. *Soil Sci. Soc. Am. J.* **2005**, *69*, 794–806. [CrossRef]

78. Jones, S.B.; Or, D. Design of porous media for optimal gas and liquid fluxes to plant roots. *Soil Sci. Soc. Am. J.* **1998**, *62*, 563–573. [CrossRef]

79. Ma, J.; Lenhart, J.H.; Tracy, K. Orthophosphate adsorption equilibrium and breakthrough on filtration media for storm-water runoff treatment. *J. Irrig. Drain. Eng.* **2011**, *137*, 244–250. [CrossRef]

80. Kumar, P.; Sudha, S.; Chand, S.; Srivastava, V.C. Phosphate removal from aqueous solution using coir-pith activated carbon. *Sep. Sci. Technol.* **2010**, *45*, 1463–1470. [CrossRef]

81. Namasivayam, C.; Sangeetha, D. Equilibrium and kinetic studies of adsorption of phosphate onto $ZnCl_2$ activated coir pith carbon. *J. Colloid Interface Sci.* **2004**, *280*, 359–365. [CrossRef]

82. Kumarasinghe, H.; Subasinghe, S.; Ransimala, D. Effect of coco peat particle size for the optimum growth of nursery plant of greenhouse vegetables. *Trop. Agric. Res. Ext.* **2016**, *18*, 51–57. [CrossRef]

83. Gai, X.; Wang, H.; Liu, J.; Zhai, L.; Liu, S.; Ren, T.; Liu, H. Effects of feedstock and pyrolysis temperature on biochar adsorption of ammonium and nitrate. *PLoS ONE* **2014**, *9*, e113888. [CrossRef]

Article

Substrate Volumetric Water Content Controls Growth and Development of Containerized Culinary Herbs

Christopher J. Currey *, Nicholas J. Flax, Alexander G. Litvin and Vincent C. Metz

Department of Horticulture, Iowa State University, 2206 Osborn Drive, Ames, IA 50011, USA; nickflax1@gmail.com (N.J.F.); aglitvin@ncsu.edu (A.G.L.); vmetz4@gmail.com (V.C.M.)
* Correspondence: ccurrey@iastate.edu; Tel.: +1-515-294-1917; Fax: +1-515-294-0730

Received: 25 September 2019; Accepted: 21 October 2019; Published: 23 October 2019

Abstract: There are no chemical plant growth retardants that may be used on containerized culinary herbs intended for consumption. Our objective was to quantify the effect of substrate moisture content on the growth of four commonly produced culinary annual herbs grown in containers in the greenhouse. Seedlings of basil (*Ocimum basilicum* L.), dill (*Anethum graveolens* L.), parsley (*Petroselinum crispum* (Mill.) Fuss), and sage (*Salvia officinalis* L.) were transplanted into 11.4 cm diameter containers filled with commercial soilless substrate comprising (by vol.) 75% sphagnum peat moss and 25% coarse perlite and amended with 3.0 kg·m^{-3} of controlled-release fertilizer. After the containers were thoroughly irrigated to container capacity, plants were placed into a sensor-controlled irrigation system, which maintained substrate volumetric water content (VWC) at 0.15, 0.28, 0.30, 0.38, or 0.45 m^3·m^{-3}. Chlorophyll fluorescence, photosynthesis, stomatal conductance, and transpiration were measured 27 d after initiating treatments, and the results showed that chlorophyll fluorescence of parsley and photosynthesis of basil increased as substrate VWC increased from 0.15 to 0.45 m^3·m^{-3}; the remaining parameters for basil, parsley, and sage were unaffected. Additionally, height, width, leaf area, and shoot dry mass of basil, dill, parsley, and sage increased as substrate volumetric water content increased from 0.15 to 0.45 m^3·m^{-3}. Our results show that growth of basil, dill, parsley, and sage can be promoted or inhibited by providing or withholding water, respectively, with no signs of stress or visual damage resulting from reduced substrate volumetric water content. Therefore, restricting irrigation and substrate volumetric water content is an effective nonchemical growth control method for containerized culinary herbs grown in peat-based substrate.

Keywords: restricted deficit irrigation; soil moisture sensors; nonchemical growth control; water use efficiency

1. Introduction

One of the primary challenges associated with growing containerized herbaceous plants is controlling shoot growth to produce plants that are proportional and aesthetically balanced to the container height. Controlling shoot growth is important to produce plants that are sized proportionally to containers for aesthetic appearance as well as to increase container density in the greenhouse and during shipping [1]. Although chemical plant growth retardants (PGRs) are commonly used to control containerized ornamental crop growth, there are currently no PGRs that are labeled for use on containerized culinary herbs [2]. Therefore, nonchemical methods of controlling containerized herb growth must be used.

There are several nonchemical growth control techniques that may be used to control containerized herb growth [3–5]. Compact cultivars are available for some herb species, including basil and dill [3], and may be more appropriately sized for container production. The concentration of mineral nutrients provided to container-grown herbs, both total and specific nutrients, also affects growth. For example, basil supplied with 200 mg·L^{-1} N from a complete, balanced water-soluble fertilizer are 33% larger than plants supplied with 50 mg·L^{-1} N from the same fertilizer [4]. Additionally, restricting P to 5 mg·L^{-1} produced basil, dill, parsley, and sage shorter than plants provided with 40 mg·L^{-1} [5]. While cultivar selection and nutrient management are useful forms of nonchemical growth control, it may be necessary to use multiple nonchemical methods of controlling growth to achieve the degree of control required in the absence of PGRs.

Reducing irrigation or substrate volumetric water content (VWC), commonly referred to as "deficit irrigation", is another effective method of controlling containerized plant growth [6–8]. The water available for plant uptake increases and growth is promoted as substrate VWC increases and, as such, restricting irrigation and reducing the substrate VWC can diminish turgor pressure and subsequent stem extension and growth [9]. For example, containerized angelonia (*Angelonia angustifolia* Benth.) and petunia (*Petunia* × *hybrid* Vilm.) bedding plant growth is promoted by substrate VWC and, by reducing VWC, compact plants of marketable quality can be produced [7,10]. Additionally, using regulated deficit irrigation can suppress stem elongation of flowering potted plants, such as poinsettia (*Euphorbia pulcherrima* Willd. ex Klotzsch), providing adequate height control during production [11]. While controlling the substrate VWC clearly has potential for use in containerized herb production, data specific to the effects of substrate VWC on containerized herb growth are lacking.

We have found some limited reports on the effects of substrate moisture on containerized perennial herb growth [8,9]. Zhen et al. [8] reported that limiting irrigation of rosemary (*Rosmarinus officinalis* L.) plants successfully controlled excessive growth. Additionally, Zhen and Burnett [9] showed that English lavender (*Lavandula angustifolia* Mill. 'Hidcote' and 'Munstead') growth diminished with decreasing substrate VWC. These data are promising for controlling containerized herb growth by limiting substrate VWC. However, we have found no other data on the use of drought stress to control excessive growth of more common containerized herb species grown with shorter production periods. Our objective was to quantify the effect of substrate VWC on the growth of four common culinary annual herbs grown in containers in the greenhouse. We hypothesized that the growth of parsley, sage, basil, and dill would be promoted by increasing substrate VWC and, as such, restricting irrigation would be an effective growth-control strategy for containerized culinary annual herb species with short growth cycles.

2. Materials and Methods

Seeds (Johnny's Selected Seed, Albion, ME, USA) of parsley (*Petroselenium crispum* (Mill.) Fuss 'Giant of Italy'), common sage (*Salvia officinalis* L.; Expt. 1), basil (*Ocimum basilicum* L. 'Italian Large Leaf'), and dill (*Anethum graveolens* L. 'Fernleaf'; Expt. 2) were individually sown in 288-cell propagation trays (PL-288-1.25; 7.1 cm^3 individual cell vol.; T.O. Plastics, Clearwater, MN, USA) filled with a soilless germination substrate comprising (by vol.) 65% fine sphagnum peat moss, 20% fine perlite, and 15% vermiculite (Propagation Mix; Sun Gro Horticulture, Agawam, MA, USA). Trays were initially hand-irrigated with clear, tempered tap water. Beginning at radicle emergence, seedlings were irrigated with tap water supplemented with a blend of water-soluble fertilizers (50 and 100 mg·L^{-1} N provided from 21N–2.2P–16.6K and 15N–2.2P–12.5K, respectively; Everris NA, Inc., Marysville, OH, USA) to provide the following (in mg·L^{-1}): 150 nitrogen, 8.6 phosphorous, 92.2 potassium, 33.3 calcium, 13.3 magnesium, 0.75 iron, 0.4 manganese and zinc, 0.2 copper and boron, and 0.5 molybdenum.

Seedlings were grown on expanded metal benches in a glass-glazed greenhouse at Iowa State University, Ames, IA (latitude 42° N) with fog cooling, radiant hot-water floor and perimeter heating, and retractable shade curtains controlled by an environmental computer (ARGUS Titan; ARGUS Control Systems, Surrey, BC, Canada). The day and night greenhouse air temperature set points were 23.0 ± 1 °C and 18.0 ± 1 °C, respectively. Aluminized shade cloth (XLS 15 Revolux; Ludvig Svensson, Kinna, Sweden) was drawn across the crop when outdoor light intensities exceeded 1000 µmol·m^{-2}·s^{-1} to avoid leaf scorch. High-pressure sodium lamps delivered a supplemental photosynthetic photon flux (PPF) of ~190 µmol·m^{-2}·s^{-1} at plant height (as measured with a quantum sensor (LI-190 SB;

LI-COR Biosciences, Lincoln, NE, USA)) when ambient light intensity was below 100 $\mu mol \cdot m^{-2} \cdot d^{-1}$ between 0600 and 2200 HR to maintain a target daily light integral (DLI) of ~12 $mol \cdot m^{-2} \cdot d^{-1}$.

Four weeks after sowing, seedlings were planted into 11.4 cm diameter containers (655 mL vol.; HC Companies, Middlefield, OH, USA) filled with soilless greenhouse substrate comprising (by vol.) 75% sphagnum peat moss and 25% coarse perlite (Sunshine® LB-2; Sun Gro Horticulture, Inc., Agawam, MA, USA) amended with 3.0 $kg \cdot m^{-3}$ controlled-release fertilizer (Florikan Plus 16.0 N–2.2 P–9.1 K with a 90 d release period; Florikan ESA, Sarasota, FL, USA). For each experimental unit, 20 plant containers were placed into two 10-cell petroleum-plastic shuttle trays adjacent to each other with individual plants spaced on 12 cm centers (69.4 plants per m^2). The inner six plant containers were measured for data gathered, while the surrounding plants were used as border plantings to simulate greenhouse practices.

An automated irrigation system controlled by soil moisture sensors was used to maintain VWC treatments similar to that described by Nemali and van Iersel [12]. Drip irrigation stakes attached to 1.9 $L \cdot h^{-1}$ pressure-compensating emitters (Netafim USA, Fresno, CA, USA) were inserted into the substrate, and plants were irrigated overhead to container capacity with clear tempered water. After overhead irrigation, capacitance moisture sensors (EC-5; Decagon Devices Inc., Pullman, WA, USA) were inserted into the substrate of the two innermost plant containers within each experimental unit. Sensors connected to a multiplexer (AM16/32B; Campbell Scientific, Logan, UT, USA) cycling measurement readings to a data logger (CR1000; Campbell Scientific, Logan, UT, USA) calculated VWC using a manufacturer-provided calibration curve specific to soilless peat-based substrates. Substrate VWC thresholds were 0.15, 0.23, 0.30, 0.38, and 0.45 $m^3 \cdot m^{-3}$, and they were chosen to represent the range of VWC to be observed in commercial production. The VWC values were maintained by the data logger controlling a solenoid valve (Orbit Irrigation Products, Inc., Bountiful, UT, USA) connected to polyethylene tubing with drip emitters for each experimental unit. Irrigation events occurred as needed when the average measured VWC of the two moisture sensors within a given experimental unit fell below its respective threshold. The data logger program was executed every 10 min to determine need. Solenoid valves corresponding to each experimental unit were controlled by a relay driver (SDM-CD16AC controller; Campbell Scientific, Logan, UT, USA) connected to the data logger. Valves opened for 10 s during each irrigation event, providing 6.2 mL of clear water per plant per event. Substrate moisture content and total irrigation volumes are presented in Figures 1 and 2, respectively.

Plants were grown in the greenhouse as previously described. The air temperature was measured every 15 s by four temperature probes (41342; R.M. Young Company, Traverse City, MI, USA) in an aspirated radiation shield (43502; R.M. Young Company, Traverse City, MI, USA), while the PPF was measured every 15 s by eight quantum sensors (LI-190SL; LI-COR Biosciences, Lincoln, NE, USA) per greenhouse. Temperature probes and quantum sensors were connected to a data logger (CR1000 Measurement and Control System; Campbell Scientific, Logan, UT, USA) with means logged every 15 min. The mean day, night, and daily temperatures and DLI are reported in Table 1.

Table 1. Mean (± standard deviation) daily light integral (DLI), average daily air temperature (ADT), and average day (DT) and night (NT) air temperature for parsley and sage (Expt. 1) or basil and dill (Expt. 2) grown in 11.4 cm diameter containers filled with a soilless substrate comprising (by vol.) 75% sphagnum peat moss and 25% coarse perlite amended with 3.0 $kg \cdot m^{-3}$ and maintained at 0.15, 0.23, 0.30, 0.38, or 0.45 $m^3 \cdot m^{-3}$ substrate volumetric water content (VWC) for four weeks.

Experiment	DLI ($mol \cdot m^{-2} \cdot d^{-1}$)	ADT (°C)	DT (°C)	NT (°C)
1	10.8 ± 0.5	23.7 ± 0.3	25.2 ± 0.3	20.7 ± 0.3
2	10.4 ± 0.7	22.9 ± 0.4	24.0 ± 0.5	20.5 ± 0.5

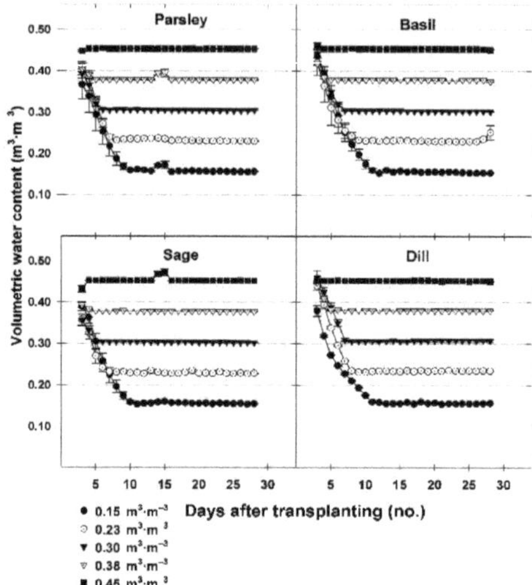

Figure 1. Substrate moisture for parsley and sage (Expt. 1) and basil and dill (Expt. 2) grown in 11.4 cm diameter containers filled with a soilless substrate comprising (by vol.) 75% sphagnum peat moss and 25% coarse perlite amended with 3.0 kg·m^{-3} controlled-release fertilizer and maintained at 0.15, 0.23, 0.30, 0.38, or 0.45 m^3·m^{-3} substrate volumetric water content for four weeks.

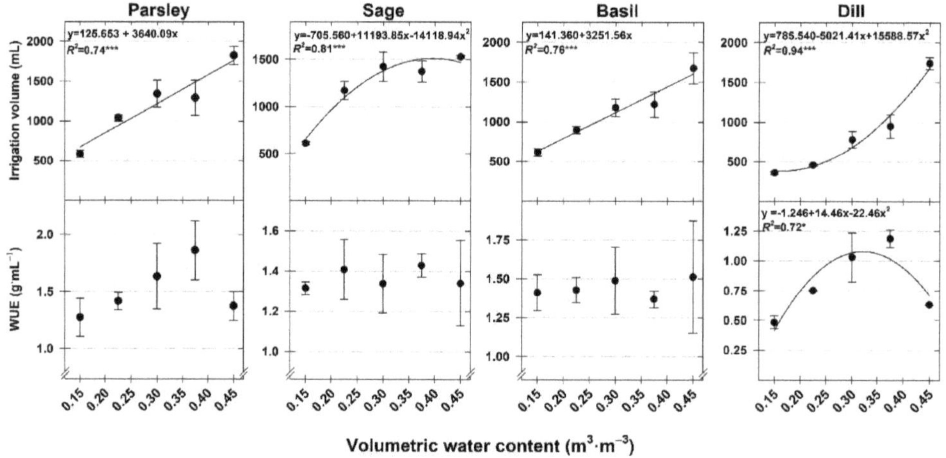

Figure 2. Total irrigation volume and water use efficiency (WUE) for parsley and sage (Expt. 1) and basil and dill (Expt. 2) grown in 11.4 cm diameter containers filled with a soilless substrate comprising (by vol.) 75% sphagnum peat moss and 25% coarse perlite amended with 3.0 kg·m^{-3} controlled-release fertilizer and maintained at 0.15, 0.23, 0.30, 0.38, or 0.45 m^3·m^{-3} substrate volumetric water content for four weeks. Regression lines are presented for significant correlations only with corresponding R^2 presented. * and *** indicate significant at $p \leq 0.05$ or 0.001, respectively.

Four weeks after transplanting seedlings, data were collected. Chlorophyll fluorescence of three plants per treatment per replication was measured on the adaxial epidermis of the most fully expanded leaf using a chlorophyll fluorescence meter (Handy Plant Efficiency Analyzer; Hansatech Instruments Ltd., Norfolk, U.K.). Using the manufacturer's clip, leaves were dark-acclimated for 15 min before measurements were taken. Fluorescence was measured by opening a shutter in the dark-acclimating clip and exposing the leaf to light with a peak wavelength of 650 nm provided by up to

3000 μmol·m^{-2}·s^{-1} for 1 s to saturate photosystem II. Chlorophyll fluorescence was expressed as a ratio of the change in chlorophyll fluorescence from initial to maximum, to maximum fluorescence (F_v/F_m).

Gas exchange measurements were conducted with a portable photosynthesis system (LI-6400XT; LI-COR Biosciences, Lincoln, NE, USA) on two plants per treatment per replication. The second most recently matured leaf placed in a 6 cm^2 leaf chamber with a light-emitting diode light source (6400-02B; red at 665 nm and blue at 470 nm) providing 400 μmol·m^{-2}·s^{-1}. The reference CO_2 concentration inside the leaf chamber was 500 μmol·mol^{-1}, and the flow of air into the chamber was set to maintain a constant mole fraction of 8.0 mmol·mol^{-1} of water inside the chamber. Leaf temperature inside the leaf chamber was maintained at 23.0 °C.

Height was measured from the substrate surface to the tallest growing point. Width was determined by measuring the widest point and 90° perpendicular and averaging these two measurements. Branch length was determined by measuring a branch at a node approximately half the total height of the plant. The number of nodes was counted. Leaf area was determined by scanning all leaves of each plant with a leaf area meter (LI-3000; LI-COR Biosciences, Lincoln, NE, USA). Shoots were severed at the substrate surface, placed in a paper bag, and dried in a forced-air oven at 67 °C for 3 d, after which shoots were weighed and the dry mass recorded. Water use efficiency (WUE) was calculated by dividing the shoot dry mass by the total irrigation volume applied per plant. Internode length was determined by dividing the height by the node number.

The experiment employed a randomized complete block design for each species. There were three blocks (replications) for each VWC for each species, with six individual plants per block. Data were analyzed using regression analyses (Sigma Plot 21.0; Systat Software, San Jose, CA, USA), with VWC concentration as the independent variable.

3. Results

3.1. Parsley

Target substrate VWC for 0.15, 0.23, 0.30, 0.38, and 0.45 were achieved 13, 8, 6, 5, and 3 d later, respectively (Figure 1). Total irrigation volume increased linearly from 587 to 1825 mL as VWC increased from 0.15 to 0.45 m^3·m^{-3} (Figure 2). The photosynthesis (P_n), conductance (g_s), and transpiration (E) of parsley was unaffected by VWC, while F_v/F_m increased from 0.82 to 0.84 as VWC increased from 0.15 to 0.45 m^3·m^{-3} (Figure 3). Height and width of parsley increased quadratically in response to VWC (Figure 4). For example, height increased by 14.8 cm as VWC increased from 0.15 to 0.38 m^3·m^{-3}, while plants grown at 0.45 m^3·m^{-3} were 1.6 cm shorter compared to those grown at 0.38 m^3·m^{-3} (Figure 4); width followed a similar trend. Increasing VWC promoted node appearance, as plants grown at 0.38 and 0.45 m^3·m^{-3} had approximately one additional node compared to those grown at 0.15 m^3·m^{-3} (Figure 4). Leaf area increased quadratically by 57.0 or 57.5 cm^2 for plants grown at 0.38 or 0.45 m^3·m^{-3}, respectively, compared to plants grown at 0.15 m^3·m^{-3} (39.2 cm^2; Figure 4). The shoot dry mass also increased quadratically from 4.5 to 14.9 g as substrate VWC increased from 0.15 to 0.45 m^3·m^{-3}, respectively. There was no significant relationship between substrate VWC and WUE of parsley (Figure 2).

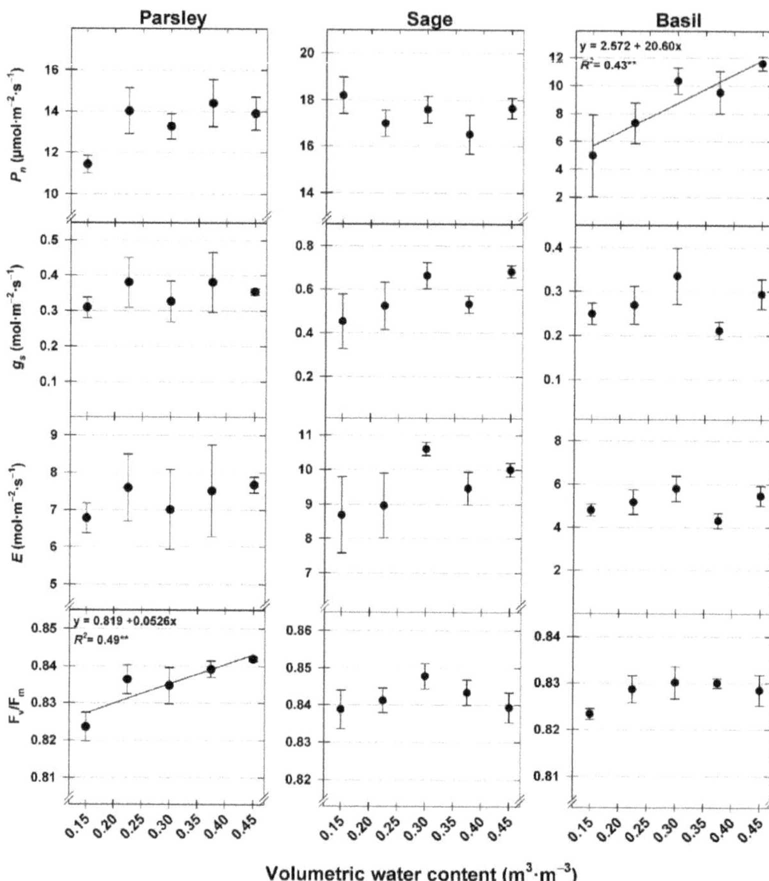

Figure 3. Photosynthesis (P_n), conductance (g_s), transpiration (E), and chlorophyll fluorescence (F_v/F_m) of parsley and sage (Expt. 1) and basil (Expt. 2) grown in 11.4 cm diameter containers filled with a soilless substrate comprising (by vol.) 75% sphagnum peat moss and 25% coarse perlite amended with 3.0 kg·m⁻³ controlled-release fertilizer and maintained at 0.15, 0.23, 0.30, 0.38, or 0.45 m³·m⁻³ substrate volumetric water content for four weeks. Regression lines are presented for significant correlations only with corresponding R^2 presented. ** indicates nonsignificant or significant at $p \leq 0.01$.

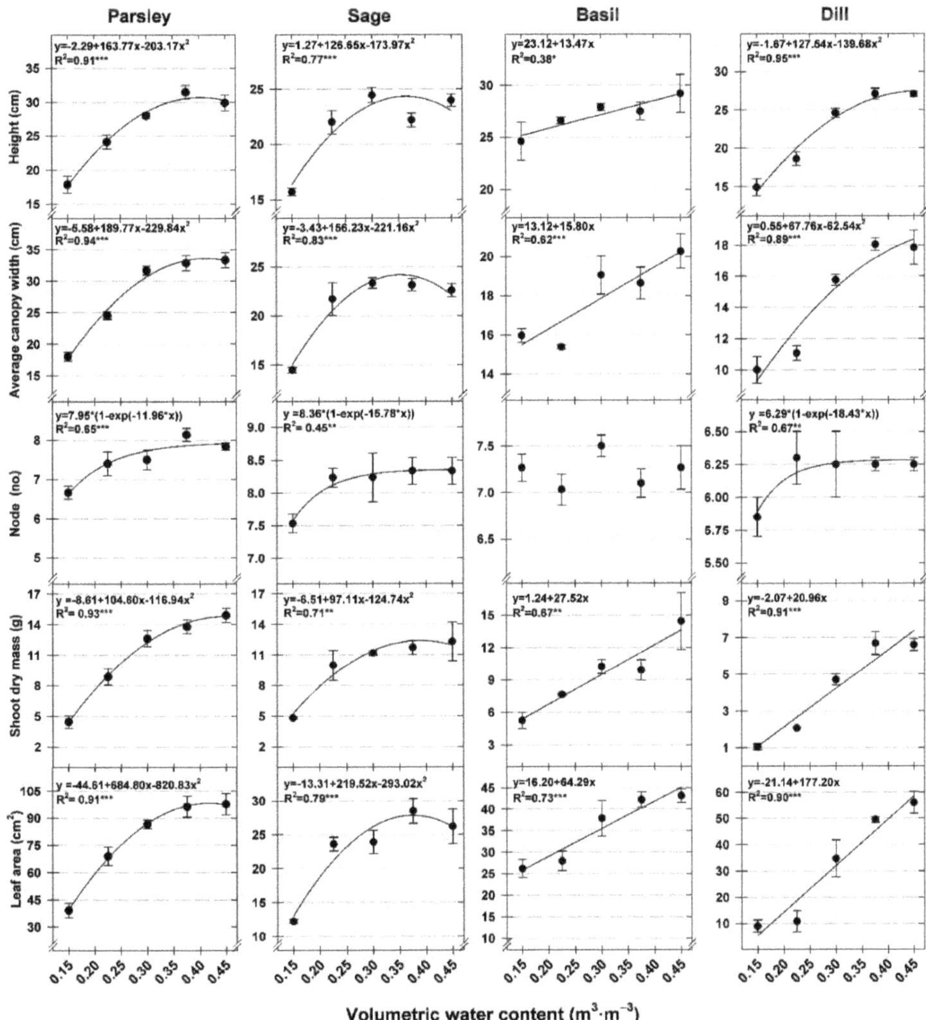

Figure 4. Height, width, node number, leaf area, and shoot dry mass of parsley and sage (Expt. 1) and basil and dill (Expt. 2) grown in 11.4 cm diameter containers filled with a soilless substrate comprising (by vol.) 75% sphagnum peat moss and 25% coarse perlite amended with 3.0 kg·m^{-3} controlled-release fertilizer and maintained at 0.15, 0.23, 0.30, 0.38, or 0.45 m^3·m^{-3} substrate volumetric water content for four weeks. Regression lines are presented for significant correlations only with corresponding R^2 presented. *, **, or *** indicates significant at $p \leq 0.05$, 0.01, or 0.001, respectively.

3.2. Sage

The time to reach target substrate conditions decreased with increasing substrate VWC, taking 10 d to reach 0.15 m^3·m^{-3} and 4 d to reach 0.45 m^3·m^{-3} (Figure 1). The total irrigation volume required to maintain substrate VWC increased from 612 to 1531 mL as VWC increased from 0.15 to 0.45 m^3·m^{-3} (Figure 2). Neither F_v/F_m nor gas exchange of sage were affected by VWC (Figure 3). The height, width, and internode length increased from 15.7 to 24.4 cm, 14.5 to 23.3 cm, and 2.0 to 3.0 cm as VWC increased from 0.15 to 0.30 m^3·m^{-3}, respectively, then decreased to 24.0 cm, 22.6 cm, and 3.0 cm, respectively, as VWC further increased up to 0.45 m^3·m^{-3} (Figures 4 and 5). Similarly, leaf area increased from 12.2 to 28.5 cm^2 as VWC increased from 0.15 to 0.38 m^3·m^{-3} (Figure 4). While node number and branch length for sage grown at 0.15 m^3·m^{-3} was 7.5 and 2.9 cm, respectively, plants grown at 0.23 to 0.45 m^3·m^{-3} had 8.2 to 8.3 nodes and branches between 6.7 and 8.9 cm long (Figure 5). Shoot dry mass increased from 4.8 to 12.3 g as VWC increased from 0.15 to 0.45 m^3·m^{-3} (Figure 4). The WUE of sage was unaffected by substrate VWC (Figure 2).

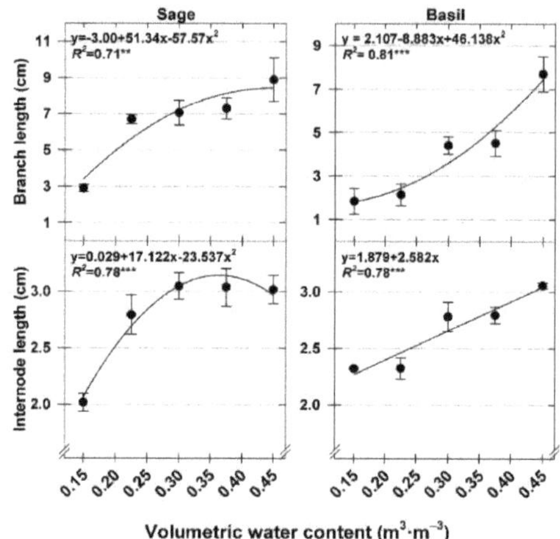

Figure 5. Branch and internode length of sage (Expt. 1) and basil (Expt. 2) grown in 11.4 cm diameter containers filled with a soilless substrate comprising (by vol.) 75% sphagnum peat moss and 25% coarse perlite amended with 3.0 kg·m^{-3} controlled-release fertilizer and maintained at 0.15, 0.23, 0.30, 0.38, or 0.45 m^3·m^{-3} substrate volumetric water content for four weeks. Regression lines are presented for significant correlations only with corresponding R^2 presented. ** or *** indicates significant at $p \leq 0.01$ or 0.001, respectively.

3.3. Basil

Increasing substrate VWC from 0.15 to 0.45 m^3·m^{-3} reduced the time from 12 to 4 d to reach VWC targets, respectively (Figure 1), whereas the amount of water required to maintain target substrate VWC increased linearly from 616 to 1674 mL (Figure 2). Although F_v/F_m, g_s, and E were unaffected by substrate VWC, P_n increased linearly from 5.0 to 11.6 µmol·m^{-2}·d^{-1} as VWC increased from 0.15 to 0.45 m^3·m^{-3} (Figure 3). Similarly, as substrate VWC increased from 0.15 to 0.45 m^3·m^{-3} the height, width, internode length, leaf area, branch length, and shoot dry mass increased by 4.6 cm, 4.3 cm, 0.7 cm, 17 cm^2, 5.9 cm, and 9.1 g, respectively (Figures 4 and 5). The WUE of basil ranged from 1.41 to 1.51 g·mL^{-1} across substrate VWC and were unaffected by treatments (Figure 2).

3.4. Dill

Substrate VWC for dill reached 0.15, 0.23, 0.30, 0.38, and 0.45 m^3·m^{-3} 13, 9, 7, 5, and 2 d after imposing treatments, respectively (Figure 1). The height and width of dill increased quadratically by 12.2 and 8.1 cm, respectively, as substrate VWC increased from 0.15 to 0.38 m^3·m^{-3} but then diminished as VWC was further increased to 0.45 m^3·m^{-3} (Figure 4). Leaf area increased linearly from 9.0 to 56.1 cm^2 as substrate VWC increased from 0.15 to 0.45 m^3·m^{-3}, respectively (Figure 4). Similarly, dill shoot dry mass increased linearly by 5.5 g as substrate VWC increased from 0.15 to 0.45 m^3·m^{-3} (Figure 4). There was no effect of substrate VWC on the number of nodes. The WUE of dill increased by 0.71 g·mL^{-1} as substrate VWC increased from 0.15 to 0.38 m^3·m^{-3} but then decreased as substrate VWC increased to 0.45 m^3·m^{-3} (Figure 2).

4. Discussion

The growth and development of containerized basil, dill, parsley, and sage is promoted with increasing substrate VWC. While the effect of substrate moisture on growth is better understood for containerized ornamental flowering crops, our results on the effect of substrate VWC on controlling growth of culinary herbs align well with the limited literature on container-grown herbs, including rosemary and English lavender [8,9]. For example, Zhen et al. [8] reported that, as substrate VWC increased from 0.05 to 0.40 $m^3 \cdot m^{-3}$, the height, width, leaf number and area, and fresh and dry mass of rosemary increased linearly. Similarly, height, width, leaf number, and area of 'Munstead' and 'Hidcote' English lavenders increased as substrate VWC increased from 0.10 to 0.40 $m^3 \cdot m^{-3}$ [9]. The effect of substrate VWC on WUE of containerized herbs was not consistent among species in the study, with parsley, sage, and basil not being affected by VWC, whereas WUE of dill increased as VWC increased up to 0.38 $m^3 \cdot m^{-3}$. This variation reflects what is seen in the literature, where WUE was found to increase with increasing substrate VWC for burkwood vibrurnum (*Viburnum* × *burkwoodii* Burkwood & Skipwith) and butterfly bush (*Buddleja davidii* Franch.); decrease with increasing substrate VWC for potato (*Solanum tuberosum* L.), salvia (*Salvia splendens* Sellow ex Roem. & Schult.), vinca (*Catharanthus roseus* (L.) G. Don), and wax begonia (*Begonia* × *semperflorens-cultorum* Hort.); or remain unaffected by substrate VWC for cheddar pink (*Dianthus gratianopolitanus* L.), columbine (*Aquilegia canadensis* L.), geranium (*Pelargonium* × *hortorum* Bailey), petunia, and rosemary [8,13–17].

The growth of basil, dill, parsley, and sage are promoted or inhibited by the provision or restriction of water to the root zone and, as such, restricting the substrate VWC to plants and growing them drier using restricted deficit irrigation is a viable nonchemical growth control method for container-grown culinary herbs. Although growing containerized herbs with restricted VWC reduces shoot mass, the harvestable or useable portion of most culinary herbs, it is important to distinguish between containerized and fresh-cut herb production. Containerized herb plants are sold as individual units (i.e., per container), not on the unit weight basis (i.e., gram) that fresh-cut culinary herbs are sold. For producers of fresh-cut herbs grown in substrate, using higher substrate VWC can promote shoot growth and yields, potentially enhancing productivity and profitability.

Although growth and development of herbs were greater at increasingly higher VWC, gas exchange was unaffected for parsley and sage (Figure 3). Under low water availability, gas exchange is reduced in most plants compared to higher availability [18]. For example, P_n and g_s of Mediterranean herbs sea beet (*Beta maritima*) and wall-rocket (*Diplotaxis ibicensis*) decreased with increasing water deficit stress [19]. Similarly, gas exchange (P_n, g_s, and E) of English lavender grown with sensor-based irrigation increased with VWC increasing from 0.10 to 0.40 $m^3 \cdot m^{-3}$ [9]. According to Yan et al. [20], annual herbs do not vary greatly in gas exchange with changing water status, suggesting limited response regulation, although the method of imposed stress may affect this. Montesano et al. [21] reported that, when irrigation was completely withheld for basil, the P_n, g_s, and E decreased after three days. However, the authors also reported that, when VWC was controlled using sensor-based irrigation and maintained 0.20, 0.30, or 0.40 $m^3 \cdot m^{-3}$, fresh mass increased with increasing VWC, whereas P_n, g_s, and E were unaffected by increasing VWC. In contrast, P_n in our study increased for basil as VWC increased from 0.15 to 0.45 $m^3 \cdot m^{-3}$; however, within 0.20 to 0.40 $m^3 \cdot m^{-3}$, P_n was similar to reports by Montesano et al. [21]. Taken together, our results align well with the literature for suppressed growth and development at lower VWC and for gas exchange under sensor-based irrigation for herbs. Drought stress reduced F_v/F_m in plants compared to well-watered conditions, which is in agreement with chlorophyll content for nontolerant species [22,23]. Nemali and van Iersel [14] reported that, as VWC increased, the quantum yield efficiency of photosynthesis increased for petunia, salvia, impatiens, and vinca, similar to parsley in this study, although basil and sage were unaffected, similar to previous reports by [9].

Sensor-based precision irrigation effectively restricted irrigation of containerized herbs in this experiment. This is especially useful for edible crops with no chemical PGRs labeled for use on them during greenhouse forcing [8] and for using drought as a nonchemical growth control method [6].

To consistently produce containerized crops at a lower substrate, VWC can be a challenge using non-sensor-controlled systems as judging the appropriate time to irrigate becomes more difficult [24,25]; automated sensor-based systems are well suited for controlling substrate VWC at desired set points [26]. Sensor-based irrigation also precisely controls substrate moisture, with minimal variation in VWC within treatment groups after initial dry down (Figure 1). However, aside from implementing precision irrigation strategies for producing containerized crops, there are other benefits when using these systems in commercial applications. Automated sensor-based irrigation is not only used to restrict irrigation for controlling height [6] but also to improve water use [24], plant growth uniformity [27], biomass [28], flower number [29], plant stress symptoms, and disease pressure [30] and can increase profitability of commercial producers compared to visual inspection- or timer-based irrigation scheduling [31].

5. Conclusions

The research presented here comprehensively quantifies the effect of substrate moisture on container-grown basil, dill, parsley, and sage regarding growth, development, and gas exchange. The growth and development of containerized culinary herbs, including height, width, node number, leaf area, and branching, were all controlled by substrate VWC, with growth and development restricted at lower VWC compared to those at higher VWC. However, while growth was suppressed when substrate VWC was lower, there were a few instances where P_n, g_s, E, or F_v/F_m were negatively impacted. Taken together, reducing substrate VWC and implementing restricted deficit irrigation is an effective growth-controlling strategy for containerized culinary herb production. Sensor-based irrigation allows for precise substrate moisture control to implement restricted deficit irrigation for controlling crop growth, although other tangible benefits may be realized in commercial production facilities. The research presented herein was performed using a round plastic container with a peat and perlite substrate. However, the different substrates that are either currently used or will be used in the future as peat alternatives [32], as well as different container shapes and sizes [33], can affect the water-holding capacity of substrates; therefore, additional work on culinary herb growth and substrate moisture content grown with different substrates and containers would be useful. While the results we have presented support the use of restricting substrate moisture to control containerized herb growth, commercial producers should conduct their own trials to determine the effectiveness of this growth-controlling technique under their unique circumstances, including the specific species and cultivars produced under specific greenhouse environmental conditions and crop culture.

Author Contributions: Conceptualization, C.J.C.; methodology, C.J.C., N.J.F., and A.G.L.; formal analysis, A.G.L.; investigation, V.C.M. and N.J.F.; writing—original draft preparation, C.J.C.; writing—review and editing, N.J.F., A.G.L. and V.C.M.; supervision, C.J.C.; funding acquisition, C.J.C.

Funding: This research was funded by the Fred C. Gloeckner Foundation.

Acknowledgments: We gratefully acknowledge Peter Lawlor for greenhouse assistance. The use of trade names in this publication does not imply endorsement by Iowa State University of products named, nor criticism of similar ones not mentioned.

Conflicts of Interest: The authors declare no conflict of interest. The funders had no role in the design of the study; in the collection, analyses, or interpretation of data; in the writing of the manuscript; or in the decision to publish the results.

References

1. Whipker, B.E.; McCall, I.; Latimer, J. Growth regulators. In *Ball Redbook, Vol. 2: Crop Production*; Nau, J., Ed.; Ball Publishing: West Chicago, IL, USA, 2011; pp. 95–105.
2. Currey, C.J. Keeping your herbs under control. *GrowerTalks* **2019**, *82*, 62, 64.
3. Currey, C.J.; Mazur, T.Z. Spinning the herb wheel. *GrowerTalks* **2018**, *82*, 66, 68–69.
4. Flax, N.J.; Currey, C.J. Controlled-release and water-soluble fertilizers affect growth and tissue nutrient concentrations of basil, dill, and parsley. *HortScience* **2016**, *51*, S297.

5. Currey, C.J.; Metz, V.C.; Flax, N.J.; Whipker, B.E. Restricting phosphorous suppresses growth of containerized culinary herbs. *HortScience* **2018**, *53*, S139–S140.

6. Alem, P.; Thomas, P.A.; Van Iersel, M.W. Controlled water deficit as an alternative to plant growth retardants for regulation of poinsettia stem elongation. *HortScience* **2015**, *50*, 565–569. [CrossRef]

7. Van Iersel, M.W.; Dove, S.; Kang, J.-G.; Burnett, S.E. Growth and water use of petunia as affected by substrate water content and daily light integral. *HortScience* **2010**, *45*, 277–282. [CrossRef]

8. Zhen, S.; Burnett, S.E.; Day, M.E.; Van Iersel, M.W. Effects of substrate water content on morphology and physiology of rosemary, canadian columbine, and cheddar pink. *HortScience* **2014**, *49*, 486–492. [CrossRef]

9. Zhen, S.; Burnett, S.E. Effects of substrate volumetric water content on English lavender morphology and photosynthesis. *HortScience* **2015**, *50*, 909–915. [CrossRef]

10. Jacobson, A.B.; Starman, T.W.; Lombardini, L. Substrate moisture content effect on growth and shelf life of *Angelonia angustifolia. HortScience* **2012**, *50*, 272–278. [CrossRef]

11. Alem, P.; Thomas, P.A.; Van Iersel, M.W. Use of controlled water deficit to regulate poinsettia stem elongation. *HortScience* **2015**, *50*, 234–239. [CrossRef]

12. Nemali, K.S.; Van Iersel, M.W. An automated system for controlling stress and irrigation in potted plants. *Sci. Hortic.* **2006**, *110*, 292–297. [CrossRef]

13. Liu, F.; Shahnazari, A.; Andersen, M.N.; Jacobsen, S.-E.; Jensen, C.R. Effects of deficit irrigation (DI) and partial root drying (PRD) on gas exchange, biomass partitioning, and water use efficiency in potato. *Sci. Hortic.* **2006**, *109*, 113–117. [CrossRef]

14. Nemali, K.S.; Van Iersel, M.W. Physiological responses to different substrate water contents: Screening for high water-use efficiency in bedding plants. *J. Am. Soc. Hortic. Sci.* **2008**, *133*, 333–340. [CrossRef]

15. Warsaw, A.L.; Fernandez, R.T.; Cregg, B.M.; Andresen, J.A. Water conservation, growth, and water use efficiency of container-grown woody ornamentals irrigated based on daily water use. *HortScience* **2009**, *44*, 1308–1318. [CrossRef]

16. Miralles-Crespo, J.; Van Iersel, M.W. A calibrated time domain transmissometry soil moisture sensor can be used for precise automated irrigation of container-grown plants. *HortScience* **2011**, *46*, 889–894. [CrossRef]

17. Flax, N.J.; Currey, C.J.; Litvin, A.G.; Schrader, J.A.; Grewell, D.; Graves, W.R. Aesthetic quality and strength of bioplastic biocontainers at different substrate volumetric water contents. *HortScience* **2018**, *53*, 483–490. [CrossRef]

18. Zhou, S.; Duursma, R.A.; Medlyn, B.E.; Kelly, J.W.G.; Prentice, I.C. How should we model plant responses to drought? An analysis of stomatal responses to water stress. *Agric. For. Meteorol.* **2013**, *182*, 204–214. [CrossRef]

19. Galmes, J.; Medrano, H.; Flexas, J. Photosynthetic limitations in response to water stress and recovery in Mediterranean plants with different growth forms. *New Phytol.* **2007**, *175*, 81–93. [CrossRef]

20. Yan, W.; Zhong, Y.; Shangguan, Z. A meta-analysis of leaf gas exchange and water status responses to drought. *Sci. Rep.* **2016**, *6*, 20917. [CrossRef]

21. Montesano, F.F.; Van Iersel, M.W.; Boari, F.; Cantore, V.; D'Amato, G.; Parente, A. Sensor-based irrigation management of soilless basil using a new smart irrigation system: Effects of set-point on plant physiological responses and crop performance. *Agric. Water Manag.* **2018**, *203*, 20–29. [CrossRef]

22. Gholamin, R.; Khayatnezhad, M. The effect of end season drought stress on the chlorophyll content, chlorophyll fluorescence parameters and yield in maize cultivars. *Sci. Res. Essays* **2011**, *6*, 5351–5357.

23. Van der Mescht, A.; De Ronde, J.A.; Rossouw, F.T. Chlorophyll fluorescence and chlorophyll content as a measure of drought tolerance in potato. *S. Afr. J. Sci.* **1999**, *95*, 407–412.

24. Choi, E.-Y.; Choi, K.-Y.; Lee, Y.-B. Non-drainage Irrigation scheduling in coir substrate hydroponic system for tomato cultivation by a frequency domain reflectometry sensor. *Eur. J. Hortic. Sci.* **2013**, *78*, 132–143.

25. Greenwood, D.J.; Zhang, K.; Hilton, H.W.; Thompson, A.J. Opportunities for improving irrigation efficiency with quantitative models, soil water sensors and wireless technology. *J. Agric. Sci.* **2010**, *148*, 1–16. [CrossRef]

26. Burnett, S.E.; Van Iersel, M.W. Morphology and irrigation efficiency of *Gaura lindheimeri* grown with capacitance sensor-controlled irrigation. *HortScience* **2008**, *43*, 1555–1560. [CrossRef]

27. Van Iersel, M.W.; Dove, S.; Burnett, S.E. The use of soil moisture probes for improved uniformity and irrigation control in greenhouses. *Acta Hortic.* **2008**, *893*, 1049–1056. [CrossRef]

28. Bacci, L.; Battista, P.; Rapi, B. An integrated method for irrigation scheduling of potted plants. *Sci. Hortic.* **2008**, *116*, 89–97. [CrossRef]

29. Cai, X.; Starman, T.; Niu, G.; Hall, C. The effect of substrate moisture content on growth and physiological responses of two landscape roses (*Rosa hybrida* L.). *HortScience* **2014**, *49*, 741–745. [CrossRef]
30. Lichtenberg, E.; Majsztrik, J.; Saavoss, M. Profitability of sensor-based irrigation in greenhouse and nursery crops. *HortTechnology* **2013**, *23*, 770–774. [CrossRef]
31. Van Iersel, M.W.; Chappell, M.; Lea-Cox, J.D. Sensors for improved efficiency of—Irrigation in—Greenhouse and nursery crop production. *HortTechnology* **2013**, *23*, 735–746. [CrossRef]
32. Gruda, N.S. Increasing sustainability of growing media constituents and stand-alone substrates in soilless culture systems. *Agronomy* **2019**, *9*, 298. [CrossRef]
33. Bilderback, T.E.; Fonteno, W.C. Effects of container geometry and media physical properties on air and water volumes in containers. *J. Environ. Hortic.* **1987**, *5*, 180–182.

Article

Microbe-Plant Growing Media Interactions Modulate the Effectiveness of Bacterial Amendments on Lettuce Performance Inside a Plant Factory with Artificial Lighting

Thijs Van Gerrewey [1,2,3,4], **Maarten Vandecruys** [3], **Nele Ameloot** [4], **Maaike Perneel** [5], **Marie-Christine Van Labeke** [6], **Nico Boon** [2] and **Danny Geelen** [1,*]

[1]   Horticell, Ghent University, Coupure Links 653, B-9000 Gent, Belgium; thijs.vangerrewey@ugent.be
[2]   Center for Microbial Ecology & Technology (CMET), Ghent University, Coupure Links 653, B-9000 Gent, Belgium; Nico.Boon@UGent.be
[3]   Urban Crop Solutions BVBA, Grote Heerweg 67, B-8791 Beveren-Leie (Waregem), Belgium; mava@urbancropsolutions.com
[4]   Agaris Belgium NV, Skaldenstraat 7a, B-9042 Gent, Belgium; nele.ameloot@agaris.eu
[5]   Cropfit, Ghent University, Coupure Links 653, B-9000 Gent, Belgium; maaike.perneel@ugent.be
[6]   Horticultural Sciences & Crop Physiology, Ghent University, Coupure Links 653, B-9000 Gent, Belgium; MarieChristine.VanLabeke@Ugent.be
*   Correspondence: danny.geelen@ugent.be; Tel.: +32(0)92646076

Received: 22 August 2020; Accepted: 21 September 2020; Published: 23 September 2020

Abstract: There is a need for plant growing media that can support a beneficial microbial root environment to ensure that optimal plant growth properties can be achieved. We investigated the effect of five rhizosphere bacterial community inocula (BCI S1–5) that were collected at three open field organic farms and two soilless farms on the performance of lettuce (*Lactuca sativa* L.). The lettuce plants were grown in ten different plant growing media (M1–10) composed of 60% *v/v* peat (black peat or white peat), 20% *v/v* other organics (coir pith or wood fiber), 10% v/v composted materials (composted bark or green waste compost) and 10% *v/v* inorganic materials (perlite or sand), and one commercial plant growing medium inside a plant factory with artificial lighting. Fractional factorial design of experiments analysis revealed that the bacterial community inoculum, plant growing medium composition, and their interaction determine plant performance. The impact of bacterial amendments on the plant phenotype relied on the bacterial source. For example, S3 treatment significantly increased lettuce shoot fresh weight (+57%), lettuce head area (+29%), root fresh weight (+53%), and NO_3-content (+53%), while S1 treatment significantly increased lettuce shoot dry weight (+15%), total phenolic content (+65%), and decreased NO_3-content (−67%). However, the effectiveness of S3 and S1 treatment depended on plant growing medium composition. Principal component analysis revealed that shoot fresh weight, lettuce head area, root fresh weight, and shoot dry weight were the dominant parameters contributing to the variation in the interactions. The dominant treatments were S3-M8, S1-M7, S2-M4, the commercial plant growing medium, S1-M2, and S3-M10. Proper selection of plant growing medium composition is critical for the efficacy of bacterial amendments and achieving optimal plant performance inside a plant factory with artificial lighting.

Keywords: plant growth-promoting rhizobacteria (PGPR); growing media; rhizosphere; lettuce; plant factory; soilless culture; plant quality; plant yield; microbiome; beneficial bacteria

---

1. Introduction

A growing world population in the course of climate change requires the food supply chain to be revised to secure future universal access to food in a sustainable way [1,2]. In controlled-environment agriculture (CEA), the recent development of state-of-the-art plant factories with artificial lighting (PFAL) allows maximizing plant growth in a resource use efficient way (water, CO_2, fertilizer, energy, etc.) [3]. Plant factories with artificial lighting can tap into new markets that are inaccessible to open-field production and conventional greenhouses by locally producing leafy greens, herbs, medicinal plants, and transplants year-round for local consumption [4].

Plant factories with artificial lighting utilize soilless culture methods [5]. Soilless culture typically requires a plant growing medium that provides a proper physicochemical and biological environment for rooting and plant growth during the seedling stage [6]. Peat, partially degraded *Sphagnum* mosses that accumulated over thousands of years under waterlogged conditions within mires, has been widely used as a plant growing medium because of its low economic cost and good performance [7,8]. However, access to peat will be limited because of sustainability and environmental concerns involving the peat production process [9–11]. Sustainable alternatives are being investigated and a variety of these are on the market (e.g., coir pith, wood fiber, composted materials, biochar, etc.) [6,8,12–16]. Nevertheless, peat will remain an essential plant growing medium constituent, for dilution purposes at any rate as it allows the blending of alternative and circular raw materials [7]. At the same time, because of the expanding world population, the demand for plant growing media is expected to increase drastically [17]. Newly developed peat-reduced plant growing media have to perform equal to or even outperform peat, to ensure universal access to food.

When selecting new plant growing medium materials, environmental factors have become as important as performance and economic cost. However, little attention is given to the microbial properties of these products and their potential to support the amendment of plant growth-promoting rhizobacteria (PGPR). Contrary to plant growing media, soil bacterial communities are widely researched [18]. Soils contain an immense diversity in bacterial communities, enabling various soil ecosystem functions [19]. However, only a minority of bacterial taxa, including Actinobacteria, Bacteroidetes, Firmicutes, and Proteobacteria, encompass the diversity present in soils [19,20]. Plants are in continuous contact with soil bacterial communities through their roots. Via rhizodeposition, plants recruit soil bacteria to the rhizosphere and endosphere that improve the capacity of the plant to adapt to the environment [20–25]. These PGPRs can stimulate germination, enhance growth, improve nutrient acquisition, promote stress resistance, and enable disease suppression [26–29].

Globally, agro-industries are starting to embrace PGPR technology but are confronted with strong variation in efficacy of PGPR application, with no benefits to considerable benefits being reported [30–33]. The underlying factors causing the differential activity are not well known. The development of bacterial amendments mainly focused on single strain PGPR products [34–37]. The complexity of bacterial communities and their interactions with environmental factors and crop specificity is suspected to play an important role in the success of the plant-microbe interaction [26,34,38,39].

Plant growing medium composition may be a determining factor in the successful amendment of microbes in a soilless environment. Rhizosphere bacteria show specific microbial substrate uptake traits that drive the assembly of the rhizosphere bacterial community [21]. In addition to plant root exudate chemistry, plant growing media could provide a source of microbial substrate allowing modulation of the rhizosphere microbiome for improved plant performance [40,41]. The role of plant growing media in beneficial plant-microbe interactions is not well studied [26]. There is evidence that plant growing media have distinct microbial features that can provide stability and resilience to crops in a diverse soilless environment. The complex biological and physicochemical interactions in organic plant growing media influence the rhizosphere microbial communities of the plant [42]. Organic plant growing media have a more diverse and sTable microbial community that decreases the susceptibility of the eggplant *Solanum melongena* to the hairy roots pathogen *Agrobacterium rhizogenes* [43]. Composts maintain a high microbial diversity that is critical to the suppression of soil-borne pathogens and improving plant performance [44–47]. Biochar amendment to peat growing media and soil may improve plant growth and disease suppressiveness [48–51]. These positive effects of biochar amendment are linked to the activity, diversity, and composition of the rhizosphere microbial community [52,53]. There is evidence that PGPR amendment can improve plant growth and decrease phytopathogen infections in soilless culture [37,54–57]. Though, the role of plant growing medium composition as a potential driver in the success of PGPR amendment is much less clear. Recent research has studied the use of plant growing medium constituents as a carrier material for bacterial inocula [33,40,41,58,59]. For example, Nadeem et al. [41] showed that the combined use of biochar, compost, and the PGPR *Pseudomonas fluorescens* alleviated the negative effect of water deficit on cucumber growth. More research has to be done on the mechanisms of action and the efficiency of using different plant growing medium constituents as a carrier for PGPR consortia.

At the start of our work, we hypothesized that plant growing medium composition plays a decisive role in the effectiveness of PGPR amendment inside a complex PFAL environment. Here we report results that show that specific microbe-plant growing medium interactions are the major determinants of performance for *Lactuca sativa* L. (lettuce). Seedlings of lettuce, a leafy green that is abundantly produced in PFALs, were grown in different plant growing media, inoculated with a few selected bacterial communities, and transferred to a PFAL. The different plant growing media were composed by varying five raw material groups: (a) peat (black peat and white peat), (b) other organics (coir pith and wood fiber), (c) composted materials (composted bark and green waste compost), (d) inorganic materials (perlite and sand), and (e) Arabic gum dosed at 1 kg·m^{-3} or 5 kg·m^{-3}. Lettuce root-associated bacterial community samples were collected from soil and soilless farms and used as an inoculum. Shoot fresh weight (FW), lettuce head area (LHA), root fresh weight (RW), shoot dry weight (DW), total phenolic content (TPC), NO_3-content, and leaf pigments were quantified.

2. Materials and Methods

2.1. Collection of Root-Associated Bacterial Communities

Lettuce root-associated bacterial community samples (S1–5) were collected at five different locations in Flanders, Belgium during the growing season: three open field organic farms and two soilless farms. An overview of all sampling locations can be found in Table 1. Sampling and extraction were performed following the method described by Barillot et al. [60]. Briefly, 30 plant and root-associated soil samples (20 cm^2 by 30 cm deep) were collected at each location, transported in polyethylene bags, and stored at 4 °C. Bulk soil was removed by manually shaking the roots. The rhizosphere fraction was collected by manually washing the roots in a sterile 0.9% NaCl solution for 10 min. Roots were subsequently washed by hand in sterile 0.9% NaCl + 0.01% Tween 80 for 10 min to obtain the rhizoplane fraction. Both fractions were homogenized on an orbital shaker (125 rpm, 90 min, room temperature). The homogenized samples were centrifuged at low speed (150 g, 10 min, room temperature) to separate soil particles and other debris from the supernatant containing bacteria. The supernatants were centrifuged at high speed (9425 g, 10 min, room temperature) to collect the bacteria in the pellet. The bacterial pellets were resuspended in tryptic soy broth (TSB) + 15% glycerol and stored at −80 °C.

Table 1. Overview of Rhizosphere Sampling Locations.

Sample	Collection Date	Location	Crop	Cultivation Method	Plant Growing Medium
S1	3 October 2017	Wachtebeke, Belgium	*Lactuca sativa* var. crispa (oakleaf)	Organic open field	Sand
S2	17 October 2017	Moerbeke-Waas, Belgium	*Lactuca sativa* var. crispa (oakleaf)	Organic open field	Loamy sand
S3	21 November 2017	Onze-Lieve-Vrouw-Waver, Belgium	*Lactuca sativa* var. crispa (lollo bionda)	Soilless	Black peat
S4	12 December 2017	Ardooie, Belgium	*Lactuca sativa* var. capitata (butterhead)	Soilless	Black peat
S5	5 June 2018	Lochristi, Belgium	*Lactuca sativa* var. crispa (lollo bionda)	Organic open field	Sand

The amount of live bacterial cells present in the rhizosphere and rhizoplane fractions was estimated using flow cytometric analysis to standardize bacterial inoculation in further analysis (see Section 2.3). The samples were diluted and stained with SYBR® Green I combined with propidium iodide (SGPI, 100 × concentrate SYBR® Green I, Invitrogen, and 50 × 20 mM propidium iodide, Invitrogen, in 0.22 μm-filtered dimethyl sulfoxide) for live-dead analysis. Staining was performed as described previously, with incubation for 13 min at 37 °C [61]. Samples were analyzed immediately after incubation on a C6+ flow cytometer (BD Biosciences, Belgium), which was equipped with four fluorescence detectors (530/30 nm, 585/40 nm, >670 nm, and 675/25 nm), two scatter detectors and a 20-mW 488-nm laser. The flow cytometer was operated with Milli-Q (Merck, Darmstadt, Germany) as sheath fluid.

2.2. Plant Growing Media Composition

Ten different experimental plant growing media were composed (M1–10; Table 2). The raw material collection took place at Agaris Belgium NV, Gent, Belgium. All plant growing media have following volumetric composition: 60% *v/v* peat, 20% *v/v* other organics, 10% *v/v* composted materials and 10% *v/v* inorganic materials. For eight plant growing media (M1, M3, M4, M5, M7, M8, M9, and M10), selection of the raw material and the Arabic gum dose was based on a 2_{III}^{5-2} fractional factorial design (Tables S1 and S2). Based on a previous study [62], two more plant growing media were composed: M2 and M6, both showing high microbial activity potential. The peat and coir based commercial plant growing medium (75% peat and 25% coir fibers, Jiffy International AS, Kristiansand, Norway) was used as a control to evaluate the performance of the experimental plant growing media. The physicochemical properties of each plant growing medium were analyzed in triplicate following Verdonck and Gabriels [63], and Gabriels et al. [64]. The data obtained is shown in Table S3.

Table 2. Composition of Plant Growing Media. Each plant growing medium consists of 4 raw material groups at different volume per volume (% v/v): peat (black peat BP or white peat WP), other organics (coir pith CP or wood fiber WF), composted materials (composted bark CB or green waste compost GC) and inorganic materials (perlite P or sand S). Arabic gum was dosed at 1 kg·m^{-3} or 5 kg·m^{-3}.

Plant Growing Medium	Peat (60% *v/v*)	Other Organics (20% *v/v*)	Composted Materials (10% *v/v*)	Inorganic Materials (10% *v/v*)	Arabic Gum (kg·m^{-3})
M1	WP	CP	CB	P	1
M2	WP	WF	CB	P	5
M3	BP	CP	CB	S	5
M4	WP	CP	GC	P	5
M5	WP	WF	CB	S	1
M6	WP	CP	CB	S	5
M7	BP	WF	CB	P	5
M8	BP	CP	GC	S	1
M9	WP	WF	GC	S	5
M10	BP	WF	GC	P	1

2.3. Plant Growth and Inoculation

Sterilized hydroponic mesh pots, with 6.5 cm height, 5 cm bottom diameter, and 7 cm top diameter, were fitted with hydroponic paper (Ellepot, Esbjerg, Denmark), filled with 200 mL of plant growing medium, and watered to saturation. Batavia lettuce seeds (Enza Zaden, Enkhuizen, The Netherlands) were sown in ten pots of each plant growing medium (Table 2). The seeds were wetted by spraying water. The pots were placed in a sterilized tray inside a growth chamber (Urban Crop Solutions, Beveren-Leie, Belgium) with a temperature of 22–23 °C, relative humidity of 60–70%, and 800 ppm CO_2-fertilization. LED light fixtures (Urban Crop Solutions, Beveren-Leie, Belgium) provided an 18 h light regime at 220 μmol.m^{-2}·s^{-1}. For the following two weeks, the pots were irrigated by hand with tap water when necessary. Two weeks after sowing six pots with uniform lettuce seedlings were selected per plant growing medium and placed in a sterilized tray fitted with an overflow drain for automated irrigation. In each tray, the six selected plants from a single plant growing medium were positioned at a distance of 18.6 cm in length and 22.2 cm in width from each other.

At this point, the bacterial community inocula (BCI S1–5) were applied to all experimental plant growing media at the base of the plant. Based on the live bacterial cell counts, determined with flow cytometric analysis (see Section 2.1), equal volumes of the collected rhizosphere and rhizoplane fractions were mixed and diluted with TSB. Application of 1 mL of inoculum provided a dose of 3.2×10^9 CFU per L plant growing medium. As a positive control treatment (PGPR), *Bacillus* sp. with plant growth-promoting properties was added as an inoculum to each plant growing medium at a dose of 3.2×10^9 CFU per L plant growing medium. As a negative control treatment (C), 1 mL of sterile TSB solution was added to every plant growing medium. Unlike the experimental plant growing media, the commercial plant growing medium was only treated with 1mL of sterile TSB solution. After inoculation, the trays were placed inside a PFAL (Urban Crop Solutions, Beveren-Leie, Belgium) for three weeks under the growing conditions as mentioned above. During these three weeks, all plants were irrigated automatically four times a day with the following nutrient solution: 14 mM NO_3^-, 2 mM PO_4^{3-}, 7 mM K^+, 4 mM Ca^{2+}, 2 mM Mg^{2+}, 635 μM SO_4^{2-}, 72 μM Fe^{2+}, 18 μM Mn^{2+}, 2 μM Zn^{2+}, 46 μM B, 0.8 μM Cu^{2+}, 1 μM Mo^{2-}, and 356 μM Si.

The experiment was split into five batches. Each batch consisted of all ten experimental plant growing media treated with one bacterial community inoculum, two randomly selected experimental plant growing media treated with the positive control, and two randomly selected experimental plant growing media treated with the negative control. The commercial plant growing medium was added to the last batch.

2.4. Plant Sample Analysis

2.4.1. Plant Sample Processing

The plants were harvested three weeks after inoculation. During harvest, top view images were taken to determine the lettuce head area (LHA) by image processing in ImageJ [65]. The harvested plants were transported in polyethylene bags to avoid excessive transpiration and stored at 4 °C until further processing. Within 24 h, the plant samples were cut at the base to separate root and shoot. Shoot fresh weight (FW) was determined by weighing the lettuce head immediately after cutting. After weighing, a section (weighing approximately 10 g) of the whole lettuce head, containing young and mature leaves, was cut out. This subsample was ground using an IKA A11 liquid nitrogen mixer (IKA, Staufen, Germany) and stored at −80 °C until further analysis. To determine the shoot dry weight (DW), the remaining shoot was placed in a paper bag and dried at 70 °C for 72 h. The difference in weight before and after drying was used to calculate the shoot dry weight of the sample. Next, the dried subsample was ground with a coffee mill (Proficook PC-KSW 1021, Clatronic International GmbH, Kempen, Germany) and stored until further analysis. If the total shoot weight was too low for obtaining both fresh and dry subsamples, priority was given to the fresh subsample. This was the case for the following treatments: S1-M6, S2-M3, PGPR-M3, PGPR-M6, and PGPR-M9.

The roots and plant growing medium of the sample were used to isolate the root-associated bacterial community following the procedure described in Section 2.1. After the second washing step, plant roots were weighed to determine root fresh weight (RW).

2.4.2. Total Phenolic Content

Total phenolic content (TPC) was ascertained following the Folin–Ciocalteu method [66]. Colorimetric TPC measurements of fresh subsample extracts in 80% methanol were carried out with a Tecan infinite plate reader (Tecan Group Ltd., Männedorf, Switzerland) at a wavelength of 765 nm. Total phenolic content was expressed as mg gallic acid equivalents (GAE) per 100 g FW.

2.4.3. Nitrate Content

The nitrate (NO_3) concentration was determined colorimetrically with salicylic acid as described by Cataldo et al. [67]. Oven-dried subsamples were used. Measurements were performed by a Tecan infinite plate reader at a wavelength of 410 nm.

2.4.4. Chlorophylls and Carotenoids

Chlorophyll a (Chl_a) a, chlorophyll b (Chl_b), and carotenoids were quantified by UV-VIS spectroscopy of a whole-pigment extract of the fresh subsamples in 80% acetone [68]. Absorption at 470 nm, 648.8 nm, and 663.2 nm wavelengths, and zero absorption at 750 nm were measured with a Tecan infinite plate reader. The amount of Chl_a, Chl_b, and carotenoids (C_{x+c}) were calculated in $\mu g.mL^{-1}$ with the following equations:

$$Chl_a = 12.25 \times A_{663.2} - 2.79 \times A_{646.8} \tag{1}$$

$$Chl_b = 21.5 \times A_{646.8} - 5.1 \times A_{663.2} \tag{2}$$

$$C_{x+c} = (1000 \times A_{470} - 1.82 \times Chl_a - 85.02 \times Chl_b)/198 \tag{3}$$

2.5. Statistical Analysis

Before subjecting the plant performance data to statistical analysis, any data points further than 1.5 times the interquartile range from the mean were considered as outliers, and were removed from the dataset. Analyses of differences between BCI means and principal component analysis (PCA) were carried out using R 3.6.1 (R Foundation for Statistical Computing, Vienna, Austria). All statistical analyses were performed at the 95% confidence level. Differences between BCI means were analyzed per plant growing medium. Quantile-quantile plots were used to check for normality of the data (*stats* package). Levene's test was used to determine the homogeneity of variance across groups (*car* package). In case the assumptions of normality and homoscedasticity were met, one-way ANOVA was used to determine significant differences between BCI means (*stats* package). As a post hoc test, a linear model was created using the *stats* package. Following, the estimated marginal means were calculated, using the Tukey's honest significance test for separation of the means at the $P < 0.05$ level (*emmeans* package). Finally, a compact letter display was created using the *multcomp* package. In case the assumptions of normality and homoscedasticity were not met, the Kruskal–Wallis test was used to compare BCI means (*stats* package). Dunn's test with Bonferroni correction was used as a post hoc method to separate the means (*FSA* package). A compact letter display of the comparison of means was produced using the *rcompanion* package.

Principal component analysis was used to determine which BCIs and plant growing media contribute most to the variation in data, and to which key performance parameters they are associated to. The data was standardized by scaling to unit variance before analysis (*stats* package). A quality of representation (cos^2) correlation circle, a contribution plot of the variables, and a contribution plot of the samples were generated using the *factoextra* package.

A 1/4 fractional factorial statistical design of experiments (DOE; 2_{III}^{5-2}) was used to simultaneously evaluate the effect of the plant growing medium raw material groups (five control factors having a high +1 and a low −1 factor level) and their interactions on the plant performance parameters. The fractional factorial design was established and analyzed in Minitab 17 (Minitab Inc., State College, Pennsylvania, United States) using main effects plots, ANOVA, and response optimization. The design was extended with an additional control factor to determine the effect of inoculation. The levels of the inoculation control factor were: negative control treatment (C) as low factor level and inoculum treatment (S1–5) as high factor level. An overview of all control factors and the final fractional factorial design can be found in Tables S1 and S2. Following decisions were made to deal with aliasing effects. (a) Three-factor and higher-order interactions are extremely rare and were omitted. (b) When aliasing occurred between the main effect and two-factor interactions, the main effect was assumed significant. (c) Aliasing between two-factor interactions was resolved by following the heredity principle: an interaction effect is likely significant when the main effects involved are also significant [69].

3. Results

3.1. Effect of Bacterial Community Inoculum and Plant Growing Medium on Shoot Fresh Weight

Both BCI ($P < 0.001$) and plant growing medium ($P < 0.001$) significantly altered FW. Lettuce FW varied from 6.03 g (S2-M3) to 74.85 g (S3-M8). Significant differences in FW were observed between BCIs in each plant growing medium (Figure 1). Bacterial community inoculum S3 significantly ($P < 0.05$) increased FW in multiple plant growing media (M5, M7, M8, M9, and M10) compared to C. For example, FW was 17.78 g in C-M7 compared to 50.83 g in S3-M7, which is more than a 2.5-fold increase. S3-M8 (74.85 g) and S3-M10 (64.63 g) were the only BCI and plant growing medium combinations that had significantly ($P < 0.05$) higher FW than the commercial plant growing medium (48.65 g). On average, inoculating the plant growing media with BCI S3 increased FW with 57% ($P < 0.001$; Figure 2f). Response optimization showed that, excluding BCI S3, the addition of a BCI was not vital to reaching maximal FW (Table S4). Moreover, BCI S3 treatment was the largest contributor to FW, compared to the plant growing medium raw material groups (Figure S1). The positive effects of S3 on FW do not occur in each plant growing medium, underlining the importance of plant growing medium composition on the effectiveness of BCI treatment.

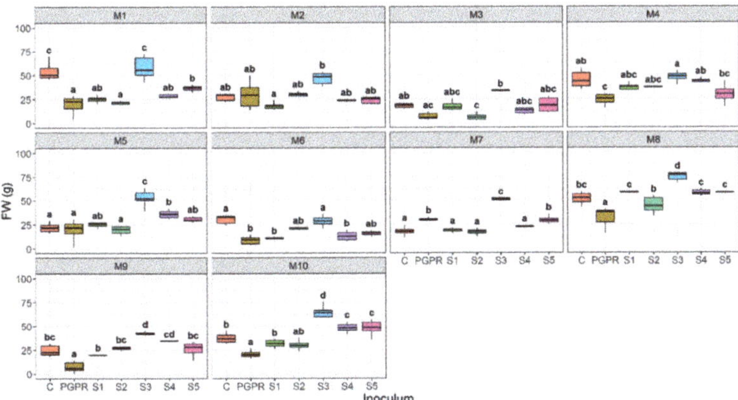

Figure 1. Boxplot of shoot fresh weight (FW; g) grouped per plant growing medium. Letters show comparison of BCI means per plant growing medium at the 95% confidence level. S indicates the bacterial community inoculum, M indicates the plant growing medium, C indicates the negative control treatment without addition of inoculum, and PGPR indicates the positive control treatment with a *Bacillus* sp. inoculum. Number of plants ≥ 3.

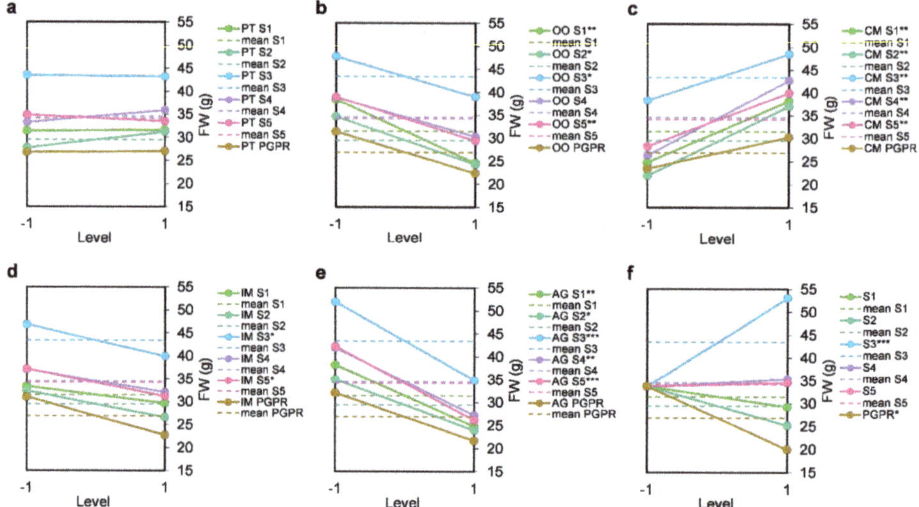

Figure 2. Main effects of plant growing medium constituents on shoot fresh weight (FW; g) under different bacterial community inoculum treatments (S1–5 and positive control PGPR). (**a**) Peat (PT; −1 = black peat and 1 = white peat); (**b**) Other organics (OO; −1 = coir pith and 1 = wood fiber); (**c**) Composted materials (CM; −1 = composted bark and 1 = green waste compost); (**d**) Inorganic materials (IM; −1 = perlite and 1 = sand); (**e**) Arabic gum (AG; −1 = 1 kg.m^{-3} and 1 = 5 kg.m^{-3}); (**f**) Bacterial inoculum (BCI; −1 = C and 1 = S1–5 or PGPR). Dashed lines indicate mean levels of FW for each bacterial treatment. Asterisks indicate level of significance: $P < 0.05$ (*), $P < 0.01$ (**) and $P < 0.001$ (***).

Surprisingly, the positive control treatment PGPR significantly decreased FW in several plant growing media (M1, M4, M6, M8, M9, and M10) compared to negative control C (Figure 1). For example, treating M9 with PGPR decreased FW with 70% compared to C. On average, the positive control treatment significantly decreased FW with 41% ($P < 0.05$; Figure 2f).

A significant ($P < 0.001$) interaction between plant growing medium and BCI was observed. Indeed, DOE analysis revealed a significant ($P < 0.05$) interaction effect between BCI S3 and the type of other organics (Figure S2). In the absence of S3, FW of lettuce grown in plant growing media containing coir pith (42.15 g) was higher than plant growing media with wood fiber (25.37 g). When treated with S3, lettuce FW increased and the difference in FW between the OO raw materials was no longer visible (coir pith: 53.42 g and wood fiber: 52.72 g). Treatment with BCI S3 negates the advantage of using coir pith over wood fiber.

Design of experiments analysis showed significant differences in FW between plant growing media, following similar trends in each BCI treatment (Figure 2). Use of coir pith increased ($P < 0.05$ in S1, S2, S3, and S5) FW compared to the use of wood fiber (+37% averaged over S1–5). Plant growing media containing green waste compost showed significantly ($P < 0.01$ in S1–5) higher FW compared to plant growing media comprising composted bark (+47% averaged over S1–5). Application of perlite instead of sand as inorganic material showed a positive trend ($P < 0.05$ in S3 and S5) in FW (+20% averaged over S1–5). The type of peat (black peat or white peat) did not significantly affect FW. Increasing the dose of Arabic gum significantly ($P < 0.05$ in S1–5) lowered FW (−35% averaged over S1–5). For the majority of the BCI treatments (S1, S2, S3, and S5) the use of coir pith, green waste compost, and a low dose of Arabic gum in the plant growing medium was required to reach maximal FW (Table S4). Additionally, the use of perlite was needed in BCI treatments S3 and S5.

A significant ($P < 0.05$ in S3 and S5) interaction effect occurred between the type of other organics and the dose of Arabic gum (Figures S2 and S3). Under a low dose of Arabic gum (1 kg·m^{-3}), the use of coir pith increased FW (59.64 g in S3) compared to wood fiber (44.22 g in S3). By increasing the amount of Arabic gum in the plant growing medium (5 kg.m^{-3}) FW dropped and the difference in FW between coir pith (35.92 g in S3) and wood fiber (33.87 g in S3) was lost.

3.2. Effect of Bacterial Community Inoculum and Plant Growing Medium on Lettuce Head Area

Lettuce head area varied significantly depending on BCI ($P < 0.001$), plant growing medium ($P < 0.001$), and BCI-plant growing medium interaction ($P < 0.001$). The BCI-plant growing medium combination S3-M8 (457.24 cm^2) exhibited the highest LHA, while S2-M3 (86.91 cm^2) showed the lowest LHA. Bacterial community inoculum treatment resulted in significant differences in LHA in each plant growing medium (Figure S4). Treatment with BCI S3 significantly ($P < 0.05$) increased LHA compared to C in the plant growing media M3, M5, M7, and M8. For example, the treatment of plant growing medium M3 with BCI S3 (305 cm^2) resulted in a more than 1.5-fold increase in LHA compared to the M3 control treatment (170 cm^2). The LHA of S3-M8 (457.24 cm^2), the highest LHA of all treatments, did not differ significantly from the LHA of the commercial plant growing medium (429.35 cm^2). The average increase in LHA under BCI S3 treatment was 29% ($P < 0.01$; Figure S5f). Response optimization showed that treatment with BCI S3 was necessary to obtain maximal LHA (Table S5). Also, S3 treatment was the largest contributor to LHA, compared to the plant growing medium raw material groups (Figure S6).

Bacterial community inoculum S2 treatment significantly ($P < 0.05$) decreased LHA compared to Fifurdecreased LHA with 51% compared to C (358 cm^2). On average, BCI S2 significantly ($P < 0.01$) decreased LHA with 33% (Figure S5f). Response optimization towards maximal LHA is reached after the removal of BCI S2 (Table S5). Additionally, BCI S2 was the largest contributor to change in LHA (absolute), compared to the plant growing medium raw material groups (Figure S7).

As also noted in lettuce FW analysis, treatment with the positive control PGPR unexpectedly decreased ($P < 0.05$) LHA compared to negative control C in M1, M3, M4, M6, M8, and M10. For example, compared to C (335 cm^2), LHA decreased by 43% when M10 was treated with PGPR (189 cm^2). On average, the application of PGPR showed a strong downward trend in LHA (-26%; Figure S5f).

Plant growing medium composition significantly affected LHA (Figure S5). The use of green waste compost resulted in significantly ($P < 0.01$ in S1–5) higher LHA compared to composted bark ($+35\%$ averaged over S1–5). Application of coir pith over wood fiber showed a positive trend ($+16\%$ averaged over S1–5) but was only significant ($P < 0.01$) under BCI S1 treatment ($+28\%$ under S1). The type of peat and inorganic material did not significantly affect LHA, though utilization of perlite resulted in a positive shift in LHA compared to sand ($+13\%$ averaged over S1–5). A high dose of Arabic gum significantly ($P < 0.05$ in S1–5) lowered LHA (-22% averaged over S1–5). For all BCI treatments (S1–5) the use of green waste compost and a low dose of Arabic gum in the plant growing medium were required to reach maximal LHA (Table S5). Also, the use of coir pith was needed under BCI treatment S1.

The treatment with BCI S1 showed a significant ($P < 0.05$) interaction effect between the type of other organics and composted materials (Figure S8). When plant growing media contained green waste compost, the application of coir pith increased LHA (382.68 cm^2) compared to wood fiber (265.09 cm^2). Changing the type of compost in the plant growing media to composted bark resulted in a decline in LHA, and the difference in LHA between coir pith (243.11 cm^2) and wood fiber (223.99 cm^2) vanished.

3.3. Effect of Bacterial Community Inoculum and Plant Growing Medium on Root Fresh Weight

Treatment of the plant growing media with BCI S3 significantly ($P < 0.05$) increased RW ($+53\%$), while both S2 and S4 significantly ($P < 0.05$) decreased RW (-53% and -18% respectively; Figure S9). Indeed, optimization of RW response towards maximum showed that the application of BCI S3 maximizes RW, while BCI S2 and S4 have to be removed to reach maximum RW (Table S6). Both BCI S2 and S3 treatment were the largest contributors to RW, compared to the plant growing medium raw material groups (Figures S10 and S11). The application of the positive control PGPR biostimulant resulted in a strong downward trend in RW (−49%; Figure S9).

Contrary to FW and LHA, the type of peat significantly ($P < 0.05$ in S1 and S4) affected RW (Figure S9). Application of white peat increased lettuce RW compared to black peat (+41% averaged over S1–5). For the remaining plant growing medium raw material groups, similar effects on RW were observed compared to FW and LHA (see Sections 3.1 and 3.2). However, the discerned trends were only marginally significant under certain BCI treatments. The interaction between the type of other organics and composted materials significantly ($P < 0.05$) affected RW under BCI S1 and S4 treatment (Figures S12 and S13). This interaction effect was also detected in the LHA DOE analysis (see Section 3.2). Bacterial community inoculum treatment S4 showed significant ($P < 0.001$) interaction with several plant growing medium raw material groups (peat, other organics, and inorganic materials): when the plant growing media were inoculated with S4, the observed differences in RW between the raw material group levels vanished.

3.4. Effect of Bacterial Community Inoculum and Plant Growing Medium on Shoot Dry Weight

Shoot dry weight was significantly affected by BCI ($P < 0.001$), plant growing medium ($P < 0.01$), and their interaction ($P < 0.001$). Shoot dry weight varied from 4.25% DW (PGPR-M10) to 7.39% DW (S1-M7). Figure 3 shows the effect of BCI treatment on lettuce DW in each plant growing medium. Compared to C, DW rose significantly ($P < 0.05$) after treatment with BCI S1 in several plant growing media (M1, M2, M4, M7, M8, and M10). For example, the treatment of plant growing media M1 and M7 with BCI S1 (S1-M1: 7.05% DW; S1-M7: 7.39% DW) resulted in a 1.3-fold increase in DW compared to C (C-M1: 5.48% DW; C-M7: 5.80% DW). Only S1-M7 (7.39% DW) and S1-M10 (7.23% DW) showed significantly ($P < 0.05$) higher DW compared to the commercial plant growing medium (6.35% DW). On average, treatment of the plant growing media with BCI S1 significantly ($P < 0.05$) increased DW (+15%; Figure S14) and BCI S1 treatment was required to optimize DW response towards maximum (Table S7). Moreover, BCI S1 treatment was the largest contributor to DW, compared to the plant growing medium raw material groups (Figure S15).

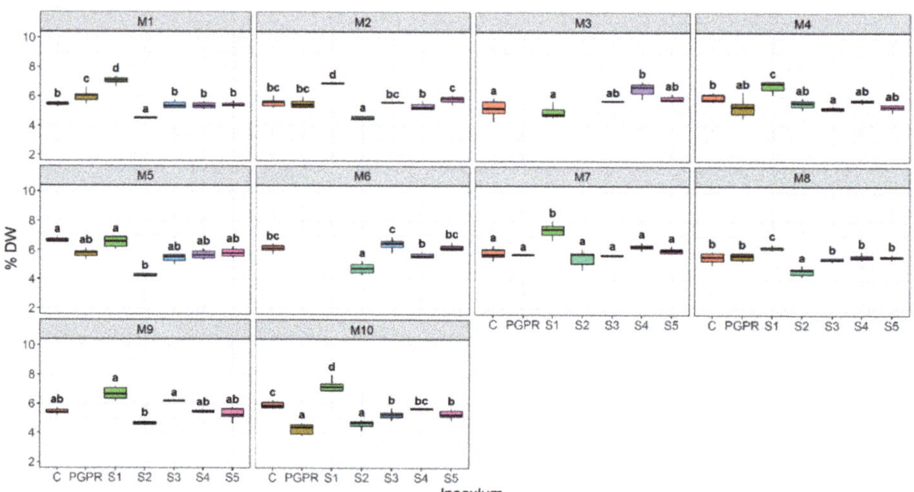

Figure 3. Boxplot of shoot dry weight (%DW) grouped per plant growing medium. Letters show comparison of BCI means per plant growing medium at the 95% confidence level. S indicates the bacterial community inoculum, M indicates the plant growing medium, C indicates the negative control treatment without addition of inoculum, and PGPR indicates the positive control treatment with a *Bacillus* sp. inoculum. Number of plants ≥ 3.

Bacterial community inoculum S2 significantly ($P < 0.05$) lowered DW compared to C in M1, M2, M5, M6, M8, and M10 (Figure 3). For instance, treating M2 with S2 (4.54% DW) decreased DW with 18.5% compared to C (5.57% DW). The average decrease in lettuce DW caused by BCI S2 treatment was 16% ($P < 0.01$) and S2 treatment was the largest contributor to DW compared to the plant growing medium raw material groups (Figure S16). Application of the positive control PGPR biostimulant resulted in a significant ($P < 0.01$) decline in DW (-5.3% on average; Figure S14). No significant effects of the plant growing medium raw material groups on lettuce DW were observed (Figure S14).

3.5. Effect of Bacterial Community Inoculum and Plant Growing Medium on Total Phenolic Content

Bacterial community inoculum treatment ($P < 0.001$), plant growing medium ($P < 0.001$), and their interaction ($P < 0.001$) impacted the TPC of lettuce. Total phenolic content levels were located between 91.50 mg GAE/100 g FW (S1-M2) and 12.73 mg GAE/100 g FW (S5-M1). Significant changes in TPC were detected between BCIs in each plant growing medium (Figure S17). Bacterial community inoculum S1 significantly ($P < 0.05$) increased TPC, compared to C, in several plant growing media (M1, M3, M5, and M10). For example, treating M1 with S1 (81.73 mg GAE/100 g FW) increased TPC with 210% compared to C-M1 (26.32 mg GAE/100 g FW). Compared to the commercial plant growing medium (43.13 mg GAE/100 g FW), TPC of lettuce grown in M1, M2, M5, and M7 was significantly ($P < 0.05$) higher when inoculated with S1. On average, BCI treatment S1 ($P < 0.001$) and S4 ($P < 0.01$) significantly increased TPC (+65% and +26% respectively), while BCI S5 significantly ($P < 0.05$) decreased TPC (-15%) (Figure S18). Response optimization of TPC towards maximum required the addition of S1 and S4 (Table S8). Furthermore, BCI S1 and S4 treatment were the largest contributors to TPC, compared to the plant growing medium raw material groups (Figures S19 and S20).

Design of experiments analysis (Figure S18) showed a significant effect ($P < 0.05$) of the OO raw material group on lettuce TPC under BCI S1, S3, and S5: the use of wood fiber increased TPC compared to coir pith (+26.5% averaged over S1–5). Remarkably, between BCI S1 and S5, an opposite effect of the CM raw material group on TPC was observed. Application of green waste compost over composted bark showed a negative trend in TPC (-35%) under S1 treatment, while TPC increased (+34.5%) under S5. These differences in TPC are caused by a significant ($P < 0.01$) interaction effect that occurred between the type of CM and inoculation with BCI S1 or S5 (Figures S21 and S22). In the C treatment, we observed no difference in TPC between lettuce grown in plant growing media containing either composted bark (33.88 mg GAE/100 g FW) or green waste compost (32.62 mg GAE/100 g FW). Treating BCI S1 to plant growing media containing composted bark resulted in a sharp increase in TPC (73.19 mg GAE/100 g FW), while lettuce TPC (36.65 mg GAE/100 g FW) in green waste compost plant growing media did not differ from C. Contrary to this, BCI S5 treatment of plant growing media containing composted bark resulted in a decrease in TPC (18.65 mg GAE/100 g FW), while TPC (38.04 mg GAE/100 g FW) in green waste compost plant growing media did not differ from C.

The interaction between BCI S4 and the OO raw materials group showed a significant effect ($P < 0.05$) on TPC (Figure S23). In the C treatment, TPC of lettuce grown in plant growing media containing wood fiber (39.16 mg GAE/100 g FW) was higher than in coir pith plant growing media (27.37 mg GAE/100 g FW). Inoculation with S4 increased lettuce TPC in coir pith plant growing media (43.19 mg GAE/100 g FW) but did not affect wood fiber plant growing media (40.47 mg GAE/100 g FW), whereby the difference in lettuce TPC between coir pith and wood fiber plant growing media was nullified.

3.6. Effect of Bacterial Community Inoculum and Plant Growing Medium on NO_3-Content

NO_3-content was significantly impacted by BCI source ($P < 0.001$), plant growing medium ($P < 0.001$), and BCI-plant growing medium interaction ($P < 0.001$). NO_3-content of all samples was well below the EU regulation limit (4000 mg/kg FW; EU 1258/2011), varying from 213 mg/kg FW (S1-M3) to 1952 mg/kg FW (S3-M8). Significant differences in lettuce NO_3-content between BCIs are shown in Figure S24. Bacterial community inoculum S3 treatment significantly ($P < 0.05$) increased

NO_3-content in M4, M5, M8, and M10 compared to C. For instance, NO_3-content of S3-M5 (1567 mg/kg FW) was close to 5-fold higher than C-M5 (322 mg/kg FW). On average, S3 significantly ($P < 0.05$) raised NO_3-content (+53%; Figure S25). Compared to C, NO_3-content significantly ($P < 0.05$) decreased in multiple plant growing media when treated with BCI S1 (M1, M3, M7, M8, and M10). For example, treating M1 with S1 (253 mg/kg FW) decreased NO_3-content with 84% compared to C-M1 (1626 mg/kg FW). Treatment of the plant growing media with BCI S1 significantly ($P < 0.01$) lowered NO_3-content with 67% on average (Figure S25). Both BCI S1 and S3 treatment were the largest contributors to NO_3-content, compared to the plant growing medium raw material groups (Figures S26 and S27). We observed no BCI-plant growing medium combinations with a significantly lower NO_3-content than the commercial plant growing medium (644 mg/kg FW). Contrary, multiple treatments (C-M1, C-M7, PGPR-M2, PGPR-M4, S2-M4, S2-M8, S3-M1, S3-M2, S3-M4, S3-M5, S3-M7, S3-M8, S3-M10, S4-M8, and S4-M10) showed significantly ($P < 0.05$) higher NO_3-content than the commercial plant growing medium.

Design of experiments analysis revealed that the plant growing medium raw material groups had no significant effect on NO_3-content (Figure S25). Treatment with BCI S5 showed a significant ($P < 0.05$) interaction effect between the type of other organics and the Arabic gum dose (Figure S28). NO_3-content of plant growing media containing coir pith was higher than wood fiber plant growing media under a low dose of Arabic gum. Contrary, an opposite shift in NO_3-content was observed under a high dose of Arabic gum. Application of BCI S1 was required to minimize NO_3-content, while BCI S3 should not be applied when minimizing NO_3-content (Table S9).

3.7. Effect of Bacterial Community Inoculum and Plant Growing Medium on Leaf Pigments

Chlorophyll a+b was significantly affected by BCI ($P < 0.001$), plant growing medium ($P < 0.01$), and their interaction ($P < 0.001$). Chlorophyll a + b varied from 11.65 mg/100 g FW (S5-M8) to 22.67 mg/100 g FW (S5-M7). Multiple treatments (C-M5, PGPR-M3, S1-M6, S2-M2, S2-M4, S2-M6, S2-M8, S2-M9, S2-M10, S5-M8, and S5-M10) showed significantly ($P < 0.05$) lower Chl_{a+b} levels than the commercial plant growing medium (21.48 mg/100 g FW). The effect of BCI treatment on Chl_{a+b} in each plant growing medium is shown in Figure S29. Overall, no clear trends in Chl_{a+b} levels, caused by BCI treatment or plant growing medium composition, were observed. However, DOE analysis revealed that BCI S2 treatment significantly ($P < 0.05$) decreased (−12%) Chl_{a+b} levels compared to C (Figure S30). Bacterial community inoculum S2 showed a significant ($P < 0.05$) interaction with the type of composted materials, where no difference in Chl_{a+b} levels was observed between BCI S2 and C in the composted bark plant growing media. However, BCI S2 treatment strongly decreased Chl_{a+b} levels in the green waste compost plant growing media compared to C (Figure S31). Contrary, BCI S4 treatment did not affect Chl_{a+b} levels in the green waste compost plant growing media compared to C. Instead, BCI S4 treatment increased Chl_{a+b} levels compared to C in the composted bark plant growing media (Figure S32). Indeed, response optimization towards maximal Chl_{a+b} showed that the combination of green waste compost with no BCI S2 application and BCI S4 application in combination with composted bark was optimal (Table S10).

When comparing all treatments, we observed that BCI treatment ($P < 0.001$), plant growing medium composition ($P < 0.001$), and their interaction ($P < 0.001$) significantly affected lettuce carotenoid content. Carotenoid levels were located between 3.12 mg/100 g FW (S2-M4) and 4.16 mg/100 g FW (S5-M7). Carotenoid content of lettuce grown in the commercial plant growing medium (4.11 mg/100 g FW) was significantly ($P < 0.05$) higher than of lettuce from S1-M8, S2-M2, S2-M4, S2-M6, S2-M8, S2-M10, S3-M8, and S3-M9. When examining the effect of BCI treatment grouped per plant growing medium (Figure S33), no clear shifts in carotenoid levels can be distinguished. Also, DOE analysis did not show any significant effects of the plant growing medium raw material groups on carotenoid content (Figure S34). However, it was revealed that BCI S2 significantly ($P < 0.05$) decreased carotenoid content compared to C (−6%). Additionally, BCI S2 treatment was the largest contributor to lettuce carotenoid content, compared to the plant growing medium raw material groups (Figure S35). A significant ($P < 0.05$)

interaction effect between composted materials and inorganic materials was observed under BCI S3 treatment (Figure S36a). When using perlite as inorganic material, carotenoid content was not affected by the type of compost. However, carotenoid content decreased when using sand in combination with green waste compost compared to composted bark. Bacterial community inoculum S3 directly interacted ($P < 0.05$) with the type of inorganic material (Figure S36b). The carotenoid content of the plant growing media containing sand was not affected by BCI S3 treatment, while BCI S3 treatment decreased carotenoid content in the plant growing media containing perlite. When treating with BCI S4, carotenoid content of the composted bark plant growing media was higher than the green waste compost plant growing media. This difference in carotenoid content was not present under C treatment (Figure S37). The type of other organics did not affect carotenoid content under a low dose of Arabic gum nor under C treatment. However, a high dose of Arabic gum or inoculation with BCI S5 increased carotenoid content in the plant growing media containing wood fiber (Figure S38a,b). No difference in carotenoid content was observed between C and BCI S5 treatment in the composted bark plant growing media. But, BCI S5 increased carotenoid levels when using green waste compost (Figure S38c). Maximizing carotenoid content depended on the BCI treatment (Table S11). Bacterial community inocula S4 and S5 were required to reach maximal carotenoid levels, while maximal carotenoid levels cannot be reached when treated with BCI S2 or S3.

3.8. Principal Component Analysis

The first two components of the PCA analysis explained 65% of the variance in the lettuce dataset (PC 1: 37.8; PC 2: 27.2%) (Figure 4). Quality of representation ($\cos^2$) values of the plant performance parameters showed that FW (96%), LHA (93%), RW (88%), and DW (71%) are well represented in PC 1 and PC 2, while the representation of NO_3 (55%), TPC (39%), Chl_{a+b} (34%), and carotenoids (44%) is low (Figure 4b). Correlation analysis of the yield and quality parameters (Table 3) demonstrated that FW, LHA, RW, and NO_3-content were significantly positively correlated to PC 1, while TPC, Chl_{a+b}, and carotenoids were significantly negatively correlated to PC 1. Correlation analysis on PC 2 revealed that LHA, RW, DW, TPC, Chl_{a+b}, and carotenoids correlated positively and NO_3-content was negatively correlated. Hereby, we observed grouping of the yield parameters (FW, LHA, and RW) along the positive PC 1 and PC 2 axis, while the quality parameters DW, TPC, Chl_{a+b}, and carotenoids were clustered towards the negative PC 1 axis and the positive PC 2 axis. NO_3-content was separated from the other yield and quality parameters along the positive PC 1 and negative PC 2 axis (Figure 4a). Shoot fresh weight (18.5%), LHA (17.8%), RW (16.8%), and DW (13.6%) were the dominant variables, contributing the most to PC 1:2 (Figure 4d). The PC 1:2 contribution values of NO_3 (10.7%), TPC (7.5%), Chl_{a+b} (6.5%), and carotenoids (8.5%) remained below the expected average contribution.

Table 3. Dimension Description of the Lettuce Yield and Quality Variables to PC 1 and PC 2 at the 95% Confidence Level.

	PC 1 (37.8%)		PC 2 (27.2%)	
	Correlation	*P* Value	Correlation	*P* Value
FW	0.961	1.71×10^{-37}	/	n.s.
LHA	0.848	2.62×10^{-19}	0.457	1.13×10^{-4}
RW	0.731	3.15×10^{-12}	0.585	2.53×10^{-7}
DW	/	n.s.	0.806	3.41×10^{-16}
TPC	−0.390	1.21×10^{-3}	0.491	2.90×10^{-5}
NO_3	0.615	3.96×10^{-8}	−0.420	4.46×10^{-4}
Chl_{a+b}	−0.347	4.28×10^{-3}	0.466	8.15×10^{-5}
Carotenoids	−0.373	2.07×10^{-3}	0.550	1.68×10^{-6}

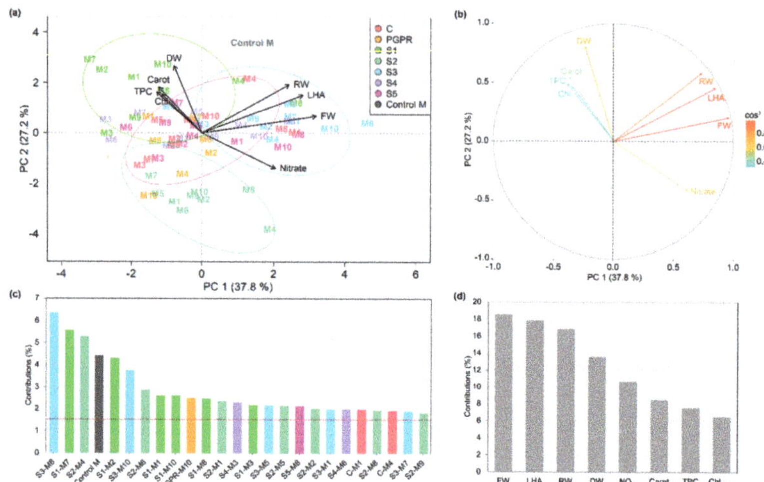

Figure 4. Principal component analysis (PCA) of the lettuce yield and quality variables under different BCI-plant growing medium treatments. (**a**) PCA biplot of individual samples to PC 1 and PC 2. Symbols indicate the type of plant growing medium (M1–10 and control M, the commercial plant growing medium) and colors indicate BCI treatment (S1–5, negative control C, and positive control PGPR). Ellipses denote 95% confidence interval of C, S1, S2, and S3. The plant performance parameters are shoot fresh weight (FW), lettuce head area (LHA), root fresh weight (RW), shoot dry weight (DW), total phenolic content (TPC), Nitrate content, chlorophyll a+b (Chl), and carotenoids (Carot); (**b**) Quality of representation (cos^2) correlation circle of variables to PC 1 and PC 2. The color gradient indicates the quality of representation of the variables; (**c**) Contribution plot of the top 25 samples to PC 1 and PC 2. Colors are the same as in a. The dashed line indicates the expected average contribution if the contribution of the samples were uniform; (**d**) Contribution plot of variables to PC 1 and PC 2. The dashed line indicates the expected average contribution if the contribution of the variables were uniform.

Principal component analysis showed grouping of the BCI-plant growing medium samples depending on BCI treatment (Figure 4a). Plant growing media treated with BCI S3 were separated from the C treatment towards FW, LHA, RW, and NO_3. Similarly, separation of BCI treatment S1 was observed towards DW, TPC, Chl$_{a+b}$, and carotenoids. Bacterial community inoculum S2 treated plant growing media clustered in the opposite direction of the plant performance parameters. We did not observe any clear grouping of BCI-plant growing medium samples based on plant growing medium type. The dominant treatments were S3-M8, S1-M7, S2-M4, the commercial plant growing medium, S1-M2, and S3-M10, each contributing more than 3% to PC 1:2 (Figure 4c).

4. Discussion

Reported evidence shows that plant growing media properties can enhance the beneficial impact of specific microbes on plant performance and stress resistance [41,42,46]. However, the role of plant growing medium composition and its interaction with rhizosphere bacterial communities in successful PGPR amendment and plant performance in soilless cultivation systems is not well understood. The presented study shows that microbe-plant growing medium interactions are important during the young plant stage for plant growth-promoting responses.

4.1. Plant Growing Medium Constituents Have Differing Effects on Lettuce Performance

The five plant growing media raw material groups peat, other organics, composted materials, inorganic materials, and Arabic gum had varied effects on the tested plant performance parameters. First, changing black peat to white peat significantly increased RW (Figure S9). Mathers et al. [70] reviewed that proper plant growing medium aeration is a vital physical characteristic influencing root growth. We observed that the air volume of the white peat growing media varied from 19.33% *v/v* (M5) to 26.33% *v/v* (M2), and for the black peat growing media from 13% *v/v* (M3) to 17% *v/v* (M7) (Table S3). Brückner [71] also reported higher air volume in white peat (24% *v/v*) compared to black peat (17% *v/v*). Thus, the positive impact of white peat on air volume improved the rooting of lettuce. Although white peat improved root weight compared to black peat, this advantage did not result in increased FW. This may be caused by the fact that after transplantation, for both white and black peat plant growing media, the roots grew out of the plant growing medium into the nutrient solution, having direct access to abundant nutrients. Since PFALs require high energy input, the production of non-salable plant parts must be minimized to reduce energy consumption [72]. The black peat growing media reduced RW of lettuce without affecting shoot FW, compared to the white peat growing media. So, the use of black peat blended with alternative materials as a plant growing medium can help minimize PFAL energy consumption through reduced lettuce root mass production. Alternatively, the use of white peat combined with alternative materials may be more advantageous for crops where the root system is the prime salable plant part.

Second, the use of perlite increased FW compared to sand, with LHA and RW showing similar trends (Figure 2, Figures S5 and S9). Similar to what we observed in the peat raw material group, the increase in plant growth likely resulted from a higher air volume and water capacity of plant growing media amended with perlite, compared to sand. Perlite is commonly amended to plant growing media to increase the air-filled pore space and water-holding capacity [73]. Contrary, sand has a small water buffer and pore volume [74]. Brückner [71] observed air volumes in sand-peat growing media ranging from 14–18% and 24–27% in perlite-peat growing media. In a previous study, we reported a higher air volume in perlite mixtures (20.5% *v/v*) compared to sand mixtures (17.8% *v/v*). Moreover, the water-holding capacity of perlite mixtures (615 g.(100 g dry matter)$^{-1}$) was double of that from sand mixtures (269 g.(100 g dry matter)$^{-1}$) [62]. The current physical analysis also showed a higher air volume and water-holding capacity of the plant growing media amended with perlite (20.8% *v/v* and 604 g.(100 g dry matter)$^{-1}$ respectively) compared to sand amendment (17.4% *v/v* and 287 g.(100 g dry matter)$^{-1}$ respectively) (Table S3).

Third, the application of green waste compost significantly increased lettuce growth (FW, LHA, RW) compared to composted bark (Figure 2,Figures S5 and S9). Spiers and Fietje [75] reported that green waste compost EC (3.43 dS·m^{-1}) was higher than bark compost EC (0.10 dS·m^{-1}). The high amount of available K$^+$ in green waste compost was mainly responsible for the high EC, with amounts reported up to 916 ppm for green waste compost compared to 19 ppm for composted bark. Previously, we also observed that plant growing media amended with green waste compost have higher EC (149 µS·cm^{-1}) than plant growing media containing bark compost (60 µS·cm^{-1}), with K$^+$ levels of 228 mg·L^{-1} and 70 mg·L^{-1} respectively [62]. In the current study, we also observed higher EC values for green waste compost growing media, varying from 130 µS·cm^{-1} (M4) to 275 µS·cm^{-1} (M8), compared to composted bark growing media, varying from 51 µS·cm^{-1} (M5) to 207 µS·cm^{-1} (M3) (Table S3). These differences in EC values between green waste compost and bark compost growing media were related to the K$^+$-content, respectively varying from 255.8 mg·L^{-1} (M8) to 335.5 mg·L^{-1} (M9), and from 84.7 mg·L^{-1} (M5) to 122.6 mg·L^{-1} (M6). The increased availability of salts, and especially K$^+$, in the plant growing media amended with green waste compost proved to be advantageous for lettuce growth.

Fourth, using wood fiber over coir pith, in the other organics raw material group, decreased all plant growth parameters tested (FW, LHA, and RW) (Figure 2, Figures S5 and S9). This reduction in growth may be caused by N-immobilization, which is a known problem in wood fiber growing media [76]. To avoid N-immobilization it is necessary to apply fertilizer from the start of plant cultivation [77]. Contrary to the commercial plant growing medium, we did not apply starter fertilizer to the experimental plant growing media. Only after transplantation to the PFAL (2 weeks after sowing), plants were irrigated regularly with nutrient solution.

These examples highlight the strong variety in which different plant growing medium constituents affect the physicochemical properties of the plant growing medium and thus plant performance. Proper selection of plant growing medium raw materials is required to achieve the desired enhancement of specific plant performance parameters.

4.2. Microbe-Plant Growing Medium Interactions and the Bacterial Source Determine Plant Performance

Plant growth-promoting rhizobacteria technology is becoming increasingly popular. However, there is still much doubt about the effectiveness of microbial amendment [31]. Design of experiments analysis revealed that bacterial amendment was the main driver affecting plant performance. However, the effectiveness of bacterial amendment and the plant performance parameters affected depended on microbe-plant growing medium interactions and the bacterial source.

Statistical analysis showed a significant interaction between the BCI and plant growing medium-class variables for all the tested plant performance parameters. Both BCI S1 and S3 positively affected plant performance. But, the observed effects did not occur in each plant growing medium (Figures 1 and 3), suggesting the potential influence of plant growing medium composition on the effectiveness of BCI treatment. Vandecasteele et al. [78] also reported that successful microbial inoculation depended on the type of plant growing medium. Biocontrol fungi showed better colonization in defibrated pure miscanthus, reed straw and flax shives compared to peat since peat did not provide the necessary compounds for fungal growth. Also, DOE analysis revealed several interaction effects between BCI treatment and plant growing medium constituents. For example, lettuce TPC was not affected by the type of compost under control treatment. However, inoculating the plant growing media with BCI S1 raised the TPC of lettuce grown in composted bark growing media while the TPC levels observed in the green waste compost growing media were unchanged (Figure S21). The higher organic matter content of the bark compost growing media, compared to the green waste compost growing media, may have provided a specific source of nutrients for the bacterial community present in S1 [62]. Overall, plant growing media without the BCI amendment did not perform as well as the commercial peat-coir based growing medium. We did not add starter fertilizer to the experimental plant growing media, while NPK levels of the commercialized plant growing medium were much higher, which may have caused retardation in growth (Table S3). However, we did observe that the BCI S3 amendment improved plant growth and even outperformed the commercial plant growing medium when amending BCI S3 to M8 and M10. This proves that specific microbe-plant growing medium combinations can create a synergistic effect that can outperform commercialized plant growing media.

Our results suggest that specific microbe-plant growing medium interactions determine plant performance. Moreover, bacterial amendment resulted in different effects on plant performance depending on the bacterial source. The BCIs were collected at separate locations. Bacterial community inoculum S1, S2, and S5 were collected at three different open field organic farms, while S3 and S4 were collected at different greenhouse soilless farms. Differences in cultivation method, fertilizer management, soil type, and crop species among others may have affected the composition of the collected root-associated bacterial communities. For instance, organic systems show greater microbial community diversity and higher microbial activity than conventional systems [79]. Roesti [80] concluded that the bacterial community structure varied between high and low fertilization strategies. Pii et al. [81] detected different microbial communities in two bulk soils. The recruitment of microbes from the soil to the rhizosphere is host-specific [82]. Rhizobia—legume interactions are well-studied, and their symbiosis is so specific that certain rhizobial species only interact with a selection of legumes [83].

All these parameters shape the bacterial community of the collected BCI samples, resulting in different effects on plant performance when amended to lettuce. For example, PCA analysis showed a grouping of the S3-plant growing medium combinations towards increased plant growth (FW, LHA, and RW) and NO_3-content (Figure 4a). Design of experiments analysis confirmed this, showing a significant increase in lettuce FW, LHA, RW, and NO_3-content under BCI S3 treatment (Figure 2, Figures S5, S9 and S25). Plant growth-promoting rhizobacteria are known to improve plant growth by enhancing nitrate uptake [84]. Because BCI S3 treatment increased lettuce NO_3-content, we suspect that BCI S3 includes certain PGPRs that improve plant growth through better nutrient acquisition.

Contrary to BCI S3, we observed a separation of BCI treatment S1 towards DW, TPC, Chl_{a+b}, and carotenoids, and away from NO_3 in the PCA analysis (Figure 4a). Indeed, DOE analysis showed a significant increase in DW and TPC, and a significant decrease in NO_3-content (Figures S14, S18 and S25). Plants are known to produce more phenolics under N-deficient conditions [85]. Also, there is evidence that PGPR treatment can induce systemic resistance against plant pathogens, and an elevated content of phenolics is suggested to play a role [86,87]. Bacterial community inoculum S1 may contain PGPRs that induce systemic resistance as suggested by the elevation in TPC.

Both BCI S1 and S3 positively affected plant performance. Meanwhile, BCI S2 treatment resulted in negative plant performance (LHA, RW, DW, Chl_{a+b}, and carotenoids) (Figure 4a), which may indicate that BCI S2 contains plant pathogenic bacteria. Surprisingly, the PGPR biostimulant (*Bacillus* sp.), which we applied as a positive control, also reduced plant performance. Design of experiments analysis even indicated a negative effect on FW and DW (Figure 2 and Figure S14). Research suggests that the amendment of several PGPRs could be more effective than individual species due to different mechanisms being used [38,88]. Moreover, PGPR application efficacy can depend on local environmental conditions and crop specificity [26]. Our results show that a complex bacterial community is a driver for successful bacterial amendment.

5. Conclusions

In summary, the reported results display the potential of bacterial enhancement of plant growing media to modulate plant performance in horticultural systems. Plant growing medium composition determines plant performance, and successful bacterial amendment can result in improved plant performance. We revealed that bacterial amendment was a key driver affecting plant performance. Not only does the effectiveness of bacterial amendment on plant performance depend on the bacterial source, but it also depends on the interaction with the plant growing medium. Further research will focus on determining how the rhizosphere bacterial community structure is associated with the observed microbe-plant growing medium interactions, and identifying the modes of action of the PGPRs affecting plant performance.

Supplementary Materials: The following are available online at http://www.mdpi.com/2073-4395/10/10/1456/s1, Figure S1: Pareto chart of the standardized effect (absolute) of the significant terms on shoot fresh weight under BCI S3 treatment, Figure S2: Interaction effects between substratum raw material groups on shoot fresh weight under BCI S3 treatment, Figure S3: Interaction effects between substratum raw material groups on shoot fresh weight under BCI S5 treatment, Figure S4: Boxplot of lettuce head area grouped per substratum, Figure S5: Main effects of substratum constituents on lettuce head area under different bacterial treatments, Figure S6: Pareto chart of the standardized effect (absolute) of the significant terms on lettuce head area under BCI S3 treatment, Figure S7: Pareto chart of the standardized effect (absolute) of the significant terms on lettuce head area under BCI S2 treatment, Figure S8: Interaction effect between other organics and composted materials on lettuce head area under BCI S1 treatment, Figure S9: Main effects of substratum constituents on root fresh weight under different bacterial treatments, Figure S10: Pareto chart of the standardized effect (absolute) of the significant terms on root fresh weight under BCI S2 treatment, Figure S11: Pareto chart of the standardized effect (absolute) of the significant terms on root fresh weight under BCI S3 treatment, Figure S12: Interaction effect between other organics and composted materials on root fresh weight under BCI S1 treatment, Figure S13: Interaction effects between substratum raw material groups on root fresh weight under BCI S4 treatment, Figure S14: Main effects of substratum constituents on shoot dry weight under different bacterial treatments, Figure S15: Pareto chart of the standardized effect (absolute) of the significant terms on shoot dry weight under BCI S1 treatment, Figure S16: Pareto chart of the standardized effect (absolute) of the significant terms on shoot dry weight under BCI S2 treatment, Figure S17: Boxplot of total phenolic content grouped per substratum, Figure S18: Main effects of substratum constituents on total phenolic content under different bacterial treatments, Figure S19: Pareto chart of the standardized effect (absolute) of the significant terms on total phenolic content under BCI S1 treatment, Figure S20: Pareto chart of the standardized effect (absolute) of the significant terms on total phenolic content (TPC) under BCI S4 treatment, Figure S21: Interaction effects between substratum raw material groups on total phenolic content (TPC; mg GAE/100 g FW) under BCI S1 treatment, Figure S22: Interaction effects between substratum raw material groups on total phenolic content under BCI S5 treatment, Figure S23: Interaction effect between other organics and BCI on total phenolic content under BCI S4 treatment, Figure S24: Boxplot of nitrate content grouped per substratum, Figure S25: Main effects of substratum constituents on nitrate content under different bacterial treatments, Figure S26: Pareto chart of the standardized effect (absolute) of the significant terms on NO_3-content under BCI S1 treatment, Figure S27: Pareto chart of the standardized effect (absolute) of the significant terms on NO_3-content under BCI S3 treatment, Figure S28: Interaction effect between other organics and Arabic gum on NO_3-content under BCI S5 treatment, Figure S29: Boxplot of chlorophyll a+b grouped per substratum, Figure S30: Main effects of substratum constituents on chlorophyll a+b content under different bacterial treatments, Figure S31: Interaction effect between composted materials and BCI on chlorophyll a+b content under BCI S2 treatment, Figure S32: Interaction effect between composted materials and BCI on chlorophyll a+b content under BCI S4 treatment, Figure S33: Boxplot of carotenoid content grouped per substratum, Figure S34: Main effects of substratum constituents on carotenoid content (mg/100 g FW) under different bacterial treatments, Figure S35: Pareto chart of the standardized effect (absolute) of the significant terms on carotenoid content under BCI S2 treatment, Figure S36: Interaction effects between substratum raw material groups on carotenoid content under BCI S3 treatment, Figure S37: Interaction effect between composted materials and BCI on carotenoid content under BCI S4 treatment, Figure S38: Interaction effects between substratum raw material groups on carotenoid content under BCI S5 treatment, Table S1: Control factors and level settings for substratum optimization, Table S2: The 2^{5-2}_{III} fractional factorial design, Table S3: Physicochemical properties of the experimental substrata and the commercial substratum, Table S4: Shoot fresh weight response optimization under each BCI treatment, Table S5: Lettuce head area response optimization under each BCI treatment, Table S6: Root fresh weight response optimization under each BCI treatment, Table S7: Shoot dry weight response optimization under each BCI treatment, Table S8: Total phenolic content response optimization under each BCI treatment, Table S9: NO_3-content response optimization under each BCI treatment, Table S10: Chlorophyll a+b content response optimization under each BCI treatment, Table S11: Carotenoid content response optimization under each BCI treatment.

Author Contributions: T.V.G.: investigation, formal analysis, data curation, visualization, writing—original draft; M.V.: supervision; N.A.: supervision; M.P.: conceptualization, project administration; M.-C.V.L.: resources, writing—review and editing; N.B.: conceptualization, funding acquisition, writing—review and editing; D.G.: conceptualization, funding acquisition, writing—review and editing, supervision. All authors have read and agreed to the published version of the manuscript.

Funding: This research was funded by the project grant VLAIO Baekeland mandate HBC.2017.0209.

Acknowledgments: The authors acknowledge Oscar Navarrete for supervision of the research and Hanne Denaeghel for assisting with the collection of the plant growing media raw materials. Seppe Top and Elsa Vancoppenolle for helping with the plant sample analysis.

Conflicts of Interest: The authors declare no conflict of interest.

Abbreviations

Controlled-environment agriculture	CEA
Plant factory with artificial lighting	PFAL
Plant growth-promoting rhizobacteria	PGPR
Tryptic soy broth	TSB
Bacterial community inoculum	BCI S1–5
Experimental plant growing media	M1–10
Peat	PT
Black peat	BP
White peat	WP
Other organics	OO
Coir pith	CP
Wood fiber	WF
Composted materials	CM
Composted bark	CB
Green waste compost	GC
Inorganic materials	IM

Perlite	P
Sand	S
Arabic gum	AG
Shoot fresh weight	FW
Lettuce head area	LHA
Root fresh weight	RW
Shoot dry weight	DW
Total phenolic content	TPC
Gallic acid equivalents	GAE
Chlorophyll a+b	Chl_{a+b}
Design of experiments	DOE
Principal component analysis	PCA

References

1. UN World Population Prospects, the 2019 Revision. Available online: https://esa.un.org/unpd/wpp/ (accessed on 8 June 2020).
2. FAO. *The State of the World's Land and Water Resources for Food and Agriculture (SOLAW)—Managing Systems at Risk*; Food and Agriculture Organization of the United Nations: Rome, Italy; Earthscan: London, UK, 2011; ISBN 9789251066140.
3. Kozai, T.; Niu, G. Plant factory as a resource-efficient closed plant production system. In *Plant Factory*; Elsevier: Amsterdam, The Netherlands, 2020; pp. 93–115. ISBN 9780128166918.
4. Kozai, T.; Niu, G. Role of the plant factory with artificial lighting (PFAL) in urban areas. In *Plant Factory*; Elsevier: Amsterdam, The Netherlands, 2020; pp. 7–34. ISBN 9780128166918.
5. Goto, E. Plant production in a closed plant factory with artificial lighting. In Proceedings of the VII International Symposium on Light in Horticultural Systems, Wageningen, The Netherlands, 15–18 October 2012; Volume 956, pp. 37–49.
6. Barrett, G.E.; Alexander, P.D.; Robinson, J.S.; Bragg, N.C. Achieving environmentally sustainable growing media for soilless plant cultivation systems—A review. *Sci. Hortic.* **2016**, *212*, 220–234. [CrossRef]
7. Schmilewski, G. The role of peat in assuring the quality of growing media. *Mires Peat* **2008**, *3*, 1–8.
8. Carlile, W.R.; Cattivello, C.; Zaccheo, P. Organic growing media: Constituents and properties. *Vadose Zone J.* **2015**, *14*, 1–13. [CrossRef]
9. Alexander, P.D.; Bragg, N.C.; Meade, R.; Padelopoulos, G.; Watts, O. Peat in horticulture and conservation: The UK response to a changing world. *Mires Peat* **2008**, *3*, 1–10.
10. Cleary, J.; Roulet, N.T.; Moore, T.R. Greenhouse gas emissions from Canadian peat extraction, 1990–2000: A life-cycle analysis. *AMBIO J. Hum. Environ.* **2005**, *34*, 456–461. [CrossRef]
11. Carlile, B.; Coules, A. Towards sustainability in growing media. In Proceedings of the International Symposium on Growing Media, Composting and Substrate Analysis, Barcelona, Spain, 17–21 October 2011; Volume 1013, pp. 341–349.
12. Kern, J.; Tammeorg, P.; Shanskiy, M.; Sakrabani, R.; Knicker, H.; Kammann, C.; Tuhkanen, E.M.; Smidt, G.; Prasad, M.; Tiilikkala, K.; et al. Synergistic use of peat and charred material in growing media—An option to reduce the pressure on peatlands? *J. Environ. Eng. Landsc. Manag.* **2017**, *25*, 160–174. [CrossRef]
13. Chrysargyris, A.; Prasad, M.; Kavanagh, A.; Tzortzakis, N. Biochar type, ratio, and nutrient levels in growing media affects seedling production and plant performance. *Agronomy* **2020**, *10*, 1421. [CrossRef]
14. Tüzel, Y.; Ekinci, K.; Öztekin, G.B.; Erdal, İ.; Varol, N.; Merken, Ö. Utilization of olive oil processing waste composts in organic tomato seedling production. *Agronomy* **2020**, *10*, 797. [CrossRef]
15. Gohardoust, M.R.; Bar-Tal, A.; Effati, M.; Tuller, M. Characterization of Physicochemical and hydraulic properties of organic and mineral soilless culture substrates and mixtures. *Agronomy* **2020**, *10*, 1403. [CrossRef]
16. Huang, L.; Gu, M.; Yu, P.; Zhou, C.; Liu, X. Biochar and vermicompost amendments affect substrate properties and plant growth of basil and tomato. *Agronomy* **2020**, *10*, 224. [CrossRef]
17. Blok, C. Growing media for food and quality of life in the period 2020–2050. *Acta Hortic.* **2020**. in press. Available online: https://peatlands.org/document/growing-media-blok-2018/ (accessed on 18 June 2020).
18. Fierer, N. Embracing the unknown: Disentangling the complexities of the soil microbiome. *Nat. Rev. Microbiol.* **2017**, *15*, 579–590. [CrossRef] [PubMed]

19. Delgado-Baquerizo, M.; Oliverio, A.M.; Brewer, T.E.; Benavent-González, A.; Eldridge, D.J.; Bardgett, R.D.; Maestre, F.T.; Singh, B.K.; Fierer, N. A global atlas of the dominant bacteria found in soil. *Science* **2018**, *359*, 320–325. [CrossRef] [PubMed]
20. Bulgarelli, D.; Schlaeppi, K.; Spaepen, S.; van Themaat, E.V.L.; Schulze-Lefert, P. Structure and functions of the bacterial microbiota of plants. *Annu. Rev. Plant. Biol.* **2013**, *64*, 807–838. [CrossRef]
21. Zhalnina, K.; Louie, K.B.; Hao, Z.; Mansoori, N.; Da Rocha, U.N.; Shi, S.; Cho, H.; Karaoz, U.; Loqué, D.; Bowen, B.P.; et al. Dynamic root exudate chemistry and microbial substrate preferences drive patterns in rhizosphere microbial community assembly. *Nat. Microbiol.* **2018**, *3*, 470–480. [CrossRef] [PubMed]
22. Bulgarelli, D.; Rott, M.; Schlaeppi, K.; Ver Loren van Themaat, E.; Ahmadinejad, N.; Assenza, F.; Rauf, P.; Huettel, B.; Reinhardt, R.; Schmelzer, E.; et al. Revealing structure and assembly cues for Arabidopsis root-inhabiting bacterial microbiota. *Nature* **2012**, *488*, 91–95. [CrossRef]
23. Edwards, J.; Johnson, C.; Santos-Medellín, C.; Lurie, E.; Podishetty, N.K.; Bhatnagar, S.; Eisen, J.A.; Sundaresan, V. Structure, variation, and assembly of the root-associated microbiomes of rice. *Proc. Natl. Acad. Sci. USA* **2015**, *112*, E911–E920. [CrossRef]
24. Venturi, V.; Keel, C. Signaling in the rhizosphere. *Trends Plant. Sci.* **2016**, *21*, 187–198. [CrossRef]
25. Zhou, D.; Huang, X.F.; Chaparro, J.M.; Badri, D.V.; Manter, D.K.; Vivanco, J.M.; Guo, J. Root and bacterial secretions regulate the interaction between plants and PGPR leading to distinct plant growth promotion effects. *Plant Soil* **2016**, *401*, 259–272. [CrossRef]
26. Rosier, A.; Medeiros, F.H.V.; Bais, H.P. Defining plant growth promoting rhizobacteria molecular and biochemical networks in beneficial plant-microbe interactions. *Plant Soil* **2018**, *428*, 35–55. [CrossRef]
27. Berg, G.; Rybakova, D.; Grube, M.; Köberl, M. The plant microbiome explored: Implications for experimental botany. *J. Exp. Bot.* **2016**, *67*, 995–1002. [CrossRef]
28. Compant, S.; Clément, C.; Sessitsch, A. Plant growth-promoting bacteria in the rhizo- and endosphere of plants: Their role, colonization, mechanisms involved and prospects for utilization. *Soil Biol. Biochem.* **2010**, *42*, 669–678. [CrossRef]
29. Lugtenberg, B.; Kamilova, F. Plant-growth-promoting rhizobacteria. *Annu. Rev. Microbiol.* **2009**, *63*, 541–556. [CrossRef]
30. Owen, D.; Williams, A.P.; Griffith, G.W.; Withers, P.J.A. Use of commercial bio-inoculants to increase agricultural production through improved phosphrous acquisition. *Appl. Soil Ecol.* **2015**, *86*, 41–54. [CrossRef]
31. Nuzzo, A.; Satpute, A.; Albrecht, U.; Strauss, S.L. Impact of soil microbial amendments on tomato rhizosphere microbiome and plant growth in field soil. *Microb. Ecol.* **2020**, *80*, 398–409. [CrossRef]
32. Compant, S.; Samad, A.; Faist, H.; Sessitsch, A. A review on the plant microbiome: Ecology, functions, and emerging trends in microbial application. *J. Adv. Res.* **2019**, *19*, 29–37. [CrossRef]
33. Backer, R.; Rokem, J.S.; Ilangumaran, G.; Lamont, J.; Praslickova, D.; Ricci, E.; Subramanian, S.; Smith, D.L. Plant growth-promoting rhizobacteria: Context, mechanisms of action, and roadmap to commercialization of biostimulants for sustainable agriculture. *Front. Plant. Sci.* **2018**, *871*, 1–17. [CrossRef]
34. Busby, P.E.; Soman, C.; Wagner, M.R.; Friesen, M.L.; Kremer, J.; Bennett, A.; Morsy, M.; Eisen, J.A.; Leach, J.E.; Dangl, J.L. Research priorities for harnessing plant microbiomes in sustainable agriculture. *PLoS Biol.* **2017**, *15*. [CrossRef] [PubMed]
35. Bona, E.; Cantamessa, S.; Massa, N.; Manassero, P.; Marsano, F.; Copetta, A.; Lingua, G.; D'Agostino, G.; Gamalero, E.; Berta, G. Arbuscular mycorrhizal fungi and plant growth-promoting pseudomonads improve yield, quality and nutritional value of tomato: A field study. *Mycorrhiza* **2017**, *27*, 1–11. [CrossRef]
36. Lingua, G.; Bona, E.; Manassero, P.; Marsano, F.; Todeschini, V.; Cantamessa, S.; Copetta, A.; D'Agostino, G.; Gamalero, E.; Berta, G. Arbuscular mycorrhizal fungi and plant growth-promoting pseudomonads increases anthocyanin concentration in strawberry fruits (*Fragaria* x *ananassa* var. Selva) in conditions of reduced fertilization. *Int. J. Mol. Sci.* **2013**, *14*, 16207–16225. [CrossRef]
37. Lee, S.; An, R.; Grewal, P.; Yu, Z.; Borherova, Z.; Lee, J. High-performing windowfarm hydroponic system: Transcriptomes of fresh produce and microbial communities in response to beneficial bacterial treatment. *Mol. Plant Microbe Interact.* **2016**, *29*, 965–976. [CrossRef]
38. Sundaramoorthy, S.; Raguchander, T.; Ragupathi, N.; Samiyappan, R. Combinatorial effect of endophytic and plant growth promoting rhizobacteria against wilt disease of *Capsicum annum* L. caused by *Fusarium solani*. *Biol. Control.* **2012**, *60*, 59–67. [CrossRef]

39. Yergeau, E.; Bell, T.H.; Champagne, J.; Maynard, C.; Tardif, S.; Tremblay, J.; Greer, C.W. Transplanting soil microbiomes leads to lasting effects on willow growth, but not on the rhizosphere microbiome. *Front. Microbiol.* **2015**, *6*, 1436. [CrossRef] [PubMed]

40. Sun, D.; Hale, L.; Crowley, D. Nutrient supplementation of pinewood biochar for use as a bacterial inoculum carrier. *Biol. Fertil. Soils* **2016**, *52*, 515–522. [CrossRef]

41. Nadeem, S.M.; Imran, M.; Naveed, M.; Khan, M.Y.; Ahmad, M.; Zahir, Z.A.; Crowley, D.E. Synergistic use of biochar, compost and plant growth-promoting rhizobacteria for enhancing cucumber growth under water deficit conditions. *J. Sci. Food Agric.* **2017**, *97*, 5139–5145. [CrossRef] [PubMed]

42. Grunert, O.; Hernandez-Sanabria, E.; Vilchez-Vargas, R.; Jauregui, R.; Pieper, D.H.; Perneel, M.; Van Labeke, M.C.; Reheul, D.; Boon, N. Mineral and organic growing media have distinct community structure, stability and functionality in soilless culture systems. *Sci. Rep.* **2016**, *6*, 18837. [CrossRef]

43. Grunert, O.; Hernandez-Sanabria, E.; Perneel, M.; Van Labeke, M.C.; Reheul, D.; Boon, N. High-throughput sequencing analysis provides a comprehensive insight into the complex bacterial relationships in horticultural growing substrates. *Acta Hortic.* **2017**, *1168*, 19–26. [CrossRef]

44. Castaño, R.; Borrero, C.; Avilés, M. Organic matter fractions by SP-MAS 13C NMR and microbial communities involved in the suppression of Fusarium wilt in organic growth media. *Biol. Control.* **2011**, *58*, 286–293. [CrossRef]

45. De Corato, U.; Viola, E.; Arcieri, G.; Valerio, V.; Zimbardi, F. Use of composted agro-energy co-products and agricultural residues against soil-borne pathogens in horticultural soil-less systems. *Sci. Hortic.* **2016**, *210*, 166–179. [CrossRef]

46. Cao, Y.; Tian, Y.; Gao, Y.; Li, J. Microbial diversity in compost is critical in suppressing plant fungal pathogen survival and enhancing cucumber seedling growth. *Compost Sci. Util.* **2018**, *26*, 189–200. [CrossRef]

47. Borrero, C.; Ordovás, J.; Trillas, M.I.; Avilés, M. Tomato Fusarium wilt suppressiveness. The relationship between the organic plant growth media and their microbial communities as characterised by Biolog®. *Soil Biol. Biochem.* **2006**, *38*, 1631–1637. [CrossRef]

48. Nieto, A.; Gascó, G.; Paz-Ferreiro, J.; Fernández, J.M.; Plaza, C.; Méndez, A. The effect of pruning waste and biochar addition on brown peat based growing media properties. *Sci. Hortic.* **2016**, *199*, 142–148. [CrossRef]

49. Sánchez-Monedero, M.A.; Cayuela, M.L.; Sánchez-García, M.; Vandecasteele, B.; D'Hose, T.; López, G.; Martínez-Gaitán, C.; Kuikman, P.J.; Sinicco, T.; Mondini, C. Agronomic evaluation of biochar, compost and biochar-blended compost across different cropping systems: Perspective from the European project FERTIPLUS. *Agronomy* **2019**, *9*, 225. [CrossRef]

50. Blok, C.; van der Salm, C.; Hofland-Zijlstra, J.; Streminska, M.; Eveleens, B.; Regelink, I.; Fryda, L.; Visser, R. Biochar for horticultural rooting media improvement: Evaluation of biochar from gasification and slow pyrolysis. *Agronomy* **2017**, *7*, 6. [CrossRef]

51. Méndez, A.; Paz-Ferreiro, J.; Gil, E.; Gascó, G. The effect of paper sludge and biochar addition on brown peat and coir based growing media properties. *Sci. Hortic.* **2015**, *193*, 225–230. [CrossRef]

52. Jaiswal, A.K.; Elad, Y.; Paudel, I.; Graber, E.R.; Cytryn, E.; Frenkel, O. Linking the belowground microbial composition, diversity and activity to soilborne disease suppression and growth promotion of tomato amended with biochar. *Sci. Rep.* **2017**, *7*, 44382. [CrossRef]

53. Kolton, M.; Graber, E.R.; Tsehansky, L.; Elad, Y.; Cytryn, E. Biochar-stimulated plant performance is strongly linked to microbial diversity and metabolic potential in the rhizosphere. *New Phytol.* **2017**, *213*, 1393–1404. [CrossRef]

54. Lee, S.; Lee, J. Beneficial bacteria and fungi in hydroponic systems: Types and characteristics of hydroponic food production methods. *Sci. Hortic.* **2015**, *195*, 206–215. [CrossRef]

55. Lee, S.; Ge, C.; Bohrerova, Z.; Grewal, P.S.; Lee, J. Enhancing plant productivity while suppressing biofilm growth in a windowfarm system using beneficial bacteria and ultraviolet irradiation. *Can. J. Microbiol.* **2015**, *61*, 457–466. [CrossRef]

56. Gül, A.; Kidoglu, F.; Tüzel, Y.; Tüzel, I.H. Effects of nutrition and *Bacillus amyloliquefaciens* on tomato (*Solanum lycopersicum* L.) growing in perlite. *Span. J. Agric. Res.* **2008**, *6*, 422–429.

57. García, J.A.L.; Probanza, A.; Ramos, B.; Palomino, M.; Mañero, F.J.G. Effect of inoculation of *Bacillus lichenformis* on tomato and pepper. *Agronomie* **2004**, *24*, 169–176. [CrossRef]

58. Vecstaudza, D.; Senkovs, M.; Nikolajeva, V.; Kasparinskis, R.; Muter, O. Wooden biochar as a carrier for endophytic isolates. *Rhizosphere* **2017**, *3*, 126–127. [CrossRef]

59. Saxena, J.; Rana, G.; Pandey, M. Impact of addition of biochar along with *Bacillus* sp. on growth and yield of French beans. *Sci. Hortic.* **2013**, *162*, 351–356. [CrossRef]
60. Barillot, C.D.C.; Sarde, C.O.; Bert, V.; Tarnaud, E.; Cochet, N. A standardized method for the sampling of rhizosphere and rhizoplan soil bacteria associated to a herbaceous root system. *Ann. Microbiol.* **2013**, *63*, 471–476. [CrossRef]
61. Van Nevel, S.; Koetzsch, S.; Weilenmann, H.U.; Boon, N.; Hammes, F. Routine bacterial analysis with automated flow cytometry. *J. Microbiol. Methods* **2013**, *94*, 73–76. [CrossRef]
62. Van Gerrewey, T.; Ameloot, N.; Navarrete, O.; Vandecruys, M.; Perneel, M.; Boon, N.; Geelen, D. Microbial activity in peat-reduced plant growing media: Identifying influential growing medium constituents and physicochemical properties using fractional factorial design of experiments. *J. Clean. Prod.* **2020**, *256*, 120323. [CrossRef]
63. Verdonck, O.; Gabriels, R. Reference method for the determination of physical and chemical properties of plant substrates. *Acta Hortic.* **1992**, 169–180. [CrossRef]
64. Gabriels, D.; Hartmann, R.; Verplancke, H.; Cornelis, W.; Verschoore, P. *Werkwijzen Voor Grondanalyses.*; University Gent: Gent, Belgium, 1998; p. 90.
65. Schindelin, J.; Rueden, C.T.; Hiner, M.C.; Eliceiri, K.W. The ImageJ ecosystem: An open platform for biomedical image analysis. *Mol. Reprod. Dev.* **2015**, *82*, 518–529. [CrossRef]
66. Liu, X.; Ardo, S.; Bunning, M.; Parry, J.; Zhou, K.; Stushnoff, C.; Stoniker, F.; Yu, L.; Kendall, P. Total phenolic content and DPPH radical scavenging activity of lettuce (*Lactuca sativa* L.) grown in Colorado. *LWT Food Sci. Technol.* 2007; *40*, 552–557. [CrossRef]
67. Cataldo, D.A.; Maroon, M.; Schrader, L.E.; Youngs, V.L. Rapid colorimetric determination of nitrate in plant tissue by nitration of salicylic acid. *Commun. Soil Sci. Plant. Anal.* **1975**, *6*, 71–80. [CrossRef]
68. Lichtenthaler, H.K.; Buschmann, C. Chlorophylls and carotenoids: Measurement and characterization by UV-VIS spectroscopy. *Curr. Protoc. Food Anal. Chem.* **2001**, *1*, F4.3.1–F4.3.8. [CrossRef]
69. Bingham, D.; Sitter, R.R. Fractional factorial split-plot designs for robust parameter experiments. *Technometrics* **2003**, *45*, 80–89. [CrossRef]
70. Mathers, H.M.; Lowe, S.B.; Scagel, C.; Struve, D.K.; Case, L.T. Abiotic factors influencing root growth of woody nursery plants in containers. *Horttechnology* **2007**, *17*, 151–162. [CrossRef]
71. Brückner, U. Physical properties of different potting media and substrate mixtures—Especially air-and water capacity. *Acta Hortic.* **1997**, *450*, 263–270.
72. Kozai, T.; Niu, G. *Challenges for the Next-Generation PFALs*; Elsevier Inc.: Amsterdam, The Netherlands, 2020; ISBN 9780128166918.
73. Evans, M.R.; Gachukia, M.M. Physical properties of sphagnum peat-based root substrates amended with perlite or parboiled fresh rice hulls. *Horttechnology* **2007**, *17*, 312–315. [CrossRef]
74. Verhagen, J.B.G.M.; Zevenhoven, M.A. Growing Media A. Available online: https://www.rhp.nl/en/home (accessed on 2 May 2018).
75. Spiers, T.; Fietje, G. Green waste compost as a component in soilless growing media. *Compos. Sci. Util.* **2000**, *8*, 19–23. [CrossRef]
76. Gruda, N. Sustainable peat alternative growing media. *Acta Hortic.* **2012**, *927*, 973–979. [CrossRef]
77. Gruda, N.; Schnitzler, W.H. Influence of wood fiber substrates and N application rates on the growth of tomato transplants. *Adv. Hortic. Sci.* **1999**, *13*, 20–24.
78. Vandecasteele, B.; Muylle, H.; De Windt, I.; Van Acker, J.; Ameloot, N.; Moreaux, K.; Coucke, P.; Debode, J. Plant fibers for renewable growing media: Potential of defibration, acidification or inoculation with biocontrol fungi to reduce the N drawdown and plant pathogens. *J. Clean. Prod.* **2018**, *203*, 1143–1154. [CrossRef]
79. Reeve, J.R.; Hoagland, L.A.; Villalba, J.J.; Carr, P.M.; Atucha, A.; Cambardella, C.; Davis, D.R.; Delate, K. Organic farming, soil health, and food quality: Considering possible links. In *Advances in Agronomy*; Elsevier Inc.: Amsterdam, The Netherlands, 2016; Volume 137, pp. 319–367. ISBN 9780128046920.
80. Roesti, D.; Gaur, R.; Johri, B.N.; Imfeld, G.; Sharma, S.; Kawaljeet, K.; Aragno, M. Plant growth stage, fertiliser management and bio-inoculation of arbuscular mycorrhizal fungi and plant growth promoting rhizobacteria affect the rhizobacterial community structure in rain-fed wheat fields. *Soil Biol. Biochem.* **2006**, *38*, 1111–1120. [CrossRef]

81. Pii, Y.; Borruso, L.; Brusetti, L.; Crecchio, C.; Cesco, S.; Mimmo, T. The interaction between iron nutrition, plant species and soil type shapes the rhizosphere microbiome. *Plant. Physiol. Biochem.* **2016**, *99*, 39–48. [CrossRef]

82. Raaijmakers, J.M.; Paulitz, T.C.; Steinberg, C.; Alabouvette, C.; Moënne-Loccoz, Y. The rhizosphere: A playground and battlefield for soilborne pathogens and beneficial microorganisms. *Plant. Soil* **2009**, *321*, 341–361. [CrossRef]

83. Wang, D.; Yang, S.; Tang, F.; Zhu, H. Symbiosis specificity in the legume—Rhizobial mutualism. *Cell. Microbiol.* **2012**, *14*, 334–342. [CrossRef] [PubMed]

84. Calvo, P.; Zebelo, S.; McNear, D.; Kloepper, J.; Fadamiro, H. Plant growth-promoting rhizobacteria induce changes in *Arabidopsis thaliana* gene expression of nitrate and ammonium uptake genes. *J. Plant. Interact.* **2019**, *14*, 224–231. [CrossRef]

85. Kováčik, J.; Bačkor, M. Changes of phenolic metabolism and oxidative status in nitrogen-deficient *Matricaria chamomilla* plants. *Plant. Soil* **2007**, *297*, 255–265. [CrossRef]

86. Muthukumar, A.; Eswaran, A.; Sangeetha, G. Induction of systemic resistance by mixtures of fungal and endophytic bacterial isolates against *Pythium aphanidermatum*. *Acta Physiol. Plant.* **2011**, *33*, 1933–1944. [CrossRef]

87. Thangavelu, R.; Palaniswami, A.; Doraiswamy, S.; Velazhahan, R. The effect *of Pseudomonas fluorescens* and *Fusarium oxysporum* f.sp. cubense on induction of defense enzymes and phenolics in banana. *Biol. Plant.* **2003**, *46*, 107–112. [CrossRef]

88. Domenech, J.; Reddy, M.S.; Kloepper, J.W.; Ramos, B.; Gutierrez-Mañero, J. Combined application of the biological product LS213 with *Bacillus*, *Pseudomonas* or *Chryseobacterium* for growth promotion and biological control of soil-borne diseases in pepper and tomato. *BioControl* **2006**, *51*, 245–258. [CrossRef]

 *agronomy*

Article

Plant Nutrient Availability and pH of Biochars and Their Fractions, with the Possible Use as a Component in a Growing Media

Munoo Prasad [1,2,*], Antonios Chrysargyris [2], Nicola McDaniel [3], Anna Kavanagh [3], Nazim S. Gruda [4] and Nikolaos Tzortzakis [2,*]

[1] Compost/AD Research & Advisory (IE, CY), Naas, W91 TYH3 Kildare, Ireland
[2] Department of Agricultural Sciences, Biotechnology and Food Science, Cyprus University of Technology, Limassol 3603, Cyprus; a.chrysargyris@cut.ac.cy
[3] Bord na Mona Research Centre, Main Street, Newbridge, W12 XR59 Kildare, Ireland; nicola.mcdaniel@bnm.ie (N.M.); annak62@outlook.ie (A.K.)
[4] Department of Horticultural Science, INRES-Institute of Crop Science and Resource Conservation, University of Bonn, 53121 Bonn, Germany; ngruda@uni-bonn.de
* Correspondence: munooprasad@yahoo.com (M.P.); nikolaos.tzortzakis@cut.ac.cy (N.T.); Tel.: +353-866012034 (M.P.); +357-25002280 (N.T.)

Received: 20 November 2019; Accepted: 17 December 2019; Published: 19 December 2019

Abstract: Biochar has the potential to be used as a growing media component, and therefore plays a role in reducing peat usage. It has unique properties apart from the ability to sequester carbon. Here we investigated the nutrient contents of four commercial biochars and their fractions. The biochars' feedstock was wood waste, except for one with paper fibres and husk. The fine or finer fractions in wood waste biochars contained higher levels of nutrients that were available to plants. The coarse fraction of the biochar derived from husk and paper fibre feedstock had a higher level of total N, P and K in contrast to the other three biochars. The pH of the finer fraction (pH of 9.08) was also higher compared with coarse fraction (pH of 8.71). It is important that when biochar a is used as a component of a peat based growing media, particle size information should be provided, as fractions from the same biochar can have different levels of total extractable nutrients and pH levels. If biochar is used to replace or reduce lime application rates of a peat-biochar mixtures, one must take into account the levels of total and extractable Ca and Mg levels, as these can vary. The variation of these elements was not only between biochars' feedstocks, even at similar pH-values, but within different fractions in the same biochar. We concluded that biochars should be characterized from the feedstock as well as from the particle size aspect, as it could have a profound effect on nutrient availability of Ca and Mg. This could lead to nutrient imbalances in cultivating plants on substrate mixtures. In addition to nutrient ratios, the suitable pH-level for a given grown species should be adjusted.

Keywords: peat replacement; particle size; calcium; magnesium; extractable nutrients

1. Introduction

Biochar is an organic carbon-rich solid by-product, which is gaining great interest in research for its utilization under the environmental and agricultural management [1–3]. Therefore, in addition to the common use as soil amendment material, biochar is being explored in terms of use for soil remediation [4–7], water filtration [8] and soilless substrates [9,10]. Recently there has been great interest in the use of biochar as a bioresource and growing media material [11–14]. Biochar and hydrothermal carbonization might play more important roles as constituents of growing media [15].

Biochar can be produced from several organic sources using pyrolysis under minimal oxygen supply, but also, for the use as growing media, from chunky timber waste; e.g., wood chips are suitable.

This has to be milled to particles sized generally less than 10 mm as a result of crushing, in order to be considered as a growing media component. A number of commercial producers supply biochars in different sizes. There are also limitations for the feedstocks used, as soft woody materials (e.g., from greenhouse crop waste), including stems, are not suitable due to excessive salinity levels [16]. In this case a hydrothermal carbonization process that requires only moderate temperatures and pressures is usually used [15].

A raft of publications has described biochar as a partial peat replacement over last three years [17–21], with several different organic materials playing an important role in decreasing the C footprint of the horticultural industry [15]. Peat is the principal material for container growing media in Europe, and peat production in Europe is more than 40 million m^3. Peat currently represents 77%–80% of the growing media annually used in the horticultural industry in Europe [22–24]. However, peat comes from peatland ecosystems, which are very important for carbon sequestration. Peatlands are most important carbon sinks and one of the most important effective eco-systems in the terrestrial biosphere. The carbon storage in peatlands in Europe is estimated to be approximately 43,000 million tons [24]. Peat has a low pH, around pH 4, and is generally almost devoid of plant nutrients. It has the advantage that due to its low pH and nutrient levels, both pH and other nutrients can be brought to pre-determined levels for crop growth [25]. Hence, when peat is used as a substrate, stored carbon is released, negatively affecting the environment and CO_2 balance [26]. For instance, Vaughn et al. [27] reported that biochar can substitute peat at levels lower than 15% (v/v), whereas higher rates derived unsatisfactory results, possible due to the high salinity and pH values, imbalance of nutrients and high C:N ratio.

Biochar has unique chemical properties. It can reduce leaching of nutrients, including nitrate [28] and P; act as a bio stimulant, especially affecting the roots and suppressing root disease [10,29,30]; reduce greenhouse gases (GHG) from growing media when ebb and flood irrigation is practiced, as anoxic conditions occur for short periods; and sequester C at the end of life of growing media, when it ultimately ends in soil [17]. At present, the costs of biochars are prohibitive. However, if during the pyrolysis process all products are used, e.g., for heat to generate electricity, bio oil for heating and using biochar as a minor component of a growing medium, and there is a gate fee for waste wood, biochar would be particularly viable. Biochar has generally a very high pH and contains nutrient elements, e.g., K, at quite high levels. These elements could be used for plant nutrition. However, only in few plant growth trials the nutrients presented in biochar were taken into account when used as a growing medium [30,31]. On the other hand, biochar's high pH was considered and the rates of lime were reduced [32,33]; or lime rate can be eliminated [34], when biochar is added. Bedussi et al. [32] used biochar that was produced by gasification instead of pyrolysis. Gasification takes place at a much higher temperature (1100 to 1200 °C) than pyrolysis (400 to 600 °C). Higher temperature has a major effect on biochar properties such as pH, total Ca and Mg levels [1,35]. Thus, products from these two above processes may not be directly comparable.

Numerous papers have evaluated biochar's positive effects on plant growth [10,11,36], but very few, such as the studies from Bedussi et al. [32], Kaudal et al. [20] and Mendez et al. [37], have characterized the biochar used in the growing media. Recently, some studies have been conducted to look at the effect of biochar particle size on physical properties of growing media [27,38]. Zaccheo et al. [33], have also shown that biochar having fine granulometry can be more effective in increasing the pH, and therefore, eliminating the need for liming of peat [32]. The nutrient properties of the particle size of biochar have been studied but only in terms of soil application [39] or for environmental use [40,41].

We are not aware of relevant studies on pH, total and extractable macronutrients and micronutrient contents of different biochar particle size fractions. Generally, quite high levels of fine fractions < 1 mm and < 2 mm have been administered, when biochar has been used in growing media [20,29,30,32,37,38]. In ad hoc trials in a few commercial biochar products with different fractions, we were surprised to find differences in pH and nutrient contents in the different fractions from the same biochar. This is in contrast to peat, as no difference in pH was reported between fine, medium and coarse peat [42]. In the present study, we studied the pH, total and available macronutrients and micronutrient contents of biochar fractions from a number of commercial biochars that have the potential for use as growing media. The nutrient availability test that we used has been found to be very strongly correlated to plant nutrient uptake in a series of publications [43], and on this basis has been accepted as a European test [44]. We particularly looked at both total and extractable (available) Ca and Mg in relation to pH in the biochar fractions.

2. Materials and Methods

For our investigations we selected four commercial grade biochars: three from Europe and one from China. One of them derived from having cereal husk and paper fibre as a feedstock (Biochar A—DU). Three were derived from wood-based materials; namely, bamboo (Biochar B—No commercial name), wood screenings (Biochar C—Verora) and forest wood (Biochar D—Carbon Terra). Following pyrolysis by different manufactures, the biochars were chunky and had been milled to approximately less than 10 mm. The four biochars were characterized for pH [45] and electrical conductivity [46] in water extract at a 1:5 (v:v) ratio. The materials were characterised by extractable NH_4-N and NO_3-N; total N, K and P [44]; bulk density; and specific surface areas of the different biochars, including pore volume and diameter that were measured with a Quantachrome Autosorb-1 surface area Analyser, using N_2 (Brunauer, Emmett and Teller—BET) sorption methods [47].

Each material was sieved to three fractions as < 1 mm, 1–2 mm and 2–4 mm (see Figure A1). We took care to ensure each entire fraction that passed through the screens for the various fractions was obtained. For each particle size fraction, 20 to 30 g of biochar was obtained, dried and mixed well prior to physic-chemical analyses. They were then analysed for total macronutrients and micronutrients, as described by Chrysargyris et al. [48]. Nitrogen was analysed using Kjeldahl method (Buchi Digest Automat K 439 and Distillation Kjelflex K 366, Switzerland). Other macro and micronutrients were ashed and the ash was digested with hydrochloric acid (2N HCl). Ash extracts were analysed using atomic absorption spectroscopy (PG Instruments AA 500 FG, Leicestershire, UK). The determinations of K and Na were made using a flame photometer (Lasany Model 1832, Lasany International, Haryana, India). Results are expressed in g kg^{-1} and mg kg^{-1} for macronutrients and micronutrients, respectively. Biochars' fractions were extracted for available nutrients using $BaCl_2$/DTPA extraction on volume basis according to EN 13651 [44]. $BaCl_2$ was used at the same concertation as the standard $CaCl_2$/DTPA, EN 13651 [44]. The extracts were analysed for macro and micronutrients. The four biochar materials by three particle sizes were assessed for pH [45] in water extracts, at 1:5 (v:v) ratios, and $BaCl_2$/DTPA extractable NH_4-N and NO_3-N [44,49]. Determination of $BaCl_2$/DTPA extractable nutrients P, K, Ca, Mg, Fe, Na, Zn, Cu, Mn and B were determined in the filtered extract by inductively coupled plasma atomic emission spectrometry (Perkin Elmer ICP-OES, Waltham, MA, USA). Sulphate was measured in the extract by separating the anions by their affinity for an anion exchange resin packed in the anion separator column. The concentration of the sulphate anion was determined by measuring the conductance as it passed through conductivity cell.

Statistical Methods

Data were statistically analysed with the IBM SPSS version 22 (IBM Corp., Armonk, NY, USA) and results are expressed as means (n = 3) ± standard errors (SEs). Differences between treatment means were compared at p = 0.05 with ANOVA, followed by Duncan's multiple range test (DMRT).

3. Results

The characterization of the examined commercial, and an unscreened biochar, showed high pH, relatively high electrical conductivity, high extractable K and very low extractable N in all four biochars (Table 1). Extractable P was significantly (p < 0.05) higher in Biochar B compared with the other Biochars. The surface area tended to be highest with Biochar D, and lowest with Biochar B. The bulk density was highest for Biochar D. The pore volume and pore size followed the same trend as the surface area (Table 1).

The total macronutrients and micronutrients of different fractions of four biochars are given in Figures 1 and 2, respectively. The pH values and extractable contents of elements of the three fractions are presented in Figure 2. The extractable macronutrients and extractable micronutrients of three fractions of four biochars are given in Figures 3 and 4, respectively, while selected correlation analysis among some elements are presented in Figure 5.

3.1. Biochar's Total Elemental Contents and Correlations

3.1.1. Total Macronutrients N, K, Ca, Mg, P and Na

The fine fraction had significantly ($p < 0.05$) higher levels of N (25.9%, 19.0% and 18.9%) for the Biochars B, C and D, respectively, compared with the coarse fraction (Figure 1), while the opposite was found in case of the Biochar type A. The coarse fraction of 2–4 mm of Biochar A showed 26.41 mg kg^{-1} N, which was 77% higher than the fine fraction, and that high N level was present only in Biochar A, which was the opposite to the other three biochars (Figure 1). The total K levels followed the same pattern as N, except that Biochar C that showed no significant effects of particle size (Figure 1). The biochar derived by husks and paper fibres, named Biochar A, revealed 17.50 mg kg^{-1} of K, which was the highest content at 2–4 mm fraction. Similar to N and K, the higher levels of total P were found in Biochar B, Biochar C and Biochar D in the finer fraction. An exception was Biochar A, which had the opposite trend with higher levels in the coarser fraction (Figure 1).

The Ca levels were significantly higher in the fine (<1 mm) and medium (1–2 mm) fractions for the examined biochars, except for Biochar B, which had no significant differences between fractions (Figure 1). Similarly, to Ca, the Mg levels were higher in the fractions of < 1 mm and 1–2 mm for the Biochar A, whereas in Biochar C and Biochar D, greater Mg contents were observed in their fractions < 1 mm. The total contents of Mg in Biochar B were the same among fractions (Figure 1). No differences were evident between the fractions for the total Na content, as presented in Figure 1.

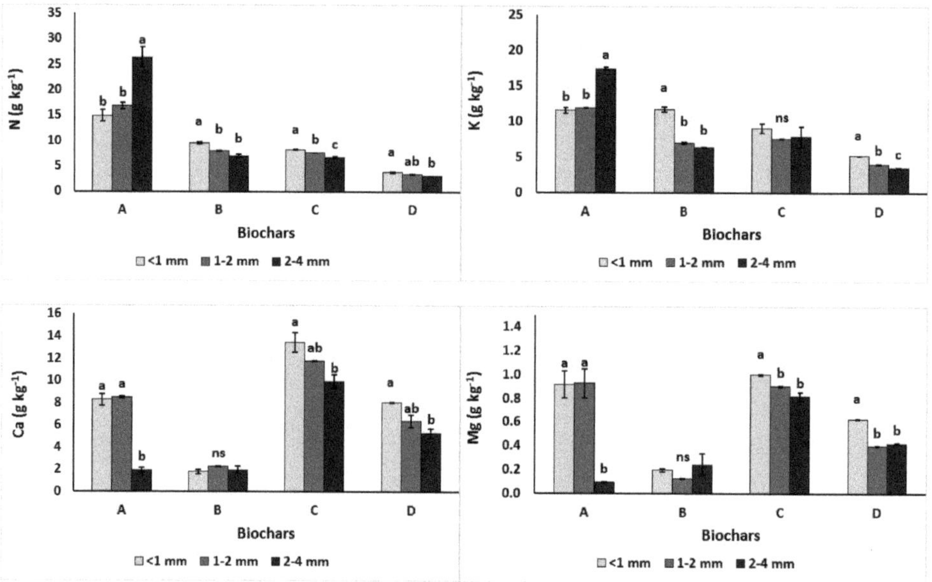

Figure 1. *Cont.*

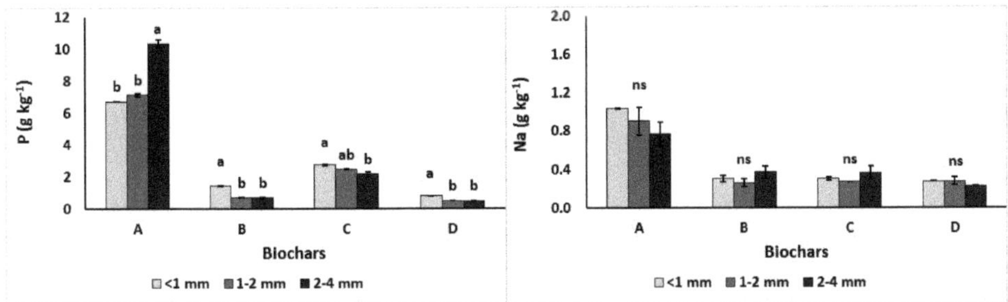

Figure 1. The effects of biochar (A, B, C and D) particle size on total macronutrients. Non-significant differences are indicated by ns, while significant differences ($p < 0.05$) among particle size for each Biochar are indicated by different Latin letters according to Duncan's Multiple Range test. Error bars show standard errors-SE.

3.1.2. Total Micronutrients Cu, Mn and Zn

Higher levels of total Cu were found in the finer fractions of two biochars—A and D. There was a clear-cut trend with the fine fraction with higher levels of Zn for the examined biochars (Figure 2). This was similar to the results that we got with extractable Zn (Figure 4). Only in Biochar C, the trend of higher levels of Mn in the fine fraction was present (Figure 2), and the results did not reflect in extractable Mn (Figure 4).

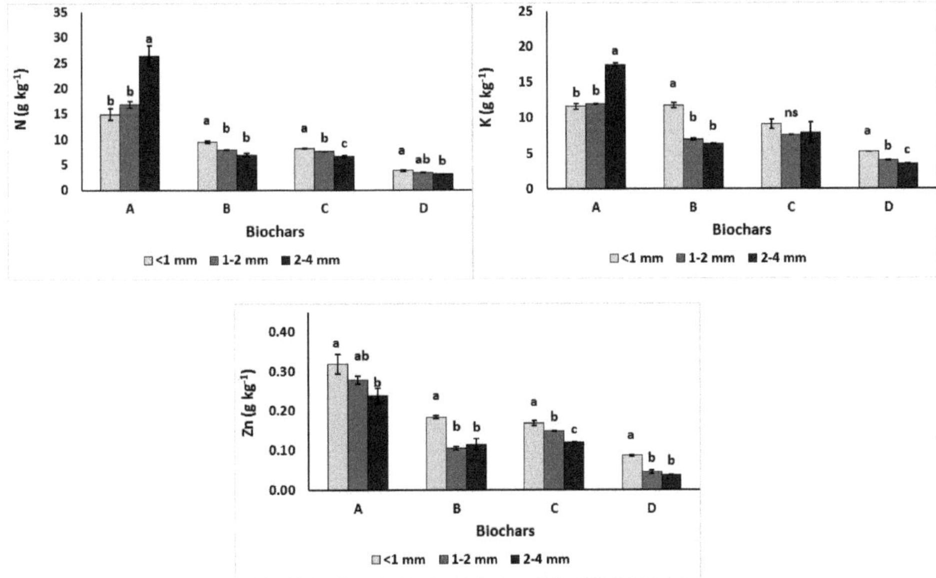

Figure 2. The effects of biochar (A, B, C, D) particle size on total micronutrients. Non-significant differences are indicated by ns, while significant differences ($p < 0.05$) among particle size for each Biochar are indicated by different Latin letters according to Duncan's Multiple Range test. Error bars show standard errors-SE.

Table 1. Characterization of commercial biochars as is (pre-screening). Values are means ± standard errors of two measurements made on pH, electrical conductivity (μS cm^{-1}) and minerals (mg L^{-1}). Means values followed by the same letter do not differ significantly at $p < 0.05$.

	Feedstock	Trade Name	pH	EC (μS cm^{-1})	NH$_4$-N (mg L^{-1})	NO$_3$-N (mg L^{-1})	N (mg L^{-1})	P (mg L^{-1})	K (mg L^{-1})	BD [1] (g L^{-1})	BET [2] (m^3 g^{-1})	Pore Vol [3] (cc g^{-2})	Pore Vol [3] (cc mL^{-1})	Pore Size [4] (Å)
Biochar A	Cereal husk/paper fibre	DU	9.56 ± 0.09 a	653 ± 7.5 a	1 ± 0.0 a	1.0 ± 0.0 a	2.0 ± 0.0 a	8.5 ± 1.5 b	891 ± 9.5 c	232	81.3	0.00592	0.00254	29.04
Biochar B	Bamboo	n/a	8.84 ± 0.08 b	571 ± 12 b	1 ± 0.0 a	0.5 ± 0.5 a	1.5 ± 0.5 a	23.0 ± 2.0 a	1158 ± 13.0 a	296	4.7	0.00021	0.00006	17.90
Biochar C	Wood screenings	Verora	9.54 ± 0.10 a	410 ± 16 c	1 ± 0.0 a	1.0 ± 0.0 a	2.0 ± 0.0 a	10.5 ± 1.5 b	659 ± 4.5 d	258	62.6	0.0045	0.0017	28.81
Biochar D	Forest wood	Carbon Terra	9.51 ± 0.04 a	637 ± 5.5 a	1 ± 0.0 a	0.5 ± 0.5 a	1.5 ± 0.5 a	5.5 ± 1.5 b	990 ± 7.0 b	389	243.0	0.146	0.056	24.03

[1] Specific surface area. [2] Total pore volumes for pores with diameters less than 31917.2. [3] Average pore diameters. [4] Å at P/Po = 0.99940.

3.2. Biochars' Extractable Elemental Contents

3.2.1. The pH and Extractable Macronutrients N, K, P, Ca, Mg, Na and SO_4

The pH was higher in the fine fraction for all four biochars (Figure 3). The pH levels revealed the same values for the medium (1–2 mm) and coarse (2–4 mm) fractions for Biochar A, Biochar B and Biochar D.

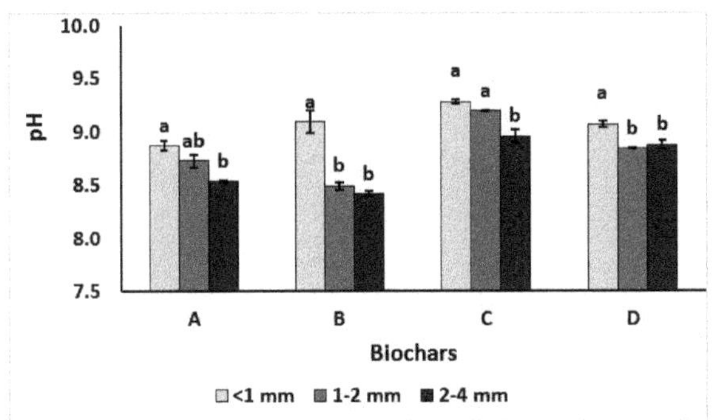

Figure 3. The effects of biochar (A, B, C, D) particle size on the pH value. Significant differences ($p < 0.05$) among particle size for each Biochar are indicated by different Latin letters according to Duncan's Multiple Range test. Error bars show standard errors-SE.

Levels of NH_4-N and NO_3-N were very low, with values < 2 mg L^{-1} (data not presented). There was a clear trend of increasing K in the fine fraction (Figure 4). The two biochars, C and D, based on wood waste material, had a similar trend. Biochar B from paper fibre and cereal husk had higher (2535 mg L^{-1}) levels of K (Figure 2).

The finer fraction (<1 mm) had significantly higher levels of extractable P in three (A, B and C) of the four biochars, while no significant differences were found among 1–2 mm and 2–4 mm fractions. Biochar D had very low values of extractable P in all the three fractions and there was no significant effect of biochar fraction (Figure 4).

Regarding extractable Ca, there was also a clear significant trend of the fine fraction containing higher levels calcium, with the woody materials, Biochars C and D, being very similar. The level of calcium in the Biochar A was high (up to 1094 mg L^{-1} at < 1 mm), while in the Biochar B, even in the fine fraction it was very low (up to 271 mg L^{-1} at < 1 mm) (Figure 4).

The fine fraction had significantly higher levels of magnesium but there was no significant difference between 1–2 mm and 2–4 mm. As with potassium the levels, in the biochar derived by bamboo (B), Mg was high, reaching 145.7 mg L^{-1} of extractable Mg (Figure 4).

Only in two types of the biochars, namely, A and D, did the fine fraction have significantly higher levels of Na, while biochars B and C did not differ among the fractions examined (Figure 4). There were no significant differences between the fine and coarser fractions among the biochars in relation to extractable SO_4 (Figure 4).

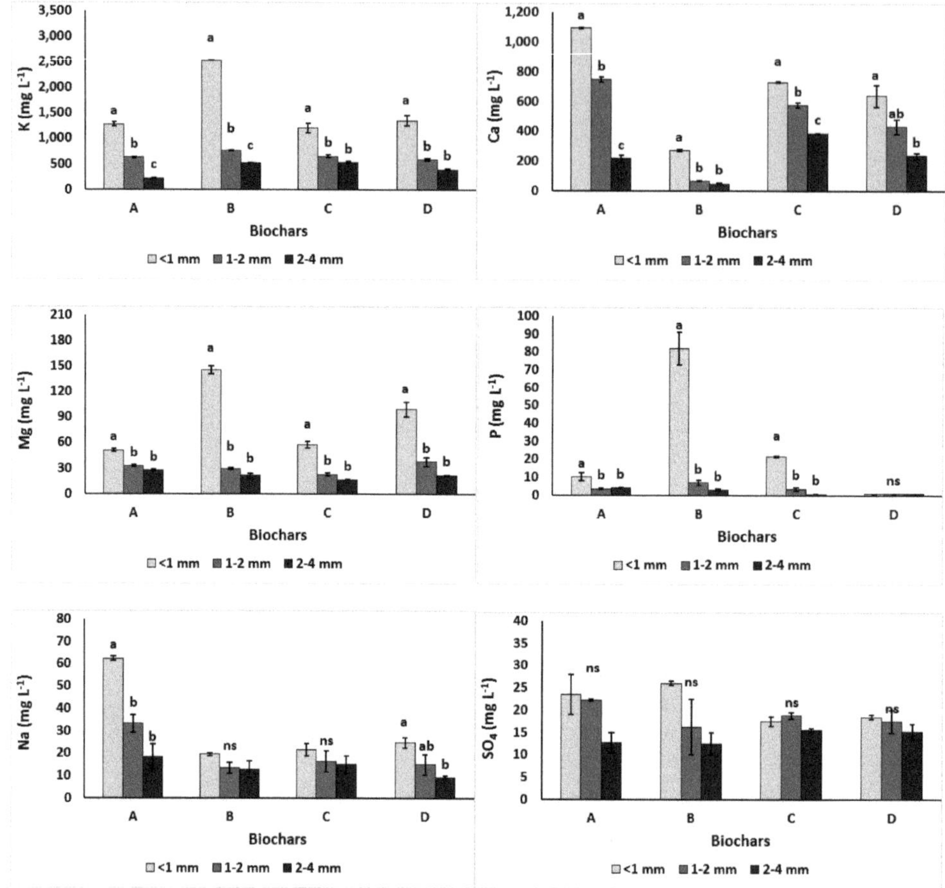

Figure 4. The effects of biochar (A, B, C, D) particle size on extractable macronutrients. Non-significant differences are indicated by ns, while significant differences ($p < 0.05$) among particle size for each Biochar are indicated by different Latin letters according to Duncan's Multiple Range test. Error bars show standard errors-SE.

3.2.2. Extractable Micronutrients (Zn, Mn, Fe, B and Cu)

There was a clear significant trend of higher levels of Zn in the fine fraction for all four biochars, but the differences between 1–2 mm and 2–4 mm were not significant (Figure 5). Biochars derived by bamboo revealed the higher (4.72 mg L^{-1}) Zn levels in the < 1 mm fraction. Manganese levels were particularly low in Biochar A and Biochar C, while both Biochar B and Biochar D revealed high (ranged from 18.82 to 20.85 mg L^{-1}) levels of Mn in their fine fractions (Figure 5).

Higher levels of extractable Cu were found in the finer fractions of two biochars, A and D, those being the same biochars with higher levels of total Cu (Figure 2). There was no clear trend regarding the four biochars in the fractions regarding the Fe and B levels (Figure 5).

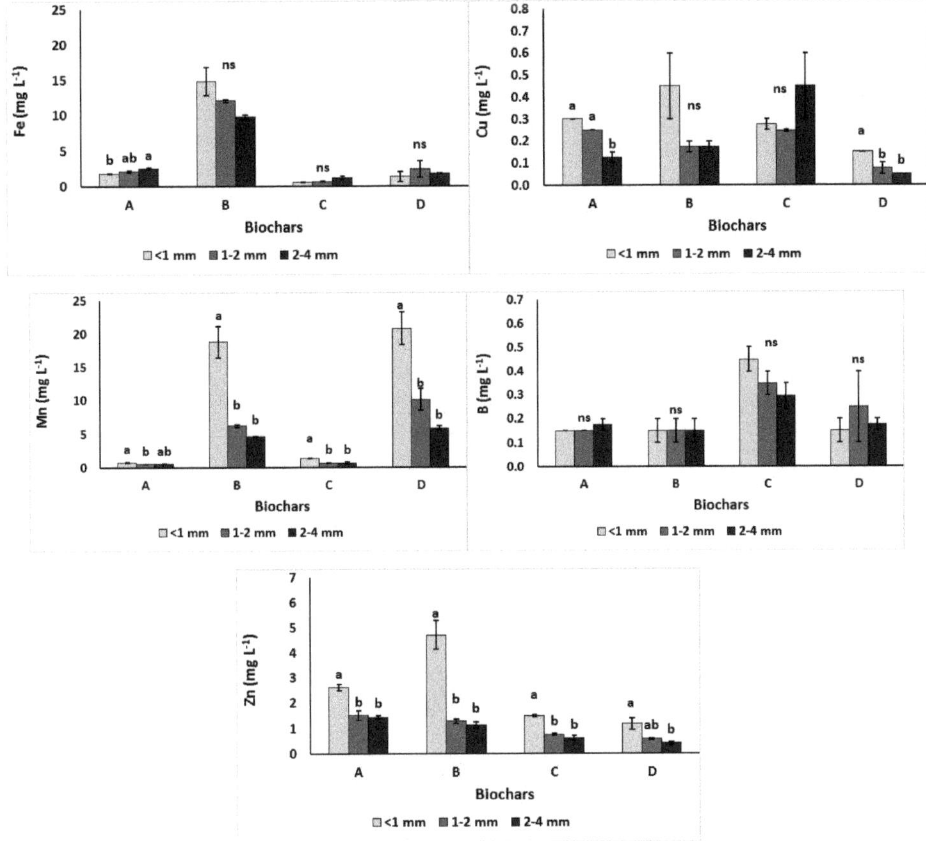

Figure 5. The effects of biochar (A, B, C, D) particle size on extractable micronutrients. Non-significant differences are indicated by ns, while significant differences ($p < 0.05$) among particle size for each Biochar are indicated by different Latin letters according to Duncan's Multiple Range test. Error bars show standard errors-SE.

3.3. Correlations

The pH value was correlated with total Ca (r = 0.518; Figure 6), and with total Mg (r = 0.364; Figure 6), respectively. However, the correlation was poor with extractable Ca (r = 0.272; Figure 6) and extractable Mg (r = 0.191; Figure 6). There was a strong correlation between total K and total N (r = 0.870) at level of $p \leq 0.01$ (Figure 7). There was a very strong correlation between total Ca and total Mg (r = 0.862) at level of $p \leq 0.001$ (Figure 7), indicating when Ca is high or low, the Mg content would follow a similar pattern. There was a very strong correlation ($p \leq 0.001$) between extractable Ca and total Ca (r = 0.711) and total Mg (r = 0.823, Figure 7). Moreover, there was a very strong correlation ($p = 0.001$) between extractable K and total Mg (r = 0.941). There was fairly a strong correlation ($p = 0.001$) between extractable Zn and extractable SO_4 (r = 0.616), extractable Fe (r = 0.602), extractable Cu (r = 0.857), respectively. Additionally, there was a strong correlation ($p \leq 0.001$) between extractable Mn and extractable Zn (r = 0.795), and a fairly good correlation (r = 0.640) between extractable Mn and extractable Cu.

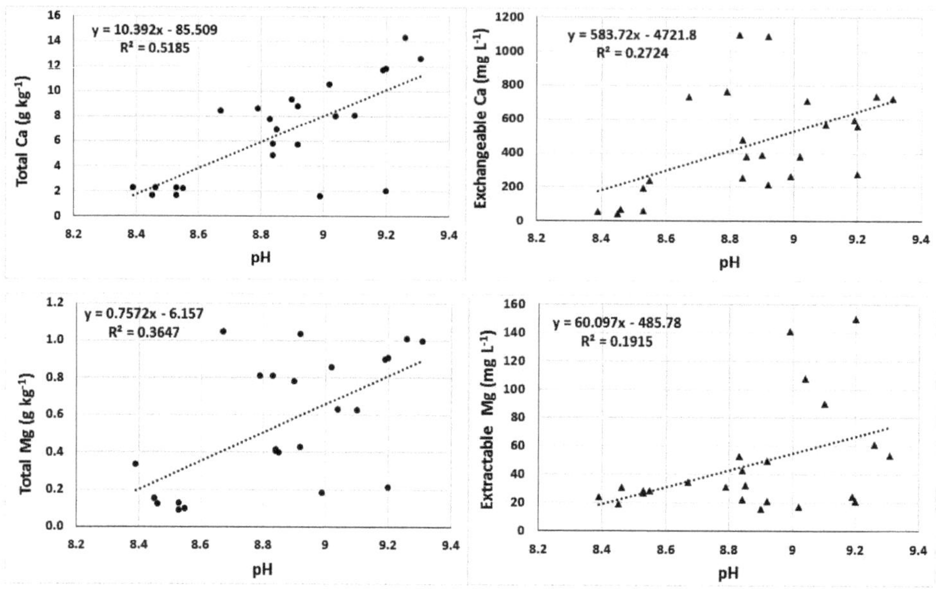

Figure 6. Correlation analysis of pH with total and extractable elements from four biochars over three fractions.

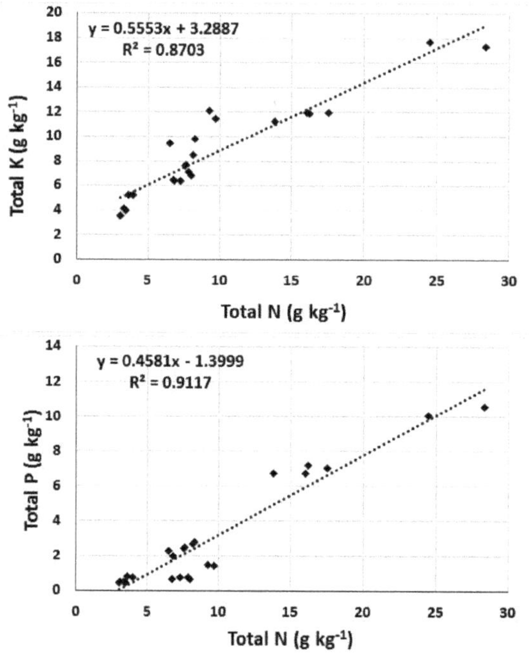

Figure 7. *Cont.*

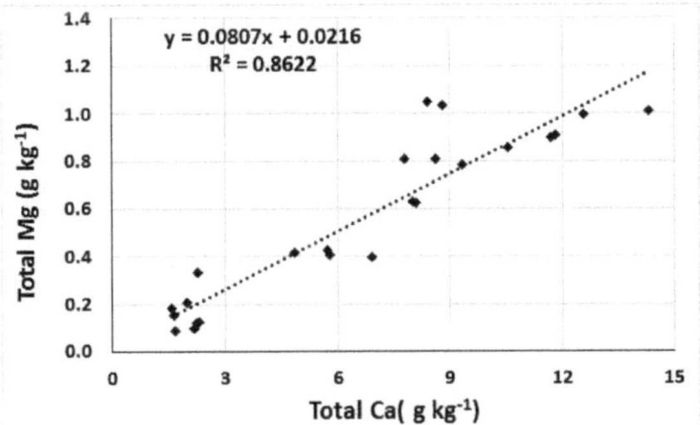

Figure 7. Correlation analysis of total elements from four biochars over three fractions.

4. Discussion

It is well known that feedstock, pyrolysis temperature and residence time have a major effect on pH and nutrient content of a biochar [1,14,35]. We investigated four commercial biochars and found a high variation between them with regard to total nutrients. This variation can be partially explained on the basis of different processing conditions and differences in feedstock. However, due to the propriety nature of the materials used in the current study, exact information on processing and feedstocks are lacking. In the present study, the levels of total N, P and K were in the same range as previously reported [20,28,32,49]. Biochar A was an exception, having particularly high nutrient levels. The feedstock of this material was different from the other three, as it was based on paper fibre and cereal husk.

The finer fraction has higher levels of total macronutrients, and to a lesser extent, micronutrients, for the three-wood based biochar. The percentage of finer fraction of a biochar is of importance through the dependence of nutrients on the particle size. Regarding the biochars used in the present study, their particle sizes have been investigated previously; they have fine particles less than 1 mm in the range of 23% to 64% [30]. Biochar B in that study had 49.5% less than 1 mm. Other authors who investigated biochars derived from woody material, reported levels of less than 1 mm of 40%–81% [29], 21%–79% [37], 38%–73% [20] and 2%–98% [38]. It is most likely that the partial removal of the fine fraction (<1 mm) could make it suitable as a component of a growing media. This could be explained on the basis that different particle size of feedstock may lead to differential heating during pyrolysis and the finer fraction could have been more carbonized than the coarser fraction. Kloss et al. [35] have shown that the levels of cations Ca, Mg and K increase with greater carbonization. They, however, found that total N levels fall as carbonization increases. However, other authors have found that N levels can increase with carbonization [41,50–52].

He et al. [41] studied the chemical properties of biochar fractions of 2–5 mm, 1–2 mm and 0.5–1 mm but also seven fractions below 0.5 mm made from pine wood. In the three fractions, which were somewhat similar to our fractions, they found an increase in total N, Ca and Mg contents in the finer fraction. The authors argued that longitudinal and transverse heterogeneity of biochar and the dominant cleavage during the preparation process may be responsible for the significant differences in properties initiated by particle size. In our investigation, N, P and K of Biochar A derived from cereal husk and paper fibre, were higher in the coarser fraction.

Generally, all the extractable nutrients were higher in the fine fraction. The fact that the availability of nutrients based on extraction with $BaCl_2/DTPA$ increases as the fraction size decreases can be explained to some extent, as it is understandable, due to the greater surface area of the fine material that allows the extractant to extract more nutrients. It may also be due to higher bulk density.

In the present study, we showed that particle size of the same biochar can have different levels on pH and total and extractable macro and micro nutrients in the different fractions (i.e., higher in the fine fraction) in the context of its use as a component of a growing media. This would ultimately affect the acidity and macro and some micronutrient content and availability of peat-biochar growing media due to fact that each fraction, particularly the fine fraction has significantly different levels of nutrients and acidity.

Numerous studies have been published showing the beneficial effects on crops due to the addition of biochar to peat; however, only in a limited number of studies has the particle size of the biochar been presented. There have been a few publications that showed the particle size of the biochar has a marked effect on physical properties of the growing media [27,32,38]. Our results indicate that in the use of biochars in growing media, particle size should be taken into serious consideration not only for physical properties, but just as importantly, chemical properties. This information should be essential when formulating nutrients/lime addition and subsequent nutrient management during cropping.

The higher nutrient availability from the four biochars in fine fractions as found in the present study, are in agreement with previous reports [39]. However, Angst and Sohi [39] examined the effect of particle size in context of a soil application of biochar, which is different from the growing media conditions. They found that the availability and the release of limited number of nutrients, namely, magnesium, potassium and phosphorus, were affected by particle size with the fine fraction material showing greater water-soluble contents of these nutrients, and better release. The fraction sizes they looked at were smaller than ours, as they studied fractions of 0.15 μm to 0.60 μm, 0.60 μm to 1.8 mm, 1.8 mm to 4 mm and > 4 mm. For K and Mg, the differences between the finer fractions were relatively small but the differences were greater between the fine fractions, that was, < 4 mm, and > 4 mm. For P the differences in the finer fraction were even less and the difference between the finer fraction and > 4 mm was less than for K and Mg [39]. The extraction solution might affect the nutrients extracted derived by different fractions of biochar. Angst and Sohi [39] carried out six extractions and for K there was little change but for Mg and P the amount and the difference between the fractions increased. They used water as an extractant, unlike our extractant which had a cation in it. One is aware that biochar has a high cation exchange capacity and also has the ability to bind nitrate nitrogen, and as such a stronger extraction, which includes a cation and contains DTPA, which may give a bigger difference between the fractions and perhaps better picture on plant availability in growing media. In any case the extractant used here has been strongly correlated the plant uptake of some macro nutrients and micronutrients in a growing media situation [30,43].

Due to the nature of biochar coming from woodchips, thereby the need for crushing and relatively easy break down of particles, variation in particle size is likely to occur between batches, and this could possibly have reflected variation in physical and chemical properties of growing media, when biochar is added to peat. Our findings point to the need to be vigilant about particle size for each batch. However, the use of dolomitic lime was added not only to neutralize the acidity of peat but also to supply essential nutrients such as calcium and magnesium. The pH values in biochar can vary according to the pyrolysis temperature, as biochars produced from sewage sludge at low temperatures (300 and 400 °C) were acidic, whereas at high temperatures (500 and 600 °C), they were alkaline [53].

The levels for the Cu, Zn and Mn were in the same range as found by Altland and Locke [28] and Bedussi et al. [32]. The average extractable K, Ca and Mg as a percentage (%) of the total K, Ca and Mg for Biochar A, Biochar B, Biochar C and Biochar D were as follows: for K—Approximately 18%, 34%, 31% and 48%, respectively; for Ca—24%,16%,16% and 19%, respectively; and for Mg—14%, 9%, 11% and 28%, respectively. Angst and Sohi [39] found greater water extractability of K compared to Mg, probably reflecting the feedstock.

It is well known that most biochars have high pH-values [31], and peats have very low pH-values, ranging from 3 to 4 [54]. Therefore, peat requires additional lime, e.g., in form of a dolomitic lime, to adjust the pH to values to around 5.5, while other substrates, e.g., wood fibres, do not need it [55]. A number of researchers suggested that biochars could be used to replace lime application [30,34]. However, surprisingly, biochar-peat blends can contain up to 80% biochar without raising the pH above 7 [34]. Unfortunately, these studies do not give any information on particle size; one can surmise it was a coarse material. Zaccheo et al. [33] reported that particle size can have an effect in increasing the pH of growing medium. Furthermore, the same authors found that the finer fraction, 0–3.3 mm, was more effective at raising the pH of a peat-biochar mixture [33]. This is in agreement with our findings, as the finer fraction had a higher pH and a higher content on Ca and Mg. However, in the same study of Zaccheo et al. [33], the premise that dolomitic lime or ordinary lime is added not only to increase the pH but also supply Ca and Mg, which are essential plant nutrients with very low levels in peat, was not taken into consideration [32]. In our study, despite the minor differences in pH, the levels of extractable Ca varied enormously between Biochar A with very high extractable Ca and Biochar B with very low extractable Ca. In addition, most biochars contain high levels of K. This could depress and reduce Mg uptake by plants, due to the cation antagonism among them. Increased K-rates, varied from 4.61–5.39 g kg^{-1} by different biochars, were also reported by Gasco et al. [31], being in agreement with the findings of the present study.

Although generally there is good relationship between pH and total calcium and magnesium content, in our case the relationship would not be good enough to predict Ca and Mg levels. This is related to the poorer relationship of pH and available and extractable Ca and Mg content in biochars. The example of a good relationship or otherwise between Ca and Mg is important, as in peat dolomitic lime is added not only to adjust pH but to supply Ca and Mg. Extractable levels of Ca are generally very low in peat. The good relationship found in our study indicates that biochar is a similar material to dolomitic lime rather than calcitic lime. The good relationship between extractable Ca and total Ca gives confidence on the supply of Ca, whereas a poor relationship would invoke doubt, either from total Ca or extractable Ca: extractable Ca gives an indication of short term availability, while total Ca gives an indication of long term availability. P levels are too low to have any significance as a biochar is added at low rates with peat. However, K would be of importance. For instance, data from Bedussi et al. [32] shows that the finer biochar from poplar using pyrogasification has higher contents of total N, P, Ca, Mg and many micronutrients. The poplar biochar had higher contents of total Ca and Mg, despite having a lower pH. In addition, Bedussi et al. [32] measured water-soluble N, P and K which were higher for these nutrients, but found lower levels of Ca and no difference for Mg. Therefore, a blanket recommendation or a suggestion that biochar can replace lime in peat-based media needs serious rethinking. We feel it is essential that the total Ca and total Mg and available Ca and available Mg, should be analysed in biochar and taken into account when recommending lime application rates if biochar has been added to peat. This is in agreement with our previous works. As the biochar component increases in a peat/biochar-mixtures, this is affecting nutrient levels in plants; for example, the levels of Mg content in a leaf dropped due to the antagonistic effect of K on Mg [30]. This is primarily due to the excessive level of K, which is a feature of most but not all biochars. Additionally, in our resent study, K, P and Cu accumulation and Mg deficiency in cabbage leaves were related to the biochar presence and feedstock [10]. In that study, the biochar's feedstock, rate and the addition of fertilizers could affect the cabbage seedling performance.

5. Conclusions

The results of the present study have shown that within the same batch of biochar, total macronutrient and micronutrient levels are different in different fraction sizes. One can clearly state from our data that the use of a biochar when added to peat and when it is considered as a substitute for lime, due to its high pH, may not be valid in many cases, as Ca levels in biochars do not always equate to a pH value. Extractable Ca and perhaps total Ca in biochars need to be considered when advising biochar rates and lime rates to peat. Moreover, the particle size of biochar is very important regarding electrical conductivity and nutrient availability, especially K. This has far reaching consequences regarding nutrient imbalances in growing media and in formulating a base dressing and a liquid feeding programme. To our knowledge, this is the first time that it has been reported for biochar in context of growing media. The variation in total nutrients in biochar fractions of the same biochar was an unexpected result for a growing media. Our recommendation is to partially take out the biochar fraction < 1 mm if the EC levels and nutrients are particularly high and use it for other purposes. The reduction of EC and extractable K would, thus, make the biochar suitable as a component of a growing medium.

Author Contributions: Conceptualization, M.P. and N.T.; data curation, A.C., M.P. and N.T.; formal analysis, M.P., N.M. and A.C.; funding acquisition, A.K. and N.T.; investigation, M.P., and N.T.; methodology, M.P., A.C. and N.T.; project administration, M.P. and N.T.; resources, N.M. and A.C.; software, A.C. and N.T.; supervision, M.P. and N.T.; validation, M.P., N.S.G. and N.T.; visualization, A.C.; writing—original draft, M.P. and A.C.; writing—review and editing, M.P., N.S.G. and N.T. All authors have read and agreed to the published version of the manuscript.

Funding: This research was funded by Bord na Mona Horticulture Ltd and Cyprus University of Technology under the project OPTIBIOCHAR and Cyprus University of Technology Open Access Author Fund.

Acknowledgments: The authors are grateful to the project OPTIBIOCHAR that has been developed under the Cooperation Programme Cyprus-Ireland, co-funded by the Bord na Mona LtD and Cyprus University of Technology.

Conflicts of Interest: The authors declare no conflict of interest.

Appendix A

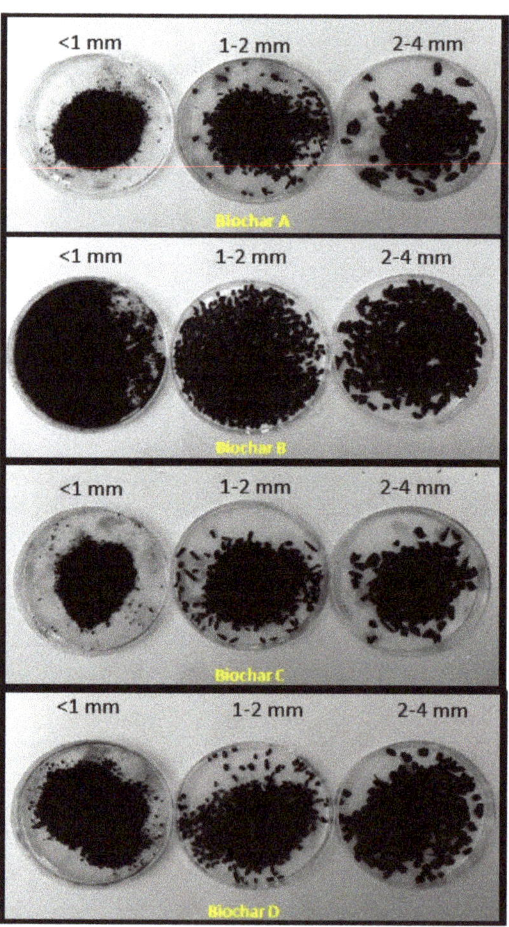

Figure A1. Biochar (A, B, C, D) particle size illustration.

References

1. Rehrah, D.; Bansode, R.R.; Hassan, O.; Ahmedna, M. Physico-chemical characterization of biochars from solid municipal waste for use in soil amendment. *J. Anal. Appl. Pyrolysis* **2016**, *118*, 42–53. [CrossRef]
2. Xu, C.Y.; Hosseini-Bai, S.; Hao, Y.; Rachaputi, R.C.N.; Wang, H.; Xu, Z.; Wallace, H. Effect of biochar amendment on yield and photosynthesis of peanut on two types of soils. *Environ. Sci. Pollut. Res.* **2015**, *22*, 6112–6125. [CrossRef] [PubMed]
3. Kerré, B.; Willaert, B.; Cornelis, Y.; Smolders, E. Long-term presence of charcoal increases maize yield in Belgium due to increased soil water availability. *Eur. J. Agron.* **2017**, *91*, 10–15. [CrossRef]
4. Gomez-Eyles, J.L.; Sizmur, T.; Collins, C.D.; Hodson, M.E. Effects of biochar and the earthworm *Eisenia fetida* on the bioavailability of polycyclic aromatic hydrocarbons and potentially toxic elements. *Environ. Pollut.* **2011**, *159*, 616–622. [CrossRef] [PubMed]
5. Han, Y.; Cao, X.; Ouyang, X.; Sohi, S.P.; Chen, J. Adsorption kinetics of magnetic biochar derived from peanut hull on removal of Cr (VI) from aqueous solution: Effects of production conditions and particle size. *Chemosphere* **2016**, *145*, 336–341. [CrossRef]
6. Komnitsas, K.; Zaharaki, D.; Pyliotis, I.; Vamvuka, D.; Bartzas, G. Assessment of pistachio shell biochar quality and its potential for adsorption of heavy metals. *Waste Biomass Valoriz.* **2015**, *6*, 805–816. [CrossRef]
7. Spokas, K.A.; Koskinen, W.C.; Baker, J.M.; Reicosky, D.C. Impacts of woodchip biochar additions on greenhouse gas production and sorption/degradation of two herbicides in a Minnesota soil. *Chemosphere* **2009**, *77*, 574–581. [CrossRef]
8. Vaccari, F.P.; Baronti, S.; Lugato, E.; Genesio, L.; Castaldi, S.; Fornasier, F.; Miglietta, F. Biochar as a strategy to sequester carbon and increase yield in durum wheat. *Eur. J. Agron.* **2011**, *34*, 231–238. [CrossRef]
9. Kim, H.S.; Kim, K.R.; Yang, J.E.; Ok, Y.S.; Kim, W., II; Kunhikrishnan, A.; Kim, K.H. Amelioration of horticultural growing media properties through rice hull biochar incorporation. *Waste Biomass Valoriz.* **2017**, *8*, 483–492. [CrossRef]
10. Chrysargyris, A.; Prasad, M.; Kavanagh, A.; Tzortzakis, N. Biochar type and ratio as a peat additive/partial peat replacement in growing media for cabbage seedling production. *Agronomy* **2019**, *9*, 693. [CrossRef]
11. Álvarez, J.M.; Pasian, C.; Lal, R.; López, R.; Díaz, M.J.; Fernández, M. Morpho-physiological plant quality when biochar and vermicompost are used as growing media replacement in urban horticulture. *Urban For. Urban Green.* **2018**, *34*, 175–180. [CrossRef]
12. Hilioti, Z.; Michailof, C.M.; Valasiadis, D.; Iliopoulou, E.F.; Koidou, V.; Lappas, A.A. Characterization of castor plant-derived biochars and their effects as soil amendments on seedlings. *Biomass Bioenergy* **2017**, *105*, 96–106. [CrossRef]
13. Huang, L.; Gu, M. Effects of biochar on container substrate properties and growth of plants—A review. *Horticulturae* **2019**, *5*, 14. [CrossRef]
14. Kaudal, B.B.; Chen, D.; Madhavan, D.B.; Downie, A.; Weatherley, A. Pyrolysis of urban waste streams: Their potential use as horticultural media. *J. Anal. Appl. Pyrolysis* **2015**, *112*, 105–112. [CrossRef]
15. Gruda, N.S. Increasing sustainability of growing media constituents and stand-alone substrates in soilless culture systems. *Agronomy* **2019**, *9*, 298. [CrossRef]
16. Blok, C.; Van Der Salm, C.; Hofland-Zijlstra, J.; Streminska, M.; Eveleens, B.; Regelink, I.; Fryda, L.; Visser, R. Biochar for horticultural rooting media improvement: Evaluation of Biochar from gasification and slow pyrolysis. *Agronomy* **2017**, *7*, 6. [CrossRef]
17. Alvarez, J.M.; Pasian, C.; Lal, R.; Lopez, R.; Fernandez, M. Vermicompost and Biochar as growing media replacement for ornamental plant production. *J. Appl. Hortic.* **2017**, *19*, 205–214.
18. Dispenza, V.; De Pasquale, C.; Fascella, G.; Mammano, M.M.; Alonzo, G. Use of biochar as peat substitute for growing substrates of *Euphorbia × lomi* potted plants. *Span. J. Agric. Res.* **2016**, *14*. [CrossRef]
19. Fryda, L.; Visser, R.; Schmidt, J. Biochar replaces peat in horticulture: Environmental impact assessment of combined biochar & bioenergy production. *Detritus* **2019**, *5*, 132–149.
20. Kaudal, B.B.; Chen, D.; Madhavan, D.B.; Downie, A.; Weatherley, A. An examination of physical and chemical properties of urban biochar for use as growing media substrate. *Biomass Bioenergy* **2016**, *84*, 49–58. [CrossRef]
21. Margenot, A.J.; Griffin, D.E.; Alves, B.S.Q.; Rippner, D.A.; Li, C.; Parikh, S.J. Substitution of peat moss with softwood biochar for soil-free marigold growth. *Ind. Crops Prod.* **2018**, *112*, 160–169. [CrossRef]

22. Gruda, N. Current and future perspective of growing media in Europe. *Acta Hortic.* **2012**, *960*, 37–43. [CrossRef]

23. Schmilewski, G. The role of peat in assuring the quality of growing media. *Mires Peat* **2008**, *3*, 1–8.

24. Prins, F. Carbon storage in peat bog areas in Europe. In Proceedings of the Workshop on CO_2 Sequestration & Biodiversity, Bratislava, Slovakia, 9–10 November 2009.

25. Armstrong, H.; Aendekerk, T.G.L. *International Substrate Manual: Analysis, Characteristics, Recommendations*; International Business Information; Elsevier: Amsterdam, The Netherlands, 2000; ISBN 9054391022.

26. Gruda, N.; Bisbis, M.; Tanny, J. Impacts of protected vegetable cultivation on climate change and adaptation strategies for cleaner production—A review. *J. Clean. Prod.* **2019**, *225*, 324–339. [CrossRef]

27. Vaughn, S.F.; Kenar, J.A.; Eller, F.J.; Moser, B.R.; Jackson, M.A.; Peterson, S.C. Physical and chemical characterization of biochars produced from coppiced wood of thirteen tree species for use in horticultural substrates. *Ind. Crops Prod.* **2015**, *66*, 44–51. [CrossRef]

28. Altland, J.E.; Locke, J.C. Effect of biochar type on macronutrient retention and release from soilless substrate. *HortScience* **2013**, *48*, 1397–1402. [CrossRef]

29. Solaiman, Z.M.; Murphy, D.V.; Abbott, L.K. Biochars influence seed germination and early growth of seedlings. *Plant Soil* **2012**, *353*, 273–287. [CrossRef]

30. Prasad, M.; Tzortzakis, N.; McDaniel, N. Chemical characterization of biochar and assessment of the nutrient dynamics by means of preliminary plant growth tests. *J. Environ. Manag.* **2018**, *216*, 89–95. [CrossRef]

31. Gascó, G.; Álvarez, M.L.; Paz-Ferreiro, J.; Miguel, G.S.; Méndez, A. Valorization of biochars from pinewood gasification and municipal solid waste torrefaction as peat substitutes. *Environ. Sci. Pollut. Res.* **2018**, *25*, 26461–26469. [CrossRef]

32. Bedussi, F.; Zaccheo, P.; Crippa, L. Pattern of pore water nutrients in planted and non-planted soilless substrates as affected by the addition of biochars from wood gasification. *Biol. Fertil. Soils* **2015**, *51*, 625–635. [CrossRef]

33. Zaccheo, P.; Crippa, L.; Cattivello, C. Liming power of different particle fractions of biochar. *Acta Hortic.* **2014**, *1034*, 363–368. [CrossRef]

34. Steiner, C.; Harttung, T. Biochar as a growing media additive and peat substitute. *Solid Earth* **2014**, *5*, 995–999. [CrossRef]

35. Kloss, S.; Zehetner, F.; Dellantonio, A.; Hamid, R.; Ottner, F.; Liedtke, V.; Schwanninger, M.; Gerzabek, M.H.; Soja, G. Characterization of slow pyrolysis biochars: Effects of feedstocks and pyrolysis temperature on biochar properties. *J. Environ. Qual.* **2012**, *41*, 990–1000. [CrossRef] [PubMed]

36. Kern, J.; Tammeorg, P.; Shanskiy, M.; Sakrabani, R.; Knicker, H.; Kammann, C.; Tuhkanen, E.M.; Smidt, G.; Prasad, M.; Tiilikkala, K.; et al. Synergistic use of peat and charred material in growing media–an option to reduce the pressure on peatlands? *J. Environ. Eng. Landsc. Manag.* **2017**, *25*, 160–174. [CrossRef]

37. Méndez, A.; Paz-Ferreiro, J.; Gil, E.; Gascó, G. The effect of paper sludge and biochar addition on brown peat and coir based growing media properties. *Sci. Hortic.* **2015**, *193*, 225–230. [CrossRef]

38. Nemati, M.R.; Simard, F.; Fortin, J.P.; Beaudoin, J. Potential use of biochar in growing media. *Vadose Zone J.* **2015**, *14*, 1–8. [CrossRef]

39. Angst, T.E.; Sohi, S.P. Establishing release dynamics for plant nutrients from biochar. *GCB Bioenergy* **2013**, *5*, 221–226. [CrossRef]

40. Li, Y.; Liao, Y.; He, Y.; Xia, K.; Qiao, S.; Zhang, Q. Polycyclic aromatic hydrocarbons concentration in straw biochar with different particle size. *Procedia Environ. Sci.* **2016**, *31*, 91–97. [CrossRef]

41. He, P.; Liu, Y.; Shao, L.; Zhang, H.; Lü, F. Particle size dependence of the physicochemical properties of biochar. *Chemosphere* **2018**, *212*, 385–392. [CrossRef]

42. Prasad, M.; Maher, M.J. Physical and chemical properties of fractioned peat. *Acta Hortic.* **1993**, *342*, 257–263. [CrossRef]

43. Alt, D.; Peters, I.; Fokken, H. Estimation of phosphorus availability in composts and compost/peat mixtures by different extraction methods. *Commun. Soil Sci. Plant Anal.* **1994**, *25*, 2063–2080. [CrossRef]

44. EN 13651-2002. *Soil Improvers and Growing Media—Extraction of Calcium Chloride/DTPA (CAT)*; British Standards Institution: London, UK, 2002.

45. EN 13037-2002. *Soil Improvers and Growing Media—Determination of pH*; British Standards Institution: London, UK, 2002.

46. EN 13038-2002. *Soil Improvers and Growing Media—Determination of Electrical Conductivity*; British Standards Institution: London, UK, 2002.

47. Hu, X.; Ding, Z.; Zimmerman, A.R.; Wang, S.; Gao, B. Batch and column sorption of arsenic onto iron-impregnated biochar synthesized through hydrolysis. *Water Res.* **2015**, *68*, 206–216. [CrossRef] [PubMed]

48. Chrysargyris, A.; Xylia, P.; Botsaris, G.; Tzortzakis, N. Antioxidant and antibacterial activities, mineral and essential oil composition of spearmint (*Mentha spicata* L.) affected by the potassium levels. *Ind. Crops Prod.* **2017**, *103*, 202–212. [CrossRef]

49. Zainul, A.; Koyro, H.-W.; Huchzermeyer, B.; Gul, B.; Khan, M.A. Impact of a biochar or a compost-biochar mixture on water relation, nutrient uptake and photosynthesis of *Phragmites karka*. *Pedosphere* **2017**. [CrossRef]

50. Al-Wabel, M.I.; Al-Omran, A.; El-Naggar, A.H.; Nadeem, M.; Usman, A.R.A. Pyrolysis temperature induced changes in characteristics and chemical composition of biochar produced from conocarpus wastes. *Bioresour. Technol.* **2013**, *131*, 374–379. [CrossRef]

51. Calvelo Pereira, R.; Kaal, J.; Camps Arbestain, M.; Pardo Lorenzo, R.; Aitkenhead, W.; Hedley, M.; Macías, F.; Hindmarsh, J.; Maciá-Agulló, J.A. Contribution to characterisation of biochar to estimate the labile fraction of carbon. *Org. Geochem.* **2011**, *42*, 1331–1342. [CrossRef]

52. Gaskin, C.; Steiner, K.; Harris, K.C.; Das, B. Bibens Effect of Low-Temperature Pyrolysis Conditions on Biochar for Agricultural Use. *Trans. ASABE* **2008**, *51*, 2061–2069. [CrossRef]

53. Hossain, M.K.; Strezov Vladimir, V.; Chan, K.Y.; Ziolkowski, A.; Nelson, P.F. Influence of pyrolysis temperature on production and nutrient properties of wastewater sludge biochar. *J. Environ. Manag.* **2011**, *92*, 223–228. [CrossRef]

54. Gruda, N.; Caron, J.; Prasad, M.; Maher, M. Growing Media. In *Encyclopedia of Soil Sciences*; Lal, R., Ed.; CRC Press Taylor & Francis Group: Boca Raton, FL, USA, 2016; pp. 1053–1058, ISBN 9781498738903.

55. Jackson, B.E.; Wright, R.D.; Gruda, N. Container medium pH in a pine tree substrate amended with peatmoss and dolomitic limestone affects plant growth. *HortScience* **2009**, *44*, 1983–1987. [CrossRef]

 *agronomy*

Article

Biochar Type and Ratio as a Peat Additive/Partial Peat Replacement in Growing Media for Cabbage Seedling Production

Antonios Chrysargyris [1], Munoo Prasad [1,2,3], Anna Kavanagh [3] and Nikos Tzortzakis [1,*]

[1]   Department of Agricultural Sciences, Biotechnology and Food Science, Cyprus University of Technology, 3603 Lemesos, Cyprus; a.chrysargyris@cut.ac.cy (A.C.); munoo.prasad@cut.ac.cy (M.P.)

[2]   Compost/AD Research & Advisory (IE/CY), Naas, W91 TYH3 Kildare, Ireland

[3]   Bord na Mona Horticulture Ltd. Research Centre, Main Street, Newbridge, W12 XR59 Kildare, Ireland; annak62@outlook.ie

*   Correspondence: nikolaos.tzortzakis@cut.ac.cy; Tel.: +357-25002280

Received: 20 September 2019; Accepted: 23 October 2019; Published: 29 October 2019

Abstract: Biochar has been proposed mainly as a soil amendment, positively affecting plant growth/yield, and to a lesser degree for growing media. In this study, four commercial grade biochars (A-forest wood; B-husks and paper fiber; C-bamboo and D-fresh wood screening), mostly wood-based materials, were selected. Initial mixtures of peat (P) with different Biochar type and ratios (0-5-10-15-20%) were selected for cabbage seedling production. Biochar material had high K content and pH $\geq$ 8.64 which resulted in increased pH of the growing media. Biochar A and C at 20% reduced cabbage seed emergence. Biochar A, B and D maintained or improved plant growth at low ratio (i.e., 5–10%) while all Biochars increased N, K and P content in leaves. Biochars A and D were further examined at 7.5% and 15% with the addition of two doses of minerals (1-fold and 1.5-fold). Biochar A and D, initially stimulated seed emergence when compared to the control. High dose of fertilizer favored plant growth in Biochar A at 7.5% and Biochar D at 15%. Leaf stomatal conductance was decreased at Biochar A+Fert at 7.5% and Chlorophyll b content was decreased at Biochar A+Fert at 15%. The presence of Biochar A increased the antioxidant activity (as assayed by 2,2-diphenyl-1-picrylhydrazyl-DPPH). Lipid peroxidation was higher in plants grown with fertilized peat and Biochar A at 15%, activating antioxidant enzymatic metabolisms. Potassium, phosphorous and copper accumulation and magnesium deficiency in cabbage leaves were related to the Biochar presence. Wooden biochar of beech, spruce and pine species (Biochar A) at 7.5% and fertilized biochar of fruit trees and hedges (Biochar D) were more promising for peat replacement for cabbage seedling production.

Keywords: biochar; peat; growth; cabbage; *Brassica*; emergence

1. Introduction

Biochar production is a process of dry pyrolysis of organic matter, whereby plant or animal-based organic materials are treated under high temperatures ranging from 450 to 600 °C, under the absence of oxygen or low oxygen conditions [1,2], while lower temperature (300 °C) for biochar production has been reported [3]. Primary material for biochar production is mainly wastes derived from intensive sectors such as agriculture, food, forest residues and wood industries with significant contribution to environmental management and recycling, decreasing the greenhouse gas (GHG) emission and sequester carbon [4–7]. Biochar (i.e., 70%) use in agriculture as an alternative container substrate adds value to the bioenergy process with significant reduction (up to 54%) of the cost for the use of peat-based substrates [8]. Moreover, biochar can substantially improve the soil adsorption capacity for heavy metals like Cd [9] and alleviate salinity stress in crops with significant protection of the environment [10]. Nowadays, attention has been focused on the potential biochar use in growing media formulation, attracting research interest [3].

Biochar is constantly receiving increasing attention through its usage for soil modifications, as it increases crop yields and retains or improves soil fertility [11]. At the same time, effective applications are questionable, as farmers need to combine biochar application with sustainable fertilizers and water input [5]. Biochar efficacy on yield increase was attributed to the application of the material in unfertile/barren lands, rather than to fertile soils [12] and the biochar co-application with fertilizers has been suggested [13].

Compared to the commonly used peat for growing media, biochar has high pH, increased surface area, excellent water and nutrient retention properties [14] and contains different forms of N and P (i.e., ammonium, orthophosphoric), considerable amount of K [2,15]. Moreover, biochar is highly resistant to biological degradation and preserve great longevity in soil [16]. Adding biochar in soil, it can assist to maintain nutrients, release and regulate contaminants, reduce the CO_2 emission to the atmosphere, boost soil physical, chemical, and biological characteristics, and enhance microbial biomass and diversity [16–19]. Biochar particle size can affect various growing media physicochemical properties, including bulk density, total pore space and available water and air for the roots [20]. However, biochar efficacy and quality (particularly surface chemical properties and the size of the pores) relies on the feedstock and the production process [21]. Substrates with low biochar rates, i.e., 10% sewage sludge biochar in lettuce [22] and 10% wood-derived biochar in pepper and tomato [23], promoted plant growth. However, higher biochar ratio had contradicting effects with either increased plant biomass and height i.e., 60–80% conifer wood biochar in *Euphorbia × lomi* [24] or suppress growth i.e., 60–100% pinewood biochar in poinsettia [25] and 80–100% pinewood biochar in tomato and basil [26].

The section of seedling and potting horticultural plant production has been improved enormously over the last few years [27–29]. The ability of handling the mixtures of growing media by choosing the mineral levels and the raw or composted material is driven to a final substrate formation with desirable physicochemical properties. Biochar has shown the potential to be added in growing media, combined with various materials such as peat [3,30], compost [8,31], coir [3] and vermicomport [32]. Little information is accessible for the physiological responses of plants following biochar applications, as biochar is mainly acting as a soil conditioner and thus mitigating the effects of climate change [33].

Peat has traditionally been used as the major growing media component in Europe, followed by coir, perlite, bark, and compost [34]. Widely used for its well-known properties (high cation exchange capacity-CEC, low nutrient levels, low pH, suitable water holding capacity and air capacity), peat production in Europe exceeds 40 million m^3 [30,35]. However, on top of the high cost of the energy used for the extraction and transportation of peat to long distances (mainly produced at northern Europe), all these procedures are adding much to carbon footprint and increase environmental constrains. Thus, there is an increased ecological concern arising from the peat extraction, including conservation policies and identification of alternative components that could be appropriate for nursery enterprises [34,36].

Based on the favorable outcomes derived from the preliminary studies with different biochars and peat mixtures, the aim of this research was: (a) to assess the impact of biochar substitution in peat on extractable nutrients, (b) to assess four commercial biochar products as a peat diluent (growing medium) as demonstrated by plant growth, physiology and nutrient content, and (c) to evaluate the fertilizer dose and biochar ratio in peat on plant metabolism and nutrient content.

2. Materials and Methods

2.1. Biochars and Plant Material

In the current study, seeds of cabbage (*Brassica oleracea* L. var. *capitata*) were used for seedling production. Four commercial grade biochars were selected, three from Europe and one from China. They were of the following feedstocks: Forest wood e.g., beech, spruce and pine from Germany (Biochar A), husks and paper fiber wood screenings from tree branches at a ratio of 1:1 (*v/v*) from Germany (Biochar B), a three-year old wild high mountain bamboo (Biochar C), and fresh wood screenings (0–20 mm) from tree and shrub cuttings mainly from urban areas and farms (fruit trees, hedges, hedgerow management) from Switzerland (Biochar D). Biochars were generated using either the Pyreg equipment for Biochar B and C at 400–600 °C, and Biochar D at 500–600 °C, or Schotteredorf process for Biochar A at 700 °C with retention time of 15–30 min. However, owing to business sensitivity, additional data about Biochar production details is not known. A high-quality professional grade H_4-H_5 on von Post scale peat (P) was used as a control and as basic material to which the biochar was added. The selected biochars were assessed for their chemical characteristics [28], as for pH [37], Electrical Conductivity (EC) in water extract at 1:5 (v:v) ratio [38], and calcium chloride/DTPA (CAT) extractable (1:5 v:v) potassium (K) and phosphorus (P), ammonium (NH_4-N), nitrate (NO_3-N), and total extractable N (NH_4-N+NO_3-N) [39]. In brief, Biochar A had pH of 9.57; EC of 0.613 mS cm^{-1}; P of 2 mg L^{-1} and K of 1087 mg L^{-1}; Biochar B had pH of 8.83; EC of 0.420 mS cm^{-1}; P of 2 mg L^{-1} and K of 376 mg L^{-1}; Biochar C had pH of 8.64; EC of 0.450 mS cm^{-1}; P of 21 mg L^{-1}, K of 755 mg L^{-1}, and NH_4-N of 1 mg L^{-1}; and Biochar D had pH of 9.55; EC of 0.410 mS cm^{-1}; P of 3 mg L^{-1} and K of 745 mg L^{-1}. Biochars had negligible amount of NO_3-N. Peat physicochemical characteristics have been reported previously [30]. In brief, peat had pH of 3.13; EC of 0.034 mS cm^{-1}; NH_4-N of 17 mg L^{-1}; NO_3-N of 3 mg L^{-1}; K of 8 mg L^{-1} and Oxygen Uptake Rate of 5.5 mmol O_2 kg^{-1} organic matter per hour.

2.2. Preparation of Growing Media

Two individual experiments were implemented in the present study. In the first experiment (Exp. I), the examined biochars mixed into the peat in different ratio. Therefore, the four biochars (A, B, C and D) were added at the rates of 0%, 5%, 10%, 15% and 20% to the peat resulting to 17 mixtures (treatments) including control treatment of peat (100% P). Then mixtures were brought to N, P and K levels (with standard fertilizers; 1-fold) to 170 mg N L^{-1} as ammonium nitrate, 70 mg P L^{-1} as triple superphosphate and 100 mg K L^{-1} as potassium sulphate respectively for the peat-biochar mixtures and limed peat (dolomitic lime at 4 g L^{-1}) and adequate amount of trace elements. The CAT extractable N, P and K that derived from the biochars were considered and the levels of fertilizers have been adjusted accordingly. There were almost insignificant amounts of N, some P and excess of K in most cases. No K was added into the mixture in case of K excess.

In the second experiment (Exp. II), the two more promising biochars and ratios were further selected for investigation with the application of additional fertilizers (1.5-fold). Therefore, the A and D biochars were selected at the rates of 0%, 7.5%, and 15% to the peat under 1-fold (N of 170 mg L^{-1}, P of 70 mg L^{-1}, and K of 100 mg L^{-1}) or 1.5-fold (N of 255 mg L^{-1}, P of 105 mg L^{-1}, and K of 150 mg L^{-1}) of fertilizers, resulting to 10 mixtures (treatments) including control (100% peat). Then mixtures were brought to adequate N, P and K levels, as described in Exp. I. The examined treatments and chemical analysis for both experiments are presented in Tables 1 and 2.

Table 1. Effects of peat (P 100) with different biochar types (A, B, C, D) and ratio (0-5-10-15-20%) on substrate chemical properties.

	pH	EC (μS cm^{-1})	NO$_3$-N (mg kg^{-1})	NH$_4$-N (mg kg^{-1})	K (mg kg^{-1})	P (mg kg^{-1})
P 100	4.97 ± 0.06 ijk	203.37 ± 2.37 a	50.13 ± 0.58 a	47.35 ± 0.55 bc	53.40 ± 1.62 k	49.71 ± 0.58 c
P-A 5%	5.01 ± 0.06 hij	143.67 ± 1.67 g	21.51 ± 1.25 e	42.37 ± 0.49 efg	84.33 ± 2.98 h	41.04 ± 1.48 f
P-A 10%	5.19 ± 0.07 gh	96.12 ± 1.12 j	16.63 ± 0.19 g	44.12 ± 1.11 de	110.85 ± 2.39 g	46.53 ± 0.54 d
P-A 15%	5.53 ± 0.11 de	96.15 ± 2.14 j	14.65 ± 0.17 i	43.82 ± 0.51 de	147.58 ± 3.57 e	38.25 ± 1.44 g
P-A 20%	5.89 ± 0.09 c	93.08 ± 1.09 j	13.84 ± 0.21 j	43.91 ± 0.53 de	213.67 ± 4.49 d	44.36 ± 1.23 e
P-B 5%	5.16 ± 0.06 ghi	167.95 ± 1.95 c	30.38 ± 1.35 b	44.28 ± 0.65 d	65.52 ± 0.76 j	44 ± 39 ± 0.51 e
P-B 10%	5.70 ± 0.05 cd	141.65 ± 1.65 gh	17.85 ± 0.26 f	41.03 ± 0.48 g	78.09 ± 1.90 i	39.89 ± 1.14 f
P-B 15%	6.46 ± 0.11 b	137.61 ± 1.60 h	17.42 ± 0.20 f	29.35 ± 0.34 h	77.63 ± 1.94 i	31.49 ± 1.45 i
P-B 20%	7.06 ± 0.13 a	149.74 ± 1.74 f	12.85 ± 0.15 k	20.45 ± 0.24 i	78.57 ± 1.35 hi	18.57 ± 0.21 j
P-C 5%	4.77 ± 0.05 k	201.34 ± 2.34 a	31.88 ± 0.37 b	48.61 ± 1.56 b	132.21 ± 2.98 f	50.49 ± 1.59 c
P-C 10%	4.92 ± 0.05 jk	181.11 ± 2.11 b	15.52 ± 0.18 h	43.04 ± 1.52 def	221.07 ± 4.57 c	43.44 ± 1.78 e
P-C 15%	5.16 ± 0.08 ghi	148.73 ± 1.73 f	10.50 ± 0.12 l	46.78 ± 0.63 c	317.57 ± 5.78 b	56.38 ± 1.65 b
P-C 20%	5.40 ± 0.07 ef	160.87 ± 1.87 d	29.06 ± 0.34 d	72.82 ± 2.82 a	367.38 ± 4.28 a	86.35 ± 3.15 a
P-D 5%	4.76 ± 0.05 k	155.87 ± 2.65 e	22.22 ± 0.26 e	43.28 ± 1.51 def	54.98 ± 1.64 k	46.26 ± 0.54 d
P-D 10%	4.87 ± 0.05 jk	121.41 ± 1.59 i	11.03 ± 0.13 l	44.71 ± 0.52 d	54.46 ± 1.21 k	43.17 ± 0.89 e
P-D 15%	5.06 ± 0.08 hij	93.08 ± 1.08 j	6.85 ± 0.28 m	43.07 ± 063 def	55.24 ± 1.76 k	37.45 ± 1.23 g
P-D 20%	5.32 ± 0.09 fg	98.14 ± 1.14 j	6.02 ± 0.07 m	41.87 ± 0.49 fg	79.97 ± 2.81 hi	34.09 ± 0.89 h

Values ($n = 2$) in columns followed by the same letter are not significantly different, $p \leq 0.05$.

Table 2. Effects of peat (P 100) with different biochar types (A, D) and ratio (7.5%, 15%) and mineral doses (with standard or with additional Fertilizers-Fert.) on substrate minerals content.

	P 100	PFert 100	P-A 7.5%	PFert-A 7.5%	P-A 15%	PFert-A 15%	P-D 7.5%	PFert-D 7.5%	P-D 15%	PFert-D 15%
NO_3-N (mg kg^{-1})	63.38 ± 1.38 c	88.42 ± 1.92 a	78.20 ± 1.71 b	81.26 ± 1.76 b	37.82 ± 0.82 f	48.04 ± 1.04 e	47.53 ± 1.03 e	58.78 ± 1.28 d	36.29 ± 0.89 f	31.69 ± 0.69 g
NH_4-N (mg kg^{-1})	20.95 ± 0.45 g	67.98 ± 1.48 b	18.51 ± 2.03 i	43.95 ± 0.95 d	33.22 ± 0.72 f	64.16 ± 1.40 c	36.80 ± 0.85 e	71.55 ± 1.55 a	32.12 ± 1.56 h	68.49 ± 1.49 ab
K (mg kg^{-1})	117.55 ± 3.55 f	178.89 ± 3.89 d	230.01 ± 5.02 c	301.55 ± 6.55 b	378.22 ± 8.23 a	373.10 ± 8.24 a	102.25 ± 2.26 f	153.34 ± 3.38 e	102.26 ± 2.21 f	148.27 ± 3.25 e
P (mg kg^{-1})	35.78 ± 0.78 g	65.93 ± 1.43 c	44.46 ± 1.96 e	71.04 ± 1.63 ab	40.38 ± 0.88 f	57.75 ± 1.82 d	44.46 ± 1.32 e	68.79 ± 1.49 bc	36.81 ± 0.98 fg	74.11 ± 1.61 a
Ca (mg kg^{-1})	695.11 ± 15.11 a	710.95 ± 15.45 a	695.12 ± 15.12 a	721.17 ± 15.67 a	718.62 ± 16.51 a	712.48 ± 15.48 a	516.73 ± 11.23 c	519.80 ± 11.35 c	481.46 ± 10.46 c	609.24 ± 13.24 b
Mg (mg kg^{-1})	482.13 ± 10.48 a	474.82 ± 12.36 a	326 ± 55 ± 7.10 b	327.72 ± 7.12 b	237.71 ± 5.16 d	234.19 ± 5.09 d	307.79 ± 6.69 b	314.64 ± 6.84 b	239.21 ± 5.23 d	274.72 ± 5.97 c
Na (mg kg^{-1})	45.49 ± 0.99 cd	46.03 ± 1.02 c	41.40 ± 0.90 de	45.49 ± 0.99 cd	42.42 ± 0.92 cde	40.89 ± 0.89 e	63.89 ± 1.39 b	62.86 ± 1.36 b	88.42 ± 1.92 a	84.85 ± 1.86 a
Fe (mg kg^{-1})	9.15 ± 0.22 bc	8.18 ± 0.18 ef	9.45 ± 0.21 b	8.43 ± 0.18 de	8.07 ± 0.17 ef	7.72 ± 0.17 f	8.28 ± 0.18 def	8.58 ± 0.18 cde	8.94 ± 0.19 bcd	14.26 ± 0.31 a
Cu (mg kg^{-1})	0.10 ± 0.00 e	0.10 ± 0.00 e	0.05 ± 0.00 f	0.05 ± 0.00 f	0.05 ± 0.00 f	0.05 ± 0.00 f	0.20 ± 0.01 c	0.25 ± 0.01 b	0.16 ± 0.01 d	0.31 ± 0.01 a
Zn (mg kg^{-1})	1.07 ± 0.02 e	0.97 ± 0.02 e	1.02 ± 0.02 e	1.17 ± 0.02 d	1.02 ± 0.02 e	0.97 ± 0.02 e	1.53 ± 0.03 c	1.58 ± 0.04 c	2.09 ± 0.05 b	2.21 ± 0.05 a
Mn (mg kg^{-1})	2.40 ± 0.05 e	2.40 ± 0.05 e	8.89 ± 0.19 b	8.88 ± 0.019 b	12.83 ± 0.28 a	12.62 ± 0.27 a	2.60 ± 0.05 de	2.66 ± 0.06 de	2.96 ± 0.07 cd	3.37 ± 0.07 c
B (mg kg^{-1})	0.61 ± 0.01 c	0.61 ± 0.01 c	0.66 ± 0.01 c	0.66 ± 0.01 c	0.68 ± 0.01 c	0.67 ± 0.01 c	0.82 ± 0.02 b	0.83 ± 0.02 b	0.97 ± 0.02 a	0.87 ± 0.02 b

Values ($n = 3$) in rows followed by the same letter are not significantly different, $p \leq 0.05$.

2.3. Seed Emergence

Cabbage seeds were sown (1 cm depth) in plastic seedling trays. Each treatment had 9 and 18 modules for Exp. I and II, respectively, of 40 cm^3 volume capacity each. Three seeds were placed in each module. Irrigation was performed daily with equal amount of water for all growing media, in order to cover the watering needs of the young seedlings. During seedling growth in the nursery, no fertilizers were applied. Max and min temperatures were 25 ± 2 °C and 20 ± 2 °C, respectively. Day light hours was L:D 16:8 with light flux density 300 µmol PAR m^{-2} s^{-1} ± 20.

A daily observation on seed emergence took place and seeds were recorded emerged when the hypocotyls were appeared. Mean emergence time (MET) was calculated as described previously [40].

2.4. Vegetative Growth and Mineral Content

Following a growing period of four to six weeks, seedlings growth-related parameters were recorded in six seedlings/treatment. Plant height and leaf number of the seedlings were measured. Leaf stomatal conductance was measured by using a ∆T-Porometer AP4 (Delta-T Devices-Cambridge, Burwell, Cambridge, UK). Leaf chlorophyll fluorescence (chlorophyll fluorometer, opti-sciences OS-30p, UK) was measured on two fully developed, light-exposed leaves per seedling. Leaves were incubated in the dark for 20 min prior to Fv/Fm measurements. Leaf chlorophyll content was assayed in six replicates/treatment either by SPAD meter or photometrically. Chlorophylls were extracted with dimethyl sulfoxide (DMSO) and Chlorophyll a (Chl a), Chlorophyll b (Chl b) and total Chlorophylls (total Chl) content was determined [28]. Seedlings were sampled above substrate surface, upper plant part was weighed (g), dried at 85 °C and then dry weight (g) was measured.

Mineral content in the upper part of the seedlings (including leaves and shoots) was determined on four replicates/treatment (two pooled plants/replicate). Plant tissue was dried to constant weight (at 65 °C for 3 day) and sub samples (~0.5 g) were ashed (at 500 °C for 5 h) and acid (2 N HCl) digested [41]. Nitrogen (N) content was determined with Kjeldahl (BUCHI, Digest automat K-439 and Distillation Kjeldahl K-360) digestion method. Phosphorus content was determined with spectrophotometer (Multiskan GO, Thermo Fisher Scientific, Waltham, MA, USA), and K, Mg, Ca, Na, Fe, Cu, Zn, and B by an atomic absorption spectrophotometer (PG Instruments AA500FG, Leicestershire, UK) for plant tissue analysis or by inductively coupled plasma atomic emission spectrometry (ICP-AES; PSFO 2.0 (Leeman Labs Inc., Mason, OH, USA) for growing media analysis. Plant mineral content were expressed in g kg^{-1} and mg kg^{-1} of dry weight, for macronutrients and micronutrients, respectively. Biochar-based media minerals were expressed in mg L^{-1}.

2.5. Total Phenolics and Antioxidant Capacity

In the Exp. II, methanolic extracts of four replicates (two pooled plants/replicate) of cabbage grown in different biochar types and ratio used for the determination of total phenolics and total antioxidant activity. The Folin–Ciocalteu method was used for the total phenolics content as described in Tzortzakis et al. [42] and results were expressed as gallic acid equivalents (mg GAE per g of fresh weight). For antioxidant capacity, two assays were used, the ferric reducing antioxidant power (FRAP) and the 2,2-diphenyl-1-picrylhydrazyl (DPPH), as described previously [43]. Results were expressed as trolox equivalents (mg trolox per g of fresh weight).

2.6. Lipid Peroxidation, Hydrogen Peroxide, and Enzymes Antioxidant Activity

Four replicates (each replicate was a poll of two plants) for each treatment were used for damage index and antioxidant enzymes activity. Lipid peroxidation and hydrogen peroxide (H$_2$O$_2$) content were assessed according to Loreto and Velikova [44] and De Azecedo Neto et al. [45]. The results were expressed as µmol H$_2$O$_2$ per g of fresh weight, while lipid peroxidation was calculated through the malondialdeyde (MDA) content (nmol of MDA per g of fresh weight).

The enzymes antioxidant activity for superoxide dismutase (SOD), for catalase (CAT) and for peroxidase activity (POD) was assayed as described previously [43]. Results were expressed as enzyme units per mg of protein. The protein content was determined by using bovine serum albumin (BSA) as a standard.

2.7. Statistical Analysis

Data were tested for normality and then statistically analyzed using analysis of variance (ANOVA) by SPSS v21.0 (SPSS Inc., Chicago, IL, USA) program. The significance of the differences between average values was based on Duncan's Multiple Range test (DMRT) at $p \leq 0.05$, following one-way ANOVA. Values are means ± standard error (SE).

3. Results

3.1. Growing Media Properties

The biochar raw material had, in general, very high pH (ranging from 8.64 to 9.57) and considerable levels of EC (ranging from 0.410 to 0.613 mS cm^{-1}). Therefore, adding biochars in ratios from 5% to 20% (Exp. I) increased the pH value of the acidic (pH of 4.97) peat-based material (Table 1). Moreover, biochar-based media had lower EC compared to the control (fertilized peat). The examined Biochars (A, B, C and D) had limited amounts of NH$_4$-N and NO$_3$-N, and this reflected the decreased levels found on the biochar-based growing media. Similar to ammonium and nitrate levels, the low (~2 mg L^{-1}) P amounts of Biochar A, B and D reflected the decreased levels of P content in the growing media. However, Biochar C had P of 21 mg L^{-1}, and as such, the P levels in the growing media increased for the ≥15% Biochar C. Interestingly, K levels of raw Biochars ranged from 376 to 1087 mg L^{-1} affected the K content in the examined biochar-based growing media, and the values increased as the Biochar ratio increased from 5% to 20% (Table 1).

Following the selection of Biochars for the Exp. II, a detailed mineral composition of the examined Biochars (A and D), ratios (7.5% and 15%) and fertilizers dose (1-fold and 1.5 fold) presented in Table 2. The additional fertilizer (1.5-fold) increased, as expected, the levels of N, K, and P at the 100% fertilized peat compared with the control (P100). Growing media containing Biochar A at 15% and Biochar D at 7.5% and 15% at both fertilizers (1-fold and 1.5-fold) levels had decreased NO$_3$-N compared to the control treatment (100% peat). The level of NH$_4$-N increased with the presence of Biochars A and D except for the Biochar A at 7.5%, with more pronounced content at the fertilized media (Table 2). Potassium levels were increased at Biochar A-based growing media and at fertilized Biochar D media (i.e., Biochar D+Fert at 7.5% and 15%) compared to 100% peat media. Increased Ca levels were found at Biochar A-based media while Biochar D-media had reduced Ca levels. Magnesium levels were decreased in both Biochars-based media. Boron, Zn, Na and Cu levels increased in case of Biochar D presence and reduced (for Cu) in case of Biochar A. Phosphorous and Mn levels increased at the Biochar-based media, while increased P levels were found also at the 100% fertilized peat. Iron content decreased in general with the presence of Biochars A and D, with exception the Biochar D+Fert at 15% media (Table 2).

3.2. Experiment I

3.2.1. Seed Emergence

In Exp. I, four biochars in four ratios were primary evaluated for cabbage seedling production. Biochar A and C at 20% reduced cabbage seed emergence compared to 100% peat (P100) as control treatment after 8 day (Figure 1A,C). Biochar B did not affect seed emergence (Figure 1B), and Biochar D (10–20%) decreased seed emergence at the first 3 day but no differences were obtained thereafter comparing with the control (Figure 1D). In general, low biochar ratios (5–10%) stimulated seed emergence for Biochar A and D compared to the control treatment for the first 3rd days.

Mean germination time is shown in Figure 1E, and it was found that Biochar C and Biochar D at ≥10% delayed the seed emergence as they had higher MET comparing to control treatment (P100). Biochar A and B did not affect the MET (Figure 1E).

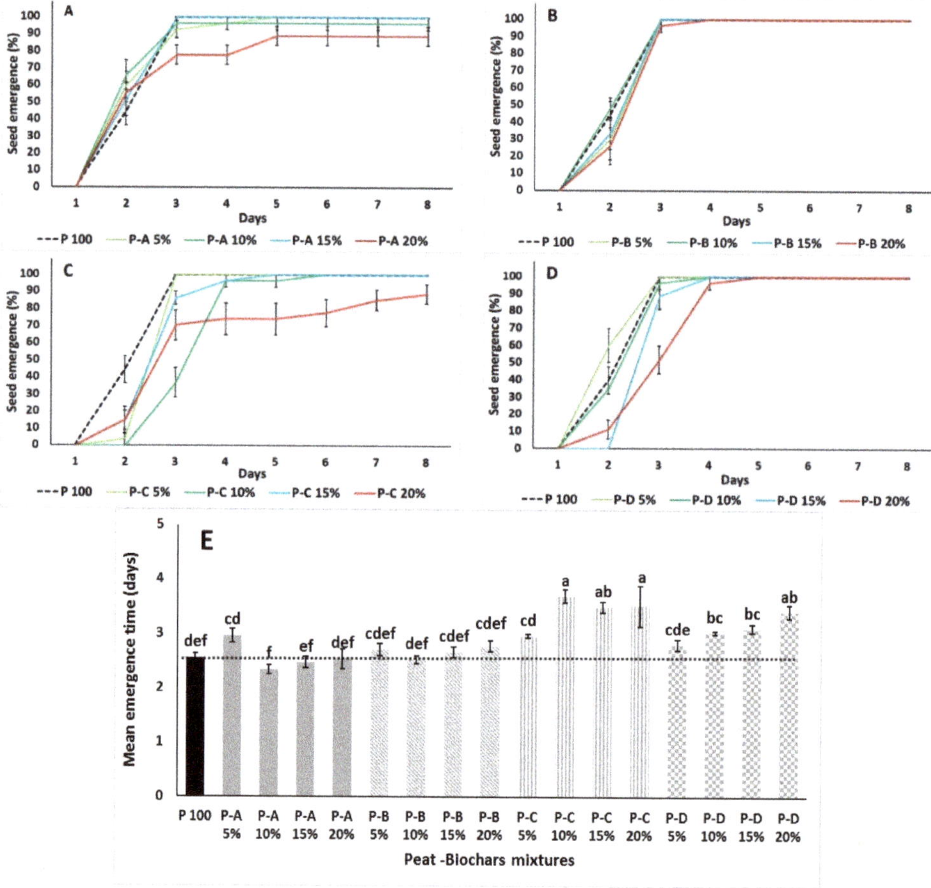

Figure 1. Cabbage cumulative seedling emergence and mean emergence time (MET) in peat with different biochar types (A, B, C, D) and ratio (0-5-10-15-20%). Biochar type is distinguished by different pattern at MET. Significant differences ($p < 0.05$) among treatments are indicated by different letters. Error bars show SE ($n = 4$). Dotted line present the levels of control treatment (100% peat). (**A**) is referring to Biochar A, (**B**) is Biochar B, (**C**) is Biochar C, (**D**) is Biochar D, (**E**) is mean emergence time for all biochars.

3.2.2. Plant Development

Biochar A and B at 5% increased but Biochar B at 20% decreased cabbage height compared to plants grown in 100% peat (Table 3). Biochar A (at 10%), Biochar B (at 5–10%), and Biochar D (at 5–10%) increased seedling fresh weight, while 20% of Biochars B and D and 10% of Biochar C decreased seedling fresh weight. Increasing Biochar ratio into the growing media, resulted in decreased plant dry weight compared to the control. The number of leaves produced did not differ among types and ratio of biochar. Biochars C and D at high levels affected negatively the cabbage root length (Table 3). Leaf chlorophyll fluorescence and content (SPAD units) were differently affected by the biochar type and ratio, with often more pronounced decreases at the higher biochars levels.

Table 3. Effects of peat (P 100) with different biochar types (A, B, C, D) and ratio (0-5-10-15-20%) on plant growth (height in cm, upper fresh weight in g, upper dry weight in g, root length in cm), chlorophyll fluorescence (Fv/Fm), chlorophyll content (SPAD units), and minerals content (N, K, P in g kg^{-1}).

	Height	Upper Fresh Weight	Upper Dry Weight	Root Length	Chlorophyll Fluorescence	SPAD	N	K	P
P 100	2.46 ± 0.11 bcd	2.62 ± 0.10 ef	0.261 ± 0.011 a	6.15 ± 0.23 abc	0.832 ± 0.042 a	27.92 ± 2.84 a	22.19 ± 0.90 g	14.70 ± 1.74 k	7.03 ± 0.19 f
P-A 5%	2.91 ± 0.07 a	3.59 ± 0.12 a	0.181 ± 0.005 cd	5.38 ± 0.41 bcd	0.701 ± 0.103 ab	24.82 ± 1.95 abc	39.11 ± 0.86 de	35.58 ± 1.19 hi	9.43 ± 0.10 de
P-A 10%	2.76 ± 0.13 ab	3.45 ± 0.11 a	0.170 ± 0.009 de	4.88 ± 0.21 bcd	0.773 ± 0.022 a	15.75 ± 1.12 ef	39.83 ± 0.28 cde	49.60 ± 0.18 de	10.01 ± 0.02 cd
P-A 15%	2.65 ± 0.11 abc	2.95 ± 0.06 cdef	0.147 ± 0.009 ef	6.13 ± 0.25 abc	0.753 ± 0.046 a	21.87 ± 2.88 abcde	34.09 ± 0.09 f	45.07 ± 2.07 efg	9.02 ± 0.11 e
P-A 20%	2.26 ± 0.08 cde	2.70 ± 0.17 ef	0.165 ± 0.012 de	6.03 ± 0.32 abcd	0.559 ± 0.049 b	27.77 ± 1.63 a	40.10 ± 1.13 cde	56.51 ± 1.09 c	9.87 ± 0.25 cde
P-B 5%	2.96 ± 0.15 a	3.35 ± 0.13 ab	0.208 ± 0.013 bc	5.75 ± 0.81 abcd	0.701 ± 0.033 ab	25.95 ± 2.36 ab	32.37 ± 0.23 f	26.45 ± 1.88 j	9.42 ± 0.32 de
P-B 10%	2.71 ± 0.16 ab	3.30 ± 0.14 abc	0.170 ± 0.013 de	5.53 ± 0.70 abcd	0.551 ± 0.074 b	18.30 ± 0.78 def	37.44 ± 2.28 e	41.22 ± 2.33 gh	10.57 ± 0.42 bc
P-B 15%	2.41 ± 0.11 bcd	2.99 ± 0.10 bcde	0.144 ± 0.007 efg	6.13 ± 0.63 abc	0.541 ± 0.077 b	18.62 ± 1.38 cdef	42.45 ± 1.87 bcd	48.15 ± 1.59 ef	10.19 ± 0.30 cd
P-B 20%	1.96 ± 0.11 e	2.14 ± 0.17 g	0.113 ± 0.009 ghi	6.36 ± 0.26 abc	0.492 ± 0.065 b	22.72 ± 1.55 abcd	38.88 ± 1.19 de	55.35 ± 2.16 cd	11.23 ± 0.05 ab
P-C 5%	2.20 ± 0.14 de	2.66 ± 0.08 ef	0.131 ± 0.005 fgh	5.30 ± 0.45 bcd	0.529 ± 0.024 b	14.47 ± 1.53 f	39.95 ± 0.49 cde	60.15 ± 5.13 c	10.63 ± 0.16 bc
P-C 10%	2.28 ± 0.11 cde	1.74 ± 0.12 h	0.091 ± 0.006 i	4.84 ± 0.75 cd	0.569 ± 0.021 b	16.97 ± 1.09 def	43.31 ± 0.67 bc	75.95 ± 1.44 b	11.63 ± 0.34 a
P-C 15%	2.65 ± 0.15 abc	2.96 ± 0.17 bcdef	0.156 ± 0.010 def	4.42 ± 0.69 d	0.561 ± 0.098 b	19.42 ± 1.97 cdef	41.37 ± 0.06 cd	74.89 ± 0.86 b	9.39 ± 0.15d e
P-C 20%	2.40 ± 0.13 bcd	2.83 ± 0.13 def	0.152 ± 0.009 def	4.50 ± 0.15 d	0.725 ± 0.070 ab	19.62 ± 1.49 cdef	45.08 ± 1.29 ab	82.17 ± 1.10 a	10.64 ± 0.66 bc
P-D 5%	2.20 ± 0.09 de	3.22 ± 0.07 abcd	0.215 ± 0.005 b	7.01 ± 0.22 a	0.751 ± 0.084 a	23.30 ± 2.03 abcd	33.21 ± 0.46 f	33.17 ± 1.42 i	9.55 ± 0.26 de
P-D 10%	2.50 ± 0.13 bcd	3.14 ± 0.04 abcd	0.164 ± 0.013 de	5.10 ± 0.50 bcd	0.584 ± 0.062 ab	17.35 ± 1.16 def	41.08 ± 1.35 cd	35.68 ± 2.01 hi	10.65 ± 0.02 bc
P-D 15%	2.65 ± 0.08 abc	2.55 ± 0.15 f	0.141 ± 0.007 efg	4.76 ± 0.31 cd	0.509 ± 0.076 b	20.35 ± 3.62 bcdef	42.74 ± 0.07 bc	42.51 ± 0.64 fg	11.83 ± 0.12 a
P-D 20%	2.73 ± 0.07 ab	2.11 ± 0.08 g	0.104 ± 0.002 hi	4.45 ± 0.29 d	0.429 ± 0.016 c	17.45 ± 1.09 def	48.09 ± 1.35 a	49.21 ± 2.21 e	10.60 ± 0.18 bc

Values ($n = 4$) in columns followed by the same letter are not significantly different, $p \leq 0.05$.

The use of biochars in the growing media significantly increased N, K and P content in cabbage leaves. Nitrogen increased (up to 80.7%, 91.3%, 103.2% and 116.7%), potassium increased (up to 284.4%, 276.5%, 459.0% and 234.8%) and phosphorus increased (up to 42.4%, 59.7%, 65.4% and 68.3%) for Biochars A, B, C and D, respectively, in relation to control (Table 3).

3.3. Experiment II

3.3.1. Seed Emergence

Following the Exp. I evaluation, two Biochars (A and D) under two ratio (7.5% and 15%) were further selected, including two mineral doses (1-fold and 1.5-fold). Biochar A and D improved seed emergence initially when compared to the control, while no differences were found after 4 day (Figure S1). Neither the biochar type nor the biochar ratio and applied fertilizers affected the mean emergence time for cabbage seeds.

3.3.2. Plant Growth and Physiology

Biochar A+Fert at 15% and Biochar D at 7.5% decreased plant height, comparing with the control treatment, while the greater plant height was found at the Biochar D+Fert at 15% (Table 4). Biochar A+Fert at 7.5% and Biochar D at 15% (independently of the fertilizers dose) increased seedling fresh weight, while Biochar A+Fert at 7.5% and Biochar D at 15% revealed increased dry weight. No differences were found on leaf number produced on biochar-based media and control (Figure S2), while the higher leaf number was found at the Biochar A+Fert at 7.5% and Biochar D+Fert at 15%.

Table 4. Effects of peat (P 100) with different biochar types (A, D) and ratio (7.5%, 15%) and mineral doses (with standard or with additional Fertilizers-Fert.) on cabbage plant growth (height in cm, upper fresh weight in g, upper dry weight in g, root length in cm).

	Height	Leaf Number	Upper Fresh Weight	Upper Dry Weight
P 100	5.40 ± 0.32 abc	3.33 ± 0.21 ab	1.962 ± 0.065 de	0.098 ± 0.005 c
PFert 100	4.98 ± 0.25 bcd	3.00 ± 0.00 b	2.230 ± 0.218 bcde	0.098 ± 0.014 c
P-A 7.5%	4.83 ± 0.31 bcd	3.00 ± 0.00 b	2.210 ± 0.074 bcde	0.124 ± 0.012 bc
PFert-A 7.5%	4.83 ± 0.23 bcd	3.50 ± 0.22 a	2.834 ± 0.147 a	0.177 ± 0.006 a
P-A 15%	4.77 ± 0.26 bcd	3.00 ± 0.00 b	2.033 ± 0.240 cde	0.107 ± 0.012 c
PFert-A 15%	4.15 ± 0.32 d	3.17 ± 0.17 ab	1.724 ± 0.180 e	0.095 ± 0.008 c
P-D 7.5%	4.20 ± 0.19 d	3.00 ± 0.00 b	2.091 ± 0.201 bcde	0.121 ± 0.009 bc
PFert-D 7.5%	4.40 ± 0.26 cd	3.00 ± 0.00 b	2.387 ± 0.301 abcd	0.110 ± 0.008 c
P-D 15%	5.52 ± 0.14 ab	3.00 ± 0.00 b	2.633 ± 0.218 abc	0.144 ± 0.007 b
PFert-D 15%	6.15 ± 0.64 a	3.50 ± 0.22 a	2.681 ± 0.0.091 ab	0.121 ± 0.005 bc

Values ($n = 6$) in columns followed by the same letter are not significantly different, $p < 0.05$.

Leaf stomatal conductance decreased at Biochar A+Fert at 7.5% and Chlorophyll b content decreased at Biochar A+Fert at 15%. No major differences were found on leaf SPAD measurements, the content of Chlorophyll a and total Chlorophylls in cabbage seedling subjected to different biochar types, ratios and fertilizer application (Table 5).

Table 5. Effects of peat (P 100) with different biochar types (A, D) and ratio (7.5%, 15%) and mineral doses (with standard or with additional Fertilizers-Fert.) on leaf chlorophyll content (SPAD units), leaf stomatal conductance (cm s^{-1}) and chlorophylls (Chl a, Chl b, total Chls) content (mg g^{-1}).

	Stomatal Conductance	SPAD	Chl a	Chl b	Total Chls
P 100	2.69 ± 0.29 ab	18.02 ± 1.99 ab	0.322 ± 0.025 abcd	0.177 ± 0.010 ab	0.439 ± 0.035 abcd
PFert 100	3.08 ± 0.65 a	18.75 ± 1.96 ab	0.311 ± 0.003 bcd	0.112 ± 0.003 b	0.422 ± 0.003 bcd
P-A 7.5%	2.08 ± 0.29 abc	15.05 ± 1.00 b	0.362 ± 0.020 ab	0.125 ± 0.006 ab	0.487 ± 0.026 ab
PFert-A 7.5%	1.11 ± 0.23 c	18.50 ± 2.55 ab	0.359 ± 0.034 ab	0.114 ± 0.009 ab	0.474 ± 0.043 abc
P-A 15%	2.07 ± 0.42 abc	15.43 ± 1.26 ab	0.274 ± 0.024 cd	0.097 ± 0.011 bc	0.371 ± 0.035 cd
PFert-A 15%	2.73 ± 0.47 ab	15.25 ± 0.53 ab	0.260 ± 0.006 d	0.081 ± 0.002 c	0.342 ± 0.008 d
P-D 7.5%	1.89 ± 0.06 bc	20.02 ± 1.25 ab	0.345 ± 0.049 abc	0.117 ± 0.015 ab	0.461 ± 0.063 abc
PFert-D 7.5%	2.00 ± 0.11 abc	18.68 ± 1.78 ab	0.319 ± 0.016 abcd	0.116 ± 0.008 ab	0.435 ± 0.025 abcd
P-D 15%	2.04 ± 0.37 abc	20.37 ± 1.21 a	0.340 ± 0.018 abcd	0.110 ± 0.004 b	0.450 ± 0.022 abcd
PFert-D 15%	1.60 ± 0.22 bc	18.18 ± 0.75 ab	0.397 ± 0.029 a	0.142 ± 0.012 a	0.539 ± 0.041 a

Values (n = 6) in columns followed by the same letter are not significantly different, $p < 0.05$.

3.3.3. Total Phenolics and Antioxidant Activity

Total phenolic content did not change much among the different treatments with the exception of Biochar D at 7.5% which revealed the highest content of phenolics (Figure 2A). Biochar A presence increased the antioxidant activity (as assayed by DPPH) of cabbage, while in case of Biochar D, DPPH increased at Biochar D at 7.5% and at Biochar D+Fert at 15% (Figure 2B). FRAP antioxidant activity revealed increased values in Biochar A at 15% and Biochar D at 7.5% (Figure 2C).

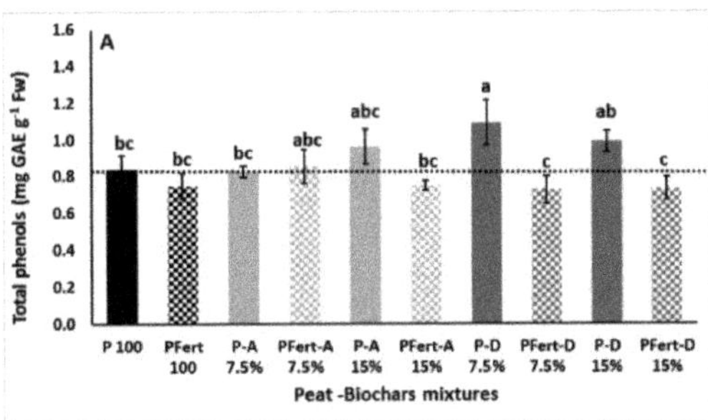

Figure 2. *Cont.*

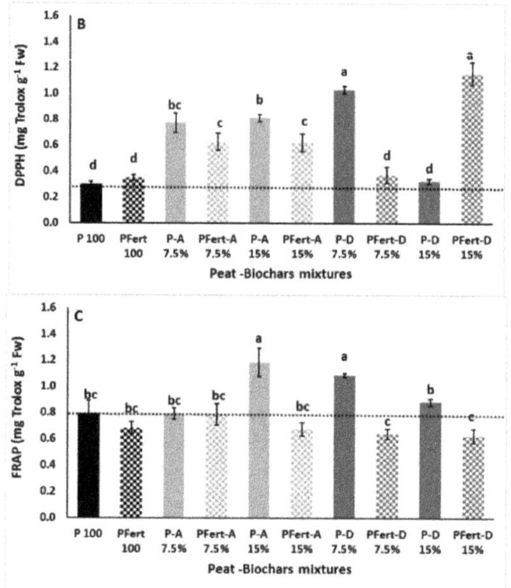

Figure 2. Effects of peat (P 100) with different biochar types (A, D) and ratio (7.5%, 15%) and mineral doses (with standard or with additional Fertilizers-Fert.) on cabbage total phenols and antioxidant activity. (**A**) total phenols, (**B**) DPPH, (**C**) FRAP. Significant differences ($p < 0.05$) among treatments are indicated by different letters. Error bars show SE ($n = 4$). Dotted line present the levels of control treatment (100% peat).

3.3.4. Lipid Peroxidation, Hydrogen Peroxide, and Enzymes Antioxidant Activity

Lipid peroxidation (MDA) increased at 100% fertilized peat compared to the non-fertilized (control) treatment (Figure 3A). Additionally, MDA increased at 7.5% of Biochar D and for Biochar A+Fert at 15% when compared to the relevant control treatments. This increase indicates cellular damage and increased stress of the plants due to the applied treatment. The production of hydrogen peroxide increased in Biochar A and D (at 7.5% and 15%), and this increase was maintained in fertilized Biochar A, but not in fertilized Biochar D (Figure 3B). In order for the plants to detoxify the increased stress, CAT antioxidant enzymatic activity was increased for Biochar A treatments (Figure 3D). SOD activity decreased for Biochar A at 15%, Biochar D at 7.5%, Biochar D+Fert at 7.5% and Biochar D+Fert at 15%, compared to the control (Figure 3C). POD activity at the fertilized peat (PFert 100) and Biochar A+Fert at 15% maintained a similar levels as the 100% peat but decreased in all other treatments (Figure 3E).

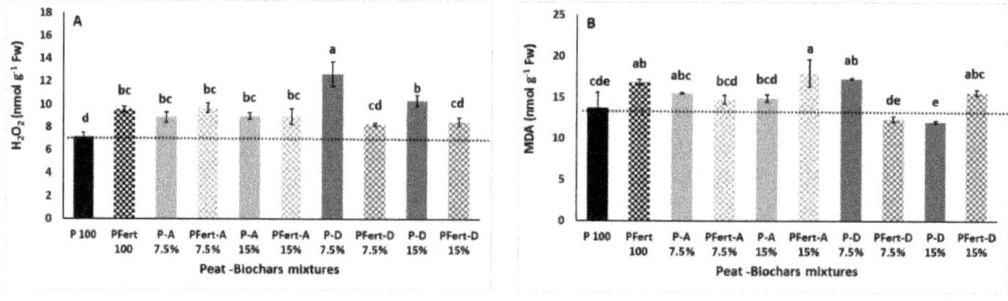

Figure 3. *Cont.*

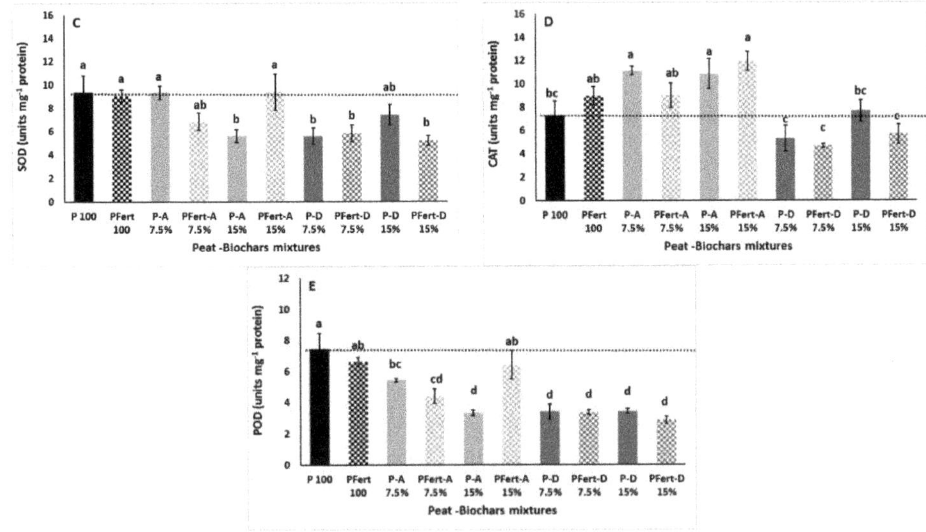

Figure 3. Effects of peat (P 100) with different biochar types (A, D) and ratio (7.5%, 15%) and mineral doses (with standard or with additional Fertilizers-Fert.) on cabbage lipid peroxidation, hydrogen peroxide and antioxidant enzymes activity. (**A**) H_2O_2, (**B**) Lipid peroxidation (MDA), (**C**) SOD, (**D**) CAT, and (**E**) POD. Significant differences ($p < 0.05$) among treatments are indicated by different letters. Error bars show SE ($n = 4$). Dotted line present the levels of control treatment (100% peat).

3.3.5. Mineral Content

The addition of Biochar A with fertilizers, as expected, increased the N accumulation in cabbage seedlings and the effects were more pronounced with 7.5% of Biochar A+Fert (Figure 4A). However, the low fertilized Biochar D reduced the N content in cabbage, while plants grown with Biochar D+Fert at 15% had increased N accumulation compared to the relevant plants grown in 100% peat. Interestingly, Biochar A increased the K accumulation in seedlings, while both Biochar ratios and fertilizer addition, increased the K accumulation. However, Biochar D needed to be fertilized and used at 15% into the mixture in order to increase the K accumulation in cabbage seedling to levels similar to the control (P-100) (Figure 4B). A similar trend to K was found for the P accumulation in the plant tissue (Figure 4C). Calcium content in cabbage increased with Biochar A at 15% (independently of the fertilizers dose) and Biochar D+Fert at 7.5% and at 15%, but was reduced with Biochar A at 7.5% (Figure 4D). Magnesium content decreased with Biochar and the effects were more pronounced in high ratio of 15% (Figure 4E). Sodium accumulation was higher with Biochar D at 15% (independent of the fertilizers dose) and lower for Biochar A at 7.5% (independent of the fertilizers dose) and 100% fertilized peat (Figure 4F). Biochar presence decreased the Fe content in cabbage while the fertilizer alleviated this effect, as Fe content was in similar levels to peat-based substrates (absence of Biochar) (Figure 4G). Copper increased with the presence of Biochar and/or fertilizers while Zn was fluctuated among the examined treatments (Figure 4H,I).

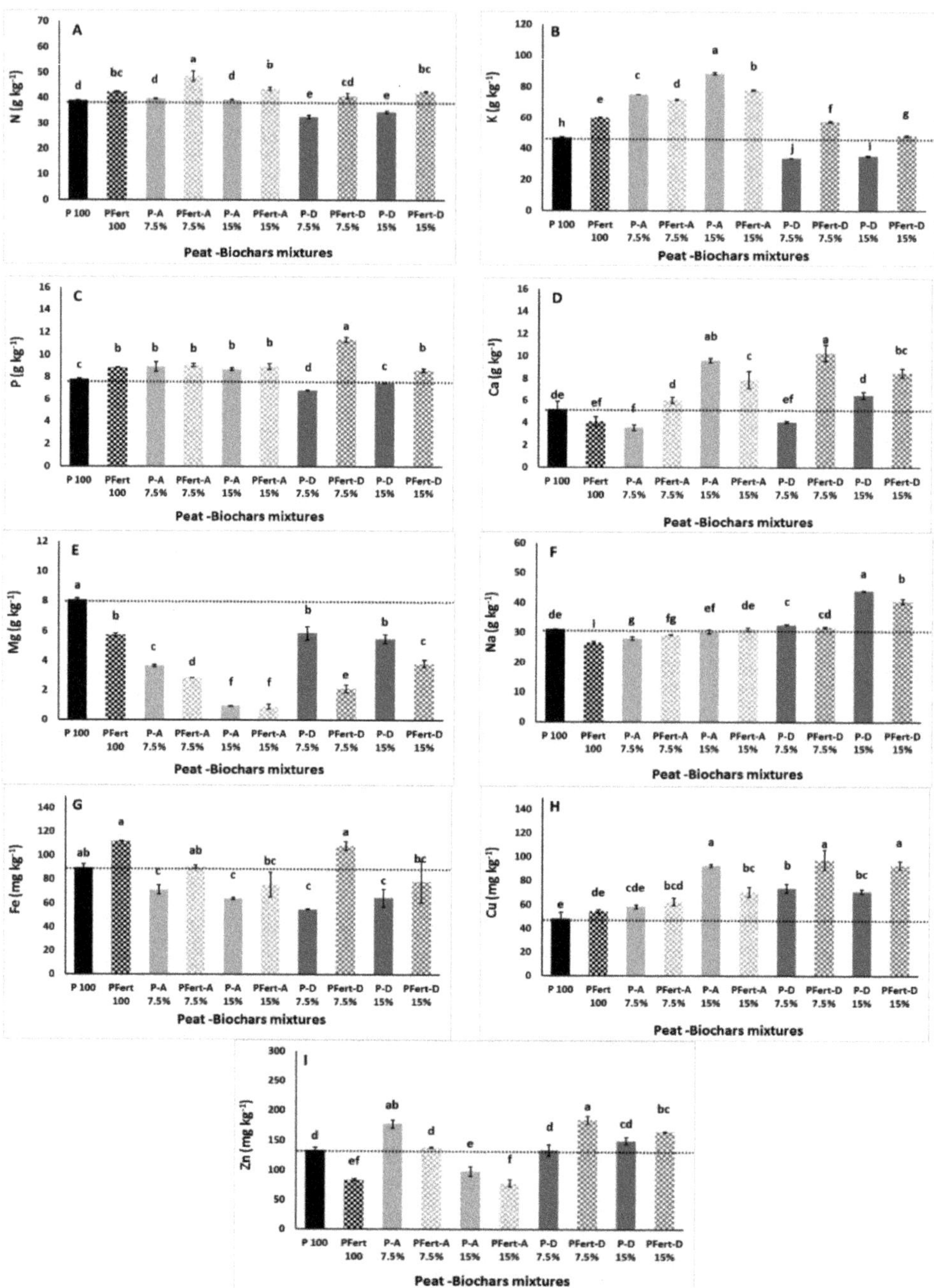

Figure 4. Effects of peat (P 100) with different biochar types (A, D) and ratio (7.5%, 15%) and mineral doses (with standard or with additional Fertilizers-Fert.) on cabbage macro (**A–F**) and micronutrient (**G–I**) content. Significant differences ($p < 0.05$) among treatments are indicated by different letters. Error bars show SE ($n = 4$). Dotted line present the levels of control treatment (100% peat).

4. Discussion

Biochar can actively restore carbon to the soil, affecting environmental parameters such as carbon footprint, and therefore, is attracting research interest for a wide range of applications in the environment, agriculture and horticulture fields [46]. In the current work, biochar application was evaluated as a growth medium amendment, as different types of biochar can have different properties and cause various effects on plants. Biochar from woody feedstock with higher lignin content and higher surface area showed different sorption abilities on metals [47]; Biochar A had the best performance in the present study, and that could be a possible explanation. However, further studies are needed before final conclusions are made. Therefore, the successful application is related to the biochar type (raw material), the ratios and to the levels of fertilizers. It is known that biochar has been effectively produced from various organic materials including municipal solid wastes (garden pruning waste), agricultural (straw, greenhouse crop residues, olive-mill waste, vineyard by-products), food waste, digestate and even sewage sludge [46,48,49]. Additionally, according to reports "not all biochars are produced in the same way". Even biochars from the same source (wood-based materials), as examined in the present study, can have different impacts on plant growth and cultivation strategies, and present biochar-specific and site-specific effects on plants [50].

Biochar has mainly been studied in applications in soil but recently, during the last 10 years, there has been a big increase in research studies and publications in the area of peat substitution by biochar [22,23,36,46,51]. Biochar addition in different ratios, as presented in Exp. I, improved growing media properties, with pH increases to more adequate levels, compared to the acidic peat-based materials, for vegetables seedling production and provided considerable amount of basic nutrients, including K and P. The increased pH in the current study is in agreement with previous reports on Biochar-based material from forest waste [6], tomato crop green and wood waste [30,52], wheat straw [53], and hardwood waste [54]. Increasing the pH due to the biochar addition is an advantage for acidic soil or growing media (as it is for peat in the current work) applications, with biochar acting as a liming material and possibly replacing the calcium oxide which is used for pH increment [55,56]. However, the low biochar ratio used in the present study, maintained pH values between 5.0 and 7.0, as the ideal substrate pH for the majority of vegetables is between 5.8 and 6.8. Additionally, biochar-based media had lower EC compared to the standard fertilized peat (control), in accordance with previous studies [30]. This outcome has very significant consequences, as materials that are commonly used for peat dilution often have elevated EC levels, such as composted green waste. Those materials could be used at a higher ratio, in combination with biochar, as high EC is very often the limiting factor for these materials to be added. The EC value is an important variable for growing media preparation and stability ranged between 1.5 and 2.0 dS m^{-1} [57]. The EC can either represent efficient nutrient support or saline conditions with adversely effects on seed germination and seedling growth [58]. However, lower initial EC values are not of consideration as substrates are commonly fertilized after plants transplanting [53]. Tailor-made fertilization is important for sustainable and successful plant growth. Therefore, increasing the fertilizers 1.5 times increased levels of minerals (i.e., K, N, P) available for the plant's growth needs. However, such nutrient enrichment can possibly create antagonistic impacts on cations such as Ca and Mg, or other effects such as increased Ca levels in Biochar A-based media and decreased Ca levels in Biochar D-media. In that case, periodical fertigation of a supplementary (hydroponic) nutrient solutions to balance the deficient levels of specific minerals could be examined. A successful case was mentioned in pot ornamental production growing in paper-waste as a substrate, supported by a hydroponic nutrient solution [29]. Previously, we had addressed the possible explanation for the decreased/low levels in nitrate and in P [30], whereas Altland and Locke [59] reported P release from biochar made from rice husks, with additional studies to be needed to explain the mechanism involved.

Seed emergence and MET in low biochar ratios (5–10%) growing media remained at similar levels with the control treatment (100% peat), while Biochars at 20% decreased seed emergence. Moreover, cabbage growing in low biochar-based media improved growth (i.e., height, fresh weight) for Biochars A, B and D. Chlorophyll fluorescence revealed low values in <15% Biochar C and in >10% in Biochars B and D, impacting the chlorophyll production, efficiency of PSII photochemistry and photosynthetic rate [28]. Increased biochar rates (i.e., 50%) resulted in decreased seed germination in myrtle and mastic seeds [6] and in tomato [30]. Solaiman et al. [1] who studied the impact of five different chars under five levels, on three plant species (wheat, mung bean, subterranean clover) indicated the early seed germination and seedling growth and this was depended on the char material and ratio. The use of biochar considerably improves seedlings' early growth [60] but some biochar may have substances that could adversely influence seed germination and early growth [1]. Seed emergence decrease was found in Biochars A and C at the rates of 20% in the present study.

Following the Exp. I, the examined Biochar ratios and types were further selected for evaluation. Biochar A and D improved seed emergence initially compared to the control. Fast and consistent seed emergence is an important issue for increased crop production, product quality, and eventually elevated profits.

In general, additional fertilizers could support plant growth with increased fresh weight at 7.5% for Biochar A and at 15% for Biochar D, observing also greater dry weight. Leaf number did not change among treatments and the decreased seedling height in case of Biochar A+Fert at 15% and Biochar D at 7.5% is not necessarily negative, as shorter (dwarf) plants are often desirable due to easy handling, transport and storage under nursery enterprises. Similar to our findings, Kim et al. [61] reported a 150% increase in shoot dry weight of kale (*Brassica olereaseae* L. var. acephala) when Biochar from rice husk was added at 5% to coir dust, perlite and verlmiculite. Vaughn et al. [62] and Steiner and Harttung [63] researched biochars for horticultural production as a substitute for peat and found no impacts on dry weight of plants. Tian et al. [51] and Mendez et al. [3] mixed biochar with compost to grow calathea and biochar with peat to grow lettuce, respectively, and revealed greater plant quality compared to those cultivated in single substrates, while Belda et al. [6] reported that the plant's response to biochar is affected by the plant species itself. No major changes were observed in plant physiology attributes in general in the present study. Leaf stomatal conductance decreased at Biochar A+Fert at 7.5% and chlorophyll b content decreased at Biochar A+Fert at 15%. The decrease in stomatal conductance and the greater water use efficiency after application of biochar shows the ability of biochar to mitigate stress from the water deficit [5].

Total phenolic content did not change much among the different treatments with the exception of the increased phenolic content in case of Biochar D at 7.5%. However, antioxidant activity increased in several cases. Interestingly, DPPH decreased when fertilizer was added with Biochar D at 7.5% but increased in case of 15% Biochar D with fertilizers, indicating the induced stress of the added minerals in the high biochar content, following MDA increment. Total phenolics and antioxidant activities increases were also found in biochar-treated *Andrographis paniculata* (kalmegh) [5]. Plants have restricted protective processes, including the production of stress response proteins and synthesis of antioxidant enzymes (includes SOD, POD and CAT) in order to overcome reactive oxygen species (ROS) accumulation [64]. The increase of MDA observed with the additionally fertilized peat (PFert 100), with Biochar A+Fert at 15% and Biochar D at 7.5% indicates cellular damage and increased stress of the plants. This was further supported with hydrogen peroxide increases and the activation of CAT antioxidant enzymes activity to detoxify the ROS accumulation [43]. The high ROS accumulation is related to intensive damage of cellular proteins, nucleic acids and lipids [65].

Although K has no direct toxicity impacts on plants, elevated K concentrations can trigger deficiencies in Mg and Ca, and plant growth reduction [66], whereas this was evident with the Biochar A mixtures that caused substantial Mg content decrease, but plant growth decrease was not observed in cabbage seedling production. Therefore, K content was increased in cabbage grown in Biochar A-based media with more pronounce effects at high ratio and/or fertilizer, while Biochar D needed to be fertilized and used in 15% into the mixture in order to obtain K levels like the control. Phosphorus accumulation followed the K trend for the examined growing media. Kim et al. [61] also reported increase of N, P and K content in kale shoots when Biochar was mixed at various ratios with the growing media. Similarly, increased K and P contents were found when *Syngonium podophyllum* was grown in different Biochar-based media, and this was related to the higher levels of these elements in the growing media [53]. Calcium content was found to be reduced with Biochar A at 7.5% indicating antagonistic effect with the K presence. However, Ca content with 7.5% Biochar A treatment was maintained to similar levels with the control, only when fertilizers at 1.5-fold were used. The high Biochar A ratio (i.e., 15%) increased the Ca content and this reduced Mg levels. In general, Mg and Fe contents were decreased with Biochar addition and the effects were more pronounced at a high ratio of 15%. The decrease SPAD units in Biochars C and D at high ratios, reflected the leaf discoloration and the decreased Mg and Fe levels, both involved in chlorophyll metabolism.

5. Conclusions

In conclusion, Biochar at a low ratio (5–10%) increased plant growth (fresh weight, height), while at 20%, it reduced cabbage seed emergence and plant height. The addition of Biochars supported the mineral accumulation in seedlings, as more available minerals could be absorbed by the plants. The production of seedlings with low height could be of benefit for nurseries, when they want to produce draft plants and where irrigation is overhead. This helps transportation and storage conditions. An increased stress occurred when a high ratio of Biochar was used (i.e., 20%), while lower ratios (5–10%) benefited plant growth-related parameters. Seeding at 20% of Biochar should be avoided as the seed emergence is decreased with higher MET. Biochars from forest wood (A) and woody feedstock (D) are quite promising materials. Finally, it seems to be preferable to use a wooden biochar of beech, spruce and pine species manufactured at 700 °C with the Schotteredorf process, and the produced Biochar (A) to be utilized at 7.5% ratio for cabbage seedling production. If fresh wooden biochar (D) of fruit trees and hedges are used, manufactured at 500–600 °C with the Pyreg equipment, then additional fertilizer is needed. However, different species need to be evaluated accordingly.

Supplementary Materials: The following are available online at http://www.mdpi.com/2073-4395/9/11/693/s1, Figure S1: Cabbage cumulative seedling emergence in peat with different biochar types (A, D) and ratio (7.5%, 15%) and mineral doses, Figure S2: Cabbage seedling production in peat with different biochar types (A, D) and ratio (7.5%, 15%) and mineral doses.

Author Contributions: Conceptualization, M.P. and N.T.; methodology, M.P. and N.T.; software, N.T.; validation, A.C., M.P. and N.T.; formal analysis, A.C. and M.P.; investigation, A.C. and A.K.; resources, A.K. and N.T.; data curation, A.C. and M.P.; writing—original draft preparation, A.C.; writing—review and editing, M.P. and N.T.; visualization, A.K.; supervision, M.P. and N.T.; project administration, M.P. and N.T.; funding acquisition, A.K. and N.T.

Funding: This research was funded by Bord na Mona Horticulture Ltd. and Cyprus University of Technology under the project OPTIBIOCHAR and Cyprus University of Technology Open Access Author Fund.

Acknowledgments: The authors are grateful to the project OPTIBIOCHAR that has been developed under the Cooperation Programme Cyprus-Ireland, co-funded by the Bord na Mona LtD and Cyprus University of Technology.

Conflicts of Interest: The authors declare no conflict of interest.

References

1. Solaiman, Z.M.; Murphy, D.V.; Abbott, L.K. Biochars influence seed germination and early growth of seedlings. *Plant Soil* **2012**, *353*, 273–287. [CrossRef]

2. Laghari, M.; Hu, Z.; Mirjat, M.S.; Xiao, B.; Tagar, A.A.; Hu, M. Fast pyrolysis biochar from sawdust improves the quality of desert soils and enhances plant growth. *J. Sci. Food Agric.* **2016**, *96*, 199–206. [CrossRef] [PubMed]

3. Méndez, A.; Paz-Ferreiro, J.; Gil, E.; Gascó, G. The effect of paper sludge and biochar addition on brown peat and coir based growing media properties. *Sci. Hortic.* **2015**, *193*, 225–230. [CrossRef]

4. Agegnehu, G.; Bass, A.M.A.M.; Nelson, P.N.P.N.; Bird, M.I.M.I. Benefits of biochar, compost and biochar-compost for soil quality, maize yield and greenhouse gas emissions in a tropical agricultural soil. *Sci. Total Environ.* **2016**, *543*, 295–306. [CrossRef] [PubMed]

5. Saha, A.; Basak, B.B.; Gajbhiye, N.A.; Kalariya, K.A.; Manivel, P. Sustainable fertilization through co-application of biochar and chemical fertilizers improves yield, quality of *Andrographis paniculata* and soil health. *Ind. Crops Prod.* **2019**, *140*, 111607. [CrossRef]

6. Belda, R.M.; Lidón, A.; Fornes, F. Biochars and hydrochars as substrate constituents for soilless growth of myrtle and mastic. *Ind. Crops Prod.* **2016**, *94*, 132–142. [CrossRef]

7. Calamai, A.; Palchetti, E.; Masoni, A.; Marini, L.; Chiaramonti, D.; Dibari, C.; Brilli, L. The influence of biochar and solid digestate on rose-scented geranium (*Pelargonium graveolens* L'Hér.) productivity and essential oil quality. *Agronomy* **2019**, *9*, 260. [CrossRef]

8. Huang, L.; Niu, G.; Feagley, S.E.; Gu, M. Evaluation of a hardwood biochar and two composts mixes as replacements for a peat-based commercial substrate. *Ind. Crops Prod.* **2019**, *129*, 549–560. [CrossRef]

9. Zhao, B.; Xu, R.; Ma, F.; Li, Y.; Wang, L. Effects of biochars derived from chicken manure and rape straw on speciation and phytoavailability of Cd to maize in artificially contaminated loess soil. *J. Environ. Manag.* **2016**, *184*, 569–574. [CrossRef]

10. Lashari, M.S.; Ye, Y.; Ji, H.; Li, L.; Kibue, G.W.; Lu, H.; Zheng, J.; Pan, G. Biochar-manure compost in conjunction with pyroligneous solution alleviated salt stress and improved leaf bioactivity of maize in a saline soil from central China: A 2-year field experiment. *J. Sci. Food Agric.* **2015**, *95*, 1321–1327. [CrossRef]

11. Lehmann, J.; Kern, D.C.; German, L.A.; McCann, J.; Martins, G.C.; Moreira, A. Soil fertility and production potential. In *Amazonian Dark Earths: Origin, Properties, Management*; Lehmann, J., Kern, D.C., Glaser, B., Woods, W.I., Eds.; Kluwer Academic Publishers: Dordrecht, The Netherlands, 2003; pp. 105–124.

12. Hussain, M.; Farooq, M.; Nawaz, A.; Al-Sadi, A.M.; Solaiman, Z.M.; Alghamdi, S.S.; Ammara, U.; Ok, Y.S.; Siddique, K.H.M. Biochar for crop production: Potential benefits and risks. *J. Soils Sediments* **2017**, *17*, 685–716. [CrossRef]

13. Sadaf, J.; Shah, G.A.; Shahzad, K.; Ali, N.; Shahid, M.; Ali, S.; Hussain, R.A.; Ahmed, Z.I.; Traore, B.; Ismail, I.M.I.; et al. Improvements in wheat productivity and soil quality can accomplish by co-application of biochars and chemical fertilizers. *Sci. Total Environ.* **2017**, *607–608*, 715–724. [CrossRef] [PubMed]

14. Vaughn, S.F.; Kenar, J.A.; Tisserat, B.; Jackson, M.A.; Joshee, N.; Vaidya, B.N.; Peterson, S.C. Chemical and physical properties of *Paulownia elongata* biochar modified with oxidants for horticultural applications. *Ind. Crops Prod.* **2017**, *97*, 260–267. [CrossRef]

15. Gul, S.; Whalen, J.K. Biochemical cycling of nitrogen and phosphorus in biochar-amended soils. *Soil Biol. Biochem.* **2016**, *103*, 1–15. [CrossRef]

16. Sohi, S.P. Carbon storage with benefits. *Science* **2012**, *338*, 1034–1035. [CrossRef]

17. Gregory, S.J.; Anderson, C.W.N.; Camps Arbestain, M.; McManus, M.T. Response of plant and soil microbes to biochar amendment of an arsenic-contaminated soil. *Agric. Ecosyst. Environ.* **2014**, *191*, 133–141. [CrossRef]

18. Jaiswal, A.K.; Frenkel, O.; Tsechansky, L.; Elad, Y.; Graber, E.R. Immobilization and deactivation of pathogenic enzymes and toxic metabolites by biochar: A possible mechanism involved in soilborne disease suppression. *Soil Biol. Biochem.* **2018**, *121*, 59–66. [CrossRef]

19. Oleszczuk, P.; Jośko, I.; Futa, B.; Pasieczna-Patkowska, S.; Pałys, E.; Kraska, P. Effect of pesticides on microorganisms, enzymatic activity and plant in biochar-amended soil. *Geoderma* **2014**, *214–215*, 10–18. [CrossRef]

20. Nieto, A.; Gascó, G.; Paz-Ferreiro, J.; Fernández, J.M.; Plaza, C.; Méndez, A. The effect of pruning waste and biochar addition on brown peat based growing media properties. *Sci. Hortic.* **2016**, *199*, 142–148. [CrossRef]

21. Wagner, A.; Kaupenjohann, M. Suitability of biochars (pyro- and hydrochars) for metal immobilization on former sewage-field soils. *Eur. J. Soil Sci.* **2014**, *65*, 139–148. [CrossRef]

22. Méndez, A.; Cárdenas-Aguiar, E.; Paz-Ferreiro, J.; Plaza, C.; Gascó, G. The effect of sewage sludge biochar on peat-based growing media. *Biol. Agric. Hortic.* **2017**, *33*, 40–51. [CrossRef]

23. Graber, E.R.; Harel, Y.M.; Kolton, M.; Cytryn, E.; Silber, A.; David, D.R.; Tsechansky, L.; Borenshtein, M.; Elad, Y. Biochar impact on development and productivity of pepper and tomato grown in fertigated soilless media. *Plant Soil* **2010**, *337*, 481–496. [CrossRef]

24. Dispenza, V.; De Pasquale, C.; Fascella, G.; Mammano, M.M.; Alonzo, G. Use of biochar as peat substitute for growing substrates of *Euphorbia* × *lomi* potted plants. *Span. J. Agric. Res.* **2016**, *14*, 21. [CrossRef]

25. Guo, Y.; Niu, G.; Starman, T.; Volder, A.; Gu, M. Poinsettia growth and development response to container root substrate with Biochar. *Horticulturae* **2018**, *4*, 1. [CrossRef]

26. Choi, H.; Son, H.; Kim, C. Predicting financial distress of contractors in the construction industry using ensemble learning. *Expert Syst. Appl.* **2018**, *110*, 1–10. [CrossRef]

27. Judd, L.A.; Jackson, B.E.; Fonteno, W.C. Advancements in root growth measurement technologies and observation capabilities for container-grown plants. *Plants* **2015**, *4*, 369–392. [CrossRef]

28. Chrysargyris, A.; Antoniou, O.; Athinodorou, F.; Vassiliou, R.; Papadaki, A.; Tzortzakis, N. Deployment of olive-stone waste as a substitute growing medium component for Brassica seedling production in nurseries. *Environ. Sci. Pollut. Res.* **2019**. [CrossRef]

29. Chrysargyris, A.; Stavrinides, M.; Moustakas, K.; Tzortzakis, N. Utilization of paper waste as growing media for potted ornamental plants. *Clean Technol. Environ. Policy* **2018**. [CrossRef]

30. Prasad, M.; Tzortzakis, N.; McDaniel, N. Chemical characterization of biochar and assessment of the nutrient dynamics by means of preliminary plant growth tests. *J. Environ. Manag.* **2018**, *216*, 89–95. [CrossRef]

31. Nadeem, S.M.; Imran, M.; Naveed, M.; Khan, M.Y.; Ahmad, M.; Zahir, Z.A.; Crowley, D.E. Synergistic use of biochar, compost and plant growth-promoting rhizobacteria for enhancing cucumber growth under water deficit conditions. *J. Sci. Food Agric.* **2017**, *97*, 5139–5145. [CrossRef]

32. Alvarez, J.M.; Pasian, C.; Lal, R.; Lopez, R.; Fernandez, M. Vermicompost and Biochar as growing media replacement for ornamental plant production. *J. Appl. Hortic.* **2017**, *19*, 205–214.

33. Álvarez, J.M.; Pasian, C.; Lal, R.; López, R.; Díaz, M.J.; Fernández, M. Morpho-physiological plant quality when biochar and vermicompost are used as growing media replacement in urban horticulture. *Urban For. Urban Green.* **2018**, *34*, 175–180. [CrossRef]

34. Landis, T.D.; Morgan, N. Growing media alternatives for forest and native plant nurseries. *Natl. Proc. For. Conserv. Nurs. Assoc. 2008* **2009**, *5814*, 26–31.

35. Li, Q.; Chen, J.; Caldwell, R.D.; Deng, M. Cowpeat as a substitute for peat in container substrates for foliage plant propagation. *Horttechnology* **2009**, *19*, 340–345. [CrossRef]

36. Gruda, N.S. Increasing sustainability of growing media constituents and stand-alone substrates in soilless culture systems. *Agronomy* **2019**, *9*, 298. [CrossRef]

37. British Standards Institution (BSI). *EN 13037, Soil Improvers and Growing Media-Determination of pH*; BSI: London, UK, 2002.

38. British Standards Institution (BSI). *EN 13038, Soil Improvers and Growing Media-Determination of Electrical Conductivity*; BSI: London, UK, 2002.

39. British Standards Institution (BSI). *EN 13651, Soil Improvers and Growing Media-Extraction of Calcium Chloride/DTPA (CAT)*; BSI: London, UK, 2002.

40. Kelepesi, S.; Tzortzakis, N.G. Olive mill wastes: A growing medium component for seedling and crop production of lettuce and chicory. *Int. J. Veg. Sci.* **2009**, *15*, 325–339. [CrossRef]

41. Chrysargyris, A.; Panayiotou, C.; Tzortzakis, N. Nitrogen and phosphorus levels affected plant growth, essential oil composition and antioxidant status of lavender plant (*Lavandula angustifolia* Mill.). *Ind. Crops Prod.* **2016**, *83*, 577–586. [CrossRef]

42. Tzortzakis, N.; Tzanakaki, K.; Conomakis, C.D. Effect of origanum oil and vinegar on the maintenance of postharvest quality of tomato. *Food Nutr. Sci.* **2011**, *2*, 974–982. [CrossRef]

43. Chrysargyris, A.; Xylia, P.; Botsaris, G.; Tzortzakis, N. Antioxidant and antibacterial activities, mineral and essential oil composition of spearmint (*Mentha spicata* L.) affected by the potassium levels. *Ind. Crops Prod.* **2017**, *103*, 202–212. [CrossRef]

44. Loreto, F.; Velikova, V. Isoprene produced by leaves protects the photosynthetic apparatus against ozone damage, quenches ozone products, and reduces lipid peroxidation of cellular membranes. *Plant Physiol.* **2001**, *127*, 1781–1787. [CrossRef]

45. De Azevedo Neto, A.D.; Prisco, J.T.; Enéas-Filho, J.; De Abreu, C.E.B.; Gomes-Filho, E. Effect of salt stress on antioxidative enzymes and lipid peroxidation in leaves and roots of salt-tolerant and salt-sensitive maize genotypes. *Environ. Exp. Bot.* **2006**, *56*, 87–94. [CrossRef]

46. Fornes, F.; Belda, R.M.; Fernández de Córdova, P.; Cebolla-Cornejo, J. Assessment of biochar and hydrochar as minor to major constituents of growing media for containerized tomato production. *J. Sci. Food Agric.* **2017**, *97*, 3675–3684. [CrossRef] [PubMed]

47. Wang, S.; Gao, B.; Zimmerman, A.R.; Li, Y.; Ma, L.; Harris, W.G.; Migliaccio, K.W. Physicochemical and sorptive properties of biochars derived from woody and herbaceous biomass. *Chemosphere* **2015**, *134*, 257–262. [CrossRef] [PubMed]

48. Shackley, S.; Ruysschaert, G.; Zwart, K.; Glaser, B. *Biochar in European Soils and Agriculture: Science and Practice*; Routledge: Oxon, UK, 2016; p. 324.

49. Ronga, D.; Francia, E.; Allesina, G.; Pedrazzi, S.; Zaccardelli, M.; Pane, C.; Tava, A.; Bignami, C. Valorization of vineyard by-products to obtain composted digestate and biochar suitable for nursery grapevine (*Vitis vinifera* L.) production. *Agronomy* **2019**, *9*, 420. [CrossRef]

50. Mukherjee, A.; Lal, R. The biochar dilemma. *Soil Res.* **2014**, *52*, 217–230. [CrossRef]

51. Tian, Y.; Sun, X.; Li, S.; Wang, H.; Wang, L.; Cao, J.; Zhang, L. Biochar made from green waste as peat substitute in growth media for *Calathea rotundifola* cv. Fasciata. *Sci. Hortic.* **2012**, *143*, 15–18. [CrossRef]

52. Dunlop, S.J.; Arbestain, M.C.; Bishop, P.A.; Wargent, J.J. Closing the loop: Use of biochar produced from tomato crop green waste as a substrate for soilless, hydroponic tomato production. *HortScience* **2015**, *50*, 1572–1581. [CrossRef]

53. Zulfiqar, F.; Younis, A.; Chen, J. Biochar or biochar-compost amendment to a peat-based substrate improves growth of *Syngonium podophyllum*. *Agronomy* **2019**, *9*, 460. [CrossRef]

54. Nair, A.; Carpenter, B. Biochar rate and transplant tray cell number have implications on pepper growth during transplant production. *Horttechnology* **2016**, *26*, 713–719. [CrossRef]

55. Sun, L.; Li, L.; Chen, Z.; Wang, J.; Xiong, Z. Combined effects of nitrogen deposition and biochar application on emissions of N_2O, CO_2 and NH_3 from agricultural and forest soils. *Soil Sci. Plant Nutr.* **2014**, *60*, 254–265. [CrossRef]

56. Bedussi, F.; Zaccheo, P.; Crippa, L. Pattern of pore water nutrients in planted and non-planted soilless substrates as affected by the addition of biochars from wood gasification. *Biol. Fertil. Soils* **2015**, *51*, 625–635. [CrossRef]

57. Abad, M.; Noguera, P.; Puchades, R.; Maquieira, A.; Noguera, V. Physico-chemical and chemical properties of some coconut coir dusts for use as a peat substitute for containerised ornamental plants. *Bioresour. Technol.* **2002**, *82*, 241–245. [CrossRef]

58. Bustamante, M.A.; Moral, R.; Paredes, C.; Pérez-Espinosa, A.; Moreno-Caselles, J.; Pérez-Murcia, M.D. Agrochemical characterisation of the solid by-products and residues from the winery and distillery industry. *Waste Manag.* **2008**, *28*, 372–380. [CrossRef] [PubMed]

59. Altland, J.E.; Locke, J.C. Effect of biochar type on macronutrient retention and release from soilless substrate. *HortScience* **2013**, *48*, 1397–1402. [CrossRef]

60. Thomas, S.C.; Gale, N. Biochar and forest restoration: A review and meta-analysis of tree growth responses. *New For.* **2015**, *46*, 931–946. [CrossRef]

61. Kim, H.S.; Kim, K.R.; Yang, J.E.; Ok, Y.S.; Kim, W.I.; Kunhikrishnan, A.; Kim, K.H. Amelioration of horticultural growing media properties through rice hull biochar incorporation. *Waste Biomass Valoriz.* **2017**, *8*, 483–492. [CrossRef]

62. Vaughn, S.F.; Kenar, J.A.; Thompson, A.R.; Peterson, S.C. Comparison of biochars derived from wood pellets and pelletized wheat straw as replacements for peat in potting substrates. *Ind. Crops Prod.* **2013**, *51*, 437–443. [CrossRef]

63. Steiner, C.; Harttung, T. Biochar as a growing media additive and peat substitute. *Solid Earth* **2014**, *5*, 995–999. [CrossRef]

64. Gupta, D.D.; Palma, J.M.; Corpas, F.J. *Reactive Oxygen Species and Oxidative Damage in Plants under Stress*; Springer International Publishing Switzerland: Cham, Switzerland, 2015; ISBN 9783319204208.

65. Sies, H.; Berndt, C.; Jones, D.P.; Sies, H.; Berndt, C.; Jones, D.P. Oxidative stress. *Annu. Rev. Biochem.* **2017**, *86*, 715–748. [CrossRef]

66. Savvas, D.; Gruda, N. Application of soilless culture technologies in the modern greenhouse industry-A review. *Eur. J. Hortic. Sci.* **2018**, *83*, 280–293. [CrossRef]

Article

Biochar Type, Ratio, and Nutrient Levels in Growing Media Affects Seedling Production and Plant Performance

Antonios Chrysargyris [1], Munoo Prasad [1,2,3], Anna Kavanagh [3] and Nikos Tzortzakis [1,*]

[1] Department of Agricultural Sciences, Biotechnology and Food Science, Cyprus University of Technology, Lemesos 3603, Cyprus; a.chrysargyris@cut.ac.cy (A.C.); Munoo.Prasad@cut.ac.cy (M.P.)

[2] Compost/AD Research & Advisory (IE/CY), Naas, W91TYH3 Kildare, Ireland

[3] Bord na Mona Horticulture Ltd. Research Centre, Main Street, Newbridge, W12XR59 Kildare, Ireland; annamarykavanagh@gmail.com

* Correspondence: nikolaos.tzortzakis@cut.ac.cy; Tel.: +357-2500-2280

Received: 21 July 2020; Accepted: 16 September 2020; Published: 18 September 2020

Abstract: Biochar can be used as an alternative component in growing media, positively affecting plant growth/yield, but also media properties. In the present study, two commercial grade biochars (BFW-forest wood; and BTS-fresh wood screening), mainly wood-based materials, were used at 7.5% and 15% (*v/v*), adding nutrient in two levels (100% and 150% standard fertilizer level-Fert). Biochar affected growing media properties, with increases on pH and changes on the nutrient content levels. Biochar BFW enhanced the emergence of seeds in comparison to the control. Increased fertilizer levels benefited plant yield in BFW and BTS at 7.5%, but not at 15%. Leaf stomatal conductance was reduced at 150% fertilized biochars (BFW + Fert and BTS + Fert) at 7.5%, while total chlorophylls increased at BTS + Fert at 7.5% and 15%. The addition of biochars decreased the antioxidant activity in the plant. Lipid peroxidation in lettuce was increased in most cases with the presence of biochars (BFW, BTS) and 150% fertilization, activating antioxidant (superoxide oxidase and peroxidase) enzymatic metabolisms. The addition of Biochars in the growing media increased the content of nutrients in seedlings, as plants could absorb more available nutrients. Biochar of beech, spruce, and pine species (BFW) at 7.5% was more promising for substituting peat to produce lettuce seedlings. However, examining different species (tomato, leek, impatiens, and geranium) with BFW at 7.5%, the results were not common, and each species needs to be evaluated further.

Keywords: biochar; peat; growth; lettuce; emergence; nursery production; container; extractable nutrients; plant nutrient content

1. Introduction

Biochar is produced with dry pyrolysis of the organic matter, in which plant or animal-based organic materials are subjected to high temperatures (450 to 600 °C), under hypoxia or anoxia environment [1,2], whereas lower temperature (300 °C) have been reported for biochar production [3]. The initial organic material for biochar production is mainly wastes coming from intensive sectors, such as forest residues and wood industries, agriculture and food, and greatly contribute to the environmental management and recycling, reducing the greenhouse gas (GHG) emission and increasing carbon sequestration [4–7]. The use of biochar in the agriculture sector as an alternative container growing media adds value to the bioenergy business process [8]. Moreover, biochar can alleviate salinity stress in crops with important environmental, agriculture, and economic benefit [9]. Biochar has demonstrated the potential for inclusion in growing media, together with different materials such as peat [3,10], compost [8,11], coir [3] and vermicompost [12].

Nowadays, there is a great deal of interest in the use of biochar as a peat replacement as evidenced by increasing scientific publications. This is supported by research and review articles [13–18]. The main objectives of this research have been to investigate if biochar could replace either totally or partly peat. Dilution rates as high as 50 to 75% mixed with peat or even at 100% have been tried with mixed results [19–21]. The objective of these peat replacement trials was often to show that at these high rates, biochar/peat mixtures performed as good as peat mostly as evidenced by plant height, fresh and dry weight [21,22]. Increased biochar ratio had conflicting impacts, either by increasing plant growth (higher biomass and height) i.e., 60–80% conifer wood biochar in *Euphorbia* × *lomi* [23] or by suppressing plant growth i.e., 60–100% pinewood biochar in poinsettia (*Euphorbia pulcherrima* L.) [20] and 80–100% pinewood biochar in tomato (*Solanum lycopersicum* L.) and basil (*Ocimum basilicum* L.) [24]. However, the researchers did not always take into consideration the economic viability of the use of biochar as a peat replacement. At the cost from \$67–177 per m^3 (bulk density of 0.3 kg L^{-1}) of biochar in the UK [25] and the cost of peat at \$25–30 per m^3 in Europe, biochar may not be a viable substitute as a peat replacement at high ratios, unless the mixtures of peat and biochar would outperform 100% professional grade peat. Handling and incorporation of biochar as a growing media constituent would also need to be considered. There are certain extraneous advantages in mixing biochar with peat due to its properties other than based on saving on peat and crop performance. For instance, the use of biochar would lead to reduction or elimination of certain inputs such as lime and certain fertilizers, e.g., potassium [2,10,26] and thus lead to savings. Use of biochar could get carbon credits, a monetary advantage, and improve the efficiency of nutrient inputs through reduced leakage of nutrients [27] to the environment, an additional savings.

There are reports that biochar may have bio-stimulant properties due to its ability to change gene expression (i.e., transcription of auxin- and brassinosteroid-related genes in *Arabidopsis*) of the plants and the presence of gibberellic acid in biochar and these could lead to changes the morphological–physiological aspects of the plants [28,29]. Prendergast-Miller et al. [30] showed that biochar is attracting roots, resulting in its partitioning between bulk and rhizosphere soil and thus, biochar directly regulates the acquisition of plant root nutrients as a source of nutrients. Fewer studies investigated the impacts of biochar on different plant growth characteristics that affect yield, such as seed germination and the architecture of shoots or roots [1,30,31]. A few papers have applied low rates of biochar at 1–15% to peat [1,16,32,33] and have found a positive response, not only on fresh and dry weight, but also morphological and physiological changes [28,29,34]. We also know from previous publications that all biochars do not behave in a similar way ("not all biochars are made equal") and crop response depends on crop species. Therefore, biochar properties are significantly affected by feedstock, temperature, and residence time. Thus, even with the same feedstock, the properties such as surface area and pore volume will vary with temperature [35]. In addition, the application of biochar and the effects of the production of greenhouse seedlings or subsequent growth of seedlings have been less reported [34,36]. In the recent years, the field of seedling and potting horticultural plants has been significantly increased [37–39]. Since biochar from different resources has different characteristics due to potential phytotoxicity, some may have adverse effects on plant growth. Phytotoxicity evaluation is important to a successful soil/soilless amendment with bioenergy by-products such as biochar [40], and seed germination testing is a reliable procedure for biochar phytotoxicity tests.

Biochar presence in soil can modify the abundance, activity, and community structure of microorganisms in the soil. Adding cotton straw biochar (i.e., 4.5 t ha^{-1}) in desert soil increased microbial respiration and carbon and nitrogen biomass with increases of activity of key enzymes involved in carbon and nitrogen transformation [41]. Gomez et al. [42] reported that low rates of biochars increased the microbial abundance with Gram-negative bacteria dominating the microbial community; however, high biochar rates (i.e., 49 t ha^{-1}) inhibited microbial activity and reduced extractable phospholipid fatty acid [43]. Therefore, biochar amendment impact is greatly depending on biochar properties and soil characteristics and the possible biochar-microorganism interaction mechanisms include toxicity and emission of volatile organic compounds acting directly on soil microorganisms or indirectly by changing soil properties and enzymatic activities [44].

Biochar application can reduce N and P uptake from the plants and additional fertilization may be required for successful plant growth [34]. Moreover, the effects of biochar on plant growth may differ with plant species since different plants may have different growing conditions or different tolerances to certain stresses. The present study was conducted with two biochars from different raw materials mixed with professional grade peat at low rates (a) to assess the impact of biochar substitution in peat on extractable nutrients and phytotoxicity for lettuce (*Lactuca sativa* L.) seed emergence, (b) to assess the impact of biochar on lettuce plant growth, physiology, and nutrient content, (c) to evaluate the fertilizer dose and biochar ratio in peat on plant metabolism and nutrient content, and (d) based on the optimum biochar applications on lettuce, four more plant species were examined for their performance on biochar application.

2. Materials and Methods

2.1. Biochars and Plant Material

In the present study, lettuce (*Lactuca sativa* L. var. Nogal; Hazera, Israel) seeds were used for seedling production. Two commercial grade biochars of the following feedstocks were used: forest wood, e.g., beech, spruce, and pine from Germany (BFW), and fresh wood screenings (0–20 mm) from tree and shrub cuttings mainly from urban areas and farms (fruit trees, hedges, hedgerow management) from Switzerland (BTS). Biochars were produced using either the Pyreg equipment (Verora, Edlibach, Switzerland) for Biochar BTS at 500–600 °C, or Schottdorf–Meiler equipment (Carbon Terra, Wallerstein, Germany) for Biochar BFW at 700 °C with retention time of 15–30 min. Nonetheless, due to business sensitivity, additional data on the specifics of Biochar production are not known. A high-quality industry standard professional grade H_4–H_5 on von Post scale fertilized/limed peat (P) from Bord na Mona (Newbridge, Ireland) was used as a reference (control) and as basic substrate to which the biochar was added. The selected biochars had roughly the same particle size to eliminate/reduce the effect on physical properties as far as possible. Biochars were evaluated for their chemical properties [38], as for pH (EN 13,037 2002) [45], Electrical Conductivity (EC) in water extract at 1:5 (*v/v*) ratio (EN 13,038 2002) [46], and barium chloride/DTPA (CAT) extractable (1:5 *v/v*) potassium (K) and phosphorus (P), ammonium (NH_4-N), nitrate (NO_3-N), and total extractable N (NH_4-N + NO_3-N) (EN 13651-2002) [47]. Barium Chloride was substituted for Calcium Chloride at the same concentration, in order to determine Ca in the extract. The surface area determined with the Brunauer, Emmett, and Teller method [48]. Peat physicochemical characteristics have been reported previously [10]. Physicochemical properties of biochars BWF and BTS and peat are presented in Table 1.

Table 1. Physicochemical properties of biochars and peat.

	Peat	BWF	BTS
pH	3.13	9.57	9.55
EC (mS cm^{-1})	0.034	0.613	0.410
NH_4-N (mg L^{-1})	17.0	<1.0	<1.0
NO_3-N (mg L^{-1})	3.0	<1.0	<1.0
K (mg L^{-1})	8.0	1087.0	745.0
P (mg L^{-1})	6.0	2.0	3.0
Oxygen Uptake Rate (mmol O_2 kg^{-1} organic matter per hour)	5.5	2.3	4.3
Particle size of < 1 mm (%)	51.0	63.9	60.5
Particle size of 1–5 mm (%)	46.0	35.6	37.0
Particle size of > 5 mm (%)	3.0	<1.0	3.0
Surface area (m^2 g^{-1})	2.4	243.2	62.3

2.2. Preparation of Growing Media

In the present study, two individual experiments were carried out. In the first experiment (Exp. I), the biochars (BFW, BTS) mixed into the peat into two ratios (7.5% and 15% *v/v*) and took the mixtures to N, P, and K levels (with standard fertilizers; 100%) to 170 mg N L^{-1} as ammonium nitrate, 70 mg P L^{-1} as triple superphosphate and 100 mg K L^{-1} as potassium sulphate respectively for the peat-biochar mixtures and limed peat (dolomitic lime at 4 g L^{-1}) plus addition of standard level of trace elements to all treatments (Table S1). The CAT extractable N, P, and K that derived from the biochars were considered and the levels of fertilizers have been adjusted appropriately. In most cases, there were nearly insignificant amounts of N, some P, and excess of K. No K has been added into the mixture of 100% fertilization in the case of K excess (K greater than that applied in standard fertilizer i.e., >100 mg K L^{-1}). The application of additional fertilizers (150%) named as "+ Fert" was also examined. Therefore, the BFW and BTS biochars were selected at the rates of 0%, 7.5%, and 15% to the peat under 100% standard rate (N of 170 mg L^{-1}, P of 70 mg L^{-1}, and K of 100 mg L^{-1}) or 150% standard rate (N of 255 mg L^{-1}, P of 105 mg L^{-1}, and K of 150 mg L^{-1}) of fertilizers, resulting in 10 mixtures (treatments) including control (100% peat). The examined treatments, their interactions, and chemical analysis for growing media are presented in Tables 2 and 3.

Table 2. Effects of different biochar types (BFW, BTS) and ratio (7.5%, 15%) and mineral doses (with standard or with additional Fertilizers-F) on lettuce plant growth, physiology, and nutrient content.

	Biochar Type (T)	Biochar Ratio (R)	Fertilizers (F)	T × R	T × F	R × F	T × R × F
Growing media							
EC (mS cm^{-1})	***	***	***	*	**	ns	ns
pH	***	ns	ns	***	ns	ns	*
Organic matter (%)	***	***	*	***	ns	ns	ns
Organic C (%)	***	***	*	***	ns	ns	ns
N (mg L^{-1})	**	***	***	***	ns	ns	ns
K (mg L^{-1})	***	***	***	***	ns	**	**
P (mg L^{-1})	ns	**	***	*	**	ns	***
Ca (mg L^{-1})	***	ns	*	ns	ns	ns	*
Mg (mg L^{-1})	ns	***	ns	*	ns	ns	ns
Na (mg L^{-1})	***	***	ns	***	ns	ns	ns
SO$_4$-S (mg L^{-1})	***	***	***	***	***	***	***
Fe (mg L^{-1})	***	***	*	***	***	***	***
Cu (mg L^1)	***	ns	***	ns	***	***	***
Zn (mg L^{-1})	***	***	ns	***	ns	ns	ns
Mn (mg L^{-1})	***	***	ns	***	ns	ns	ns
B (mg L^{-1})	***	**	ns	**	ns	ns	ns
Lettuce plant							
Plant height (cm)	*	ns	***	*	ns	**	**
Leaf Number	ns	ns	**	ns	ns	ns	*
Chlorophyll fluorescence (Fv/Fm)	***	***	ns	ns	**	ns	***
Stomatal conductance (cm s^{-1})	ns	ns	ns	*	*	**	ns
Fresh weight (g)	ns	ns	***	ns	ns	ns	ns
Dry weight (g)	ns	ns	***	ns	ns	*	ns
Chlorophyll a (mg g^{-1} Fw)	*	ns	*	*	*	ns	ns
Chlorophyll b (mg g^{-1} Fw)	*	ns	**	*	*	ns	ns
Total Chlorophylls (mg g^{-1} Fw)	*	ns	*	*	*	ns	ns
Total phenolics (mg GAE g^{-1} Fw)	ns	ns	ns	*	ns	ns	ns
FRAP (mg trolox g^{-1} Fw)	*	ns	ns	**	ns	ns	ns
DPPH (mg trolox g^{-1} Fw)	ns	ns	ns	*	ns	ns	ns
H$_2$O$_2$ (nmol g^{-1} Fw)	ns	ns	***	***	***	ns	ns
MDA (nmol g^{-1} Fw)	***	***	**	***	***	***	**
SOD (units mg^{-1} protein)	*	ns	ns	ns	ns	ns	*
CAT (units mg^{-1} protein)	ns	***	*	***	**	**	*
POD (units mg^{-1} protein)	ns	***	***	ns	ns	ns	ns
N (g kg^{-1})	***	***	**	ns	***	***	***
K (g kg^{-1})	***	***	ns	***	***	***	**
P (g kg^{-1})	***	ns	***	***	***	***	***
Ca (g kg^{-1})	***	***	***	ns	ns	**	ns
Mg (g kg^{-1})	***	***	***	*	ns	**	ns
Na (g kg^{-1})	***	ns	***	***	***	***	ns
Fe (mg kg^{-1})	ns	ns	*	*	ns	ns	ns
Cu (mg kg^{-1})	***	***	***	***	*	*	ns
Zn (mg kg^{-1})	***	***	*	***	***	***	***

*, **, *** Significant difference at $p \leq$ 5%, 1%, and 0.1% following three-way ANOVA. ns: non-significant.

Table 3. Effects of peat (P) with different biochar types (BFW, BTS) and ratio (7.5%, 15%) and mineral doses (with standard or with additional Fertilizers-Fert.) on substrate minerals content.

	P	P + Fert	BFW 7.5%	BFW + Fert 7.5%	BFW 15%	BFW + Fert 15%	BTS 7.5%	BTS + Fert 7.5%	BTS 15%	BTS + Fert 15%
EC (μS cm^{-1})	417.14 ± 16.34 c	576.02 ± 25.98 a	329.85 ± 17.45 d	478.90 ± 20.90 b	209.53 ± 4.27 e	236.15 ± 13.65 e	318.90 ± 15.41 d	502.45 ± 22.35 b	252.66 ± 11.16 e	361.15 ± 02.18 d
pH	4.79 ± 0.01 c	4.89 ± 0.05 c	5.08 ± 0.06 b	4.86 ± 0.6 c	5.42 ± 0.03 a	5.41 ± 0.05 a	4.86 ± 0.06 c	4.96 ± 0.10 bc	4.57 ± 0.02 d	4.38 ± 0.07 e
Organic matter (%)	96.32 ± 0.18 a	94.84 ± 0.02 c	95.71 ± 0.11 b	95.76 ± 0.17 ab	95.57 ± 0.10 b	96.01 ± 0.02 ab	94.74 ± 0.09 c	94.76 ± 0.12 c	93.43 ± 0.34 d	93.51 ± 0.26 d
Organic C (%)	55.87 ± 0.10 a	55.01 ± 0.01 c	55.51 ± 0.06 b	55.55 ± 0.10 ab	55.43 ± 0.06 b	55.69 ± 0.01 ab	54.95 ± 0.05 c	54.96 ± 0.07 c	54.20 ± 0.20 d	54.24 ± 0.15 d
N (mg L^{-1})	85.16 ± 2.66 e	157.93 ± 4.93 a	87.10 ± 2.72 e	126.44 ± 3.94 b	71.74 ± 2.24 f	113.60 ± 3.61 c	85.16 ± 2.66 e	131.61 ± 4.11 b	46.45 ± 1.45 g	101.21 ± 3.20 d
K (mg L^{-1})	118.70 ± 3.70 f	180.64 ± 5.64 d	232.25 ± 7.25 c	304.23 ±9.51 b	381.92 ± 9.54 a	372.96 ± 11.24 a	103.22 ± 3.22 f	154.83 ± 4.64 e	103.21 ± 3.16 f	147.42 ± 4.74 e
P (mg L^{-1})	36.13 ± 1.13 e	66.58 ± 2.08 b	44.85 ± 1.40 d	71.74 ± 2.24 ab	40.77 ± 1.27 de	58.35 ± 1.85 c	44.78 ± 1.28 d	69.16 ± 2.16 ab	37.16 ± 1.16 e	74.87 ± 2.17 a
Ca (mg L^{-1})	701.91 ± 21.91 a	717.54 ± 22.41 a	701.84 ± 21.73 a	728.23 ± 22.73 a	725.65 ± 22.65 a	719.42 ± 21.53 a	521.79 ± 16.29 c	524.88 ± 16.38 c	486.17 ± 15.17 c	615.12 ± 11.87 b
Mg (mg L^{-1})	486.85 ± 15.21 a	479.46 ± 14.96 a	329.74 ± 10.29 b	330.93 ± 9.56 b	240.04 ± 7.49 d	236.59 ± 7.49 d	310.56 ± 9.69 b	317.72 ± 9.92 b	241.54 ± 7.54 d	277.54 ± 8.79 c
Na (mg L^{-1})	45.93 ± 1.43 c	46.45 ± 1.45 c	41.80 ± 1.28 c	45.93 ± 1.43 c	42.83 ± 1.33 c	41.31 ± 1.31 c	64.51 ± 2.01 b	63.48 ± 1.98 b	89.28 ± 2.78 a	85.71 ± 1.95 a
SO$_4$-S (mg L^{-1})	8.77 ± 0.27 b	8.26 ± 0.26 b	6.71 ± 0.21 c	6.71 ± 0.21 c	6.70 ± 0.20 c	6.72 ± 0.22 c	6.71 ± 0.21 c	6.71 ± 0.20 c	6.19 ± 0.19 c	13.94 ± 0.44 a
Fe (mg L^{-1})	9.25± 0.29 bc	8.26 ± 0.21 cde	9.54 ± 0.30 b	8.51 ± 0.23 cde	8.15 ± 0.25 de	7.79 ± 0.21 e	8.36 ± 0.26 cde	8.64 ± 0.26 bcde	9.03 ± 0.28 bcd	13.98 ± 0.45 a
Cu (mg L^{-1})	0.11 ± 0.00 e	0.10 ± 0.00 e	0.04 ± 0.00 f	0.04 ± 0.00 f	0.04 ± 0.00 f	0.04 ± 0.00 f	0.21 ± 0.01 c	0.26 ± 0.01 b	0.15 ± 0.00d	0.31 ± 0.01 a
Zn (mg L^{-1})	1.08 ± 0.03 cd	0.98 ± 0.03 d	1.03 ± 0.03 d	1.18 ± 0.04 c	1.03 ± 0.03 d	0.98 ± 0.03 d	1.55 ± 0.05 b	1.55 ± 0.05 b	2.11 ± 0.06 a	2.22 ± 0.07 a
Mn (mg L^{-1})	2.42 ± 0.07 d	2.43 ± 0.07 d	8.98 ± 0.28 b	8.97 ± 0.027 b	12.95 ± 0.39 a	12.75 ± 0.37 a	2.63 ± 0.08 d	2.68 ± 0.08 cd	2.99 ± 0.09 cd	3.41 ± 0.11 c
B (mg L^{-1})	0.62 ± 0.02 c	0.62 ± 0.02 c	0.67 ± 0.02 c	0.66 ± 0.02 c	0.67 ± 0.02 c	0.67 ± 0.02 c	0.83 ± 0.03 b	0.83 ± 0.02 b	0.98 ± 0.03 a	0.88 ± 0.03 b

Values (n = 3) in rows followed by the same letter are not significantly different, $p \leq 0.05$.

In the second experiment (Exp. II), the most promising treatment of biochar was further selected for investigation and examined in four different plant species of high importance and marketability interest for seedling production. Tomato (*Solanum lycopersicum* cv. Fi Akron), leek (*Allium porrum* cv. F1 Stamford), geranium (*Pelargonium* × *hortorum* cv. Fi Horizon), and impatiens (*Impatiens walleriana* cv. F1 New Guinea Divine Orange) seeds were used. Plant growth, nutrients, and physiology-related attributes were examined.

2.3. Seed Emergence

Both Exp. I and Exp. II investigated the emergence of seeds. Seeds were sown in plastic seedling trays (1 cm depth). Each treatment had 18 modules, each with a volume of 40 cm^3. Each module was seeded with three seeds. Irrigation took place daily with equal amount of potable water for all growing media. No fertilizers were applied during seedling growth. The recorded maximum and minimum temperatures were 25 ± 2 °C and 20 ± 2 °C, respectively. Day light hours was L:D 16:8 with light flux density 300 μmol PAR m^{-2} s^{-1} ± 20.

Seed emergence was observed daily and seeds were marked emerged with the hypocotyl's appearance. Mean emergence time (MET) was calculated as described previously [49].

2.4. Plant Growth and Nutrient Content

In six seedlings/treatments, seedlings were recorded with growth-related parameters following 4–6 weeks of plant growth. The seedling height and the number of leaved produced were recorded. The stomatal conductance of leaves was measured with a ΔT-Porometer AP4 (Delta-T Devices-Cambridge, Burwell, Cambridge, UK) [38]. Leaf chlorophyll fluorescence (chlorophyll fluorometer, opti-sciences OS-30p, UK) was measured on two fully developed, light-exposed leaves per seedling. Following leaves incubation in the dark for 20 min, the Fv/Fm ratio was measured [38]. Leaf chlorophyll content was assayed in six replicates/treatment. Chlorophyll was extracted with dimethyl sulfoxide (DMSO) and Chlorophyll a (Chl a), Chlorophyll b (Chl b) and total Chlorophylls (total Chl) content was determined [38]. Seedlings were sampled above substrate surface, the upper plant part was weighed (g), dried at 85 °C and then the dry weight (g) was recorded.

Nutrient content was measured in the seedling's upper part (including leaves and shoots) on four replicates/treatment (two pooled plants/replicate). Dried plant tissue (at 65 °C for 3 d) was used (~0.5 g) and was ashed (at 500 °C for 5 h) and acid (2 N HCl) digested [50]. Nitrogen (N) content was determined with Kjeldahl (BUCHI, Digest automat K-439 and Distillation Kjeldahl K-360) digestion method. Phosphorus content was determined spectrophotometrically (Multiskan GO, Thermo Fisher Scientific, Waltham, MA, USA), and K, Mg, Ca, Na, Fe, Cu, and Zn by an atomic absorption spectrophotometer (PG Instruments AA500FG, Leicestershire, UK) for plant tissue analysis or by inductively coupled plasma atomic emission spectrometry (ICP-AES; PSFO 2.0, Leeman Labs INC., Mason, OH, USA) for growing media analysis [10,38]. Plant nutrient content was expressed in g kg^{-1} and mg kg^{-1} of dry weight, for macronutrients and micronutrients, respectively, and growing media mineral content was expressed in mg L^{-1}.

2.5. Total Phenols Content and Antioxidant Capacity

In the Exp. I, methanolic extracts of four replicates (two pooled plants/replicate) of lettuce grown in different biochar type and ratio were used for total phenols content and total antioxidant activity determination. Total phenols content was determined as previously described [51] and results were expressed as gallic acid equivalents (mg GAE per gram of fresh weight). For antioxidant capacity, two assays were employed, the ferric reducing antioxidant power (FRAP) and the 2,2-diphenyl-1-picrylhydrazyl (DPPH), as described previously [52]. Results were expressed as trolox equivalents (mg trolox per gram of fresh weight).

2.6. Lipid Peroxidation, Hydrogen Peroxide, and Enzyme Antioxidant Activity

In the Exp. I, four replicates (two pooled plants/replicate) of each treatment were used for the evaluation of damage index and antioxidant enzymatic activity. Lipid peroxidation (assayed through the malondialdeyde-MDA content) and hydrogen peroxide (H_2O_2) content were measured [53,54]. Results were expressed as μmol H_2O_2 per gram of fresh weight, and nmol of MDA per gram of fresh weight.

The enzymes antioxidant activity for superoxide dismutase (SOD), for catalase (CAT) and for peroxidase activity (POD) was assayed as described previously [52]. Results were expressed as enzyme units per mg of protein. The protein content was determined with bovine serum albumin (BSA), as a standard.

2.7. Statistical Analysis

A three-factor (Biochar type, Biochar rate and Fertilizer) factorial experiment was carried out. Results were statistically analyzed with a three-way analysis of variance (ANOVA) with the IBM SPSS v.22 software for Windows. The Duncan's Multiple Range test (DMRT) was used for comparing means in case of the effect of factors and their interaction, at $p \leq 0.05$, following one-way ANOVA. Mean values ± standard error (SE) of three biological replications ($n = 3$) for growing media and of four biological replications ($n = 4$) for plant-related analysis were used.

3. Results

Table 2 presents the effects of biochar type, biochar ratio, fertilizer, and their interaction on growing media and plant-related parameters. Biochar type affected significantly growing media parameters (EC, pH, organic matter, organic carbon, K, Ca, Na, SO_4, Fe, Cu, Zn, Mn, and B at $p < 0.001$; N at $p < 0.01$) and plant (Fv/Fm, MDA, N, K, P, Ca, Mg, Na, Cu, and Zn at $p < 0.001$; height, chlorophylls, FRAP, and SOD at $p < 0.05$). Biochar ratio affected significantly growing media parameters (EC, organic matter, organic carbon, N, K, Mg, Na, SO_4, Fe, Zn, and Mn at $p < 0.001$; P and B at $p < 0.01$) and plant (Fv/Fm, MDA, CAT, POD, N, K, Ca, Mg, Cu, and Zn at $p < 0.001$). Fertilizer affected significantly growing media parameters (EC, organic matter, organic carbon, N, K, P, SO_4, and Cu at $p < 0.001$; Ca, Fe at $p < 0.05$) and plant-related parameters (height, fresh weight, dry weight, H_2O_2, POD, P, Ca, Mg, Na, and Cu at $p < 0.001$; leaf number, chlorophyll b, MDA, and N at $p < 0.01$; chlorophyll a, total chlorophyll, CAT, Fe, and Zn at $p < 0.05$).

Considering the interaction of the examined factors, Biochar type × Biochar ratio (T × R) affected significantly growing media (pH, organic matter, organic carbon, N, K, Na, SO_4, Fe, Zn, and Mn at $p < 0.001$; B at $p < 0.01$; EC, P and Mg at $p < 0.05$) and plant (H_2O_2, MDA, CAT, K, P, Na, Cu, and Zn at $p < 0.001$; FRAP at $p < 0.01$; height, stomatal conductance, chlorophylls, total phenols, DPPH, Mg, and Fe at $p < 0.05$). Biochar type × Fertilizer (T × F) affected significantly growing media (SO_4, Fe, and Cu at $p < 0.001$; EC and P at $p < 0.01$) and plant (H_2O_2, MDA, N, K, P, Na, and Zn at $p < 0.001$; Fv/Fm and CAT, at $p < 0.01$; stomatal conductance, chlorophylls, and Cu at $p < 0.05$). Biochar rate × Fertilizer (R × F) affected significantly growing media (SO_4, Fe, and Cu at $p < 0.001$; K, at $p < 0.01$) and plant (MDA, N, K, P, Na, and Zn, at $p < 0.001$; height, stomatal conductance, CAT, Ca, Mg, at $p < 0.01$; dry weight, and Cu, at $p < 0.05$). Biochar type × Biochar rate × Fertilizer (T × R × F) affected significantly growing media (P, SO_4, Fe, and Cu at $p < 0.001$; K at $p < 0.01$; pH and Ca at $p < 0.05$) and plant (Fv/Fm, N, P, and Zn at $p < 0.001$; height, MDA, and K at $p < 0.01$; leaf number, SOD, and CAT at $p < 0.05$).

3.1. Growing Media Properties

The growing media properties from different mixtures based on different biochars types (BFW or BTS), ratios (7.5% and 15%), and fertilizer level (100% and 150%) are shown in Table 3. The addition of NPK-fertilizer at a level of 150% increased the EC and the levels of N, K, and P at the 100% fertilized peat (P + Fert) in comparison to the control (P). Fertilized substrates (+ Fert) of 150% revealed higher

EC values compared to the 100%. The addition of BFW and BTS decreases the EC, more with the former and more at the higher rate of biochar. This was present at both rates of fertilizer (Table 3). BTS-based media had lower organic matter compared with the BFW. Adding BFW at 15% increased pH value compared to lower ratio (i.e., 7.5% BFW) or BTS-based media. The adding of BFW and BTS at 15% into the growing media decreased N content comparing to the control (peat). This was also evidenced at the 150% fertilized BFW and BTS even at lower ratio, i.e., at 7.5%, but also at the 15%. Potassium was increased at BFW-based growing media (independently of the fertilization), but decreased at the 150% fertilized BTS media (i.e., BTS + Fert at 7.5% and 15%) in comparison to the relevant control (peat or peat + Fert, respectively). Phosphorus increased at 7.5% of BFW and BTS compared to peat, increased at BTS + Fert at 15%, and decreased at BFW + Fert at 15% compared to the peat + Fert treatment. The addition of BTS into the growing media decreased the Ca, but increased the Na and B levels, independently of the fertilization and/or biochar ratios. Magnesium and sulfur levels were decreased in BFW- and BTS-based media. The addition of BFW decreased Cu while the addition of BTS increased Cu levels compared to the control. The Zn levels were increased in BTS-based media and in case of BFW + Fert at 7.5%. Iron content decreased in BFW at 15%, but increased in BTS + Fert at 15%. Manganese levels were increased in BFW-based media compared with the relevant control (peat or peat + Fert, respectively) (Table 3).

3.2. Experiment I

3.2.1. Seed Emergence

Biochar BFW at 7.5% increased seed emergence after 4 days compared to control (peat). Biochar BTS did not change the emergence of lettuce seeds (Figure 1). Neither the type nor the ratio of the examined biochars and applied fertilizers (100% or 150%) affected the mean emergence time for lettuce seeds (data not shown).

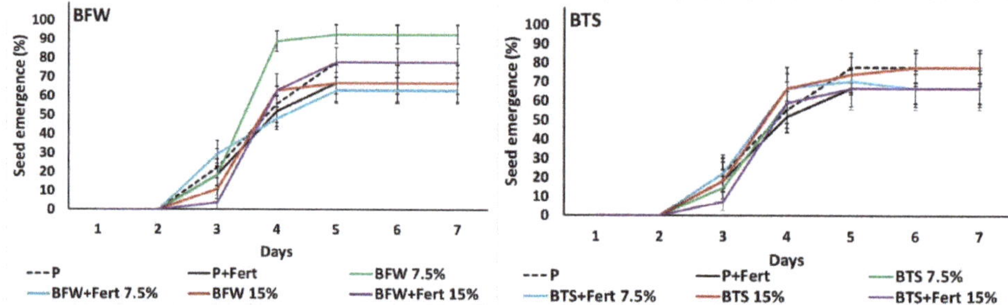

Figure 1. Lettuce cumulative seedling emergence in peat (P) with different biochar types (BFW, BTS) and ratio (7.5%, 15%) and mineral doses (with standard or with Fertilizers-Fert.). Error bars show SE ($n = 4$).

3.2.2. Plant Growth and Physiology

Biochar BFW and BTS either at 7% or at 15% reduced plant height, when compared with the control treatment (Table 4). The tallest seedlings were found at the BTS + Fert at 7.5% treatment. Fertilization at 150% and biochar type were affecting upper seedling fresh weight as BTS + Fert at 15% and BFW + Fert at 7.5% and at 15% decreased seedling fresh weight compared with the peat + Fert treatment. BFW + Fert at 15% and BTS + Fert at 15% decreased dry weight when compared to control and 7.5% of 150% fertilized Biochars (BFW + Fert and BTS + Fert). The number of leaves produced was similar in plants grown on biochar-based media and control (Figure S1), while the higher number of leaves was obtained at the BTS + Fert at 7.5% and relevant control (Peat + Fert) (Table 4).

Table 4. Effects of peat (P) with different biochar types (BFW, BTS) and ratio (7.5%, 15%) and mineral doses (with standard or with additional Fertilizers-Fert.) on lettuce plant growth (height in cm, upper fresh weight in g, upper dry weight in g, root length in cm).

	Height	Leaf Number	Upper Fresh Weight	Upper Dry Weight
P	9.11 ± 0.71 abc	5.33 ± 0.21 ab	0.908 ± 0.153 bcd	0.090 ± 0.026 bc
P + Fert	9.61 ± 0.55 ab	5.50 ± 0.42 a	1.588 ± 0.241 a	0.207 ± 0.054 a
BFW 7.5%	6.50 ± 0.53 de	4.16 ± 0.31 b	0.504 ± 0.053 d	0.046 ± 0.002 c
BFW + Fert 7.5%	8.30 ± 0.58 bcd	5.16 ± 0.47 ab	1.061 ± 0.047 bc	0.240 ± 0.051 a
BFW 15%	6.06 ± 0.42 e	4.66 ± 0.21 ab	0.603 ± 0.015 cd	0.085 ± 0.002 bc
BFW + Fert 15%	8.01 ± 0.49 bcde	5.00 ± 0.25 ab	0.889 ± 0.083 cd	0.093 ± 0.018 bc
BTS 7.5%	6.42 ± 0.41 de	4.50 ± 0.42 ab	0.570 ± 0.009 d	0.091 ± 0.000 bc
BTS + Fert 7.5%	10.83 ± 1.06 a	5.50 ± 0.62 a	1.341 ± 0.274 ab	0.159 ± 0.011 ab
BTS 15%	7.01 ± 0.59 de	4.83 ± 0.40 ab	0.625 ± 0.063 cd	0.089 ± 0.009 bc
BTS + Fert 15%	7.56 ± 0.47 cde	4.66 ± 0.33 ab	0.853 ± 0.186 cd	0.101 ± 0.018 bc

Values ($n = 6$) in columns followed by the same letter are not significantly different, $p < 0.05$.

The stomatal conductance of leaves was increased at BFW + Fert at 15% (Table 5). Leaf chlorophyll fluorescence decreased with the biochars (BFW or BTS) presence at both 7.5 and 15% ratios. However, fertilization at 150% increased chlorophyll fluorescence only in the case of BFW at 7.5%. Chlorophyll a content increased at BTS at 7.5% and at BTS (at 7.5 and 15%) + Fert compared to relevant controls (100% fertilized peat in the first case and 150% fertilized peat in the latter). Chlorophyll b content was also increased at BTS + Fert, which resulted in increased total chlorophylls content at BTS at 7.5% and 15% (Table 5).

Table 5. Effects of peat (P) with different biochar types (BFW, BTS) and ratio (7.5%, 15%) and mineral doses (with standard or with additional Fertilizers-Fert.) on lettuce leaf stomatal conductance (cm s^{-1}), chlorophyll fluorescence (Fv/Fm), and chlorophylls (Chl a, Chl b, total Chls) content (mg g^{-1}).

	Stomatal Conductance	Chlorophyll Fluorescence	Chl a	Chl b	Total Chls
P	13.64 ± 0.60 ab	0.903 ± 0.006 b	0.621 ± 0.003 d	0.161 ± 0.001 c	0.782 ± 0.001 c
P + Fert	11.41 ± 1.28 bc	0.878 ± 0.007 c	0.695 ± 0.035 cd	0.184 ± 0.008 c	0.879 ± 0.044 c
BFW 7.5%	13.70 ± 2.12 ab	0.871 ± 0.005 c	0.638 ± 0.093 cd	0.165 ± 0.027 c	0.804 ± 0.121 c
BFW + Fert 7.5%	8.41 ± 0.68 c	0.919 ± 0.002 a	0.681 ± 0.055 cd	0.177 ± 0.018 c	0.858 ± 0.074 c
BFW 15%	11.01 ± 0.91 bc	0.865 ± 0.003 c	0.795 ± 0.032 bcd	0.202 ± 0.007 bc	0.997 ± 0.040 bc
BFW + Fert 15%	17.26 ± 1.03 a	0.868 ± 0.001 c	0.822 ± 0.086 bcd	0.214 ± 0.022 bc	1.037 ± 0.109 bc
BTS 7.5%	16.13 ± 2.29 a	0.865 ± 0.003 c	0.830 ± 0.045 bc	0.207 ± 0.009 bc	1.036 ± 0.054 bc
BTS + Fert 7.5%	7.13 ± 1.70 c	0.838 ± 0.001 d	1.139 ± 0.108 a	0.308 ± 0.030 a	1.446 ± 0.138 a
BTS 15%	9.33 ± 1.67 bc	0.825 ± 0.007 d	0.661 ± 0.037 cd	0.168 ± 0.008 c	0.830 ± 0.045 c
BTS + Fert 15%	7.33 ± 1.19 c	0.838 ± 0.002 d	0.939 ± 0.048 b	0.246 ± 0.013 b	1.185 ± 0.061 b

Values ($n = 6$) in columns followed by the same letter are not significantly different, $p < 0.05$.

3.2.3. Total Phenol Content and Antioxidant Activity

Total phenol content decreased at 7.5% BFW and 15% BTS, independently of the fertilization scheme (Figure 2A). BFW presence at 7.5%, independently of the fertilization, decreased the antioxidant activity (as assayed by DPPH and FRAP) of lettuce, while in the case of BTS, DPPH and FRAP were decreased at 100% fertilized BTS at 7.5% and 15% (Figure 2B,C).

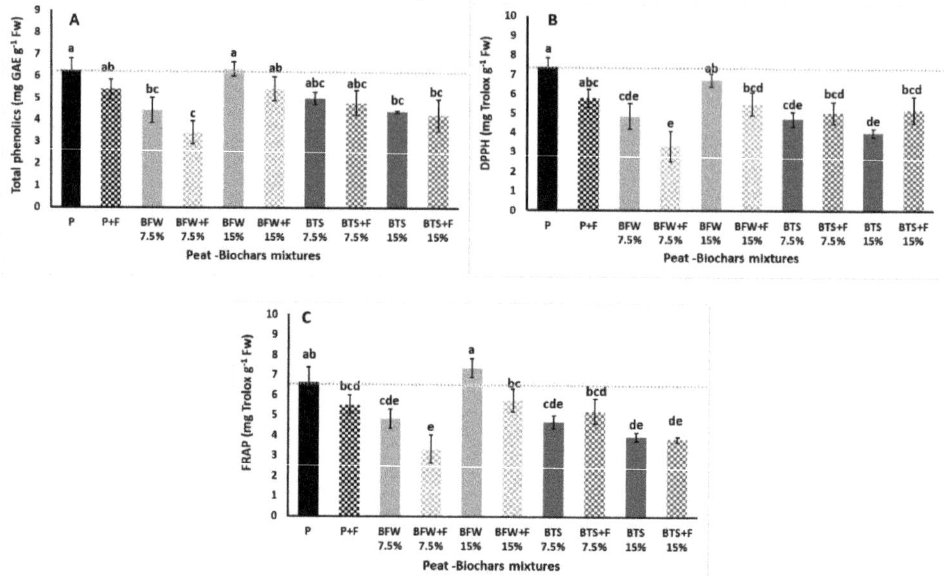

Figure 2. Effects of peat (P) with different biochar types (BFW, BTS) and ratio (7.5%, 15%) and mineral doses (with standard or with additional Fertilizers-F.) on lettuce total phenols and antioxidant activity. (**A**) total phenols, (**B**) DPPH, and (**C**) FRAP. Significant differences ($p < 0.05$) among treatments are indicated by different letters. Error bars show SE ($n = 4$). Dotted line presents the levels of control treatment (100% peat).

3.2.4. Lipid Peroxidation, Hydrogen Peroxide, and Enzyme Antioxidant Activity

The 150% fertilized peat (peat + Fert) revealed increases in the plant lipid peroxidation (MDA) when compared to the standard rate of fertilized (control) treatment (Figure 3A). Moreover, MDA increased at 7.5% and 15% BFW and at BFW + Fert at 15% in comparison to peat. In the case of BTS, MDA content increased at 7.5% and 15% BTS as well as the BTS + Fert at 7.5%, but MDA decreased at BTS + Fert at 15% compared to relevant controls. The MDA increases were followed by the increased trend of production of hydrogen peroxide in most cases (Figure 3B). Antioxidant enzymes have fluctuated among the treatments, so that the plants can detoxify the elevated stress. SOD activity increased for BTS + Fert at 7.5% and BTS at 15%, when compared to the relevant control (Figure 3C). CAT antioxidant enzymatic activity was decreased for BFW at 15% (independently of the fertilization) treatments and for BTS at 15% (Figure 3D). POD activity at the 150% fertilized peat (P + Fert) increased compared to the 100% peat (Figure 3E). POD activity decreased for the BFW at 15%, BTS at 7.5%, BTS at 15% compared to peat. Fertilized (at 150%) Biochars (BFW and BTS) decreased POD activity compared with the relevant control (peat + Fert).

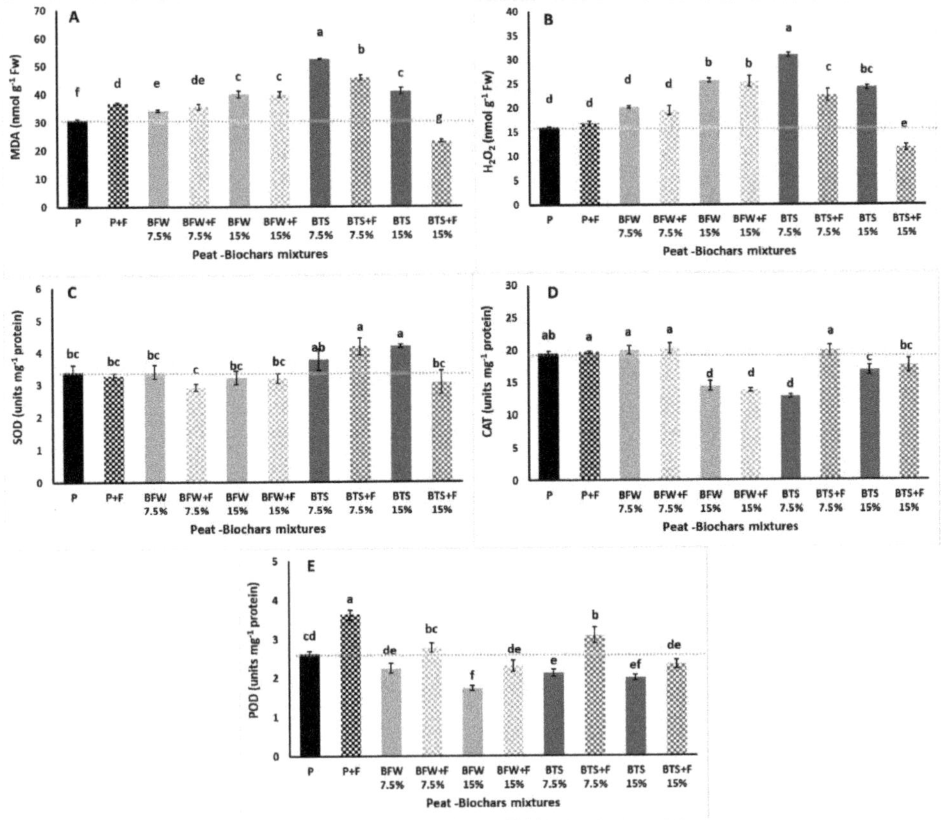

Figure 3. Effects of peat (P) with different biochar types (BFW, BTS) and ratio (7.5%, 15%) and mineral doses (with standard or with additional Fertilizers-F) on lettuce lipid peroxidation, hydrogen peroxide and antioxidant enzymes activity. (**A**) Lipid peroxidation (MDA), (**B**) H_2O_2, (**C**) superoxide dismutase (SOD), (**D**) catalase (CAT), and (**E**) peroxidase activity (POD). Significant differences ($p < 0.05$) among treatments are indicated by different letters. Error bars show SE ($n = 4$). Dotted line presents the levels of control treatment (100% peat).

3.2.5. Nutrient Content

The addition of 150% fertilizers in peat, increased the N, P, Mg, and Na but decreased Ca and Cu accumulation in lettuce seedlings (Figure 4A–I). The BTS at 7.5% increased further the N accumulation in lettuce, while BFW at 15%, BTS at 15%, and 150% fertilized BFW and BTS at both rations decreased the N content in lettuce in comparison to the plants grown in control (Figure 4A). BFW significantly increased the accumulation of K in seedlings, as both the Biochar ratios and the fertilizer presence, increased the K content in plants. Nonetheless, BTS decreased in general the K content in lettuce and was necessary to be fertilized at 150% and used at 15% into the growing media so to increase the K content in lettuce seedling to levels comparable to the control (peat) (Figure 4B). The P content decreased at 15% BFW and BTS at 7.5% and 15% compared to peat, while BFW + Fert and BTS + Fert resulted in decreased P content in lettuce, independently of the ratio of 7.5% or 15% used (Figure 4C). A similar trend to P was observed in the plant tissue for the Mg accumulation (Figure 4E). Calcium content in lettuce was accumulated less with BFW at 7.5% and at 15% and BTS at 7.5% compared to peat, but Ca content was increased with BTS + Fert at 15% (Figure 4D). Sodium was accumulated more in plants grown at BTS at 15% (no matter the dosage of fertilizer) and lesser at BFW at 15% (independent of the fertilizers levels) and BFW + Fert at 7.5% compared to the relevant controls (Figure 4F). Biochar presence at 7.5% for BTS and at 15% for BFW and BTS decreased the Fe content in lettuce, while the 150% fertilization alleviated this effect, as Fe content was in comparable levels to peat-based growing media (without Biochar) (Figure 4G). Copper levels were increased with the adding of BTS and/or fertilizers, but decreased with the BFW at 7.5% (Figure 4H). A similar tendency to Cu was observed for Zn accumulation with exception the decreased Zn content with the BTS at 7.5% compared to the control treatment (Figure 4I).

3.3. Experiment II

3.3.1. Seed Emergence

In tomato, no differences were found on seed emergence percentage and MET (Figure S2). The first seed emergence took place on day 4, while all seeds were emerged on the 5th day. In leek, seed emergence percentage increased (up to 60%) in Biochar-based media when compared to the control (100% peat) and significant differences were found after the 11th day (Figure S2). There were not any delays on the emergence time as the MET was similar for control and Biochar-based media. In geranium, a slight increase on seed emergence was found in Biochar-based media however, the effects ended up not to be significant at the end (Figure S2). No differences were found on MET among the examined growing media. In impatiens, the first seed emergence took place on the 9th day and the emergence was completed on the day 16 (Figure S2). Seed emergence on the Biochar-based media was significantly increased up to the 12th day compared with the relevant emergence on the control treatment. MET was decreased in biochar-based growing media. In tomato, seed emergence decreased in biochar-based media after the 4th day, while MET was the same for peat and biochar-based media (Figure S2).

3.3.2. Effects on Plant Growth, Physiology, and Nutrient Content

In tomato, BFW + Fert at 7.5% increased plant height and K accumulation, but decreased leaf stomatal conductance and P levels compared to plants grown in peat (control) (Figure 5). In leak, Biochar presence increased seedling dry weight, and the levels of chlorophylls (Chlorophyll a, Chlorophyll b and total Chlorophylls), but decreased the nutrient accumulation by decreasing N, K, and P levels. In geranium, plants grown in Biochar-enriched growing media revealed higher plant height, dry weight, and K content compared to the control treatment. In impatiens, Biochar presence increased leaf number, P, and K content, but decreased chlorophyll fluorescence, Chlorophylls content, and N content compared to the peat (control) (Figure 5).

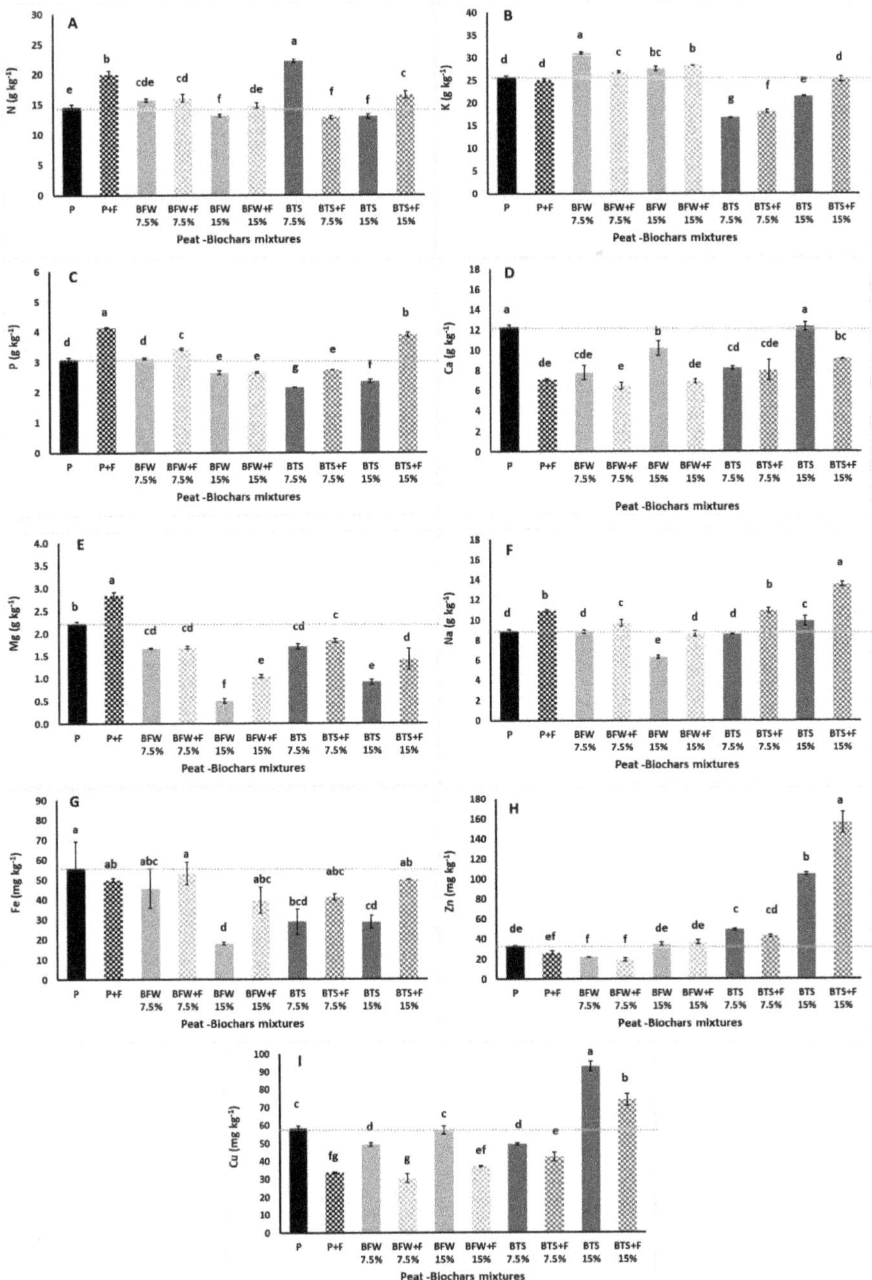

Figure 4. Effects of peat (P) with different biochar types (BFW, BTS) and ratio (7.5%, 15%) and mineral doses (with standard or with additional Fertilizers-F) on lettuce macro- and micronutrient content. Significant differences ($p < 0.05$) among treatments are indicated by different letters. Error bars show SE ($n = 4$). Dotted line presents the levels of control treatment (100% peat).

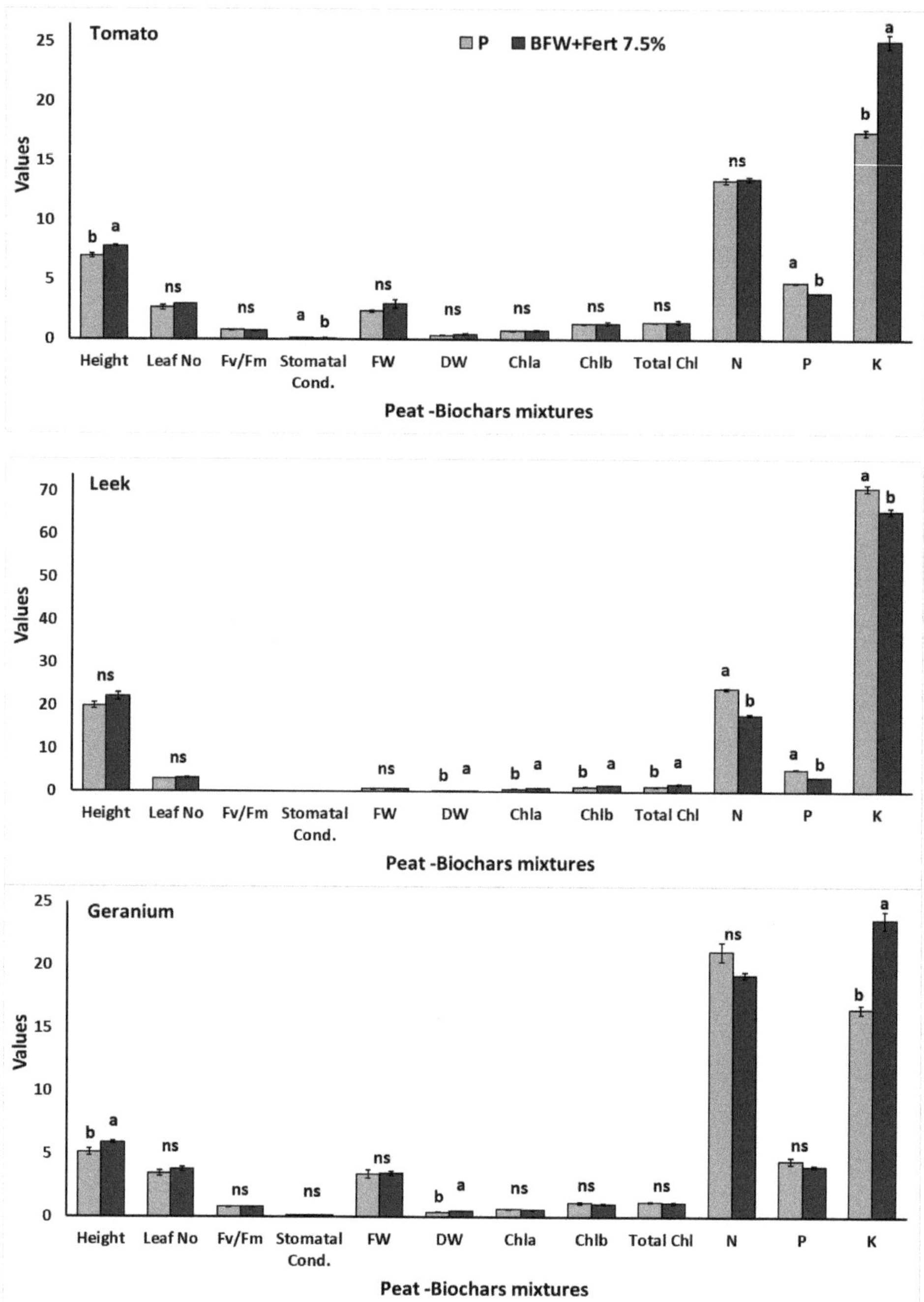

Figure 5. *Cont.*

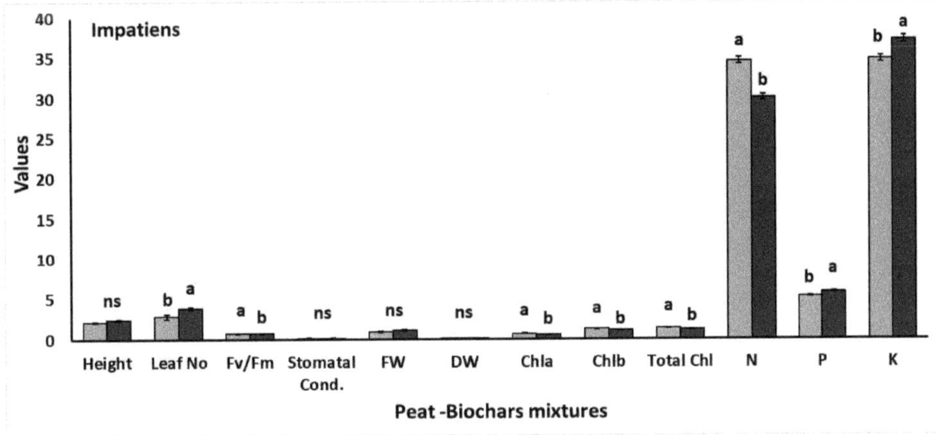

Figure 5. Effects of peat (P; light grey) with BFW at 7.5% (dark grey) with additional Fertilizers-Fert. on tomato, leek, geranium, and impatiens plant growth, physiology, and nutrient content-related parameters. Significant differences ($p < 0.05$) among treatments are indicated by different letters. Error bars show SE ($n = 4$). ns: not significant.

4. Discussion

In the present study, biochar type and ratio as well as fertilization levels were examined for lettuce seedling production and affected lettuce growth and physiology-related attributes with the levels of fertilization impacting plant performance. However, optimized biochar and fertilizer application do not have the same impacts on different plant species and selection should be on a plant-species based strategy. Biochar production is of different organic materials including urban wastes (garden pruning waste), agricultural waste (straw, residue from greenhouse crops, olive-mill waste, by-products from vineyards), food waste, digestate, and even sewage sludge [55–57], and different sources of organic material result in different biochar quality, as presented in Table 1.

In the present study, the increased pH is in line with numerous studies on Biochar-based material derived from forest waste [5], tomato crop green and wood waste [10,58], wheat straw [14], and hardwood waste [59]. The increase of pH following biochar addition is beneficial for acidic soil or growing media (as is the case for peat in the present work), with biochar serving as a liming agent and likely replacing the calcium oxide used to increase the pH [60,61]. However, there is need for caution, as we have shown that biochars of similar pH can have different levels of extractable Ca [62]. The low ratio of biochars employed in the present study retained pH values between 5.0 and 7.0, since the ideal substrate pH for peat substrate is between 5.0 and 5.5 [63]. Moreover, biochar-based media had lower EC than standard fertilized peat (control) and was related to rate of application with the higher rate (15%) to be more effective than 5% and the biochar with higher surface area being more effective, in accordance with previous reports [10]. This finding has very serious implications, as materials widely used for peat dilution have usually high levels of EC, such as compost from green waste [64]. Those materials could be used in combination with biochar at a higher ratio since high EC is commonly a limiting factor to be added to these materials. Wang et al. [65] also found that Biochar derived from woody feedstock with higher lignin content and greater surface area revealed different metal sorption capabilities. However, additional studies on biochar type and rate applied on other crops, and at different growth stage apart from seedlings, are required before final conclusions are made. We had previously discussed the possible reason for decreased/low levels of nitrate and of P [10], while Altland and Locke [66] documented P release of biochar from rice husks, with further studies required to understand the mechanism involved. There are various reasons for the retention of nitrate and phosphate in biochar amended growing media. Biochar provides refugee for soil microorganism to influence the binding of carbon and anions [67]. Another explanation could be that due to improvement of the root milieu due to addition of biochar e.g., due to soluble organic carbon and this leads in increased microbial activity. Accelerated metabolism of soil biota turns the inorganic nitrogen into organic form, hence less available N and less N uptake [68]. The reduction of phosphate availability and uptake could be due to increased availability of Ca and Mg, due to addition of biochar [69]. Finally, increased surface area and net surface charge may also be responsible [70]. Kammann et al. [71] hypothesized that surface ageing plus non-conventional ion-water bonding in micro- and nano-pores enhanced the capture of nitrates in the biochar particles. Amending (N-rich) bio-waste with biochar may promote its agronomic value and reduce nutrient losses from bio-wastes and agricultural soils.

Biochar BFW at 7.5% stimulated seed emergence compared to the control, but this was not evidenced at the BTS. One possible explanation for that is that BFW had increased K levels compared with BTS, and priming effect of extra K on seed germination is already documented in Chinese cabbage [72]. Under nursery conditions, the consistent and fast emergence of seeds is required to meet the increased demands of healthy plant material, delivered to the producers.

Lettuce seedling height was decreased with the presence of BFW or BTS, independently of the biochar ratio, which can be of benefit for a nursery, as shorter plants can be handled, transported, and stored easier than taller plants. BFW and BTS were nutrient rich (EC ranged from 0.41–$0.61\,\mathrm{mS\,cm^{-1}}$) materials, but in the present study, they were used in low levels (7.5–15% *v/v*) from one hand, while the fertilized peat used, was a more nutrient rich (higher EC) component compared to biochar-based growing media. Therefore, the reduced plant height might be attributed to the decreased available nutrients to the roots and/or due to the different particle size/surface area and extractable nutrients, i.e., Ca at the biochar-based media [62]. Therefore, the decreased plant height cannot be considered of disadvantage at all. Based on that, plant fresh weight was not affected by the biochar type and ratio used. However, when fertilization took place, plant fresh biomass decreased in both BFW and BTS, dry weight decreased at 15% fertilized biochar's media, indicating an increased stress condition due to the overloaded fertilizers particularly K, without affecting the number of leaves produced per plant. High ratio of biochar (up to 20%) and different biochar type can affect the cabbage seedling production, as biochar from bamboo or from husks and paper fiber wood screenings affected negatively plant growth and successful cabbage seedling production [34].

Growing media with low biochar rates, for example 10% sewage-sludge derived biochar in lettuce [73] and 10% wood-derived biochar in pepper (*Capsicum annuum* L.) and tomato [74], promoted plant growth. Similar to our findings based on the low biochar ratio used (i.e., 7.5%), Kim et al. [75] reported increased dry weight of kale (*Brassica olereaseae* L.) shoots with the use of low biochar levels (i.e., 5% rice husk biochar) into a coir dust-based media. However, other studies reported no effects of biochar on plant dry weight in tomato and marigold (*Tagetes erecta* L.) [22] and sunflower (*Helianthus annuus* L.) [21]. This fluctuation of results among studies might be related to the different biochar sources (wood, straw) and method of production, thus quality, the ratio used (from 5% up to 75%), and the plant species [5]. Mendez et al. [3] mixed biochar with peat to grow lettuce revealed higher quality plants in comparison to those cultivated in single substrate. Changes in chlorophyll fluorescence is affecting the efficiency of PSII photochemistry and the plant photosynthetic performance [38], and this was evidenced at biochar-based media that revealed decreased leaf chlorophyll fluorescence and plant height. Leaf stomatal conductance increased at BFW+Fert at 15%. *Andrographis paniculate* (kalmegh) herb grown in biochar-based media mitigated drought stress altering plant metabolism, decreasing the stomatal conductance and increased the water use efficiency under such conditions [7].

Total phenols content did not differ much among the examined treatments, except for the decreased total phenols content in BFW at 7.5% and BTS at 15%. Lettuce antioxidant activity (DPPH, FRAP) decreased in most cases of BWF and BTS presence indicating the reduced capacity of the plant to withstand oxidative stress and which is less appreciated and accepted by consumers and markets/industry, who are seeking added value products of high antioxidant status [76]. In contrast, total phenols content and antioxidant activity were increased in the case of high biochar ratio and/or fertilizer for cabbage, as reported by Chrysargyris et al. [34]. Even though total phenols and antioxidants remained low, lettuce plants were subjected to oxidative stress with the presence of biochars (BFW, BTS) and/or fertilizers as revealed by the increased levels of MDA, indicating lipid peroxidation and cellular damage. This was further supported with the increased levels of H_2O_2 and the activation of SOD and POD antioxidant enzymes to detoxify the reactive oxygen species (ROS) accumulation [52]. Plants responded to oxidative stress by activation of protecting mechanism, producing stress response proteins and antioxidant enzymes (including SOD, POD, and CAT) to resolve the accumulation of ROS [77].

Increased K levels are not directly toxic to plants however, the increased K levels can cause antagonism and resulted in Mg and Ca deficiencies with decreases in plant growth [78]. In the present study, decreased Mg levels in lettuce were evidenced in biochar-based media, those media that had lower Mg and higher K content compared to control media (peat). Similarly, lower Mg was recorded when Biochars was added to a peat growing media [10]. Moreover, K accumulated in lettuce produced in BFW-based media with more pronounced effects at high ratio and/or fertilizer, whereas BTS had to be fertilized at 150% and mixed at 15% to obtain K levels such as control. Similarly, when *Syngonium podophyllum* was grown in various Biochar-based media, increased K content was found and this was related to the higher level of this element in the growing media [14]. Fertilized (150%) biochar-based media and BFW and BTS at high ratio, decreased N and P content in lettuce, being in accordance with findings of our previous studies [10]. In contrast, Kim et al. [75] documented increase contents of N and P in kale shoots when Biochar was added at different ratios with the growing media. Calcium content was decreased in lettuce grown in BFW (at 7.5–15%) and BTS (at 7.5%) -based media and fertilization at 150% were needed to overcome this decrease. Biochar presence in general decreased the Fe content in lettuce but fertilization alleviated this effect, as the content of Fe was similar to that of peat-based substrates (without Biochar). Copper and Zn content in lettuce were increased by adding BTS and/or fertilizers in the growing media with more profound effect at the higher ratio of 15%, and this is reflecting the increased Cu and Zn levels into the growing media.

Following selection of BFW at 7.5% with 150% fertilization, seed emergence was improved in leek and impatiens, but not in tomato and geranium. In tomato, geranium, and impatiens BFW + Fert at 7.5% increased several plant-growth related parameters and nutrient accumulation, mainly of K compared to the control. However, in some cases, such as tomato, leaf stomatal conductance decreased in plants grown in biochar-based media.

5. Conclusions

In conclusion, Biochar increased plant growth (biomass, height) at a low ratio (7.5%), while it reduced the emergence of lettuce seed and plant height at 15%. The addition of Biochars provided nutrients in the seedlings, because the plants could absorb more available nutrients. Production of low-weight seedlings may be beneficial to nurseries when they want to produce dwarf plants and overhead irrigation. This helps to conditions for transport and storage. It seems better to use a wooden biochar of beech, spruce, and pine species produced at 700 °C with the Schotteredorf process and to use the resulting Biochar (BFW) at a ratio of 7.5% for the production of lettuce seedlings. Different species however need to be assessed accordingly. These results showed clearly the ability of biochars to reduce EC depending on rate of application and biochar surface area. This finding is very significant as most materials used to dilute peat e.g., composted greenwaste have high EC and the limiting factor on the rate of peat dilution are their high EC. These results also clearly showed that any investigation into the suitability of adding biochar to peat as peat replacement and/or biostimulant, must consider the nutrient implications of this addition to the plant. In addition, bringing the nutrient to the levels based on calculation e.g., peat growing media, may not be enough as there is a strong interaction between biochar and N and P availability as based on our substrate analysis and plant nutrient uptake data. This area of work needs attention when experiments are conducted to evaluate biochar as an addition to peat and to other growing media.

Supplementary Materials: The following are available online at http://www.mdpi.com/2073-4395/10/9/1421/s1, Table S1: Growing media composition. Figure S1: Lettuce seedling production in peat with different biochar types (BFW, BTS) and ratio (7.5%, 15%) and nutrient doses. Figure S2. Effects biochar BFW at 7.5% with additional Fertilizers-Fert. on tomato, leek, geranium, and impatiens seed emergence and mean emergence time.

Author Contributions: Conceptualization, M.P. and N.T.; methodology, M.P. and N.T.; software, N.T.; validation, A.C., M.P., and N.T.; formal analysis, A.C. and M.P.; investigation, A.C. and A.K.; resources, A.K. and N.T.; data curation, A.C. and M.P.; writing—original draft preparation, A.C.; writing—review and editing, M.P. and N.T.; visualization, A.K.; supervision, M.P. and N.T.; project administration, M.P. and N.T.; funding acquisition, A.K. and N.T. All authors have read and agreed to the published version of the manuscript.

Funding: This research was funded by Bord na Mona Horticulture Ltd. and Cyprus University of Technology under the project OPTIBIOCHAR.

Acknowledgments: The authors are grateful to the project OPTIBIOCHAR that has been developed under the Cooperation Programme Cyprus-Ireland, co-funded by the Bord na Mona LtD and Cyprus University of Technology. Moreover, we are grateful to the reviewer for their substantial and critical contribution during the review/evaluation process.

Conflicts of Interest: The authors declare no conflict of interest.

References

1. Solaiman, Z.M.; Murphy, D.V.; Abbott, L.K. Biochars influence seed germination and early growth of seedlings. *Plant Soil* **2012**, *353*, 273–287. [CrossRef]
2. Laghari, M.; Hu, Z.; Mirjat, M.S.; Xiao, B.; Tagar, A.A.; Hu, M. Fast pyrolysis biochar from sawdust improves the quality of desert soils and enhances plant growth. *J. Sci. Food Agric.* **2016**, *96*, 199–206. [CrossRef]
3. Méndez, A.; Paz-Ferreiro, J.; Gil, E.; Gascó, G. The effect of paper sludge and biochar addition on brown peat and coir based growing media properties. *Sci. Hortic. Amst.* **2015**, *193*, 225–230. [CrossRef]
4. Agegnehu, G.; Bass, A.M.A.M.; Nelson, P.N.P.N.; Bird, M.I.M.I. Benefits of biochar, compost and biochar-compost for soil quality, maize yield and greenhouse gas emissions in a tropical agricultural soil. *Sci. Total Environ.* **2016**, *543*, 295–306. [CrossRef]
5. Belda, R.M.; Lidón, A.; Fornes, F. Biochars and hydrochars as substrate constituents for soilless growth of myrtle and mastic. *Ind. Crop. Prod.* **2016**, *94*, 132–142. [CrossRef]
6. Calamai, A.; Palchetti, E.; Masoni, A.; Marini, L.; Chiaramonti, D.; Dibari, C.; Brilli, L. The influence of biochar and solid digestate on rose-scented geranium (*Pelargonium graveolens* L'Hér.) productivity and essential oil quality. *Agronomy* **2019**, *9*, 260. [CrossRef]
7. Saha, A.; Basak, B.B.; Gajbhiye, N.A.; Kalariya, K.A.; Manivel, P. Sustainable fertilization through co-application of biochar and chemical fertilizers improves yield, quality of Andrographis paniculata and soil health. *Ind. Crop. Prod.* **2019**, *140*, 111607. [CrossRef]
8. Huang, L.; Niu, G.; Feagley, S.E.; Gu, M. Evaluation of a hardwood biochar and two composts mixes as replacements for a peat-based commercial substrate. *Ind. Crop. Prod.* **2019**, *129*, 549–560. [CrossRef]
9. Lashari, M.S.; Ye, Y.; Ji, H.; Li, L.; Kibue, G.W.; Lu, H.; Zheng, J.; Pan, G. Biochar-manure compost in conjunction with pyroligneous solution alleviated salt stress and improved leaf bioactivity of maize in a saline soil from central China: A 2-year field experiment. *J. Sci. Food Agric.* **2015**, *95*, 1321–1327. [CrossRef]
10. Prasad, M.; Tzortzakis, N.; McDaniel, N. Chemical characterization of biochar and assessment of the nutrient dynamics by means of preliminary plant growth tests. *J. Environ. Manag.* **2018**, *216*, 89–95. [CrossRef]
11. Nadeem, S.M.; Imran, M.; Naveed, M.; Khan, M.Y.; Ahmad, M.; Zahir, Z.A.; Crowley, D.E. Synergistic use of biochar, compost and plant growth-promoting rhizobacteria for enhancing cucumber growth under water deficit conditions. *J. Sci. Food Agric.* **2017**, *97*, 5139–5145. [CrossRef]
12. Alvarez, J.M.; Pasian, C.; Lal, R.; Lopez, R.; Fernandez, M. Vermicompost and Biochar as growing media replacement for ornamental plant production. *J. Appl. Hortic.* **2017**, *19*, 205–214. [CrossRef]
13. Alvarez, J.M.; Pasian, C.; Lal, R.; Lopez-Nuñez, R.; Fernández, M. A biotic strategy to sequester carbon in the ornamental containerized bedding plant production: A review. *Span. J. Agric. Res.* **2018**, *16*, 4. [CrossRef]
14. Zulfiqar, F.; Younis, A.; Chen, J. Biochar or Biochar-Compost Amendment to a Peat-Based Substrate Improves Growth of Syngonium podophyllum. *Agronomy* **2019**, *9*, 460. [CrossRef]

15. Blok, C.; Van Der Salm, C.; Hofland-Zijlstra, J.; Streminska, M.; Eveleens, B.; Regelink, I.; Fryda, L.; Visser, R. Biochar for horticultural rooting media improvement: Evaluation of Biochar from gasification and slow pyrolysis. *Agronomy* **2017**, *7*, 6. [CrossRef]
16. Gu, M.; Li, Q.; Steele, P.H.; Niu, G.; Yu, F. Growth of "Fireworks" gomphrena grown in substrates amended with biochar. *J. Food Agric. Environ.* **2013**, *11*, 819–821.
17. Fryda, L.; Visser, R.; Schmidt, J. Biochar Replaces Peat in Horticulture: Environmental Impact Assessment of Combined Biochar & Bioenergy Production. *Detritus* **2019**, *5*, 132–149.
18. Huang, L.; Gu, M. Effects of biochar on container substrate properties and growth of plants—A review. *Horticulturae* **2019**, *5*, 14. [CrossRef]
19. Margenot, A.J.; Griffin, D.E.; Alves, B.S.Q.; Rippner, D.A.; Li, C.; Parikh, S.J. Substitution of peat moss with softwood biochar for soil-free marigold growth. *Ind. Crop. Prod.* **2018**, *112*, 160–169. [CrossRef]
20. Guo, Y.; Niu, G.; Starman, T.; Volder, A.; Gu, M. Poinsettia Growth and Development Response to Container Root Substrate with Biochar. *Horticulturae* **2018**, *4*, 1. [CrossRef]
21. Steiner, C.; Harttung, T. Biochar as a growing media additive and peat substitute. *Solid Earth* **2014**, *5*, 995–999. [CrossRef]
22. Vaughn, S.F.; Kenar, J.A.; Thompson, A.R.; Peterson, S.C. Comparison of biochars derived from wood pellets and pelletized wheat straw as replacements for peat in potting substrates. *Ind. Crop. Prod.* **2013**, *51*, 437–443. [CrossRef]
23. Dispenza, V.; De Pasquale, C.; Fascella, G.; Mammano, M.M.; Alonzo, G. Use of biochar as peat substitute for growing substrates of *Euphorbia × lomi* potted plants. *Span. J. Agric. Res.* **2016**, *14*, 21. [CrossRef]
24. Choi, H.; Son, H.; Kim, C. Predicting financial distress of contractors in the construction industry using ensemble learning. *Expert Syst. Appl.* **2018**, *110*, 1–10. [CrossRef]
25. Shackley, S.; Hammond, J.; Gaunt, J.; Ibarrola, R. The feasibility and costs of biochar deployment in the UK. *Carbon Manag.* **2011**, *2*, 335–356. [CrossRef]
26. Gul, S.; Whalen, J.K. Biochemical cycling of nitrogen and phosphorus in biochar-amended soils. *Soil Biol. Biochem.* **2016**, *103*, 1–15. [CrossRef]
27. Altland, J.E.; Locke, J.C. Biochar affects macronutrient leaching from a soilless substrate. *HortScience* **2012**, *47*, 1136–1140. [CrossRef]
28. Viger, M.; Hancock, R.D.; Miglietta, F.; Taylor, G. More plant growth but less plant defence? First global gene expression data for plants grown in soil amended with biochar. *GCB Bioenergy* **2015**, *7*, 658–672. [CrossRef]
29. French, E.; Iyer-Pascuzzi, A.S. A role for the gibberellin pathway in biochar-mediated growth promotion. *Sci. Rep.* **2018**, *8*, 1–10. [CrossRef]
30. Prendergast-Miller, M.T.; Duvall, M.; Sohi, S.P. Biochar-root interactions are mediated by biochar nutrient content and impacts on soil nutrient availability. *Eur. J. Soil Sci.* **2014**, *65*, 173–185. [CrossRef]
31. Abiven, S.; Hund, A.; Martinsen, V.; Cornelissen, G. Biochar amendment increases maize root surface areas and branching: A shovelomics study in Zambia. *Plant Soil* **2015**, *395*, 45–55. [CrossRef]
32. Hilioti, Z.; Michailof, C.M.; Valasiadis, D.; Iliopoulou, E.F.; Koidou, V.; Lappas, A.A. Characterization of castor plant-derived biochars and their effects as soil amendments on seedlings. *Biomass Bioenergy* **2017**, *105*, 96–106. [CrossRef]
33. Bachmann, R.T.; Adawiyah, S.; Krishnan, T.; Khoo, B.; Sian, T.S.; Richards, T. Partial substitution of peat moss with biochar for sustainable cultivation of *Durio zibethinus* L. in nurseries. *Arab. J. Geosci.* **2018**, *11*, 426. [CrossRef]
34. Chrysargyris, A.; Prasad, M.; Kavanagh, A.; Tzortzakis, N. Biochar type and ratio as a peat additive/partial peat replacement in growing media for cabbage seedling production. *Agronomy* **2019**, *9*, 693. [CrossRef]
35. Kloss, S.; Zehetner, F.; Dellantonio, A.; Hamid, R.; Ottner, F.; Liedtke, V.; Schwanninger, M.; Gerzabek, M.H.M.H.; Soja, G. Characterization of Slow Pyrolysis Biochars: Effects of Feedstocks and Pyrolysis Temperature on Biochar Properties. *J. Environ. Qual.* **2012**, *41*, 990–1000. [CrossRef]
36. Yu, P.; Li, Q.; Huang, L.; Niu, G.; Gu, M. Mixed hardwood and sugarcane bagasse biochar as potting mix components for container tomato and basil seedling production. *Appl. Sci.* **2019**, *9*, 4713. [CrossRef]
37. Judd, L.A.; Jackson, B.E.; Fonteno, W.C. Advancements in root growth measurement technologies and observation capabilities for container-grown plants. *Plants* **2015**, *4*, 369–392. [CrossRef]

38. Chrysargyris, A.; Antoniou, O.; Athinodorou, F.; Vassiliou, R.; Papadaki, A.; Tzortzakis, N. Deployment of olive-stone waste as a substitute growing medium component for *Brassica* seedling production in nurseries. *Environ. Sci. Pollut. Res.* **2019**, *26*, 35461–35472. [CrossRef]

39. Chrysargyris, A.; Stavrinides, M.; Moustakas, K.; Tzortzakis, N. Utilization of paper waste as growing media for potted ornamental plants. *Clean Technol. Environ. Policy* **2019**, *21*, 1937–1948. [CrossRef]

40. Busch, D.; Stark, A.; Kammann, C.I.; Glaser, B. Genotoxic and phytotoxic risk assessment of fresh and treated hydrochar from hydrothermal carbonization compared to biochar from pyrolysis. *Ecotoxicol. Environ. Saf.* **2013**, *97*, 59–66. [CrossRef]

41. Liao, N.; Li, Q.; Zhang, W.; Zhou, G.; Ma, L.; Min, W.; Ye, J.; Hou, Z. Effects of biochar on soil microbial community composition and activity in drip-irrigated desert soil. *Eur. J. Soil Biol.* **2016**, *72*, 27–34. [CrossRef]

42. Gomez, J.D.; Denef, K.; Stewart, C.E.; Zheng, J.; Cotrufo, M.F. Biochar addition rate influences soil microbial abundance and activity in temperate soils. *Eur. J. Soil Sci.* **2014**, *65*, 28–39. [CrossRef]

43. Ameloot, N.; Sleutel, S.; Case, S.D.C.; Alberti, G.; McNamara, N.P.; Zavalloni, C.; Vervisch, B.; delle Vedove, G.; De Neve, S. C mineralization and microbial activity in four biochar field experiments several years after incorporation. *Soil Biol. Biochem.* **2014**, *78*, 195–203. [CrossRef]

44. Gorovtsov, A.V.; Minkina, T.M.; Mandzhieva, S.S.; Perelomov, L.V.; Soja, G.; Zamulina, I.V.; Rajput, V.D.; Sushkova, S.N.; Mohan, D.; Yao, J. The mechanisms of biochar interactions with microorganisms in soil. *Environ. Geochem. Health* **2019**, *42*, 2495–2518. [CrossRef]

45. EN 13037-2002, 2002 Soil Improvers and Growing Media-Determination of pH.

46. EN 13038-2002, 2002 Soil Improvers and Growing Media-Determination of Electrical Conductivity.

47. EN 13651-2002, 2002. Soil Improvers and Growing Media-Extraction of Calcium Chloride/DTPA (CAT).

48. Sinha, P.; Datar, A.; Jeong, C.; Deng, X.; Chung, Y.G.; Lin, L.C. Surface Area Determination of Porous Materials Using the Brunauer-Emmett-Teller (BET) Method: Limitations and Improvements. *J. Phys. Chem. C* **2019**, *123*, 20195–20209. [CrossRef]

49. Kelepesi, S.; Tzortzakis, N.G. Olive mill wastesA growing medium component for seedling and crop production of lettuce and chicory. *Int. J. Veg. Sci.* **2009**, *15*, 325–339. [CrossRef]

50. Chrysargyris, A.; Panayiotou, C.; Tzortzakis, N. Nitrogen and phosphorus levels affected plant growth, essential oil composition and antioxidant status of lavender plant (*Lavandula angustifolia* Mill.). *Ind. Crop. Prod.* **2016**, *83*, 577–586. [CrossRef]

51. Tzortzakis, N.G.; Tzanakaki, K.; Economakis, C.D.C.D. Effect of origanum oil and vinegar on the maintenance of postharvest quality of tomato. *Food Nutr. Sci.* **2011**, *2*, 974–982. [CrossRef]

52. Chrysargyris, A.; Xylia, P.; Botsaris, G.; Tzortzakis, N. Antioxidant and antibacterial activities, mineral and essential oil composition of spearmint (*Mentha spicata* L.) affected by the potassium levels. *Ind. Crop. Prod.* **2017**, *103*, 202–212. [CrossRef]

53. Loreto, F.; Velikova, V. Isoprene produced by leaves protects the photosynthetic apparatus against ozone damage, quenches ozone products, and reduces lipid peroxidation of cellular membranes. *Plant Physiol.* **2001**, *127*, 1781–1787. [CrossRef]

54. De Azevedo Neto, A.D.; Prisco, J.T.; Enéas-Filho, J.; De Abreu, C.E.B.; Gomes-Filho, E. Effect of salt stress on antioxidative enzymes and lipid peroxidation in leaves and roots of salt-tolerant and salt-sensitive maize genotypes. *Environ. Exp. Bot.* **2006**, *56*, 87–94. [CrossRef]

55. Fornes, F.; Belda, R.M.; Fernández de Córdova, P.; Cebolla-Cornejo, J. Assessment of biochar and hydrochar as minor to major constituents of growing media for containerized tomato production. *J. Sci. Food Agric.* **2017**, *97*, 3675–3684. [CrossRef] [PubMed]

56. Shackley, S.; Ruysschaert, G.; Zwart, K.; Glaser, B. *Biochar in European Soils and Agriculture: Science and Practice*; Routledge: London, UK, 2016.

57. Ronga, D.; Francia, E.; Allesina, G.; Pedrazzi, S.; Zaccardelli, M.; Pane, C.; Tava, A.; Bignami, C. Valorization of Vineyard By-Products to Obtain Composted Digestate and Biochar Suitable for Nursery Grapevine (*Vitis vinifera* L.) Production. *Agronomy* **2019**, *9*, 420. [CrossRef]

58. Dunlop, S.J.; Arbestain, M.C.; Bishop, P.A.; Wargent, J.J. Closing the loop: Use of biochar produced from tomato crop green waste as a substrate for soilless, hydroponic tomato production. *HortScience* **2015**, *50*, 1572–1581. [CrossRef]

59. Nair, A.; Carpenter, B. Biochar rate and transplant tray cell number have implications on pepper growth during transplant production. *Horttechnology* **2016**, *26*, 713–719. [CrossRef]

60. Sun, L.; Li, L.; Chen, Z.; Wang, J.; Xiong, Z. Combined effects of nitrogen deposition and biochar application on emissions of N_2O, CO_2 and NH_3 from agricultural and forest soils. *Soil Sci. Plant Nutr.* **2014**, *60*, 254–265. [CrossRef]

61. Bedussi, F.; Zaccheo, P.; Crippa, L. Pattern of pore water nutrients in planted and non-planted soilless substrates as affected by the addition of biochars from wood gasification. *Biol. Fertil. Soils* **2015**, *51*, 625–635. [CrossRef]

62. Prasad, M.; Chrysargyris, A.; McDaniel, N.; Kavanagh, A.; Gruda, N.S.; Tzortzakis, N. Plant nutrient availability and pH of biochars and their fractions, with the possible use as a component in a growing media. *Agronomy* **2020**, *10*, 10. [CrossRef]

63. Bunt, A. *Media and Mixes for Container-Grown Plants: A Manual on the Preparation and Use of Growing Media for Pot Plants*, 1st ed.; Unwin Hyman Ltd.: London, UK, 1988.

64. Reyes-Torres, M.; Oviedo-Ocaña, E.R.; Dominguez, I.; Komilis, D.; Sánchez, A. A systematic review on the composting of green waste: Feedstock quality and optimization strategies. *Waste Manag.* **2018**, *77*, 486–499. [CrossRef]

65. Wang, S.; Gao, B.; Zimmerman, A.R.; Li, Y.; Ma, L.; Harris, W.G.; Migliaccio, K.W. Physicochemical and sorptive properties of biochars derived from woody and herbaceous biomass. *Chemosphere* **2015**, *134*, 257–262. [CrossRef]

66. Altland, J.E.; Locke, J.C. Effect of biochar type on macronutrient retention and release from soilless substrate. *HortScience* **2013**, *48*, 1397–1402. [CrossRef]

67. Liang, B.; Lehmann, J.; Solomon, D.; Kinyangi, J.; Grossman, J.; O'Neill, B.; Skjemstad, J.O.; Thies, J.; Luizão, F.J.; Petersen, J.; et al. Black Carbon Increases Cation Exchange Capacity in Soils. *Soil Sci. Soc. Am. J.* **2006**, *70*, 1719–1730. [CrossRef]

68. Maroušek, J.; Kolář, L.; Vochozka, M.; Stehel, V.; Maroušková, A. Biochar reduces nitrate level in red beet. *Environ. Sci. Pollut. Res.* **2018**, *25*, 18200–18203. [CrossRef] [PubMed]

69. Takaya, C.A.; Fletcher, L.A.; Singh, S.; Anyikude, K.U.; Ross, A.B. Phosphate and ammonium sorption capacity of biochar and hydrochar from different wastes. *Chemosphere* **2016**, *145*, 518–527. [CrossRef] [PubMed]

70. Chintala, R.; Mollinedo, J.; Schumacher, T.E.; Papiernik, S.K.; Malo, D.D.; Clay, D.E.; Kumar, S.; Gulbrandson, D.W. Nitrate sorption and desorption in biochars from fast pyrolysis. *Microporous Mesoporous Mater.* **2013**, *179*, 250–257. [CrossRef]

71. Kammann, C.I.; Schmidt, H.P.; Messerschmidt, N.; Linsel, S.; Steffens, D.; Müller, C.; Koyro, H.W.; Conte, P.; Stephen, J. Plant growth improvement mediated by nitrate capture in co-composted biochar. *Sci. Rep.* **2015**, *5*, 11080. [CrossRef]

72. Yan, M. Seed priming stimulate germination and early seedling growth of Chinese cabbage under drought stress. *S. Afr. J. Bot.* **2015**, *99*, 88–92. [CrossRef]

73. Méndez, A.; Cárdenas-Aguiar, E.; Paz-Ferreiro, J.; Plaza, C.; Gascó, G. The effect of sewage sludge biochar on peat-based growing media. *Biol. Agric. Hortic.* **2017**, *33*, 40–51. [CrossRef]

74. Graber, E.R.; Harel, Y.M.; Kolton, M.; Cytryn, E.; Silber, A.; David, D.R.; Tsechansky, L.; Borenshtein, M.; Elad, Y. Biochar impact on development and productivity of pepper and tomato grown in fertigated soilless media. *Plant Soil* **2010**, *337*, 481–496. [CrossRef]

75. Kim, H.S.; Kim, K.R.; Yang, J.E.; Ok, Y.S.; Kim, W., Il; Kunhikrishnan, A.; Kim, K.H. Amelioration of Horticultural Growing Media Properties Through Rice Hull Biochar Incorporation. *Waste Biomass Valorization* **2017**, *8*, 483–492. [CrossRef]

76. Chrysargyris, A.; Nikolaidou, E.; Stamatakis, A.; Tzortzakis, N. Vegetative, physiological, nutritional and antioxidant behaivor of spearmint (*Mentha spicata* L.) in response to differnet nitrogen supply in hydroponics. *J. Appl. Res. Med. Aroma.* **2017**, *6*, 52–61.

77. Gupta, D.D.; Palma, J.M.; Corpas, F.J. *Reactive Oxygen Species and Oxidative Damage in Plants under Stress*; Springer: Berlin/Heidelberg, Germany, 2015; ISBN 9783319204208.

78. Savvas, D.; Gruda, N. Application of soilless culture technologies in the modern greenhouse industry—A review. *Eur. J. Hortic. Sci.* **2018**, *83*, 280–293. [CrossRef]

 *agronomy*

Article

Utilization of Olive Oil Processing Waste Composts in Organic Tomato Seedling Production

Yüksel Tüzel [1,*], Kamil Ekinci [2], Gölgen Bahar Öztekin [1], İbrahim Erdal [2], Nurhan Varol [3] and Özen Merken [4]

[1] Department of Horticulture, Faculty of Agriculture, Ege University, 35100 Bornova-Izmir, Turkey; golgen.oztekin@ege.edu.tr

[2] Departments of Agricultural Machinery and Technology Engineering and Soil Science and Plant Nutrition, Faculty of Agriculture, Isparta University of Applied Sciences, 32200 Isparta, Turkey; kamilekinci@isparta.edu.tr (K.E.); ibrahimerdal@isparta.edu.tr (İ.E.)

[3] Ministry of Agriculture and Forestry, Olive Research Institute, 35100 Bornova-Izmir, Turkey; varol.nurhan@gmail.com

[4] Viticulture Research Institute, 45125 Yunusemre, Manisa, Turkey; ozen.merken@gthb.gov.tr

* Correspondence: yuksel.tuzel@ege.edu.tr; Tel.: +90-232-3111398

Received: 6 May 2020; Accepted: 31 May 2020; Published: 4 June 2020

Abstract: Olive oil byproducts show differences according to the olive oil extraction systems, which are called olive mill solid wastes, olive oil wastewater and olive oil wastewater sludge. Three different kinds of composts, including two-phase and three-phase olive mill solid wastes, and olive oil wastewater sludge were produced with separated dairy manure, poultry manure, and straw. The composts obtained from two-phase and three-phase olive mill solid wastes and olive oil wastewater sludge were named as two-phase, three-phase, and water sludge composts, respectively. They were separately enriched by rock phosphate and potassium salt. These composts were mixed with peat in a ratio of 0%, 25%, 50%, 75%, and 100% (*v/v*). Tomato seeds were sown in all mixtures on 3 February 2016. All the seeds were sown into 2 trays and each plug included 2 replicates. The trays were left in a germination room for 3 days, then moved to a heated greenhouse which is specialized for growing seedlings, and the seedlings were grown there for 3 weeks. The results showed that increasing compost ratios in the growing medium and also the enrichment of the growing medium increased organic matter content, electrical conductivity, and macro and micro nutrient concentrations. The germination period lasted longer with increasing compost ratios. The shoot length was lower at a compost ratio of over 50% excluding water sludge compost, which reacted to over 75%. The highest plant dry weights were obtained in the plants grown on the media with compost ratios of 50%, 25%, and 25% for water sludge compost, enriched two-phase compost, and enriched three-phase compost, respectively. We concluded that the composts obtained from two-phase and three-phase olive mill solid wastes and olive oil waste water sludge can be used without any need of enrichment and a ratio of 25% was found appropriate in most of the measured properties.

Keywords: *Solanum lycopersicum*; olive oil waste; two-phase; three-phase; water sludge

1. Introduction

Turkey is one of the most important olive- and olive oil-producing countries among the Mediterranean countries with a production of 1,500,467 tonnes of olive in 2018 [1]. The share of organic olive production among the total production in 2018 was 14.22% [2]. Two-phase or three-phase olive oil processing systems are used for the extraction of olive oil and both systems generate large amounts of by-products, which are called two-phase and three-phase olive mill solid wastes, olive oil waste water and olive oil wastewater sludge [3]. In Turkey, a survey study showed that all the shares of producers running three-phase, two-phase, and traditional cold stone pressed olive oil production systems were 71%, 27%, and 2%, respectively [4].

Olive oil production produces a large amount of solid and liquid wastes each year. Three-phase olive mill solid wastes contain broken seeds of olive. Olive mill wastewater contains 83%–96% water, 3.5%–15% organic matter, and 0.5%–2.0% mineral salts, depending on factors such as olive varieties, growing conditions, soil and climatic conditions, extraction methods, etc. [5]. Both effluents pose environmental problems since they exhibit highly phytotoxic and antimicrobial properties mainly due to phenols and they are not easily biodegradable [6–8]. Therefore, olive processing wastes have been considered as soil and water pollutants and cannot be used directly for agricultural purposes [7]. Within the framework of the measures taken by the Ministry of Environment and Urbanization of the Republic of Turkey, it is recommended that factory owners accumulate olive mill wastewater in lagoons or open ponds, evaporate their water, and utilize olive oil waste water sludge as the least risky solution for the environment. Additionally, factories should convert their processing systems into a two-phase system. Chowdhury et al. [9] reported that two-phase systems produce a lignocellulosic olive humid husk, which is a watery solid by-product with high contents of water (56.6%–74.5%) and phenols (0.62%–2.39%).

As a result, it is necessary to utilize solid and liquid wastes from both systems. Numerous researchers indicate that composting of olive oil production wastes with manure and some other organic materials is the best way of recycling as agricultural material [10,11]. The composted olive oil processing solid waste can be utilized as organic inputs for soil fertility and plant nutrition in agricultural production.

Fertilization is the most important input necessary for the conservation and maintenance of soil fertility in crop production in organic agriculture. On the other hand, with the growth in agricultural production, the amount of organic wastes arising from agriculture-based industry is increasing day by day. By composting these resources, it is possible to obtain organic raw materials that are beneficial to the soil and to protect the environment [12]. At the same time, rational input can be provided in organic agriculture for plant nutrition. Cegarra et al. [11] stated that the final form of composted olive oil processing solid waste has a higher organic matter content and remarkable mineral elements without toxic elements.

Several studies were carried out on the applications of compost obtained from olive oil processing wastes in agricultural production. Raviv et al. [13] applied composts produced from solid and liquid wastes of olive oil mill on tomato seedlings. Michailides et al. [14] produced compost from three-phase olive pomace waste and olive leaves and tested it on lettuce yield. Killi and Kavdır [15] carried out a study on the effects of compost produced from three-phase pomace waste on tomato yield. Diacono and Montemurro [16] conducted a study on the effects of composts obtained from two-phase olive pomace on the yield of organic emmer crop. However, none of these studies carried out a comparative study as to the effects of composts obtained from two-phase and three-phase olive mill solid wastes and olive oil waste water sludge separately on the growth performance of *Solanum lycopersicum* L. seedlings in growing medium.

The purposes of this study were to evaluate composts obtained from two-phase and three-phase olive mill solid wastes and olive oil waste water sludge, to determine the effects of enrichment of composts, and to compare different compost rates on organic tomato seedling production.

2. Materials and Methods

This study was conducted during the years of 2015 and 2016. Composts were produced at Composting Facility in Olive Research Institute, Ministry of Agriculture and Forestry. Then, they were tested in seedling production at the Horticulture Department of the Faculty of Agriculture at Ege University, Izmir, Turkey (38°27′17″ N, 27°14′17″ E). Organically certified seeds of tomato cultivar 'Şencan-9' (provided from Ataturk Central Horticultural Research Institute, Yalova, Turkey) were used for the study. Compost materials were obtained from the mixture of olive oil processing wastes from two and three phase systems (two-phase and three-phase olive mill solid wastes, olive oil waste water and olive oil waste water sludge) with separated dairy manure, poultry manure, and wheat straws. All input materials were obtained from the organically certified farms. The optimized mixing ratios for 3 different kinds of composts determined at Composting Laboratory at the Department of Agricultural Machinery and Technology Engineering in Isparta University of Applied Sciences (Table 1) were produced based on dry weight (Table 1) were produced (based on dry weight) [17].

Table 1. The optimized mixing ratios for 3 different kinds of composts used in this research.

	2P	3P	WS
	Mixing Ratios (%)		
Two-phase olive mill solid wastes	60	-	-
Three-phase olive mill solid wastes	-	46	-
Olive oil waste water		1	-
Olive oil waste water sludge	-	-	20
Separated dairy manure	23	27	53
Poultry manure	10	21	21
Wheat straws	7	5	6
C/N ratio	30.17	25.26	20.16

An aerated static pile composting method was used for composting the wastes (Figure 1). Piles with a width of 2 m, a length of 3 m and a height of 1.50 m were formed. Rutgers aeration strategies [18] were performed for aeration of piles for 360 days, which is in agreement with those reported in the study of Chowdhury et al. [9]. Although the composting process was monitored for temperature, pH, electrical conductivity (EC), moisture, and organic matter contents, C, N, and heavy metals ratios, and total phosphorus, they are not reported here. At the later stages, 0.38 kg of cotton seed meal per one kg of dry matter of the initial compost was added to each compost pile to enrich the composts at day 330 of composting (maturation and stabilization stages). Additionally, 0.16 kg of rock phosphate and 0.02 kg of raw potassium salt [19] per one kg of dry matter of the initial compost was added to each compost pile for the enrichment of composts at day 360 of composting. Composting lasted for 425 days including the maturation and stabilization periods. This prolonged period was due to the enrichment (E) process of composts. Therefore, the enriched versions of each compost were labeled as E2P, E3P, and EWS. Powder sulfur was applied at the fourth month of composting to reduce the pH value in the piles. For this purpose, 8 g of powder elemental sulfur was applied to one kg of dry compost [20].

Figure 1. Aerated static pile composting system.

Peat provided from Denizli local peat bogs (Turkey) and composts (2P, 3P, WS, E2P, E3P, and EWS) were used as organic substrates in the growing media with compost ratios (%, *v/v*) of 0%, 25%, 50%, 75%, and 100% with local peat. Neither lime nor any nutrient was added into the peat.

Tomato seeds were sown in all growing media on 3 February 2016. All the seeds were sown into 2 trays with 128 plugs in each. Each plug included 2 replicates. After sowing, the trays were left in a germination room at a day/night temperature of 24/24 °C and 80% relative humidity (RH) under dark conditions for 3 days, then moved to a heated greenhouse (15/24 °C and 70%RH) which is specialized for growing seedlings and the seedlings were grown there for 3 weeks. The seedlings were fertilized with liquid farmyard manure (Botanica, Camli Yem Besicilik, Izmir, Turkey) (2 cc L^{-1}, EC:1.32 dS m^{-1}, pH:4.6) every day with a boom system based on the previous results of Tuzel et al. [21]. In this period, the germination rate and germination period of the seeds were noted. The germination rate was calculated by counting the number of germinated seeds in the cells and expressed as %. The germination period was determined as the number of days required for 50% seed emergence.

When the seedlings were ready for planting in a month, they were harvested from each replicate containing 20 seedlings of treatments in order to measure shoot and root biomass. The roots were washed and cleaned from the growing medium and separated from the shoots. The root and shoot (stem and leaf) samples were weighed for fresh weight (g) and dried for 48 h in a thermo-ventilated oven at 65 °C. Then, these dried samples were weighed for dry weight (g) and dry matter was calculated as (%). The longest root length (from top to bottom) was measured with a tape meter and the average result was expressed in cm. The distance between the starting point of the roots and the tip of plant leaves was measured again with a tape meter (cm) and the values were used as seedling height. Stem diameter was also measured above the root collar of the seedlings between nodium with digital caliper (mm).

Minolta colorimeter (CR-400, Minolta Co., Tokyo, Japan) was used to determine leaf color as CIE L* a* b*. The obtained values of a* and b* were used to calculate hue angle $[h° = \tan^{-1} (b*/a*)]$ and chroma $[C* = \sqrt{(a*^2 + b*^2)}]$, which determine the saturation and the essential components of the color (red, yellow, blue, and green), respectively [22]. The total chlorophyll index was measured with a chlorophyll meter (SPAD-502Plus, Konica Minolta, Chiyoda-Tokyo, Japan) and expressed as SPAD.

In order to determine plant nutrient concentrations, the seedlings were harvested after the experiment period over the soil surface. Then, they were washed with tap water and distilled water to clean surface residues, dried at 65 °C until constant weight, and were grounded. The samples were wet digested with a microwave digestion system and then filtered up to 50 mL with de-ionized water for P, K, Ca, Mg, Cu, Zn, and Mn measurement. Except for P, other nutrients in the supernatant were measured using an atomic absorption spectrophotometer (AAnalyst 400, Perkin Elmer, Waltham, MA, USA). Phosphorus was determined calorimetrically using the spectrophotometer (TU1880 Double Beam UV-VIS, PG Instruments, Leicestershire, UK). In order to determine the N concentration, the samples were wet digested in 250 mL macro-Kjeldahl tubes using concentrated H_2SO_4 and Khjeldahl tablet at 350–400 °C. After digesting the samples with NaOH (40%), NH_4-N was fixed in H_3BO_3 (2%) and titrated with 0.1 N H_2SO_4 [23]. The same procedures and methods were applied to determine the mineral compositions of composts and peat used in the growing media and their mixtures as in plant analysis. The organic matter content of the dry samples of materials was analyzed after incinerating the samples at 550 °C as recommended by the US Department of Agriculture and the US Composting Council [24]. The pH and EC of the fresh samples were extracted by shaking at 180 rpm for 20 min at a solid:water ratio of 1:10 (*w/v*) [25], and were measured using pH (pH 720, WTW, Weilheim, Germany) and EC (Multi 340i, WTW, Weilheim, Germany) meters.

The experimental design was randomized blocks with 4 replicates ($n = 20$). A factorial analysis was performed with the composts (WS, EWS, 2P, E2P, 3P, E3P) and ratios (0%, 25%, 50%, 75%, and 100% with local peat and the interaction between these 2 factors. The data were subjected to analysis of variance to determine any statistically significant differences by using the JMP statistical analysis package program (SAS Institute, Cary, NC, USA). The Tukey range test was conducted at a 5% significance level.

3. Results

3.1. Physical and Chemical Properties of Substrates

Some physical and chemical characteristics of the seedling growing media were determined before seed sowing (Tables 1–3). The organic matter, content of the media was 38.45% at the initial stage. However, when the compost ratio was increased from 0% to 100% in the growing media at the start, the organic matter contents increased with the rate of 38.49%, 28.32%, 41.40%, 19.25%, 62.21% and 67.70% for WS, EWS, 2P, E2P, 3P and E3P, respectively. The highest organic matter (64.48%) was determined for E3P with a compost ratio of 100%. EC of the local peat was 1.11 dS m^{-1} before seed sowing. By the use of composts, EC values increased dramatically in particular when the composts were enriched and used with 75% and/or more. The pH of the growing media changed between 5.60 and 7.38. The pH decreased with an increasing compost ratio in the growing medium (Table 2).

Table 2. Initial organic matter, EC, and pH values of the growing media.

Composts	Compost Ratios in Peat (%)	Organic Matter (%)	EC (dS m^{-1})	pH
WS	0	38.45 h	1.11 n	7.38 a
	25	42.11 fgh	1.77 lmn	7.13 a
	50	47.01 d–h	2.26 k–n	6.69 abc
	75	45.93 d–h	3.26 f–k	6.28 bcd
	100	53.25 b–e	4.49 ef	5.60 d
EWS	0	38.45 h	1.11 n	7.38 a
	25	38.58 h	2.75 h–m	6.77 abc
	50	44.08 e–h	3.69 f–i	7.13 a
	75	42.63 fgh	4.30 efg	7.01 ab
	100	49.34 d–g	6.23 cd	6.64 a
2P	0	38.45 h	1.11 n	7.38 a
	25	38.30 h	1.65 mn	7.28 a
	50	44.29 e–h	2.35 j–n	7.33 a
	75	49.34 d–g	2.19 k–n	6.92 abc
	100	54.37 bcd	3.59 f–j	6.16 cd
E2P	0	38.45 h	1.11 n	7.38 a
	25	38.30 h	2.00 k–n	6.96 ab
	50	42.53 fgh	2.98 g–l	7.02 ab
	75	48.90 d–g	4.30 efg	6.95 ab
	100	45.85 d–h	7.39 c	7.06 a
3P	0	38.45 h	1.11 n	7.38 a
	25	40.48 gh	2.06 k–n	7.13 a
	50	47.57 d–h	2.99 g–l	7.06 a
	75	48.89 d–g	2.45 i–m	7.35 a
	100	62.37 ab	3.79 e–h	6.78 abc
E3P	0	38.45 h	1.11 n	7.38 a
	25	43.71 e–h	3.67 f–j	7.19 a
	50	50.95 c–f	5.04 de	7.33 a
	75	59.80 abc	9.08 b	7.25 a
	100	64.48 a	11.14 a	7.04 ab
		*	***	***

Means within each column followed by the same letters are not significantly different according to the Tukey test. * and ***: significant at 0.01 < p ≤ 0.05 and p ≤ 0.001, respectively.

The main and interaction effects of the treatments on the N, P, K, and Mg concentration of the growing medium before seed sowing were found to be significantly different. The initial N concentration (0.81%) of the growing media increased due to the increase in the compost ratio from 0% to 100% at the start with the rate of 34.57%, 70.37%, 16.05%, 53.09%, 61.73% and 111.11% and for WS, EWS, 2P, E2P, 3P and E3P, respectively. Higher compost ratios produced higher P and K concentrations of the media. The average Ca concentrations of WS, EWS, 2P, E2P, 3P, and E3P were 2.18%, 2.33%, 2.26%, 2.51%, 3.10%, and 2.25% at the start, while the Mg concentration changed between 0.45% and 0.82% (Table 3).

Table 3. Macro nutrient concentrations of the growing medium before seed sowing.

Composts	Compost Ratios in Peat (%)	N (%)	P (%)	K (%)	Ca (%)	Mg (%)
WS	0	0.81 i	0.12 f	0.62 e	2.33	0.80 a
	25	0.88 hi	0.22 ef	0.63 e	1.92	0.82 a
	50	0.94 ghi	0.28 de	0.63 e	3.52	0.82 a
	75	1.07 e–i	0.36 cd	0.71 de	2.85	0.81 a
	100	1.09 e–i	0.41 bc	0.95 b–e	2.02	0.80 a
EWS	0	0.81 i	0.12 f	0.62 e	2.33	0.80 a
	25	1.03 f–i	0.67 a	0.77 cde	1.97	0.69 a–d
	50	1.10 e–i	0.69 a	0.76 cde	1.84	0.76 ab
	75	1.42 a–d	0.72 a	1.04 bcd	2.09	0.75 abc
	100	1.38 b–e	0.74 a	1.12 b	2.17	0.66 a–d
2P	0	0.81 i	0.12 f	0.62 e	2.33	0.80 a
	25	0.80 i	0.15 f	0.68 e	2.73	0.82 a
	50	0.95 ghi	0.22 ef	0.75 cde	2.32	0.78 ab
	75	0.95 ghi	0.24 ef	0.69 e	2.66	0.77 ab
	100	0.94 ghi	0.37 cd	0.90 b–e	2.18	0.77 ab
E2P	0	0.81 i	0.12 f	0.62 e	2.33	0.80 a
	25	0.99 f–i	0.67 a	0.72 de	2.08	0.68 a–d
	50	1.23 d–g	0.73 a	0.89 b–e	1.97	0.68 a–d
	75	1.28 d–g	0.76 a	1.08 bc	3.31	0.72 a–d
	100	1.24 c–g	0.75 a	1.51 b	2.32	0.77 a–b
3P	0	0.81 i	0.12 f	0.62 e	2.33	0.80 a
	25	0.94 ghi	0.31 cde	0.84 b–e	3.05	0.81 a
	50	1.00 f–i	0.43 bc	0.89 b–e	2.66	0.79 ab
	75	1.08 e–i	0.39 bcd	0.89 b–e	2.52	0.79 ab
	100	1.31 b–f	0.50 b	1.11 b	3.87	0.72 a–d
E3P	0	0.81 i	0.12 f	0.62 e	2.33	0.80 a
	25	1.19 d–h	0.74 a	0.81 b–e	2.46	0.45 de
	50	1.59 ab	0.74 a	0.75 cde	2.71	0.52 b–e
	75	1.56 abc	0.75 a	1.77 a	2.57	0.49 cde
	100	1.71 a	0.74 a	1.80 a	2.69	0.64 a–d
		***	*	**	ns	***

Means within each column followed by the same letters are not significantly different according to the Tukey test. ns, *, ** and ***: nonsignificant, significant at $0.01 < p \le 0.05$, $0.001 < p \le 0.01$ and $p \le 0.001$, respectively.

The type of composts and ratios also affected the Zn, Mn, and Cu concentrations of the growing media at the start of the experiment. The Zn concentration varied between 68.2 and 432.4 mg kg^{-1}, the Mg and Cu concentration varied between 107.8–287.8 mg kg^{-1} and 36.6–55.0 mg kg^{-1} before seed sowing (Table 4).

3.2. Germination Period and Rate

The number of days from seed sowing until germination was 4.25 days in local peat (0%) and increased in all composts with increasing compost ratios in the growing medium particularly in the enriched treatments. The use of a compost ratio of 25% in the growing medium shortened the number of days compared with other compost ratios, but extended 11.8%, 17.6%, 5.9%, 17.6%, 5.9% and 111.8% in WS, EWS, 2P, E2P, 3P and E3P, respectively, compared to local peat, while the extension rate was 41.2%, 252.9%, 117.6%, 194.1%, 152.9%, and 264.7% for a compost ratio of 100% compared with local peat (Table 5). The germination rate also showed the same tendency and decreased with increasing compost ratios, but the ratio changed dramatically in the enriched growing medium (Table 5).

Table 4. Micro nutrient concentrations of the growing medium before seed sowing.

Composts	Compost Ratios in Peat (%)	Zn (mg kg^{-1})	Mn (mg kg^{-1})	Cu (mg kg^{-1})
WS	0	68.2 d	136.0 def	49.7
	25	108.3 bcd	184.1 b–f	37.3
	50	199.1 bcd	205.0 a–d	39.1
	75	179.6 bcd	206.4 a–d	39.4
	100	240.6 bcd	230.9 abc	45.9
EWS	0	68.2 d	136.0 def	49.7
	25	123.9 bcd	108.5 ef	50.5
	50	150.8 bcd	123.4 def	50.8
	75	263.8 abc	135.8 def	52.1
	100	228.9 bcd	152.3 c–f	49.7
2P	0	68.2 d	136.0 def	49.7
	25	86.7 cd	149.5 c–f	37.3
	50	88.9 cd	110.1 ef	40.0
	75	88.4 cd	116.5 ef	41.3
	100	108.9 bcd	107.8 f	40.2
E2P	0	68.2 d	136.0 def	49.7
	25	140.7 bcd	154.5 c–f	41.9
	50	171.0 bcd	186.2 b–f	52.7
	75	247.5 a–d	241.0 ab	52.1
	100	432.4 a	287.8 a	46.8
3P	0	68.2 d	136.0 def	49.7
	25	137.3 bcd	194.6 b–e	36.6
	50	162.6 bcd	185.9 b–f	38.6
	75	97.7 bcd	164.9 b–f	38.5
	100	147.1 bcd	160.2 b–f	39.9
E3P	0	68.2 d	136.0 def	49.7
	25	132.9 bcd	120.5 def	55.0
	50	215.8 bcd	131.6 def	46.7
	75	134.8 bcd	167.7 b–f	45.2
	100	282.2 ab	168.3 b–f	45.4
		*	**	ns

Means within each column followed by the same letters are not significantly different according to the Tukey test. ns, * and **: nonsignificant, significant at $0.01 < p \leq 0.05$ and $0.001 < p \leq 0.01$, respectively.

Table 5. Effects of composts with local peat on germination period and the rate of *Solanum lycopersicum*.

Compost Ratios in Peat (%)	Composts						Mean$_{ratio}$
	WS	EWS	2P	E2P	3P	E3P	
	Germination Period (day)						
0	4.25 j	4.25 j	4.25 j	4.25 j	4.25 j	4.25 j	4.25 E
25	4.75 ij	5.00 ij	4.50 j	5.00 ij	4.50 j	9.00 fg	5.46 D
50	6.00 hi	9.00 fg	5.25 hj	9.00 fg	6.50 h	12.75 b	8.08 C
75	8.00 g	11.25 cd	8.00 g	11.75 bd	9.50 ef	15.50 a	10.67 A
100	6.00 hi	15.00 a	9.25 fg	12.50 bc	10.75 de	0.00 k *	8.92 B
Mean$_{compost}$	5.80 D	8.90 A	6.25 D	8.50 AB	7.10 C	8.30 B	
	Germination Rate (%)						
0	94.92 ab	94.92 ab	94.92 ab	94.92 ab	94.92 ab	94.92 ab	94.92 A
25	94.14 ab	90.23 bc	91.41 ac	95.31 ab	92.58 ab	92.97 ab	92.77 AB
50	91.18 ac	94.92 ab	91.02 ac	93.36 ab	94.92 ab	79.30 e	90.78 B
75	96.88 a	85.94 cd	92.58 ab	81.64 de	91.02 ac	70.70 f	86.46 C
100	94.53 ab	54.02 g	94.53 ab	71.88 f	76.95 ef	14.45 h	67.73 D
Mean$_{compost}$	94.33 A	84.01 D	92.89 AB	87.42 C	90.08 BC	70.47 E	

* "0" is accepted as germination rates lower than 50%. Means within each column followed by the same letters are not significantly different according to the Tukey test. Capital letters show significant differences in mean values of composts and compost ratios in peat; lowercase letters indicate significant differences in interaction.

3.3. Seedling Growth

The effects of the treatments on the lengths of shoots and roots and stem diameter were found to be significantly different (Table 5). The shoot length changed between 16.33 and 4.65 cm. A compost ratio of up to 50% in the growing medium promoted the shoot length, but an increasing compost ratio had an impact on shoot growth excluding the compost ratio of 75% in WS. The shoot length sharply decreased in E2P and E3P. The root length was similar in the treatments, but it decreased by 35% in E3P. The stem diameter also showed similarities to the other measured parameters and decreased in the enriched treatments with an increasing compost ratio (Table 6).

Table 6. Effects of treatments on growth.

Compost Ratios in Peat (%)	Composts						
	WS	EWS	2P	E2P	3P	E3P	Mean$_{ratio}$
Shoot Length (cm)							
0	4.95 mn	4.95 mn	4.95 mn	4.95 mn	4.95 mn	4.95 mn	4.95 D
25	13.03 d	15.66 ab	11.10 eg	15.13 bc	14.48 c	6.43 kl	12.63 A
50	14.38 c	16.33 a	10.40 gi	12.23 de	14.73 bc	5.25 mn	12.22 A
75	15.65 ab	11.19 eg	9.93 hi	7.00 jk	11.83 ef	4.65 n	10.04 B
100	11.15 eg	10.89 fh	7.83 j	5.85 lm	9.48 i	4.74 mn	8.32 C
Mean$_{compost}$	11.83 A	11.80 A	8.84 C	9.03C	11.09 B	5.20 D	
Root Length (cm)							
0	7.03 de	7.03 de	7.03 de	7.03 de	7.03 de	7.03 de	7.03
25	7.70 ad	7.21 ce	7.78 ad	7.28 be	8.00 ac	6.63 e	7.43
50	7.28 be	7.95 ac	7.83 ad	7.33 be	8.03 ac	4.48 f	7.15
75	7.25 be	8.33 a	8.28 a	7.38 be	8.33 a	3.73 fg	7.21
100	8.08 ab	8.40 a	7.95 ac	7.38 be	8.00 ac	2.93 g	7.12
Mean$_{compost}$	7.47 BC	7.78 AB	7.77 AB	7.28 C	7.88 A	4.96 D	
Stem Diameter (mm)							
0	1.17 jk	1.17 jk	1.17 jk	1.17 jk	1.17 jk	1.17 jk	1.17 E
25	2.14 ce	2.46 a	1.83 h	2.46 a	2.37 ab	1.33 ij	2.10 A
50	2.41 ab	2.23 bd	1.88 fh	2.05 dg	2.46 a	1.01 k	2.00 B
75	2.31 ac	1.94 eh	1.84 gh	1.38 ij	2.06 dg	0.77 l	1.72 C
100	2.20 bd	1.86 fh	1.52 i	1.09 k	2.08 df	0.79 l	1.59 D
Mean$_{compost}$	2.04 A	1.93 B	1.65 C	1.63 C	2.02 AB	1.01 D	

Means within each column followed by the same letters are not significantly different according to the Tukey test. Capital letters show significant differences in mean values of composts and compost ratios in peat; lowercase letters indicate significant differences in interaction.

The treatments affected the dry weights of the roots significantly. Although root dry/fresh weights increased with compost ratios in the growing medium, this tendency did not continue with increasing ratios. Particularly, the values of E2P and E3P with compost ratios of over 25% showed less root growth (Figure 2).

The main and interaction effects of the treatments on shoot growth were also found to be significantly different. The results showed that the highest dry weights were in WS, while the lowest values were determined for the seedlings grown in E3P. Increasing doses of compost ratios of more than 25% and enrichment had negative effects on seedling dry weights (Figure 3).

3.4. Chlorophyll Index

The treatments affected the chlorophyll index values (SPAD) significantly. However, there was a slight reduction in WS, 2P, and 3P with increasing compost ratios, whereas the chlorophyll index increased in the enriched compost treatments (Table 7).

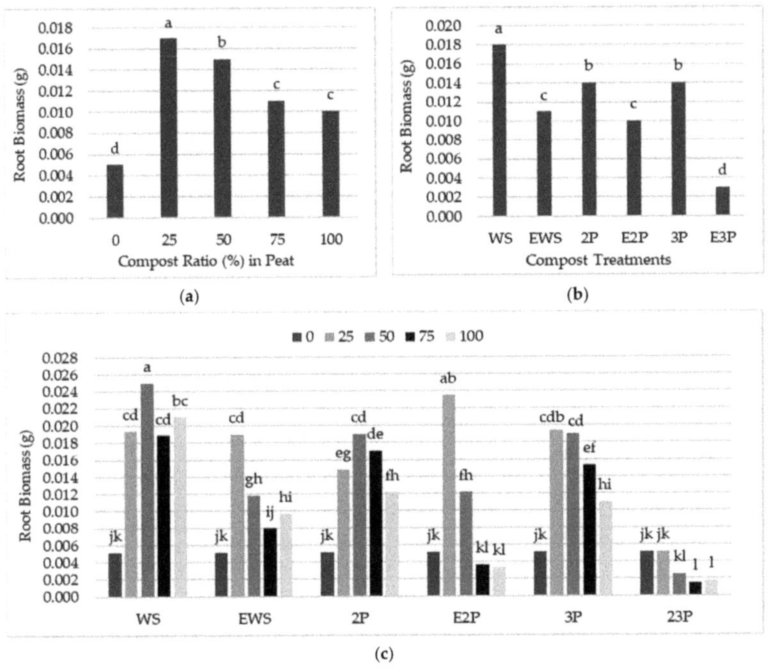

Figure 2. Main (**a**,**b**) and interaction (**c**) effects of the treatments on root dry weight. Means within each column followed by the same letters are not significantly different according to the Tukey test.

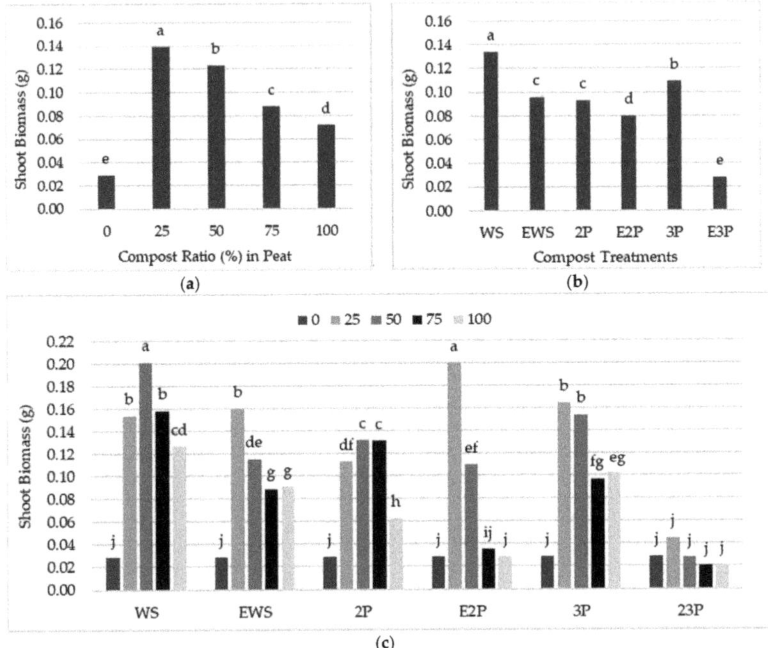

Figure 3. Main (**a**,**b**) and interaction (**c**) effects of the treatments on shoot dry weight. Means within each column followed by the same letters are not significantly different according to the Tukey test.

Table 7. Effects of the treatments on chlorophyll index values (SPAD).

Compost Ratios in Peat (%)	Composts						
	WS	EWS	2P	E2P	3P	E3P	Mean$_{ratio}$
0	30.58 d–h	30.58 d–h	30.58 d–h	30.58 d–h	30.58 d–h	30.58 d–h	30.58 BC
25	27.45 f–i	32.69 c–g	30.94 d–h	34.81 b–e	30.03 e–i	41.63 a	32.92 A
50	26.21 hi	36.16 ad	29.18 e–i	33.23 c–f	25.43 hi	39.97 ab	31.70 ABC
75	26.06 hi	34.33 b–e	25.84 hi	34.12 b–e	25.86 hi	34.48 b–e	30.11 C
100	24.18 i	38.69 abc	26.86 ghi	34.9 b–e	29.75 e–i	37.31 abc	31.96 AB
Mean$_{compost}$	26.90 C	34.49 B	28.68 C	33.55 B	28.33 C	36.79 A	

Means within each column followed by the same letters are not significantly different according to the Tukey test. Capital letters show significant differences in mean values of composts and compost ratios in peat; lowercase letters indicate significant differences in interaction.

3.5. Leaf Color

The main effects of composts and compost ratios on the "L*" value of leaf color were significant. The lowest "L*" was in the growing medium composed of local peat. Additionally, "L*" was lower in the enriched composts. The compost ratios only affected the "a*" value and the treatments showed significant difference when compared with peat usage. However, the "b*" value was affected by the main and interaction effect of the treatments and the b* values of the enriched composts were lower. The value of "h" changed according to the compost ratios and peat usage and 2P with a compost ratio of 75% gave the lowest hue value. However, "C*" had the same tendency with the "b*" value (Table 8).

Table 8. Effects of treatments on leaf color.

Compost Ratios in Peat (%)	Composts						
	WS	EWS	2P	E2P	3P	E3P	Mean$_{ratio}$
				L*			
0	28.91	28.91	28.91	28.91	28.91	28.91	28.91 B
25	47.92	39.63	48.18	40.61	48.30	39.59	44.04 A
50	49.75	37.79	47.57	40.28	49.41	39.66	44.07 A
75	49.34	38.85	48.97	43.19	51.35	40.52	45.37 A
100	49.04	39.14	48.39	40.14	42.91	42.29	43.65 A
Mean$_{compost}$	44.99 A	36.86 B	44.40 A	38.63 B	44.17 A	38.19 B	
				a*			
0	−0.56	−0.56	−0.56	−0.56	−0.56	−0.56	−0.56 A
25	−14.51	−15.96	−12.61	−15.60	−17.43	−15.33	−15.24 B
50	−17.15	−16.73	−16.17	−16.52	−17.69	−15.09	−16.56 B
75	−18.24	−15.31	−15.78	−17.12	−18.03	−16.34	−16.80 B
100	−16.89	−15.41	−15.38	−15.86	−15.47	−14.07	−15.51 B
Mean$_{compost}$	−13.47	−12.79	−12.10	−13.13	−13.84	−12.28	
				b*			
0	8.38 h	8.38 h	8.38 h	8.38 h	8.38 h	8.38 h	8.38 D
25	28.86 a–f	23.47 efg	24.60 c–g	21.52 g	29.82 a–e	21.05 g	24.89 BC
50	31.00 abc	20.64 g	32.38 ab	22.22 fg	33.76 a	20.56 g	26.76 AB
75	31.13 abc	20.20 g	33.37 a	25.42 b–g	34.22 a	21.84 fg	27.70 A
100	32.26 ab	20.14 g	30.80 a–d	21.44 g	23.83 d–g	19.09 g	24.59 C
Mean$_{compost}$	26.32 A	18.57 B	25.90 A	19.80 B	26.00 A	18.18 B	

Table 8. *Cont.*

Compost Ratios in Peat (%)	Composts						Mean$_{ratio}$
	WS	EWS	2P	E2P	3P	E3P	
				$h°$			
0	173.75	173.75	173.75	173.75	173.75	173.75	173.75 A
25	116.71	124.30	117.08	125.99	120.31	126.07	121.74 B
50	119.04	129.08	116.53	126.62	117.68	126.41	122.56 B
75	120.36	127.19	115.41	123.99	117.77	126.79	121.92 B
100	117.71	127.45	116.44	126.51	123.18	126.44	122.96 B
Mean$_{compost}$	129.52	136.54	127.84	135.37	130.54	135.89	
				$C*$			
0	9.45 h	9.45 h	9.45 h	9.45 h	9.45 h	9.45 h	9.45 C
25	32.30 a–e	28.39 c–g	27.64 d–g	26.58 efg	34.54 abc	26.04 efg	29.25 B
50	35.43 ab	26.59 efg	36.19 ab	27.69 d–g	38.13 a	25.53 efg	31.59 A
75	36.08 ab	25.35 fg	36.94 ab	30.65 b–f	38.69 a	27.27 efg	32.50 A
100	36.45 ab	25.36 fg	34.44 a–d	26.67 efg	28.43 c–g	23.76 g	29.18 B
Mean$_{compost}$	29.94 A	23.03 B	28.93A	24.21 B	29.85 A	22.41 B	

Means within each column followed by the same letters are not significantly different according to the Tukey test. Capital letters show significant differences in mean values of composts and compost ratios in peat; lowercase letters indicate significant differences in interaction.

3.6. Nutrient Concentration

Individual effects of composts and compost ratio with local peat and their interactions on the N and P concentrations of the seedlings showed a similar effect. Based on the interactions, both nutrient concentrations containing enriched composts with compost ratios of 50%, 75%, and 100% for EWS and E2P and with compost ratios of 25%, 50%, and 75% for E3P were higher than those of the composts without enrichment. The mean plant nutrient concentrations of compost rates significantly varied from 2.71% (a compost ratio of 25%) to 3.54% (a compost ratio of 75%) for N, and from 0.15% (0%) to 0.76% for P (Table 9). As for the plant Ca concentration obtained from composts × compost ratio interactions, increasing the compost ratios resulted in a decrease of Ca in plant tissue. This result implies that 100% local peat as seedling substrate had the highest Ca concentration. These results can also be obtained from the compost ratio comparison. The mean values showed that the Ca concentrations obtained from E3P were higher than those obtained from other composts. The plant Mg concentrations showed a similar tendency to Ca. Namely, except for E3P with a compost ratio of 25%, all the other plant Mg concentrations measured from the plugs with 100% local peat were higher. Furthermore, higher compost ratios generally led to a decrease in the plant Mg concentrations. The same trend was recorded from the means of compost ratios. While the lowest Mg concentrations were determined from 2P, there was not a significant variation among the means of the other composts (Table 9).

Table 9. Effects of the treatments on macro element concentrations of leaves.

Compost Ratios in Peat (%)	Composts						Mean$_{ratio}$
	WS	EWS	2P	E2P	3P	E3P	
				N (%)			
0	3.05 c–f	3.05 c–f	3.05 c–f	3.05 c–f	3.05 c–f	3.05 c–f	3.05 AB
25	1.73 ef	3.46 b–e	2.01 ef	2.98 c–f	1.90 ef	4.19 a–d	2.71 B
50	1.90 ef	4.79 abc	1.27 f	4.81 abc	1.87 ef	4.41 a–d	3.17 AB
75	2.12 ef	5.02 ab	1.27 f	5.21 ab	2.22 ef	5.39 a	3.54 A
100	2.21 ef	4.40 a–d	2.33 ef	5.33 a	2.70 def	3.48 b–e	3.41 A
Mean$_{compost}$	2.20 B	4.14 A	1.98 B	4.28 A	2.35 B	4.10 A	

Table 9. *Cont.*

Compost Ratios in Peat (%)	Composts						
	WS	EWS	2P	E2P	3P	E3P	Mean$_{ratio}$
	P (%)						
0	0.15 g	0.15 g	0.15 g	0.15 g	0.15 g	0.15 g	0.15 D
25	0.28 fg	0.71 bc	0.20 g	0.69 bc	0.32 fg	0.84 ab	0.50 C
50	0.41 d–g	0.83 ab	0.37 efg	0.84 ab	0.59 b–e	0.83 ab	0.64 B
75	0.68 bcd	0.84 ab	0.51 c–f	0.78 ab	0.71 bc	1.05 a	0.76 A
100	0.61 b–e	0.81 ab	0.63 b–e	0.81 ab	0.75 bc	0.69 bc	0.72 AB
Mean$_{compost}$	0.42 BC	0.67 A	0.37 C	0.65 A	0.50 B	0.71 A	
	Ca (%)						
0	6.70 a	6.70 a	6.70 a	6.70 a	6.70 a	6.70 a	6.70 A
25	2.69 b–g	2.88 b–f	3.18 bcd	2.86 b–f	3.05 b–e	3.36 bc	3.00 B
50	2.49 c–g	2.36 d–g	2.68 b–g	2.25 d–g	2.49 c–g	2.84 b–f	2.52 C
75	2.10 efg	1.87 g	2.28 d–g	2.39 d–g	2.32 d–g	2.81 b–g	2.30 C
100	2.02 fg	1.96 fg	2.77 b–g	2.84 b–f	2.26 d–g	3.56 b	2.57 C
Mean$_{compost}$	3.20 BC	3.15 C	3.52 B	3.41 BC	3.36 BC	3.85 A	
	Mg (%)						
0	0.93 b	0.93 b	0.93 b	0.93 b	0.93 b	0.93 b	0.93 A
25	0.78 bc	0.78 bc	0.70 bc	0.76 bcd	0.75 bcd	1.20 a	0.84 B
50	0.73 bcd	0.76 bcd	0.71 bcd	0.76 bcd	0.73 bcd	0.76 bc	0.74 C
75	0.70 b–e	0.77 bc	0.67 cde	0.78 bc	0.72 bcd	0.63 cde	0.71 C
100	0.57 cde	0.77 bc	0.45 e	0.58 cde	0.72 bcd	0.51 de	0.60 D
Mean$_{compost}$	0.74 AB	0.80 A	0.70 B	0.76 AB	0.77 AB	0.81 A	

Means within each column followed by the same letters are not significantly different according to the Tukey test. Capital letters show significant differences in mean values of composts and compost ratios in peat; lowercase letters indicate significant differences in interaction.

The plant Zn concentrations were significantly affected by individual factors and their interactions (Table 10). The Zn concentrations increased with increasing compost ratios. The Zn concentrations of the seedlings grown on the enriched composts were usually higher than those of the other composts without enrichment and the highest values were measured from E3P with a compost ratio of 75% and E3P with a compost ratio of 100% with the values of 325 and 226 mg kg^{-1} Zn in seedling tissue. Compared to the control (0%), the plant Zn concentrations showed more than threefold increment with increasing compost ratios up to 75%. The means of composts showed that Zn levels determined from the enriched compost were higher than those obtained from non-enriched composts. The highest Zn concentration was measured from the plants growing on E3P. The individual effects of composts and compost ratio showed a significant effect on the Mn and Cu concentrations (Table 10). While the seedling Mn concentrations increased with the compost ratio, the plant Cu concentrations decreased, but no significant differences were observed among compost ratios between 25% and 100%. The results show that the enriched composts seemed to be more effective than the non-enriched composts on the plant Mn concentrations. Additionally, WS was statistically in the same group. The Mn concentrations obtained from 2P and 3P substrates had the lowest values. Similarly, the plant Cu concentrations measured from the enriched composts were higher than those measured from the non-enriched composts and the highest Cu concentration was determined from the plant grown on E3P. WS had the lowest effect on the plant Cu concentration.

Table 10. Effects of the treatments on micro element concentrations of leaves.

Compost Ratios in Peat (%)	Composts						
	WS	EWS	2P	E2P	3P	E3P	Mean$_{ratio}$
	Zn (mg kg^{-1})						
0	44 g	44 g	44 g	44 g	44 g	44 g	44 C
25	49 fg	91 c–g	76 d–g	95 c–g	67 efg	103 c–g	80 B
50	67 efg	143 b–g	67 efg	160 b–f	86 c–g	188 bc	118 AB
75	84 c–g	170 b–e	64 efg	155 b–f	100 c–g	325 a	150 A
100	101 c–g	150 b–g	100 c–g	186 bcd	125 b–g	226 ab	148 A
Mean$_{compost}$	69 D	120 BC	70 D	128 B	85 CD	177 A	
	Mn (mg kg^{-1})						
0	32	32	32	32	32	32	32 C
25	12	15	14	18	18	33	18 C
50	10	50	15	52	26	60	36 C
75	70	83	40	62	52	89	66 B
100	114	104	90	102	69	76	93 A
Mean$_{compost}$	49 AB	57 A	38 B	53 A	39 B	58 A	
	Cu (mg kg^{-1})						
0	28	28	28	28	28	28	28 A
25	9	14	12	12	8	21	13 B
50	7	15	11	19	11	24	14 B
75	7	14	9	16	16	22	14 B
100	6	14	7	20	14	24	14 B
Mean$_{compost}$	11 C	17 AB	13 BC	19 AB	15 ABC	24 A	

Means within each column followed by the same letters are not significantly different according to the Tukey test. Capital letters show significant differences in mean values of composts and compost ratios in peat; lowercase letters indicate significant differences in interaction.

4. Discussion

Seedlings are grown in a limited volume of containers, however, materials and rates utilized in formulations of growing medium affect the physical, chemical and/or biological properties of medium [26], which is also directly linked with seedling quality. Growing medium provides physical support, aeration, supply of water, and nutrients [27]. In our experiments, the enrichment of the growing medium and also increasing the compost ratio increased organic matter content, electrical conductivity, and macro and micro element concentrations. The origin of compost also affects the nutritional features of growing medium. Furthermore, olive oil processing wastes are rich in nutrients with a higher electrical conductivity [28,29]. Although there were slight changes in organic matter content before planting, P, Ca, Mg, Zn, and Mn decreased during the seedling growth due to plant consumption. However, the increase in N was most probably due to the ongoing mineralization affected by the composition and the characteristics of the material, temperature, and water content [30].

The germination rate changed between 14.45% and 96.88% and decreased by the enrichment of the growing medium in particular in EWS and E2P when the compost ratio was 75% and over, while the germination rate declined in E3P after a compost ratio of 50% and with the increasing compost ratio in the growing medium. However, the germination period also lasted longer with the enrichment of the growing medium and increasing compost ratios. Sánchez–Monedero et al. [31] also reported a lower germination rate and a delay in seedling emergence when the relative proportion of the compost increased in the growing medium, leading to higher EC. The rate and duration of germination are affected by the physical and chemical properties of the growing medium, the rate of ingredients, the requirement of crop species, and crop management including irrigation, fertigation, and the use of beneficial microorganisms as well as environmental conditions [32].

In terms of germination rate, two composts made from olive pomace waste and green waste were used as growing medium components at four ratios (20%, 45%, 70%, 90%, *v/v*) and compost made of green waste with ratio 20% and 45% and olive pomace waste with ratio of 20% showed the best performances [29]. Perez-Murcia et al. [33] tested the addition of increasing quantities of composted sewage sludge to peat (0%, 15%, 30%, and 50%, *v/v*), and increasing sewage sludge treatments (especially 30% and 50%) reduced the germination of lettuce and broccoli, but in cauliflower seedlings, an increment of germination was observed for the 15% and 30% treatments compared with the control. A compost ratio of 25% for composted rose oil processing [34] and for olive oil production wastes [35] was found appropriate in terms of the rate and duration of germination for organic tomato seedling production which is in harmony with our results.

Healthy seedling growth is a prerequisite for the success of crop production [36]. The shoot length was lower in compost ratios over 50% excluding WS, which reacted to over 75%. Shoot length and stem diameter decreased by the enrichment of the growing medium over 50% compost rate in EWS and E2P. The longest root lengths were also affected by the enrichment of medium excluding WS and EWS which could be also be related to the washing process. The development of shoot, root, and stem was the poorest in E3P. The nutrient contents of the growing media were higher in the ones with higher compost ratios and the enriched ones (Table 2), but the EC values were also high in those ones. The highest average EC value was in E3P treatment, resulting in the poorest shoot, root and stem development.

Tomato is moderately sensitive to salinity and salinity threshold of tomatoes is 2.5 dS m^{-1} [37]. Increasing salinity in the rhizosphere restricts root cell growth and increases root lesion, resulting in a reduction in root elongation rate and lateral root growth. Additionally, a reduction in photosynthesis and tissue expansion and the inhibition of cell division affect leaf and shoot growth [38]. Maggio et al. [39] found that high EC (approx. 9.6 dS m^{-1}) caused a sharp increase in the values of root and shoot abscisic acid (ABA), which coincided with the reduction of stomatal resistance to ABA, a different partitioning of Na ions between young and mature leaves, and the increase of root to shoot ratio [39]. In our experiment, morphological measurements (a decrease in shoot length, stem diameter, shoot and root biomass with an increasing compost ratio and enrichment process, poor growth particularly in under E2P and E3P) and SPAD readings, which showed the greenness or the relative chlorophyll concentration of leaves and the highest root to shoot dry matter ratio (in E3P), confirm the effect of salt stress on the seedlings.

The highest plant dry weights were measured from the plants grown on the media with compost ratios of 50% and 25% for WS and E2P, respectively. The variation of the results could be explained in terms of the chemical composition of the composts [40–42]. However, some other properties such as humic and fulvic acid and some other hormones like substances may also have positive effects on plant growth, and thus dry weight [43]. The decrease of dry weight with an increase higher than 50% in compost ratio either enriched or not might be due to the toxicity of some fenolic compounds on plant growth [44,45]. In order to prevent the toxic effect of WS, it was reported to follow the changes occurring in phenols and biotoxicity during composting. Moreover, Zenjari et al. [46] indicated that toxicity disappeared after 2 months of composting. Many studies conducted with different plants grown on different composting materials proposed rates of WS in composting between 25% and 67% [31,47]. The enrichment of 2P (E2P) with P and Ca due to different materials, especially rock phosphate, may have a positive effect on plant growth and dry weight.

The results show that all the composts, either enriched or not, and compost ratios had significantly different effects on most of the plant nutrient concentrations. If a general evaluation is made for the plant N, P, and Zn concentrations, it can be clearly seen that these nutrient concentrations in plants grown on the enriched composts were higher than the non-enriched composts. A number of studies showed that pre-mixing rock phosphate with agro-wastes followed by composting increased the P availability to plants [48–51]. Local peat seems to be the best medium in terms of the plant Ca and Mg concentrations. However, it is quite clear that the dilution effect played a very important role especially for Ca, as dry weights obtained from 100% local peat containing plug were quite low when compared to most of the media. It is well-documented in the literature that nutrients are diluted in plant tissues with plant growth and concentrated with growth retention [41,52].

Among the tested compost ratios, a ratio of 25% was found appropriate in most of the measured properties. However, compost ratios could be increased by up to 50% in the case of water sludge use. Previous research results also propose a rate starting from 25% up to 67% in different crops (such as poinsettia with olive mill wastes [53]; tomato with municipal solid waste compost [47]; broccoli, onion, and tomato with sweet sorghum bagasse, pine bark, and either urea or brewery sludge [31]; lettuce, chard, broccoli, and coriander with exhausted grape marc and cattle or poultry manure [54]). The chemical and physical properties of compost affect the compost ratio in the growing medium [47] and nitrogen has the greatest effect on transplant growth [55]. In our experiment, the higher EC level of the growing medium when enriched and/or included higher compost ratio affected plant growth starting from the seed germination stage. These results are in harmony with the results of our experiments conducted with composts containing rose oil processing wastes [34] and olive oil production wastes [35].

Peat is the most common substrate in seedling production. Although peat-based growing media are allowed in organic production, peat substitution in plant nursery activity and, in particular, in organic seedling production is a debated issue [56] since peat utilization contradicts numerous fundamental principles of organic agriculture. EGTOP (Expert Group for Technical Advice on Organic Production) advises that its use in growing media should be limited to a maximum of 80% by volume, as normally 20%–30% of peat by volume in growing media for professional use could be replaced by compost [57]. Our results showed that composts based on olive mill wastes and olive oil wastewater sludge could be used in the growing medium of vegetable seedlings and there is no need to enrich the medium, which results in a much higher electrical conductivity and higher costs.

Future studies should focus on the enrichment of composts with the effective microorganisms to improve soil fertility and facilitate the nutrient uptake from the soil.

5. Conclusions

In conclusion, the composts obtained from two-phase and three-phase olive mill solid wastes and olive oil wastewater sludge can be used without any need of enrichment and a ratio of 25% was found appropriate in most of the measured properties. However, compost ratios could be increased by up to 50% in the case of water sludge compost use.

Author Contributions: All the authors contributed to the study conception and design. Material preparation, data collection and analysis were performed by Y.T., K.E., G.B.Ö., İ.E., N.V. and Ö.M. The manuscript was written by Y.T., K.E., G.B.Ö. and İ.E. All the authors read and approved the final manuscript. All authors have read and agreed to the published version of the manuscript.

Funding: This research was carried out in the framework of the project "Developing of Input Production Methods for Utilization in Organic Plant Production", which was approved by The Scientific and Technological Research Council of Turkey (TUBİTAK) with project number 11-G−055.

Acknowledgments: The authors are grateful to Gulay Beşirli for organic seed supply.

Conflicts of Interest: The authors declare no conflict of interest.

References

1. TurkStat. Turkish Statistical Institute Data Basis. 2020. Available online: http://www.tuik.gov.tr (accessed on 31 January 2020).
2. Republic of Turkey Ministry of Agriculture and Forestry. Organic Agricultural Production Data. 2020. Available online: http://www.tarim.gov.tr/Konular/Bitkisel-Uretim/Organik-Tarim/Istatistikler (accessed on 31 January 2020).
3. Roig, A.; Cayuela, M.L.; Sánchez-Monedero, M.A. An overview on olive mill wastes and their valorisation methods. *Waste Manag.* **2006**, *26*, 960–969. [CrossRef]

4. Hocaoglu, S.M.; Haksevenler, B.H.G.; Basturk, I.; Talazan, P.; Aydoner, C. Assessment of technology modification for olive oil sector through mass balance: A case study for Turkey. *J. Clean. Prod.* **2018**, *188*, 786–795. [CrossRef]

5. Tunalıoglu, R.; Bektas, T. The problem of olive mill waste water in the Turkish olive industry. *Zeytin Bilimi* **2010**, *1*, 65–71.

6. Ramos-Cormenzana, A.; Monteoliva-Sanchez, M.; Lopez, M.J. Bioremediation alpechin. *Int. Biodeterior. Biodegrad.* **1995**, *35*, 249–268. [CrossRef]

7. Galiatsatou, P.; Metaxas, M.; Arapoglou, D.; Kasselouri-Rigopoulou, V. Treatment of olive waste water with activated carbons from agricultural byproducts. *Waste Manag.* **2002**, *22*, 803–812. [CrossRef]

8. Ammar, E.; Nasri, M.; Medhioub, K. Isolation of Enterobacteria able to degrade simple aromatic compounds from the wastewater from olive oil extraction. *World J. Microbiol. Biotechnol.* **2005**, *21*, 253–259. [CrossRef]

9. Chowdhury, A.K.M.M.B.; Akratos, C.S.; Vayenas, D.V.; Pavlou, S. Olive mill waste composting: A review. *Int. Biodeter. Biodegrad.* **2013**, *85*, 108–119. [CrossRef]

10. Madejón, E.; Galli, E.; Tomati, U. Composting of wastes produced by low water consuming olive mill technology. *Agrochimica* **1998**, *42*, 135–146.

11. Cegarra, J.; Amor, J.B.; Gonzálvez, J.; Bernal, M.P.; Roig, A. Characteristics of a new solid olive-mill-byproduct ("alperujo") and its suitability for composting. In Proceedings of the International Composting Symposium, Halifax/Dartmouth, NS, Canada, 19–23 September 1999; Volume 1, pp. 124–140.

12. Gruda, N.S. Increasing sustainability of growing media constituents and stand-alone substrates in soilless culture systems. *Agronomy* **2019**, *9*, 298. [CrossRef]

13. Raviv, M.; Aviani, I.; Laor, Y.; Medina, S.H.; Krassnov, A. Co-composting of solid and liquid olive mill wastes: Management aspects and the horticultural value of the resulting composts. *Bioresour. Technol.* **2010**, *101*, 6699–6706.

14. Michailides, M.; Christou, G.; Akratos, C.S.; Tekerlekopoulou, A.G.; Vayenas, D.V. Composting of olive leaves and pomace from a three-phase olive mil plant. *Int. Biodeterior. Biodegrad.* **2011**, *65*, 560–564. [CrossRef]

15. Killi, D.; Kavdır, Y. Effects of olive solid waste and olive solid waste compost application on soil properties and growth of *Solanum lycopersicum*. *Int. Biodeterior. Biodegrad.* **2013**, *82*, 157–165. [CrossRef]

16. Diacono, M.; Montemurro, F. Olive pomace compost in organic emmer crop: Yield, soil properties, and heavy metals' fate in plant and soil. *J. Soil Sci. Plant. Nutr.* **2019**, *19*, 63–70. [CrossRef]

17. Beşirli, G.; Tüzel, Y.; Türemiş, N.; Ekinci, K.; Varol, N. *Final Report. Developing of Input Production Methods for Utilization in Organic Plant Production*; TUBITAK Project No. 111G055, Final Report; Tübitak: Ankara, Turkey, 2018; 80p.

18. Finstein, M.S.; Miller, F.C. Principles of Composting Leading to Maximization of Decomposition Rate, Odor Control and Cost Effectiveness. In *Composting of Agricultural and Other Wastes*; Gasser, J.K.R., Ed.; Elsevier Applied Science Publications: Barking/Essex, UK, 1985; pp. 13–26.

19. OG. *Regulation on the Principles of Organic Farming and Their Implementation*; The Official Gazette of Republic of Turkey; No: 27676; Republic of Turkey, Official Gazette: Turkey, 2010. (In Turkish)

20. Roig, A.; Cayuela, M.L.; Sanchez-Monedero, M.A. The use of elemental sulphur as organic alternative to control pH during composting of olive mill wastes. *Chemosphere* **2004**, *57*, 1099–1105. [CrossRef] [PubMed]

21. Tuzel, Y.; Oztekin, G.B.; Tan, E. Use of different growing media and nutrition on organic seedling production. *Acta Hort.* **2015**, *1107*, 165–175. [CrossRef]

22. McGuire, R.G. Reporting of objective color measurements. *HortScience* **1992**, *27*, 1254–1255. [CrossRef]

23. Jones, J.B., Jr.; Wolf, B.; Mills, H.A. *Plant Analysis Handbook. A Practical Sampling, Preparation, Analysis, and Interpretation Guide*; Micro-Macro Publishing Inc.: Athens, Greece, 1991.

24. USCC. *Test Methods for the Examination of Composting and Composts*; US Government Printing Office, The US Composting Council, 2002. Available online: http://tmecc.org/tmecc/index.html (accessed on 15 April 2020).

25. Şevik, F.; Tosun, İ.; Ekinci, K. The effect of FAS and C/N ratios on co-composting of sewage sludge, dairy manure and tomato stalks. *Waste Manag.* **2018**, *80*, 450–456. [CrossRef]

26. Reis, M.; Coelho, L. Compost mixes as substrates for seedling production. *Acta Hort.* **2007**, *747*, 283–292. [CrossRef]

27. Landis, T.D.; Douglass, F.J.; Wilkinson, K.M.; Luna, T. Tropical Nursery Manual: A guide to starting and operating a nursery for native and traditional plants. In *Growing Media*; Wilkinson, K.M., Landis, T.D., Haase, D.L., Daley, B.F., Dumroese, R.K., Eds.; Agriculture Handbook 732; U.S. Department of Agriculture, Forest Service: Washington, DC, USA, 2014; 376p.

28. Altieri, R.; Esposito, A. Evaluation of the fertilizing effect of olive mill waste compost in short-term crops. *Int. Biodeterior. Biodegrad.* **2010**, *64*, 124–128. [CrossRef]

29. Ceglie, F.G.; Elshafie, H.; Verrastro, V.; Tittarelli, F. Evaluation of olive pomace and green waste composts as peat substitutes for organic tomato seedling production. *Compost Sci. Util.* **2011**, *9*, 293–300. [CrossRef]

30. Cabrera, M.L.; Kissel, D.E.; Vigil, M.F. Nitrogen mineralization from organic residues: Research opportunities. *J. Environ. Qual.* **2005**, *34*, 75–79. [CrossRef] [PubMed]

31. Sanchez-Monedero, M.A.; Roig, A.; Cegarra, J.; Pilar Bernal, M.; Noguera, P.; Abad, M.; Antón, A. Composts as media constituents for vegetable transplant production. *Compost Sci. Util.* **2004**, *12*, 161–168. [CrossRef]

32. Tuzel, Y.; Oztekin, G.B.; Aktan, H.; Yolageldi, L. *Improvement of Organic Seedling Production Methods*; TUBITAK Project No 111G151, Final Report; Tübitak: Ankara, Turkey, 2017; 119p.

33. Perez-Murcia, M.D.; Moreno-Caselles, J.; Moral, R.; Perez-Espinosa, A.; Paredes, C.; Rufete, B. Utilization of composted sewage sludge as horticultural growth media: Effects on germination and trace element extraction. *Commun. Soil Sci. Plant Anal.* **2005**, *36*, 571. [CrossRef]

34. Oztekin, G.B.; Ekinci, K.; Tuzel, Y.; Merken, O. Effects of Composts obtained from two different composting methods on organic tomato seedling production. *Acta Hort.* **2017**, *1164*, 209–216. [CrossRef]

35. Tuzel, Y.; Varol, N.; Öztekin, G.B.; Ekinci, K.; Merken, O. Effects of composts obtained from olive oil production wastes on organic tomato seedling production. *Acta Hort.* **2017**, *1164*, 217–224. [CrossRef]

36. Atif, M.J.; Jellani, G.; Malik, M.H.A.; Saleem, N.; Ullah, H.; Khan, M.Z.; İkram, S. Different growth media effect the germination and growth of tomato seedlings. *Sci. Tech. Dev.* **2016**, *35*, 123–127.

37. Tanji, K.K.; Kielen, N.C.; Food and Agriculture Organization of The United Nations. *Agricultural Drainage Water Management in Arid and Semi-Arid Areas*; FAO Irrigation and Drainage Paper; FAO: Rome, Italy, 2003; ISBN 92-5-104839-8.

38. Zhang, P.; Senge, M.; Dai, Y. Effects of salinity stress on growth, yield, fruit quality and water use efficiency of tomato under hydroponics system. *Rev. Agric. Sci.* **2016**, *4*, 46–55. [CrossRef]

39. Maggio, A.; Raimondi, G.; Martino, A.; De Pascale, S. Salt stress response in tomato beyond the salinity tolerance threshold. *Environ. Exp. Bot.* **2007**, *59*, 276–282. [CrossRef]

40. Aktas, H.; Daler, S.; Ożen, O.; Gencer, K.; Bayindur, D.; Erdal, I. The effect of some growing substrate media on yield and fruit quality of eggplant (*Solanum melongena* L.) grown and irrigated by drip irrigation system in green house. *Infrastruktura i Ekologia Terenów Wiejskich* **2013**, *1*, 5–11.

41. Erdal, I.; Kaya, M.; Kucukyumuk, Z. Effects of zinc and nitrogen fertilizations on grain yield and some parameters affecting zinc bioavailability in lentil seeds. *Legume Res.* **2014**, *37*, 55–61. [CrossRef]

42. Erdal, I.; Aktas, H.; Daler, S.; Gencer, K. Effects of Different Growing Mediums on Nutrient Concentration of Eggplant in Soilless Culture (Chapter: 38). In *Technological Advancement for Vibrant Agriculture*, 2nd ed.; Amitava, R., Ed.; Athens Institute for Education and Research: Athens, Greece, 2014; pp. 417–424.

43. Çelik, C.; Yüksek, L.T.; Kimya, A.B.D. *Zeytin Karasuyundan Humik (HA) ve Fulvik (FA) Asitlerin Eldesi ve Karakterizasyonu*; Fen Bilimleri Enstitüsü, Çukurova Üniversitesi: Adana, Turkey, 2010; 50p.

44. Niaounakis, M.; Halvadakis, C.P. *Olive-Mill Waste Management: Literature Review and Patent Survey*; Typothito-George Dardanos: Athens, Grecee, 2004; ISBN 960-402-123-0.

45. Piotrowska, A.; Iamarino, G.; Rao, M.A.; Gianfreda, L. Short-term effects of olive mill waste water (OMW) on chemical and biochemical properties of a semiarid Mediterranean soil. *Soil Biol. Biochem.* **2006**, *38*, 600–610. [CrossRef]

46. Zenjari, B.; El Hajjouji, H.; Ait Baddi, G.; Bailly, J.R.; Revel, J.C.; Nejmeddine, A.; Hafidi, M. Eliminating toxic compounds by composting olive mill wastewater–straw mixtures. *J. Hazard. Mater.* **2006**, *13*, 433–437. [CrossRef] [PubMed]

47. Herrera, F.; Castillo, J.E.; Chica, A.F.; López Bellido, L. Use of municipal solid waste compost (MSWC) as a growing medium in the nursery production of tomato plants. *Bioresour. Technol.* **2008**, *99*, 287–296. [CrossRef] [PubMed]

48. Singh, C.P.; Amberger, A. Solubilization and availability of P during decomposition of rock phosphate enriched straw and urine. *Biol. Agric. Hortic.* **1991**, *7*, 261–269. [CrossRef]

49. Singh, C.P.; Amberger, A. The effect of rock phosphate enriched compost on the yield and phosphorus nutrition of rye grass. *Am. J. Altern. Agric.* **1995**, *10*, 82–87. [CrossRef]
50. Biswas, D.R.; Narayanasamy, G. Mobilization of phosphorus from rock phosphate through composting using crop residue. *Fert. News* **2002**, *47*, 53–56.
51. Vassilev, N.; Vassileva, M. Biotechnological solubilization of rock phosphate on media containing agro-industrial wastes. *Appl. Microbiol. Biotechnol.* **2003**, *61*, 435–440. [CrossRef]
52. Erdal, I.; Bozkurt, M.A.; Mesut, K.M.; Karaca, S.; Saglam, M. Effects of humic acid and phosphorus applications on growth and phosphorus uptake of corn plant (*Zea mays* L.) grown in a calcareous soil. *Turk. J. Agric. For.* **2000**, *24*, 663–668.
53. Papafotiou, M.; Phsyhalou, M.; Kargas, G.; Chatzipavlidis, I.; Chronopoulos, J. Olive-mill wastes compost as growing medium component for the production of poinsettia. *Sci. Hort.* **2004**, *102*, 167–175. [CrossRef]
54. Bustamente, M.A.; Paredes, C.; Moral, R.; Agullo, E.; Perez-Murcia, M.D.; Abad, M. Composts from distillery wastes as peat substitutes for transplant production. *Resour. Conserv. Recycl.* **2008**, *52*, 792–799. [CrossRef]
55. Zandstra, J.W.; Liptay, A. Nutritional effects on transplant root and shoot growth—A review. *Acta Hort.* **1999**, *504*, 23–32. [CrossRef]
56. Clark, S.; Cavigelli, M. Suitability of composts as potting media for production of organic vegetable transplants. *Compost Sci. Util.* **2005**, *13*, 150–156. [CrossRef]
57. EGTOP. Final Report on Greenhouse Production (Protected Cropping). 7th Plenary Meeting of 19–20 June 2013. Available online: https://ec.europa.eu/agriculture/organic/sites/orgfarming/files/docs/body/final_report_egtop_on_greenhouse_production_en.pdf (accessed on 7 January 2013).

Article

Biochar and Vermicompost Amendments Affect Substrate Properties and Plant Growth of Basil and Tomato

Lan Huang [1], Mengmeng Gu [2], Ping Yu [3], Chunling Zhou [4] and Xiuli Liu [5,*]

[1] Institute of Urban Agriculture, Chinese Academy of Agricultural Sciences, Chengdu 610000, China; huanglan_92@163.com

[2] Department of Horticultural Sciences, Texas A&M AgriLife Extension Service, College Station, TX 77843, USA; mgu@tamu.edu

[3] Department of Horticultural Sciences, Texas A&M University, College Station, TX 77843, USA; yuping520@tamu.edu

[4] School of Landscape Architecture & Forestry, Qingdao Agricultural University, Qingdao 266109, China; zhou_chl@sina.com

[5] School of Landscape Architecture, Beijing Forestry University, Beijing 100083, China

* Correspondence: showlyliu@bjfu.edu.cn; Tel.: +86-136-1104-0888

Received: 12 December 2019; Accepted: 1 February 2020; Published: 4 February 2020

Abstract: The suitability of biochar (BC) as a container substrate depends on the BC mix ratio and plant species. Mixes with mixed hardwood BC (20%, 40%, 60%, and 80%, by volume) and vermicompost (VC; 5%, 10%, 15%, and 20%, by volume) were evaluated as container substrates on basil (*Ocimum basilicum* L.) and tomato (*Solanum lycopersicum* L. 'Roma') plants compared to a commercial peat-based substrate (CS). The CS made up the rest of the volume when BC and VC did not add up to 100%. The total porosity of all mixes with BC, VC, and CS (BC:VC:CS mixes) was similar to the control. Mixes with 80% BC had lower container capacity than the control. At 9 weeks after transplanting, the leachate pH of all the BC:VC:CS mixes was higher than that of the control, except for mixes of 20%BC and 5%VC with the rest (75%) being CS (20BC:5VC:75CS) and 20BC:10VC:70CS with tomato plants. The soil plant analysis development (SPAD) readings in BC:VC:CS mixes were similar to or higher than the control except for tomato plants in 80BC:5VC:15CS, 80BC:15VC:5CS, and 80BC:20VC:0CS mixes. Plants in BC:VC:CS mixes had similar growth indexes and total dry weight with respect to those in 100% CS, with the root DW of basil plants in 60BC:15VC:25CS being the highest among all treatments. Therefore, the BC (20%, 40%, 60%, or 80%, by volume) and VC (5%, 10%, 15%, or 20%, by volume) mixes had the potential to replace CS for container-grown plants, with the estimate wholesale price for 80BC:5VC:15CS was only 61.6% that of the control.

Keywords: container; growing media; nursery production; carbon; peat moss; bioenergy

1. Introduction

Biochar (BC), attracting increasing interests in recent years for its use in agriculture, can be used to replace some components of commonly used container substrates [1–3]. Biochar could be made from the pyrolysis [4,5] or gasification [6,7] of biomass. The main purpose of the bioenergy production process, pyrolysis and gasification, is to produce syngas or bio-oil [4,8,9], and BC is the by-product. The application of BC in other fields including agriculture provides extra benefits to the bioenergy producers. Biochars can be made from green waste [10,11], wheat straw [12,13], wood [13,14], and rice hull [15], and are renewable and quickly generated [16]. Biochars produced from various raw materials or production conditions would be different and thus cause diverse effects when being incorporated in container substrates [17,18]. Meanwhile, BCs made from the same feedstock but with different fraction sizes could have different pH and nutrient levels [19].

The incorporation of BC in container substrates has many benefits. Ecological issues caused by extracting the most commonly used horticultural substrate constituent peat from peatlands has increased the necessity of using alternative growing media components including BC in the near future [20]. Research has shown that peatlands have been drained for peat use in agriculture for a long time, which led to the loss of carbon to the atmosphere [21]. Drained peatlands would cause the release of 1.91 Gt CO_2-eq. contemporary annual greenhouse gas emission, and peatland rehabilitation is strongly needed [22]. Due to the environmental concern of using peat, the use of BC in containers substituting peat could be a more sustainable choice for the horticultural industry. Biochar in container substrates could increase water-holding capacity [23] and reduce nutrient leaching [24]. Furthermore, the incorporation of BC in peat-based substrate could increase substrate electrical conductivity (EC) and mineral nutrients uptake [25]. Many of the BCs used were alkaline and thus could be used to raise the pH of acidic substrate [10,26,27].

Although BC in container substrates has a lot of benefits, different plant species and BC mix ratios in the container may lead to different results. Kadota and Niimi [28] concluded that mixing 10% or 30% (by volume) BC to the basal medium (substrate with peat, vermiculite, soil, and sand at the ratio of 2:1:1:1, by volume) caused enhanced zinnia (*Zinnia linearis* Benth) shoot growth but no positive effects on marigold (*Tagetes patula* L.) or scarlet sage (*Salvia splendens* Ker Gawl.). Mixes with potato anaerobic digestate and acidified wood pellet BC (1:1, by volume) increased tomato (*Solanum lycopersicum* L.) dry weight (DW) but decreased marigold DW compared to those in the 1:1 peat:vermiculite control [29]. Mixes of 50% or 70% (by volume) sugarcane bagasse BC with the rest being bark-based container substrates led to decreased tomato total DW but no negative effects on basil DW compared to the control [30]. In addition to plant species, different percentages of BC mixed with other substrates components also led to diverse results. Gu et al. [31] has shown that the BC rate in pine bark mixes was positively correlated with gomphrena (*Gomphrena globosa* L.) fresh weight (FW) and DW. Housley et al. [32] found that the pansy (*Viola hybrida* Schur) aboveground DW was increased in the mixes with pine bark, coir, clinker ash, and coarse sand when incorporated with 2.5% (*w/w*) Sydney blue gum (*Eucalyptus saligna* Sm.) wood chip BC, while suppressed when incorporated with 10% (*w/w*) BC, compared to the control. Webber et al. [33] also showed that the amendment of 50% (by volume) pneumatic bagasse BC could increase squash (*Cucurbita pepo* L.) plant DW, but the amendment of 25% (by volume) BC caused no negative effect on plant DW in comparison with the control.

Vermicompost (VC) is produced by using worms to digest and thus break down organic matter, such as sewage sludge [34], animal waste [35–37], and crop residues [38]. Vermicompost is finely textured and rich in nutrients [39,40], and it has good water-holding capacity [41]. Beneficial effects have been shown in a lot of studies in which VC was used in containers with other substrates. Atiyeh et al. [40] concluded that VC addition in container substrates would enhance plant growth. The swamp rose mallow (*Hibiscus moscheutos* L.) grown in containers with VC showed improved plant DW [42]. Vermicompost mixed with coir at a ratio of 2:1 (*w/w*) as container substrate increased Swiss chard (*Beta vulgaris* L.) plant height and FW [43]. The substitution of peat-based growing media with VC (10%, 20%, 30%, 40%, and 50%, by volume) and BC (4%, 8%, and 12%, by volume) did not cause any negative effect on the shoot DW of petunia (*Petunia hybrida* E.Vilm.) and pelargonium (*Pelargonium peltatum* L.) except pelargonium in mixes of 4% BC and 50% VC [44]. It was also shown that the plant size, flower production, and root growth capacity of petunia and pelargonium in BC (4%, 8%, and 12%, by volume) and VC (10%, 20%, and 30%, by volume) mixes was similar to or higher than those in the peat-based substrate control [45]. However, VC made from different parent materials could have different properties [46]. The addition of VC in container substrate may not always cause positive effects on plant growth. Liu et al. [47] have shown that the DW of vegetative and flower organs and growth index of pepper (*Capsicum annuum* L.) in BC (70%, 80%, and 90%, by volume) mixes with VC were lower than those in the Sunshine #1 Mix, which is a peat moss-based substrate. Therefore, although a specific percentage of certain VCs could be used as container substrates to grow plants, caution is also needed due to the complexity of VC.

Tomato (*Solanum lycopersicum* L.) and basil (*Ocimum basilicum* L.) are two widely consumed plants in the horticultural industry. Tomato is an important source of various antioxidant vitamins including ascorbic acid, tocopherols, vitamin C, and carotenoids [48]. Tomato plants are considered as "heavy feeder", requiring medium to high fertility [49]. Basil is referred to as the "king of the herbs" [50]. The essential oil of basil is used in various food products, perfumes, insecticides, medicines, and industrial products [51,52]. Basil is sensitive to high fertility [53]. It was shown that high fertility (500 mg N L^{-1}) reduced the basil leaf area, when compared to the ones growing with 100 mg N L^{-1} [54].

Few research studies have investigated combinations of BC with VC as container substrates. Due to the high cost of VC and the proneness to use more BC in containers to replace the commonly used peat-based substrate, low percentages of VC (5%, 10%, 15%, and 20%, by volume) and wide range of percentages of BC (20%, 40%, 60%, and 80%, by volume) were used in this experiment. The purpose of this experiment was to test the potential of the mixed hardwood BC and VC mixes as replacements for a commercial peat-based substrate (CS). The specific objectives were to (1) investigate the physical and chemical properties of the BC and VC mixes; and (2) compare the impacts of different mixes of BC with VC on container-grown basil and tomato plants to 100% CS.

2. Materials and Methods

2.1. Plant Materials and Container Substrates Treatments

Tomato 'Roma' (Morgan County Seeds, Barnett, MO, USA) and basil seeds (Johnny's Selected Seeds, Winslow, ME, USA) were sown in commercial propagation mix (Propagation mix; Sun Gro® Horticulture, Agawam, MA, USA) in plug trays on 28 October 2016. One tomato seed and four basil seeds were sown per cell (hexagon with side length of 2.6 cm; height: 4.2 cm; volume: 20 mL). Uniform basil and tomato seedlings were selected and transplanted into the experimental substrates in pots (depth: 10.8 cm; top diameter: 15.5 cm; bottom diameter: 11.3 cm; volume: 1330 mL) on 16 November 2016 after true leaves emerged. Each container contained one tomato seedling or four basil seedlings. Sixteen BC and VC mixes were formulated by mixing four rates of BC (20%, 40%, 60%, and 80%, by volume; a by-product of fast pyrolysis of mixed hardwood, Proton Power, Inc., Lenior City, TN, USA) with four rates of VC (5%, 10%, 15%, and 20%, by volume; Pachamama earthworm castings; Lady Bug Brand, Conroe, TX, USA). The CS (BM7 35BKS; Berger, Saint-Modeste, QC, Canada) made up the rest of the volume when the BC and VC did not add up to 100%. The CS was used as the control (Figure 1). The CS (Berger BM7 35BKS) used in this research consisted of 55% coarse peat moss, 35% pine bark, and 10% horticultural perlite. The wholesale price for the mixed hardwood BC was $65.4 per cubic meter (Personal Communication). The wholesale price was approximately $176.6 per cubic meter for the CS (Berger BM7 35BKS) and $607.4 per cubic meter for the VC [55]. The estimated wholesale price for the 17 different substrates per cubic meter was shown in Figure 2. The estimated wholesale price for mixes of 20% BC and 10% VC by volume with the rest (70%) being the CS (20BC:10VC:70CS), 20BC:15VC:65CS, 40BC:15VC:45CS, 20BC:20VC:60CS, 40BC:20VC:40CS, and 60BC:20VC:20CS was higher than the 100% CS (control), while the other BC:VC:CS mixes were all cheaper than the control. The cheapest treatment (80BC:5VC:15CS) was only 61.6% of the price of the CS. The nutrient concentration (N, P, K, Ca, Mg, S, B, Ca, Cu, Fe, Mn, Na, and Zn) of the CS, BC, and VC were tested by the Texas A&M AgriLife Extension Service Soil, Water and Forage Testing Laboratory in College Station, TX, USA and shown in Table 1.

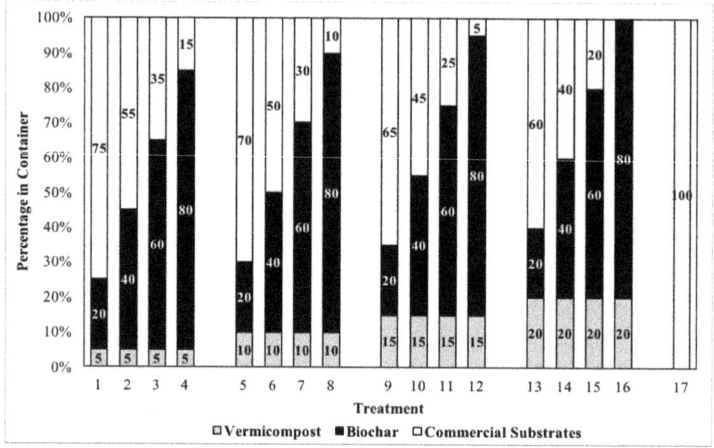

Figure 1. Seventeen formulated substrates including mixes of biochar (20%, 40%, 60%, or 80%, by volume) with vermicompost (5%, 10%, 15%, or 20%, by volume) and the 100% commercial peat-based substrates, Berger BM7 35BKS.

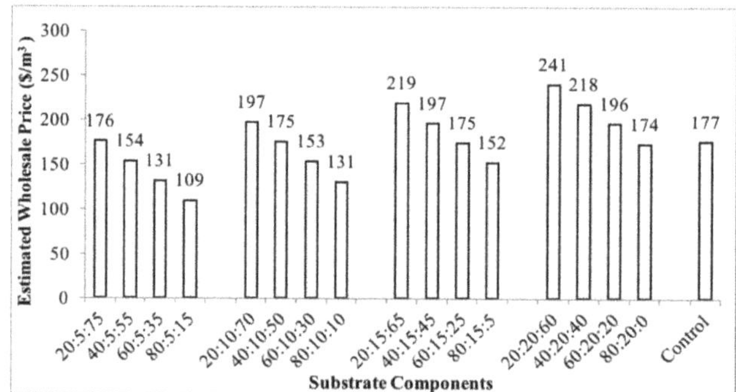

Figure 2. The estimated wholesale price ($) for the 17 formulated substrates per cubic meter. The ratios on the X-axis indicate the percentage ratio of biochar to vermicompost to commercial substrate (by volume). The control was 100% commercial substrate (Berger BM7 35BKS).

Table 1. Nutrient analysis of the commercial substrate (CS, Berger BM7 35BKS), biochar (BC), and vermicompost (VC).

Substrate	N	P	K	Ca	Mg	S	Fe	B	Cu	Mn	Na	Zn
	(%)						(mg kg⁻¹)					
CS	0.70 b[z]	540 b	1265 c	25108 a	4237 a	1744 b	1508 b	11 b	17 b	98 c	953 a	46 b
BC	0.23 c	456 b	6362 a	27507 a	1299 b	231 b	2039 b	15 b	9 b	905 a	107 c	13 b
VC	2.43 a	4901 a	3714 b	25841 a	3819 a	5996 a	4835 a	42 a	165 a	374 b	351 b	385 a

[z] Means within a column under each mean factor followed by the same letter are not significantly different according to the Tukey's HSD test at $p < 0.05$ ($n = 4$).

The pH of the CS, BC, and VC was measured by using a handheld pH-EC meter (HI 98129, Hanna Instruments, Woonsocket, RI, USA), and the EC was measured by using the Bluelab Combo Meter (Bluelab Corporation Limited, Tauranga, New Zealand) according to the pour-through extraction method [56]. The pH of the CS was 6.06, and the EC was 1.3 dS m⁻¹, respectively. The pH of the BC was 11.18 and the EC was 2.0 dS m⁻¹, respectively. The pH of the VC was 4.8 and the EC was 6.7 dS m⁻¹, respectively. The total porosity, container capacity, air space, and bulk density of the BC were 84.7%,

60.3%, 24.4%, and 0.15 g cm^{-3}, respectively. Particle size distribution of the BC was determined by passing 40 g BC through 2.8, 2, 1, 0.425, and 0.25 mm sieves, and the weight was measured to determine the percentage of each particle size. Percentages of the BC particles ranging from greater than 2.8 mm, 2.0 mm to 2.8 mm, 1.0 mm to 2.0 mm, 0.425 mm to 1.0 mm, 0.25 mm to 0.425 mm, and smaller than 0.25 mm in diameter were 47.9%, 19.4%, 19.4%, 9.1%, 2.0%, and 2.2% (*w/w*), respectively.

Six replications of the 17 treatments (16 BC:VC:CS mixes plus control) were arranged in randomized complete blocks in the greenhouse located on Texas A&M University campus, College Station, TX, USA to control the environmental variance. The temperature, humidity, and dew point in the greenhouse were monitored using Watchdog (Spectrum Technologies Inc., Paxinos, PA, USA). During the experimental period, the average greenhouse temperature, relative humidity, and dew point were 20.5 °C, 76.0%, and 15.4 °C, respectively. The basil plants were irrigated with 200 mg nitrogen (N) L^{-1} (20N-4.4 P-16.6K) Peters® Professional (Everris NA Inc., Dublin, OH, USA) nutrient solution. The tomato plants were irrigated with 200 mg N L^{-1} (20N-4.4P-16.6K) Peters® Professional nutrient solution from 0 to 3 weeks after transplanting (WAT) and changed to 300 mg N L^{-1} from 4 WAT. The total N in Peters® Professional contains 8.1% ammoniacal N and 11.9% nitrate N. The pH of the 200 mg N L^{-1} (20N-4.4P-16.6K) Peters® Professional nutrient solution was 6.1, and the EC was 1.0 dS m^{-1}. The pH of the 300 mg N L^{-1} (20N-4.4P-16.6K) Peters® Professional nutrient solution was 5.9, and the EC was 1.3 dS m^{-1}.

2.2. Substrate Physical Properties and Substrate Leachate pH

Four replications of each substrate were tested to determine physical properties including the bulk density, total porosity, air space, and container capacity of the 17 substrates using the porometers of the North Carolina State University Horticultural Substrates Laboratory [57]. The substrate leachate pH was measured at 0, 2, 4, 6, and 9 WAT using a handheld pH-EC meter (HI 98129, Hanna Instrument, Woonsocket, RI, USA) according to the pour-through extraction method using the same amount of leachate for each test [56].

2.3. Plant Growth and Development

The plant growth index (GI) of each plant was measured at 0, 2, 4, 6, and 9 WAT, respectively. The height of the plant was measured from the medium surface to the highest point of the plant. The widest plant canopy width and its perpendicular width were measured. The plant GI was determined by the following formula: *GI = plant height/2 + (plant width 1 + plant width 2)/4* [58]. The leaf chlorophyll content of each plant was measured as soil plant analysis development (SPAD) values at 2, 4, 6, and 9 WAT, respectively using a portable SPAD 502 Plus Chlorophyll Meter (Spectrum Technologies, Inc., Plainfield, IL, USA). Plant leaves were too small to measure SPAD at 0 WAT. The leaf greenness of each plant was determined using the average of readings from three mature leaves.

At the end of nine WAT, plants were harvested to measure DW. For each tomato plant, the stems, leaves, root, and combined fruits and flowers were harvested separately. For each basil plant, the shoot and root were harvested separately. All the plant parts were oven-dried at 80 °C to constant weight before the DW measurements. The total DW of each plant was calculated by adding the DW of all parts of the plant.

2.4. Statistical Analysis

Data were analyzed with one-way analysis of variance (ANOVA) using JMP Statistical Software (version Pro 12.2.0; SAS Institute, Cary, NC, USA) to test the effect of different substrates on the physical and chemical properties and plant growth. The type of substrate was the main factor. Tukey's Honestly Significant Difference (HSD) tests were used for the comparison of means among treatments at $p < 0.05$. Tomato and basil plants were treated as independent studies and were not compared.

3. Results and Discussion

3.1. Physical Properties of the Container Substrates

The total porosity of all the BC:VC:CS mixes was similar to the control (Figure 3a). There was no difference between the container capacity of the mixes of BC (20% or 40%) with VC (5%, 10%, or 15%) and the control (Figure 3b). The container capacity of 80% BC mixes were significantly lower than the control, since BC had lower container capacity (60.3%) than that of CS (70.7%). The air space of 80BC:5VC:15CS, 60BC:20VC:20CS, and 80BC:20VC:0CS was higher than that of the control due to the high incorporation rate of BC with large particle size (Figure 3c), which increased the macropores and thus the air space. The fraction of BC with size greater than 2.8 mm (47.9%) was higher than that of CS (25.4%) and VC (1.2%). The past research showed the variable results of the substrates' physical properties after BC incorporation. Tian et al. [10] reported that the total porosity and container capacity of peat substrate with or without 50% (by volume) green waste BC were similar, while others found that the total porosity and container capacity of the substrates increased with the increasing BC rate [59–61]. Guo et al. [62] found that the air space increased as the pine wood BC rate increased. Yu et al. [63] indicated that the air space increased with the increasing mixed hardwood BC incorporation rate from 10% to 100% (by volume), but the trend was totally opposite for sugarcane bagasse BC. Another research showed that the substitution of peat with 10% (by volume) sewage sludge BC caused no difference on the air space in comparison with the 100% peat substrate control [64]. The effect of BC incorporation on a substrate's physical properties is BC-specific. Container substrates hold water in the micropores between or inside the container substrate components' particles [65]. The container capacity would be increased if the incorporation of BC leads to a higher fraction of micropores. Air space is the proportion of air-filled macropores after the water drains [65]. Air space is closely related to the particle size distributions of BC and the other substrate components, and the changed interporosity after BC incorporation affects air space. The effect of BC on total porosity is related to container capacity and air space, since total porosity is the sum of container capacity and air space.

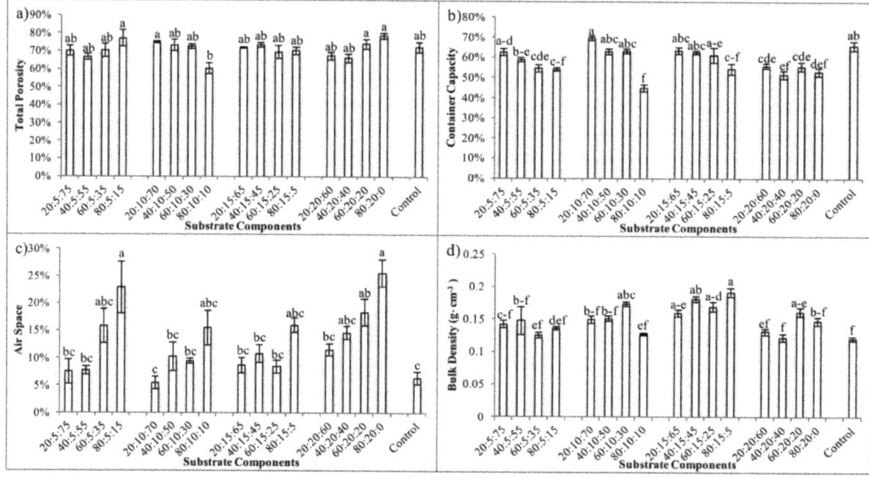

Figure 3. Total porosity (**a**), container capacity (**b**), air space (**c**), and bulk density (**d**) (mean ± standard error) of the 17 different formulated substrates. The ratios on the X-axis indicate the percentage ratio of biochar to vermicompost to commercial substrate (by volume). The control was 100% commercial substrate (Berger BM7 35BKS). Means indicated by the same letter are not significantly different according to Tukey's Honestly Significant Difference (HSD) test at $p < 0.05$ ($n = 4$).

The bulk density of 15% VC mixes, 60BC:10VC:30CS, and 60BC:20VC:20CS was higher than that of the control (Figure 3d). The increased bulk density could be due to the high bulk density of VC (0.38 g cm^{-3}) and BC (0.15 g cm^{-3}) compared to the control (0.10 g cm^{-3}). Similar to our results, a lot of research has shown that substitution of the commonly used substrate with BC could increase bulk density [13,23,29].

3.2. Substrate Leachate pH

Compared to the control, the substrate leachate pH in all BC:VC:CS mixes was increased, except for those of the 20BC:5VC:75CS and 20BC:10VC:70CS mixes with tomato plants at 9 WAT (see Table S1). The increased pH was probably due to the high pH of the BC (11.18) used in this experiment. The liming effect of BC was found in a lot of research [66–68]. In addition, substrate leachate pH tended to slightly decrease during the study (see Table S1), which was possibly due to the acidifying effect of the fertilizer 20N-4.4P-16.6K Peters® Professional (a potential acidity of 188 kg calcium carbonate equivalent per 1000 kg of the fertilizer). Therefore, the reason for the similar pH of mixes of 20BC:5VC:75CS and 20BC:10VC:70CS with tomato plants at 9 WAT with the control could be due to the low percentage of the BC incorporation rate and the relative large amount of nutrient solution applied to the tomato plants for 9 weeks 'washing down' the substrate leachate pH.

3.3. Plant Growth and Development

For basil, BC:VC:CS mixes caused no negative effect on the SPAD readings in comparison with the control at 2, 4, 6, or 9 WAT (see Table S2). For tomato, the SPAD readings of the plants in BC:VC:CS mixes were similar to or higher than those in the control at 2, 4, and 6 WAT, while at 9 WAT, the leaf SPAD readings of tomato plants grown in the 80BC:5VC:15CS, 80BC:15VC:5CS, and 80BC:20VC:0CS mixes were lower than those in the control (see Table S2). Similarly, Liu et al. [47] found that SPAD readings of bell pepper leaves in mixes of BC (70%, 80%, and 90%, by volume) with the rest being VC were lower than those in commercial substrates. The decreased leaf SPAD readings at 9 WAT could be caused by two reasons. First, it was shown that leaf SPAD readings was closely related to leaf N concentration, and lower SPAD readings indicated lower leaf N concentration [69]. The decreased SPAD readings could be due to the BC's ability to immobilize N [70]. Second, the increased substrate pH after the incorporation of the BC with high pH (11.18) could reduce iron (Fe) availability, causing decreased leaf greenness. It was shown that shoot Fe concentration was lower at substrate with higher pH [71], and leaf SPAD readings were significantly correlated with Fe availability, since Fe is essential for the chlorophyll synthesis [72]. The leaf chlorosis (as measured by chlorophyll concentration) could be more severe with less iron concentration [73]. The reason for the decreased leaf SPAD readings of the tomato plants only shown at 9 WAT was due to the nutrient deficiency of the leaves caused by the strong nutrient sink (fruits and flowers) at that stage, since all the tomato plants had flowers and fruits at 9 WAT.

However, the possible N binding of BC and reduced Fe availability caused by increased pH did not decrease the plant GI and DW of either tomato or basil plants in this research. The GIs of both basil and tomato plants grown in BC:VC:CS mixes were similar to those in the control at 9 WAT (see Figure S1). All basil plants grown in BC:VC:CS mixes had similar shoot and total DWs in comparison with the control (see Figure S2). The root DWs of basil plants in BC:VC:CS mixes were similar to or higher than those in 100% CS, with those in 60BC:15VC:25CS being the highest among all treatments (Figure 4). Similarly, all tomato plants grown in BC:VC:CS mixes had similar DWs (the combined flower and fruit, leaf, stem, root, and total DW) with respect to the control (see Figure S3). The reasons for the enhanced plant growth could be the VC's extra nutrient supply and the BC's nutrient-holding ability. Similar results were reported by Huang et al. [74], who indicated that tomato and basil plant growth in mixes of BC (60% or 70%, by volume) with either 5% VC or chicken manure compost with the rest being CS was similar to or higher than the control. Alvarez et al. [44] also found that the amendment of VC

(10%, 20%, 30%, 40%, and 50%, by volume) and BC (4%, 8%, and 12%, by volume) to a peat-based substrate did not adversely affect the petunia shoot DW.

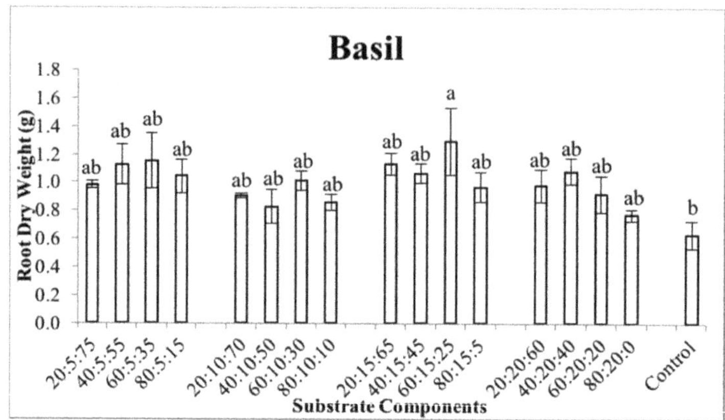

Figure 4. Root dry weight (mean ± standard error) per basil plant harvested at 9 weeks after transplanting. The ratios on the X-axis indicate the percentage ratio of biochar to vermicompost to commercial substrate (by volume). The control was 100% commercial substrate (Berger BM7 35BKS). Means indicated by the same letter are not significantly different according to Tukey's HSD test at $p < 0.05$ ($n = 6$).

4. Conclusions

The mixes of mixed hardwood BC (20%, 40%, 60%, or 80%, by volume) made from fast pyrolysis and VC (5%, 10%, 15%, or 20%, by volume) used in this study had the potential to replace the CS to grow basil and tomato plants. Our results found difference in the substrate leachate pH between the 100% CS and BC:VC:CS mixes except for 20BC:5VC:75CS and 20BC:10VC:70CS with tomato plants at 9 WAT, which indicated the liming effect of the mixed hardwood BC used in this research. At 9 WAT, the leaf SPAD readings of tomato plants grown in 80BC:5VC:15CS, 80BC:15VC:5CS, and 80BC:20VC:0CS mixes were lower than those in the control, which was possibly due to the binding ability of BC or reduced Fe availability caused by increased substrate pH after BC incorporation. The growth index and total dry weight of basil and tomato plants in BC:VC:CS mixes were similar to those in the CS. Considering the cost of the alternative substrates, all the BC:VC:CS mixes (except for 20BC:10VC:70CS, 20BC:15VC:65CS, 40BC:15VC:45CS, 20BC:20VC:60CS, 40BC:20VC:40CS, and 60BC:20VC:20CS) in this experiment could be selected as the suitable ones to grow plants, with the 80BC:5VC:15CS being the cheapest and most recommended. This study is important for the future use of mixtures of BC with VC in container substrate for greenhouse and nursery plant production to provide a sustainable and environmentally friendly way to substitute peat use in agriculture and add value to the bioenergy process by using the by-product BC. Using the BC:VC:CS mixes with wholesale prices cheaper than the 100% CS could provide more economical ways to grow plants and benefit the growers. Tomato and basil plants were used as model plants in this study. Since results were similar for these two plants with different optimal growing conditions, these results could be applicable to many other plants. The results in this study can be only suitable for the specific mixed hardwood BC made from pyrolysis and VC (Pachamama earthworm castings; Lady Bug Brand) due to the complexity of BC and VC. More mixed hardwood BC incorporation percentages and other potential amendment candidates need to be tested for economic viability.

Supplementary Materials: The following are available online at http://www.mdpi.com/2073-4395/10/2/224/s1, Table S1: The leachate pH of the 17 substrates with basil and tomato plants at 0, 2, 4, 6, and 9 weeks after transplanting (WAT), Table S2: The SPAD reading of basil and tomato leaves in 17 substrates at 2, 4, 6, and 9 weeks after transplanting (WAT), Figure S1: Cumulative growth index (mean ± standard error) per basil (a) or tomato (b) plant grown in 17 substrates at 0, 2, 4, 6, and 9 weeks after transplanting (WAT), Figure S2: Shoot (a) and total (b) dry weight (mean ± standard error) per basil plant harvested at 9 weeks after transplanting, and Figure S3: Leaves (a), stem (b), root (c), combined flower and fruit (d), and total (e) dry weight (mean ± standard error) per tomato plant harvested at 9 weeks after transplanting.

Author Contributions: Conceptualization, L.H., M.G., and X.L.; Data curation, L.H.; Formal analysis, L.H.; Investigation, L.H., P.Y., C.Z., and X.L.; Resources, M.G.; Supervision, M.G. and X.L.; Writing—original draft, L.H.; Writing—review and editing, M.G., P.Y., C.Z., and X.L. All authors have read and agreed to the published version of the manuscript.

Funding: This research received no external funding.

Acknowledgments: The authors thank the Agriculture Women Excited to Share Opinions, Mentoring, and Experiences (AWESOME) faculty group of the College of Agriculture and Life Sciences at Texas A&M University for assistance with editing the manuscript.

Conflicts of Interest: The authors declare no conflict of interest.

References

1. Webber, C.L., III; White, P.M., Jr.; Gu, M.; Spaunhorst, D.J.; Lima, I.M.; Petrie, E.C. Sugarcane and pine biochar as amendments for greenhouse growing media for the production of bean (*Phaseolus vulgaris* L.) seedlings. *J. Agric. Sci.* **2018**, *10*, 58–68. [CrossRef]

2. Choi, H.-S.; Zhao, Y.; Dou, H.; Cai, X.; Gu, M.; Yu, F. Effects of biochar mixtures with pine-bark based substrates on growth and development of horticultural crops. *Horti. Environ. Biotechnol.* **2018**, *59*, 345–354. [CrossRef]

3. Peng, D.; Gu, M.; Zhao, Y.; Yu, F.; Choi, H.-S. Effects of biochar mixes with peat-moss based substrates on growth and development of horticultural crops. *Hortic. Sci. Technol.* **2018**, *36*, 501–512.

4. Gvero, P.M.; Papuga, S.; Mujanic, I.; Vaskovic, S. Pyrolysis as a key process in biomass combustion and thermochemical conversion. *Therm. Sci.* **2016**, *20*, 154. [CrossRef]

5. Rahman, A.A.; Sulaiman, F.; Abdullah, N. Influence of washing medium pre-treatment on pyrolysis yields and product characteristics of palm kernel shell. *J. Phys. Sci.* **2016**, *27*, 53.

6. Hansen, V.; Müller-Stöver, D.; Ahrenfeldt, J.; Holm, J.K.; Henriksen, U.B.; Hauggaard-Nielsen, H. Gasification biochar as a valuable by-product for carbon sequestration and soil amendment. *Biomass Bioenergy* **2015**, *72*, 300–308. [CrossRef]

7. Altland, J.E.; Locke, J.C. Gasified rice hull biochar is a source of phosphorus and potassium for container-grown plants. *J. Environ. Hortic.* **2013**, *31*, 138–144.

8. Laird, D.A. The charcoal vision: A win–win–win scenario for simultaneously producing bioenergy, permanently sequestering carbon, while improving soil and water quality. *Agron. J.* **2008**, *100*, 178–181.

9. Zheng, W.; Sharma, B.; Rajagopalan, N. *Using Biochar as a Soil Amendment for Sustainable Agriculture*; Illinois Sustainable Technology Center: Champaign, IL, USA, 2010.

10. Tian, Y.; Sun, X.; Li, S.; Wang, H.; Wang, L.; Cao, J.; Zhang, L. Biochar made from green waste as peat substitute in growth media for *Calathea rotundifola* cv. Fasciata. *Sci. Hortic.* **2012**, *143*, 15–18. [CrossRef]

11. Buss, W.; Graham, M.C.; Shepherd, J.G.; Mašek, O. Risks and benefits of marginal biomass-derived biochars for plant growth. *Sci. Total Environ.* **2016**, *569*, 496–506. [CrossRef]

12. Xu, G.; Zhang, Y.; Sun, J.; Shao, H. Negative interactive effects between biochar and phosphorus fertilization on phosphorus availability and plant yield in saline sodic soil. *Sci. Total Environ.* **2016**, *568*, 910–915. [CrossRef] [PubMed]

13. Vaughn, S.F.; Kenar, J.A.; Thompson, A.R.; Peterson, S.C. Comparison of biochars derived from wood pellets and pelletized wheat straw as replacements for peat in potting substrates. *Ind. Crops Prod.* **2013**, *51*, 437–443. [CrossRef]

14. Hansen, V.; Hauggaard-Nielsen, H.; Petersen, C.T.; Mikkelsen, T.N.; Müller-Stöver, D. Effects of gasification biochar on plant-available water capacity and plant growth in two contrasting soil types. *Soil Till. Res.* **2016**, *161*, 1–9. [CrossRef]

15. Locke, J.C.; Altland, J.E.; Ford, C.W. Gasified rice hull biochar affects nutrition and growth of horticultural crops in container substrates. *J. Environ. Hortic.* **2013**, *31*, 195–202.
16. Yu, F.; Steele, P.H.; Gu, M.; Zhao, Y. Using Biochar as Container Substrate for Plant Growth. U.S. Patent 9,359,267 B2, 7 June 2016.
17. Huang, L.; Gu, M. Effects of biochar on container substrate properties and growth of plants—A review. *Horticulturae* **2019**, *5*, 14. [CrossRef]
18. Chrysargyris, A.; Prasad, M.; Kavanagh, A.; Tzortzakis, N. Biochar type and ratio as a peat additive/partial peat replacement in growing media for cabbage seedling production. *Agronomy* **2019**, *9*, 693. [CrossRef]
19. Prasad, M.; Chrysargyris, A.; McDaniel, N.; Kavanagh, A.; Gruda, N.S.; Tzortzakis, N. Plant nutrient availability and pH of biochars and their fractions, with the possible use as a component in a growing media. *Agronomy* **2020**, *10*, 10. [CrossRef]
20. Gruda, N.S. Increasing sustainability of growing media constituents and stand-alone substrates in soilless culture systems. *Agronomy* **2019**, *9*, 298. [CrossRef]
21. Hatala, J.A.; Detto, M.; Sonnentag, O.; Deverel, S.J.; Verfaillie, J.; Baldocchi, D.D. Greenhouse gas (CO_2, CH_4, H_2O) fluxes from drained and flooded agricultural peatlands in the Sacramento-San Joaquin Delta. *Agric. Ecosyst. Environ.* **2012**, *150*, 1–18. [CrossRef]
22. Leifeld, J.; Menichetti, L. The underappreciated potential of peatlands in global climate change mitigation strategies. *Nat. Commun.* **2018**, *9*, 1071. [CrossRef]
23. Dumroese, R.K.; Heiskanen, J.; Englund, K.; Tervahauta, A. Pelleted biochar: Chemical and physical properties show potential use as a substrate in container nurseries. *Biomass Bioenergy* **2011**, *35*, 2018–2027. [CrossRef]
24. Crutchfield, E.F.; McGiffen, M.E., Jr.; Merhaut, D.J. Effects of biochar on nutrient leaching and begonia plant growth. *J. Environ. Hortic.* **2018**, *36*, 126–132.
25. Zulfiqar, F.; Younis, A.; Chen, J. Biochar or biochar-compost amendment to a peat-based substrate improves growth of *Syngonium podophyllum*. *Agronomy* **2019**, *9*, 460. [CrossRef]
26. Dispenza, V.; De Pasquale, C.; Fascella, G.; Mammano, M.M.; Alonzo, G. Use of biochar as peat substitute for growing substrates of *Euphorbia × lomi* potted plants. *Span. J. Agric. Res.* **2016**, *14*, e0908. [CrossRef]
27. Nieto, A.; Gascó, G.; Paz-Ferreiro, J.; Fernández, J.; Plaza, C.; Méndez, A. The effect of pruning waste and biochar addition on brown peat based growing media properties. *Sci. Hortic.* **2016**, *199*, 142–148. [CrossRef]
28. Kadota, M.; Niimi, Y. Effects of charcoal with pyroligneous acid and barnyard manure on bedding plants. *Sci. Hortic.* **2004**, *101*, 327–332. [CrossRef]
29. Vaughn, S.F.; Eller, F.J.; Evangelista, R.L.; Moser, B.R.; Lee, E.; Wagner, R.E.; Peterson, S.C. Evaluation of biochar-anaerobic potato digestate mixtures as renewable components of horticultural potting media. *Ind. Crops Prod.* **2015**, *65*, 467–471. [CrossRef]
30. Yu, P.; Huang, L.; Li, Q.; Lima, I.M.; White, P.M.; Gu, M. Effects of mixed hardwood and sugarcane biochar as bark-based substrate substitutes on container plants production and nutrient leaching. *Agronomy* **2020**, *10*, 156. [CrossRef]
31. Gu, M.; Li, Q.; Steele, P.H.; Niu, G.; Yu, F. Growth of 'fireworks' gomphrena grown in substrates amended with biochar. *J. Food Agric. Environ.* **2013**, *11*, 819–821.
32. Housley, C.; Kachenko, A.; Singh, B. Effects of eucalyptus saligna biochar-amended media on the growth of *Acmena smithii*, *Viola* var. *hybrida, and Viola × wittrockiana*. *J. Hortic. Sci. Biotechnol.* **2015**, *90*, 187–194. [CrossRef]
33. Webber, C.L., III; White, P.M., Jr.; Spaunhorst, D.J.; Lima, I.M.; Petrie, E.C. Sugarcane biochar as an amendment for greenhouse growing media for the production of cucurbit seedlings. *J. Agric. Sci.* **2018**, *10*, 104. [CrossRef]
34. Mitchell, M.; Hornor, S.; Abrams, B. Decomposition of sewage sludge in drying beds and the potential role of the earthworm, *Eisenia foetida*. *J. Environ. Qual.* **1980**, *9*, 373–378. [CrossRef]
35. Edwards, C. Production of feed protein from animal waste by earthworms. *Philos. Trans. R. Soc. Lond. B Biol. Sci.* **1985**, *310*, 299–307.
36. Chan, P.L.; Griffiths, D. The vermicomposting of pre-treated pig manure. *Biol. Wastes* **1988**, *24*, 57–69. [CrossRef]
37. Hartenstein, R.; Bisesi, M.S. Use of earthworm biotechnology for the management of effluents from intensively housed livestock. *Outlook Agric.* **1989**, *18*, 72–76. [CrossRef]

38. Manna, M.; Jha, S.; Ghosh, P.; Ganguly, T.; Singh, K.; Takkar, P. Capacity of various food materials to support growth and reproduction of epigeic earthworms on vermicompost. *J. Sustain. For.* **2005**, *20*, 1–15. [CrossRef]

39. Sinha, R.K.; Agarwal, S.; Chauhan, K.; Valani, D. The wonders of earthworms & its vermicompost in farm production: Charles darwin's 'friends of farmers', with potential to replace destructive chemical fertilizers. *Agric. Sci.* **2010**, *1*, 76.

40. Atiyeh, R.; Subler, S.; Edwards, C.; Bachman, G.; Metzger, J.; Shuster, W. Effects of vermicomposts and composts on plant growth in horticultural container media and soil. *Pedobiologia* **2000**, *44*, 579–590. [CrossRef]

41. Edwards, C.A.; Burrows, I. Potential of earthworm composts as plant growth media. In *Earthworms in Waste and Environmental Management*; Edwards, C.A., Neuhauser, E.F., Eds.; SPB Academic Publishing: The Hague, The Netherlands, 1988.

42. McGinnis, M.; Bilderback, T.; Warren, S. Vermicompost amended pine bark provides most plant nutrients for *Hibiscus moscheutos* 'Luna Blush'. *Acta Hortic.* **2011**, *891*, 249–256. [CrossRef]

43. Abbey, L.; Young, C.; Teitel-Payne, R.; Howe, K. Evaluation of proportions of vermicompost and coir in a medium for container-grown Swiss chard. *Int. J. Veg. Sci.* **2012**, *18*, 109–120. [CrossRef]

44. Alvarez, J.; Pasian, C.; Lal, R.; López, R.; Fernández, M. Vermicompost and biochar as substitutes of growing media in ornamental-plant production. *J. Appl. Hortic.* **2017**, *19*, 205–214.

45. Álvarez, J.M.; Pasian, C.; Lal, R.; López, R.; Díaz, M.J.; Fernández, M. Morpho-physiological plant quality when biochar and vermicompost are used as growing media replacement in urban horticulture. *Urban For. Urban Green.* **2018**, *34*, 175–180. [CrossRef]

46. Szczech, M.; Smolińska, U. Comparison of suppressiveness of vermicomposts produced from animal manures and sewage sludge against *Phytophthora nicotianae* Breda de Haan var. *nicotianae*. *J. Phytopathol.* **2001**, *149*, 77–82. [CrossRef]

47. Liu, R.; Gu, M.; Huang, L.; Yu, F.; Jung, S.-K.; Choi, H.-S. Effect of pine wood biochar mixed with two types of compost on growth of bell pepper (*Capsicum annuum* L.). *Hortic. Environ. Biotechnol.* **2019**, *60*, 313–319. [CrossRef]

48. Abushita, A.A.; Hebshi, E.A.; Daood, H.G.; Biacs, P.A. Determination of antioxidant vitamins in tomatoes. *Food Chem.* **1997**, *60*, 207–212. [CrossRef]

49. Jones, J.B., Jr. Tomato plant culture. In *The Field, Greenhouse, and Home Garden*; CRC Press: Boca Raton, FL, USA, 2007.

50. Makri, O.; Kintzios, S. Ocimum sp. (basil): Botany, cultivation, pharmaceutical properties, and biotechnology. *J. Herbs Spices Med. Plants* **2008**, *13*, 123–150. [CrossRef]

51. Bernstein, N.; Sela, S.; Dudai, N.; Gorbatsevich, E. Salinity stress does not affect root uptake, dissemination and persistence of salmonella in sweet-basil (*Ocimum basilicum*). *Front. Plant Sci.* **2017**, *8*, 675. [CrossRef]

52. Grayer, R.J.; Kite, G.C.; Goldstone, F.J.; Bryan, S.E.; Paton, A.; Putievsky, E. Infraspecific taxonomy and essential oil chemotypes in sweet basil, *Ocimum basilicum*. *Phytochemistry* **1996**, *43*, 1033–1039. [CrossRef]

53. Sharafzadeh, S.; Alizadeh, O. Nutrient supply and fertilization of basil. *Adv. Environ. Biol.* **2011**, *5*, 956–960.

54. Tesi, R.; Chisci, G.; Nencini, A.; Tallarico, R. Growth response to fertilisation of sweet basil (*Ocimum basilicum* L.). *Acta Hortic.* **1995**, *390*, 93–96. [CrossRef]

55. BWI Companies, Inc. Available online: https://www.bwicompanies.com/ (accessed on 2 February 2020).

56. LeBude, A.; Bilderback, T. *The Pour-Through Extraction Procedure: A Nutrient Management Tool for Nursery Crops*; AG-717-W:2009; North Carolina Cooperative Extension: Raleigh, NC, USA, 2009.

57. Fonteno, W.; Hardin, C.; Brewster, J. *Procedures for Determining Physical Properties of Horticultural Substrates Using the NCSU Porometer*; Horticultural Substrates Laboratory, North Carolina State University: Raleigh, NC, USA, 1995.

58. Guo, Y.; Niu, G.; Starman, T.; Volder, A.; Gu, M. Poinsettia growth and development response to container root substrate with biochar. *Horticulturae* **2018**, *4*, 1. [CrossRef]

59. Zhang, L.; Sun, X.-Y.; Tian, Y.; Gong, X.-Q. Biochar and humic acid amendments improve the quality of composted green waste as a growth medium for the ornamental plant *Calathea insignis*. *Sci. Hortic.* **2014**, *176*, 70–78. [CrossRef]

60. Fan, R.; Luo, J.; Yan, S.; Zhou, Y.; Zhang, Z. Effects of biochar and super absorbent polymer on substrate properties and water spinach growth. *Pedosphere* **2015**, *25*, 737–748. [CrossRef]

61. Méndez, A.; Paz-Ferreiro, J.; Gil, E.; Gascó, G. The effect of paper sludge and biochar addition on brown peat and coir based growing media properties. *Sci. Hortic.* **2015**, *193*, 225–230. [CrossRef]

62. Guo, Y.; Niu, G.; Starman, T.; Gu, M. Growth and development of Easter lily in response to container substrate with biochar. *J. Hortic. Sci. Biotechnol.* **2018**, *94*, 80–86. [CrossRef]

63. Yu, P.; Li, Q.; Huang, L.; Niu, G.; Gu, M. Mixed hardwood and sugarcane bagasse biochar as potting mix components for container tomato and basil seedling production. *Appl. Sci.* **2019**, *9*, 4713. [CrossRef]

64. Méndez, A.; Cárdenas-Aguiar, E.; Paz-Ferreiro, J.; Plaza, C.; Gascó, G. The effect of sewage sludge biochar on peat-based growing media. *Biol. Agric. Hortic.* **2017**, *33*, 40–51. [CrossRef]

65. Landis, T.D. Growing media. *Contain. Grow. Med.* **1990**, *2*, 41–85.

66. Ducey, T.F.; Novak, J.M.; Johnson, M.G. Effects of biochar blends on microbial community composition in two coastal plain soils. *Agriculture* **2015**, *5*, 1060–1075. [CrossRef]

67. Streubel, J.; Collins, H.; Garcia-Perez, M.; Tarara, J.; Granatstein, D.; Kruger, C. Influence of contrasting biochar types on five soils at increasing rates of application. *Soil Sci. Soc. Am. J.* **2011**, *75*, 1402–1413. [CrossRef]

68. Chan, K.; Van Zwieten, L.; Meszaros, I.; Downie, A.; Joseph, S. Agronomic values of greenwaste biochar as a soil amendment. *Soil Res.* **2008**, *45*, 629–634. [CrossRef]

69. Li, Y.; Alva, A.; Calvert, D.; Zhang, M. A rapid nondestructive technique to predict leaf nitrogen status of grapefruit tree with various nitrogen fertilization practices. *HortTechnology* **1998**, *8*, 81–86. [CrossRef]

70. Nair, A.; Carpenter, B. Biochar rate and transplant tray cell number have implications on pepper growth during transplant production. *HortTechnology* **2016**, *26*, 713–719. [CrossRef]

71. Dickson, R.W.; Fisher, P.R.; Padhye, S.R.; Argo, W.R. Evaluating calibrachoa (*Calibrachoa× hybrida* cerv.) genotype sensitivity to iron deficiency at high substrate ph. *HortScience* **2016**, *51*, 1452–1457. [CrossRef]

72. Radhamani, R.; Kannan, R.; Rakkiyappan, P. Leaf chlorophyll meter readings as an indicator for sugarcane yield under iron deficient typic haplustert. *Sugar Tech.* **2016**, *18*, 61–66. [CrossRef]

73. Wallihan, E.F. Relation of chlorosis to concentration of iron in citrus leaves. *Am. J. Bot.* **1955**, *42*, 101–104. [CrossRef]

74. Huang, L.; Niu, G.; Feagley, S.E.; Gu, M. Evaluation of a hardwood biochar and two composts mixes as replacements for a peat-based commercial substrate. *Ind. Crops Prod.* **2019**, *129*, 549–560. [CrossRef]

 *agronomy*

Article

Effects of Mixed Hardwood and Sugarcane Biochar as Bark-Based Substrate Substitutes on Container Plants Production and Nutrient Leaching

Ping Yu [1], Lan Huang [2], Qiansheng Li [3], Isabel M. Lima [4], Paul M. White [5] and Mengmeng Gu [2,*]

[1] Department of Horticultural Sciences, Texas A&M University, 2133 TAMU, College Station, TX 77843, USA; yuping520@tamu.edu

[2] Institute of Urban Agriculture, Chinese Academy of Agricultural Sciences, Chengdu 610000, China; huanglan_92@163.com

[3] Department of Horticultural Sciences, Texas A&M AgriLife Extension Service, 2134 TAMU, College Station, TX 77843, USA; qianshengli@tamu.edu

[4] USDA, Agriculture Research Service, New Orleans, LA 70124, USA; Isabel.Lima@ars.usda.gov

[5] USDA, Agriculture Research Service, Houma, LA 70360, USA; Paul.White@ars.usda.gov

* Correspondence: mgu@tamu.edu; Tel.: +1-979-845-8567

Received: 31 December 2019; Accepted: 20 January 2020; Published: 22 January 2020

Abstract: Biochar (BC) has the potential to replace bark-based commercial substrates in the production of container plants. A greenhouse experiment was conducted to evaluate the potential of mixed hardwood biochar (HB) and sugarcane bagasse biochar (SBB) to replace the bark-based commercial substrate. A bark-based commercial substrate was incorporated with either HB at 50% (vol.) or SBB at 50% and 70% (vol.), with a bark-based commercial substrate being used as the control. The total porosity (TP) and container capacity (CC) of all SBB-incorporated mixes were slightly higher than the recommended value, while, the others were within the recommended range. Both tomato and basil plants grown in the BC-incorporated mixes had a similar or higher growth index (GI), leaf greenness (indicated by soil-plant analyses development), and yield than the control. The leachate of all mixes had the highest NO_3–N concentration in the first week after transplantation (1 WAT). All BC-incorporated mixes grown with both tomato and basil had similar NO_3–N concentration to the control (except 50% SBB at 1 and 5 WAT, and 50% HB at 5 WAT with tomato plants; 50% SBB at 5 WAT with basil plants). In conclusion, HB could replace bark-based substrates at 50% and SBB at 70% for both tomato and basil plant growth, without negative effects.

Keywords: biochar; NO_3–N; plant; substrate; container; production

1. Introduction

Both tomato and basil are important crops and 95% of tomato and basil are produced in soilless cultivation systems using different horticultural growing substrates [1]. Tomato is one of the most important horticulture crops, with a total production estimated to be at 164 MT worldwide [2]. Tomato can be grown in coconut fiber, and perlite alone or in mixture with peat, and produce good yields [3]. Additionally, 50% coco–peat mixed with 50% perlite was recommended for tomato seedling production [4]. Basil is an annual herb that is commercially important for its medical and culinary purposes [5,6]. Basil plants can be grown in 75% sphagnum peat moss mixed with 25% coarse perlite [7]. Additionally, the mix of 60% sphagnum peat and 10% biochar with compost, has proven to be suitable for basil production [8].

Container plant production has become a major source of N leaching and runoff that can be a potential contamination source [9,10]. Container plant production requires a large amount of fertilizer, with nitrogen as the key component, making container plant production a major source of N leaching or runoff [9]. The leachate of N can be a potential contamination source for surface and underground water, resulting in environmental and health concerns [11]. NO_3–N, the main form for plants absorption, contributes in large to the N leaching and runoff in soilless production systems.

Bark has become one of the most commonly used container organic components in horticulture [12]. The reason for bark being commonly used in horticulture is because it has suitable properties for container plants to grow well and it is easy to get access to [13,14]. Compared to peat moss, another most commonly used container component, bark, is a byproduct of the forestry industry, is less expensive because it is available locally and does not require extra shipping costs [15,16]. In the USA, Douglas fir bark is mainly used in the pacific northwest, while pine bark is mainly used in the southwest [17,18].

Although bark has been a good container component, besides peat moss, its inconstant and unpredictable supply in recent years has limited its usage in horticulture industry [16,19,20]. Bark supply competes with many other markets, including alternatives of industrial fuel, timber production, housing and paper market, all of which prevent bark from being a constant source for the horticulture industry [20–22]. Since the supply of bark is fluctuating and unpredictable, it would be beneficial for the horticulture industry to explore less expensive and more constant alternatives with similar properties [16,22].

Biochar (BC), a by-product from thermochemical biomass decomposition under an oxygen-depleted or oxygen-limited environment [23–25] with specific time and temperature conditions and from certain carbon-rich raw materials, can be a potential alternative to common substrates for plant growth, as has been documented in many trials [16,26–29]. Research has shown that BC can increase water and nutrient holding capacity, ameliorate substrate acidity, and provide suitable environments for plants [30–32]. It, thus, improves greenhouse crop growth, yield, and quality, under appropriate conditions [32–36].

Biochar has been considered to be a sustainable component of a growing substrate because it can be derived from various agriculture by-products, such as green waste [33], wood, straw [31,37–40], bark [41], rice hull [42], and wheat straw [31,43]. Additionally, due to the significant variation in pyrolysis conditions, the BC properties could vary significantly, and there is no universal standard for BC addition to plant production and BC's effects on container substrates vary, as a result [28]. Research on BC as a substrate amendment is still in its infant stage [29]. In this present study, a trial was conducted to determine whether two types of BCs had the potential to be a replacement of bark-based substrate amendments for container plant production.

2. Materials and Methods

2.1. Plant Material

Plant seeds (tomato, *Solanum lycopersicum* 'Red Robin™', Fred C. Gloeckner and Company Inc., Harrison, NY, USA; basil, *Ocimum basilicum*, Johnny's Selected Seeds, Winslow, ME, USA) were sown in 72-cell plug trays (one seed per cell, cell dimension: 5 cm*4 cm*4 cm, depth/length/width; volume: 55 mL) with a commercial germination substrate (BM2 Berger, Saint-Modeste, Quebec, Canada), on 26 February 2019. After the first pair of true leaves expanded, uniform seedlings were transplanted into 6-inch azalea pots (dimension: 10.8 cm* 15.5 cm*11.3 cm, depth/top/bottom diameter; volume: 1330 mL) with a commercial growing substrate (Jolly Gardener, Oldcastle Lawn & Garden Inc., Atlanta, GA, USA) that was incorporated with either sugarcane bagasse biochar (SBB, American Biocarbon LLC White Castle, LA, USA) at two different rates (50% and 70%; by vol.), or with mixed hardwood biochar (HB, Proton Power Inc. Lenoir City, TN, USA) at 50% (by vol.), on 27 March 2019.

The composition used in this study was chosen because a previous study had showed that 70% of HB can be successfully incorporated with peat moss based commercial substrates and with composts for tomato and basil production [29], and 50% of SBB can be used for petunia growth (not published). We wanted to do further tests of HB with different compositions, on tomato and basil, using tests of

SBB with different plant species. The main components for the commercial growing substrate was aged pine bark (55%; by vol.), the other ingredients in the substrate were Canadian sphagnum peat moss, perlite, and vermiculite. The commercial substrate was used as the control. The pH of SBB and of HB were 5.9 and 10.1, respectively (Table 1). The SBB and HB had electrical conductivity (EC) of 753 µS/cm and 1,058 µS/cm, respectively [44]. During transplanting, slow-release fertilizer Osmocote Plus (15N-4P-10K, Scotts-Sierra Horticultural Products Company, Marysville, OH, USA) was surface-dressed at the rate of 4.8 g/pot for basil and 7.7 g/pot for tomato. All mixes were placed in a greenhouse at Texas A&M University, College Station, TX, USA. The average greenhouse temperature, relative humidity, and dew point were 23.7 °C, 82%, and 19.6 °C, respectively.

Table 1. The pH, electrical conductivity (EC), total porosity (TP), container capacity (CC), air space (AS), and bulk density (BD) of biochars and the substrate mixes used in this study.

Composition	pH	EC µS/cm	TP%	CC %	AS %	BD g/cm^3
SBB	5.9	753	74	71	3	0.11
HB	10.1	1058	87	66	20	0.13
50%SBB + 50%CS	6.3	2073	81	75	7	0.13
50%HB + 50%CS	7.5	1370	78	62	17	0.13
70%SBB + 30%CS	6.4	1830	89	76	13	0.14
CS	6.5	1819	97	85	12	0.15
Suitable range ^{Z}	-	-	50–80	45–65	10–30	0.19–0.7

Note: SBB = Sugarcane Bagasse Biochar; HB = Mixed hardwood Biochar; and CS = Commercial bark-based growing mix; ^{Z} Recommended physical properties of container substrate by Yeager et al. [45].

2.2. Measurements

2.2.1. Potting Mix Physical and Chemical Properties

Mix physical properties—total porosity (TP), container capacity (CC), air space (AS), and bulk density (BD)—were measured according to North Carolina State University Horticultural Substrates Laboratory Porometer [46]. The leachate EC and pH were measured every other week, starting at one week after transplantation (1 WAT), with a portable EC/pH meter (Hanna Instrument, Woonsocket, RI, USA), according to the pour-through method [47].

Nutrient leachate was collected whenever EC and pH were measured and was stored in the refrigerator (4 °C) until analysis. A HQ440d Benchtop Meter and ISENO3181 nitrate electrode (Hach Company, Loveland, CO, USA) were used for leachate NO_3–N measurements.

2.2.2. Plant Growth

Plant height and two widest canopy widths (width 1: horizontal, width 2: perpendicular) were measured at 1, 3, 5, and 7 WAT. The plant growth index (GI) was calculated according to the formula: GI = plant height/2 + (width 1 + width 2)/4 [26]. Plants' leaf greenness was measured at 1 WAT with a portable soil-plant analyses development (SPAD) meter, (SPAD 502 Plus Chlorophyll Meter, Spectrum Technologies, Inc., Plainfield, IL, USA). Each plant's leaf greenness was determined by taking averages of readings from three random mature leaves. Plant stem, leaf, and fruit were harvested separately. After being dried at 80 °C in an oven until a consistent weight was reached, their dry weights (shoot dry weight (SDW), leaf dry weight (LDW), fruit dry weight (FDW)) were measured. Plant roots were washed under running water, after harvest. Root length, root surface area, average root diameter, and the number of root tips were measured by using a root scanner (WinRHIZO, Regent Instruments Canada Inc., Quebec, Canada). Root dry weights (RDW) were determined after being dried at 80 °C in an oven, until a constant weight was reached. Total dry weights (TDW) were calculated by adding up the SDW, LDW, FDW, and RDW.

2.3. Statistical Analysis

This experiment was designed as a completely randomized block design with six replications for each mix. A one-way analysis of variance using JMP Statistical Software (version Pro 14.2.0; SAS Institute, Cary, NC, USA) was used for data analysis. All the means were separated by using Dunnett's test when treatments were significantly different from control at $p < 0.05$. A principle component analysis (PCA) was conducted to evaluate the relationship between the selected variables and were treated using R programing software (version 3.5.1).

3. Results

3.1. Potting Mix Physical and Chemical Properties

Most of the mixes' physical properties were within the recommended range [45], except for the SBB-incorporated mixes, which had a slightly higher TP and CC than the recommended value (Table 1). The 50% SBB mix had a slightly lower AS, as compared to the recommended value. All the mixes had slightly lower BD in comparison to the recommended value and the commercial mix had the lowest BD among all the mixes.

Tomato and basil plants grown in all BC-incorporated pots had similar EC as compared to the control, throughout the experiment, except for the tomato plants in 50% HB at 1 WAT (Figure 1). The 50% HB mixes with tomato plants had a significantly higher pH than the control at 1, 3, and 7 WAT (Figure 2A). The SBB-incorporated mix with tomato plants (50% at 1 WAT, 70% SBB at 7 WAT) had a significantly lower pH, compared to the control. Plants in all the other BC-incorporated mixes had a similar pH, throughout the experiment. Basil plants grown in 50% HB mixes had a significantly higher pH compared to the control, throughout the experiment (Figure 2B). However, basil plants grown in SBB-incorporated mixes (50% and 70%, at 5 and 7 WAT) had a significantly lower pH, compared to the control.

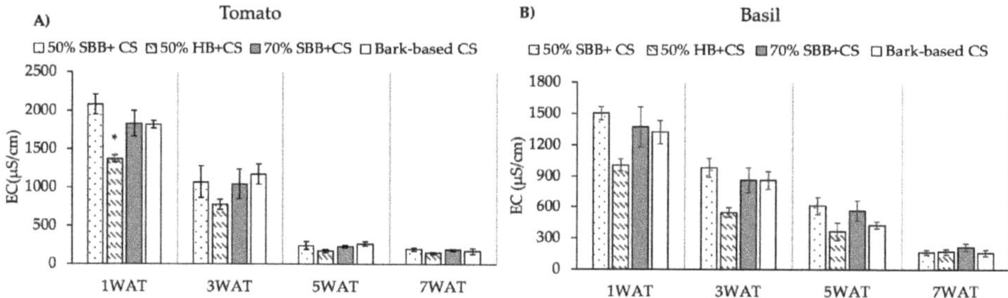

Figure 1. The EC (mean ± standard error) of potting mixes with 50% sugarcane bagasse biochar (SBB), 50% mixed hardwood biochar (HB), and 70% SBB (by vol.) mixed with bark-based commercial substrate (CS) with tomato (**A**) and basil (**B**) plants at 1, 3, 5, and 7 week(s) after transplanting (WAT). *indicated significant differences from CS using Dunnett's test at $p \leq 0.05$.

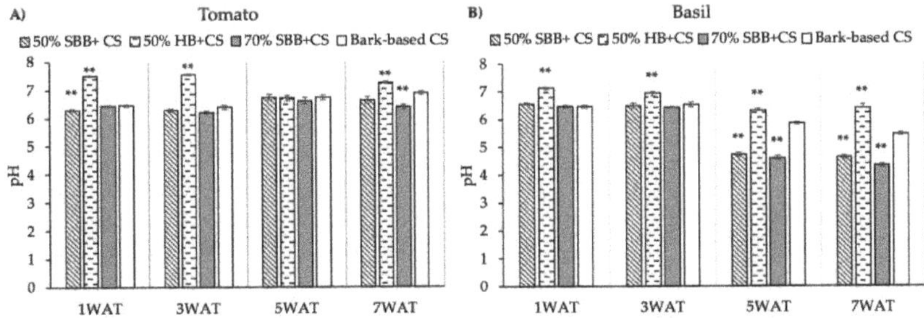

Figure 2. The pH (mean ± standard error) of container mixes, with 50% sugarcane bagasse biochar (SBB), 50% mixed hardwood (HB), and 70% SBB (by vol.) mixed with bark-based commercial substrate (CS) grown with tomato (**A**) and basil (**B**) plants at 1, 3, 5, and 7 week(s) after transplantation (WAT). **indicated significant differences from CS using Dunnett's test at $p \leq 0.01$.

3.2. Leachate NO$_3$–N

The leachate of all BC-incorporated mixes (both with tomato and basil plants) had a similar or higher NO$_3$–N concentration compared to the control. The leachate NO$_3$–N concentration generally decreased from 1 WAT to 7 WAT, for each mix (Figure 3).

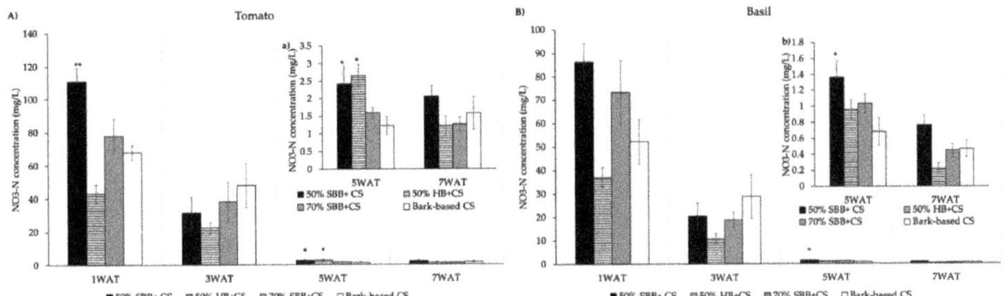

Figure 3. Leachate NO$_3$–N (mean ± standard error) of tomato (**A**) and basil (**B**) plants grown in container mixes with 50% (by vol.) sugarcane bagasse biochar (SBB), 50% mixed hardwood biochar (HB), and 70% SBB mixed with bark-based commercial substrate (CS). (**A,B**) Amplified figure for tomato (**a**) and basil (**b**) from 5 WAT to 7 WAT. *, **indicated significant differences from CS using Dunnett's test at $p \leq 0.05$ and $p \leq 0.01$, respectively.

3.3. Plant Growth

In the BC-incorporated mixes, both tomato and basil plants had a similar or higher GI, in comparison to the control, throughout the experiment (Figure 4). Tomato plants in all BC-incorporated mixes had similar SDW and FDW (yield), compared to the control, however, tomato plants in SBB-incorporated mixes had significantly lower TDW, RDW, and LDW compared to the control (Figure 5A). Basil plants grown in all BC-incorporated mixes had similar RDW, SDW (except 50% HB), LDW, FDW, and TDW to the control (Figure 5B). The SPAD of tomato and basil plants grown in all BC-incorporated mixes was no different from the control (Figure 6).

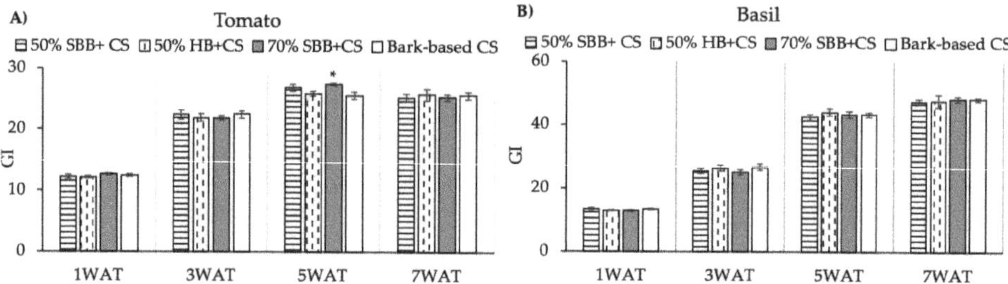

Figure 4. Growth index (mean ± standard error) of plants tomato (**A**) and basil (**B**) grown in container mixes with 50% sugarcane bagasse biochar (SBB), 50% mixed hardwood biochar (HB), and 70% SBB (by vol.) mixed with bark-based commercial substrate (CS) at 1, 3, 5, and 7 week(s) after transplantation (WAT). *indicated significant differences from CS, using Dunnett's test at $p \leq 0.05$.

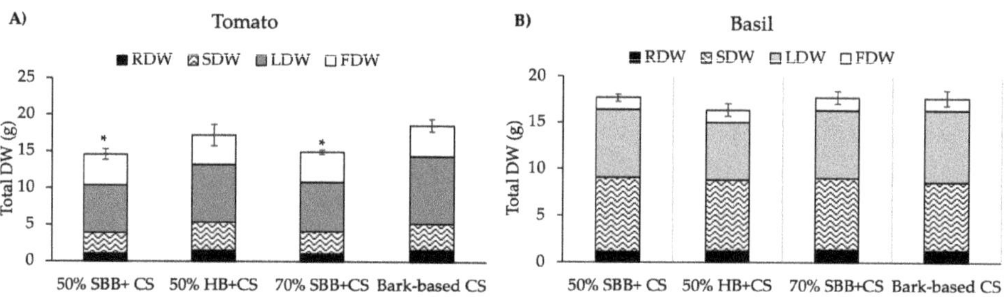

Figure 5. Total dry weight (Total DW = root dry weight (RDW) + shoot dry weight (SDW) + leave dry weight (LDW) + fruit dry weight (FDW); mean ± standard error) of tomato (**A**) and basil (**B**) grown in container mixes with 50% sugarcane bagasse biochar (SBB), 50% mixed hardwood biochar (HB), and 70% SBB (by vol.) mixed with bark-based commercial substrate (CS). *indicated significant differences on the total DW from CS using Dunnett's test at $p \leq 0.05$.

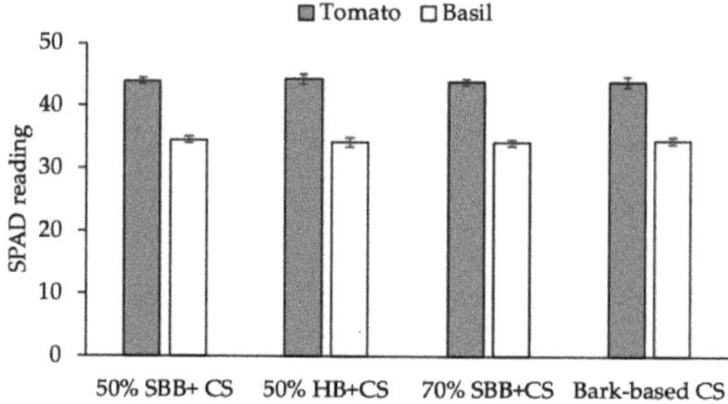

Figure 6. The soil-plant analyses development (SPAD) (mean ± standard error) of tomato and basil grown in container mixes with 50% sugarcane bagasse biochar (SBB), 50% mixed hardwood biochar (HB), and 70% SBB (by vol.), mixed with bark-based commercial substrate (CS).

Similar root length, average root diameter, and number of root tips were observed between tomato plants grown in all BC-incorporated mixes and the control (except 50% SBB), however, significantly smaller root surface area of tomato plants grown in all SBB-incorporated mixes were noticed (Table 2). Basil plants grown in all BC-incorporated mixes had significantly shorter root length but bigger diameter than the control. Basil plants in all BC-incorporated mixes had similar root surface area to the control, yet those in 50% BC-incorporated mixes had significantly fewer root tips than the control (Table 2).

Table 2. The root development (mean ± standard error) of plants grown in potting mixes with 50% sugarcane bagasse biochar (SBB), 50% mixed hardwood biochar (HB), and 70% SBB (by vol.) mixed with bark-based commercial substrate (CS). *, **, and ***indicated significant differences from CS using Dunnett's test at $p \le 0.05$, $p \le 0.01$, and $p \le 0.001$, respectively.

Mixes	Root Length (cm)	Root Surface Area (cm^2)	Average Root Diameter (mm)	Number of Root Tips
		Tomato		
50%SBB + 50%CS	1214 ± 60	442 ±37 *	1.2 ± 0.1	2650 ± 94 *
50%HB + 50%CS	1454 ± 67	557 ± 24	1.2 ± 0.1	3349 ± 171
70%SBB + 30%CS	1234 ± 74	421 ± 25 *	1.1 ± 0.1	2970 ± 196
CS	1324 ± 40	543 ± 19	1.3 ± 0.1	3227 ± 157
		Basil		
50%SBB + 50%CS	1415 ± 48 ***	819 ± 18	1.9 ± 0.1 ***	3092 ± 166 **
50%HB + 50%CS	1887 ± 117 *	866 ± 23	1.5 ± 0.1 *	3006 ± 149 **
70%SBB + 30%CS	1850 ± 115 *	870 ± 19	1.5 ± 0.1 *	3528 ± 222
CS	2240 ± 74	832 ± 26	1.2 ± 0.0	4003 ± 80

4. Discussion

4.1. Potting Mix Physical and Chemical Properties

Despite the fact that BC can have various effects on substrate properties contingent on the types of feedstocks and the pyrolysis conditions of BC [28,48], many types of BC have been proven to be suitable replacements for commercial growing substrates, without negatively affecting the plant [28,35]. Biochar from fast pyrolysis (pinewood, 450 °C), for instance, could replace commercial substrate at up to 80%, providing suitable properties for the poinsettia and Easter lily growth [26,27]. Biochar from fast pyrolysis (mixed hardwood) could be suitable for tomato and basil plant growth, due to the proper properties it created [29]. Sugarcane bagasse BC and pinewood BC mixes had similar physical properties to commercial growing mix, allowing them to be acceptable for bean and cucurbit seedlings production, even though some of the TP and CC in the SBB-incorporated mixes were slightly higher than the recommended values [44]. Adding pruning residue BC (fast pyrolysis, 500 °C) to soilless mixes can render appropriate physical properties for vegetable production [35,49]. In this study, even though 50% SBB and 70% SBB mixes had slightly higher TP (81%, 89%, respectively) and CC (75%, 76%, respectively) than the recommended value (TP 50%–80% and CC 45%–65%) [45], the growth of tomato and basil plants was not affected, as observed in Webber's study [44].

Different initial BC pH (HB: 10.05, SBB: 5.94) resulted in differences in pH levels in the different BC mixes. Mixes with HB (50%, by vol.) and commercial bark-based substrates (initial pH: 6.81) had a pH lower than the initial HB but higher than the initial commercial bark-based substrate. The same was true for all SBB mixes. Since SBB had an acidic initial pH, adding 30% to 50% of the commercial substrate (pH: 6.81) resulted in mixes with a pH that was lower than the commercial substrate but was higher than the SBB.

4.2. Biochar Effects on Leachate NO_3–N

Plant species, plant stage, and substrate properties can influence NO_3–N leaching [9,50,51]. Tomato, as a heavy feeder fertilizer crop, require more nutrients throughout the growing season than other lighter feeder fertilizer crops, such as snapdragon and bedding plants [52,53]. As a result of administering the same amount of fertilizer to different plant species due to their divergent nutrient requirements, the final NO_3–N leaching varies. Additionally, the nutrients demand for plant at different stages also vary. During the growing period, plants' requirement for nutrients presents a skewed "s" curve—vegetative periods need less nutrient yet when entering the flowering/fruit-set period, the demand for nutrients increases dramatically [54]. Nitrate leaching can be also affected by soil or substrate texture and normally, coarse textured mixtures lead to more nitrate leaching [55]. Substrate properties affecting nitrate leaching can explain why leachate from 50% HB (in both case of tomato and basil) had the lowest NO_3–N concentration (except tomato at 5 WAT), among all mixes.

4.3. Biochar Effects on Plants Growth

Biochar can have positive, null, and negative effects on plant growth [26,56,57], contingent on plant species, BC types, incorporation rates, and the interactions of both. For instance, pinewood BC had positive effects on bell pepper growth [58], similar results were reported on Easter lily, poinsettia, and "Firework" *Gomphrena*. Mixed hardwood BC can positively affect tomato and basil plants growth [16,26,27,29]. The null and negative effects of BC (from tomato crop waste or wood pellet) on tomato plant growth have also been reported [56,57]. This study obtained similar results to some previous studies that found that BC does not negatively affect plant growth at high incorporation rates [16,26,27,29].

There are few studies with detailed information on BC–root systems [59]. Since roots are essential parts for water and nutrients uptake, plants with better roots were desired [59,60], and the effects of BC on root development is an eventuality. In this study, tomato plants grown in all the BC-incorporated mixes had similar root length, root surface area (except 50% and 70% SBB), average root diameter, and number of tips, in comparison to the control. Basil plants had similar root surface area to the control, which can explain why plants grown in BC-incorporated mixes performed as well as those in the control.

4.4. Treatment Factors Determined Plants and Mix Properties

As the effect of biochar on plants and mix properties can be complex and difficult to explain, given the fact that two types of biochars and multiple variables were included in this study, a principal component analysis (PCA) was used to depict variables shaped by different biochars with tomato (Figure 7A) and basil (Figure 7B) plants. For tomato plants, 88.9% of the variability was explained by the first two components (Figure 7A). PC1 accounted for 65.8% variance, with SBB differing from HB and CS. Sugarcane bagasse biochar was associated more with yield (FDW) and NO_3–N leaching, while CS and HB was related more to plant growth (RDW, LDW, and GI). PC2 accounted for 23.1% variance, distinguishing the CS and BC mixes. Commercial substrate tended to be affiliated with plant biomass, however, BC mixes appeared to be related to nutrient leaching. For basil plants, the first two components explained 77.1% of the variability (Figure 7B). PC1 accounted for 42.9% variance, SBB 50% differing from HB and CS mixes. A 50% sugarcane bagasse biochar mix showed a greater association with NO_3–N leaching and SDW, while CS, 70% SBB, and HB showed a greater relation to plant growth, including root parameters (RDW, root length (RL), root tip (RT), and root surface area (RSA)) and chemical properties of the mixes (EC, pH). PC2 accounted for 34.2% variance, distinguishing between the CS and BC mixes. Commercial substrates tended to affiliated with plant biomass, however, BC mixes appeared to be related to the chemical properties of the mixes (EC, pH, NO_3–N).

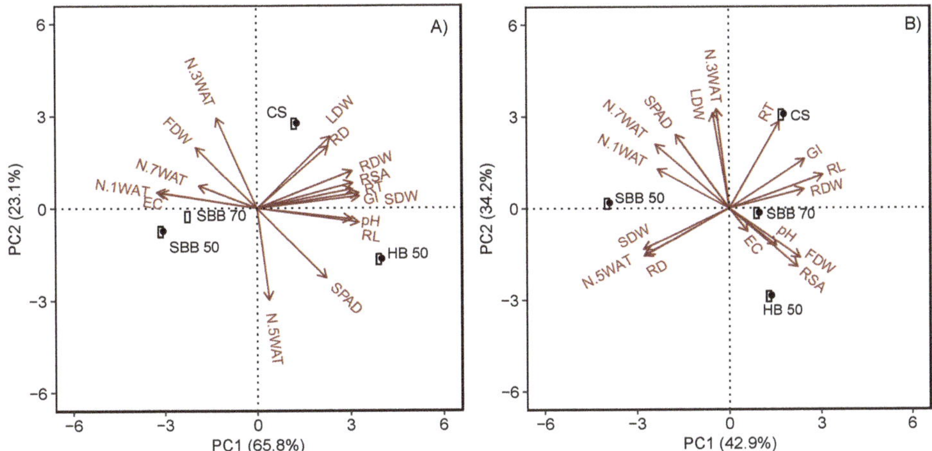

Figure 7. Principal component analysis (PCA) depicting the relationships between selected variables and treatment factors with tomato (**A**) and basil (**B**). Selected variables are displayed by arrows and include plant growth parameters—SPAD, growth index (GI), fruit dry weight (FDW), leave dry weight (LDW), shoot dry weight (SDW), root length (RL), root dry weight (RDW), root diameter (RD), root surface area (RSA), and number of root tips (RT); substrate chemical parameters were pH, EC, and NO_3–N leachate at different weeks. Treatment factors are displayed by filled grey circles: 50% sugarcane bagasse biochar (SBB 50), 50% mixed hardwood biochar (HB 50), 70% SBB (SBB 70) mixed with bark-based commercial substrate, and bark-based commercial substrate (CS).

5. Conclusions

The mixed hardwood biochar and sugarcane bagasse biochar used in this experiment could be used as bark-based substrate amendments for container plant production. The mixed hardwood biochar could replace the bark-based substrate at 50% and the sugarcane bagasse biochar at 70%, as growing mixes for tomato and basil production. More than 5.4 M ft^3 container substrates were used in horticulture industry in 2017 and the current container substrate major components—peat moss and bark are causing serious environmental concerns [61]. As can be seen from the results of this study, if mixed hardwood biochar or sugarcane bagasse biochar was chosen for greenhouse production, around 1.35 M ft^3 fewer peat moss or 1.94 M ft^3 fewer bark could be used annually (assuming container substrate contains 50% peat moss or bark).

Author Contributions: P.Y. conducted the experiments, collected and analyzed the data, and wrote the manuscript with assistance from all other authors, mainly M.G., L.H. and Q.L. provided technical advice and assistance when the study was conducted. I.M.L. and P.M.W. revised and improved the manuscript. All authors have read and agreed to the published version of the manuscript.

Funding: This research received no external funding.

Acknowledgments: We would like to thank American Biocarbon LLC White Castle, Louisiana, Proton Power Inc. Lenoir City, Tennessee, for supplying the biochar for the experiment; Elizabeth Pierson and Kevin Crosby, for supplying experimental instruments; Patricia Goodson; all the students in CEHD 603 fall 2019 writing class; Charles L. Webber III and the Agriculture Women Excited to Share Opinions, Mentoring and Experiences (AWESOME) faculty group of the College of Agriculture and Life Sciences at Texas A&M University for assistance with editing the manuscript. The author would also like to thank the Texas A&M University Open Access to Knowledge Fund (OAK Fund, supported by the University Libraries and the Office of the Vice President for Research) for partially covering the open access publication fees.

Conflicts of Interest: The authors declare no conflict of interest.

References

1. Grunert, O.; Hernandez-Sanabria, E.; Vilchez-Vargas, R.; Jauregui, R.; Pieper, D.H.; Perneel, M.; Van Labeke, M.-C.; Reheul, D.; Boon, N. Mineral and organic growing media have distinct community structure, stability and functionality in soilless culture systems. *Sci. Rep.* **2016**, *6*, 18837. [CrossRef] [PubMed]
2. Rodríguez-Ortega, W.M.; Martínez, V.; Nieves, M.; Simón, I.; Lidón, V.; Fernandez-Zapata, J.; Martinez-Nicolas, J.; Cámara-Zapata, J.M.; García-Sánchez, F. Agricultural and physiological Responses of tomato plants Grown in Different Soilless Culture systems with saline WATer under Greenhouse Conditions. *Sci. Rep.* **2019**, *9*, 6733. [CrossRef] [PubMed]
3. Kılıc, P.; Erdal, I.; Aktas, H. Effect of different substrates on yield and fruit quality of tomato grown in soilless culture. *Infrastruktura i Ekologia Terenów Wiejskich* **2018**. [CrossRef]
4. Sedaghat, M.; Kazemzadeh-Beneh, H.; Azizi, M.; Momeni, M. Optimizing Growing Media for Enhancement to Vegetative Growth, Yield and Fruit Quality of Greenhouse Tomato Productionin Soilless Culture System. *World J. Agric. Sci* **2017**, *13*, 82–89.
5. Mairapetyan, S.; Alexanyan, J.; Tovmasyan, A.; Daryadar, M.; Stepanian, B.; Mamikonyan, V. Productivity, biochemical indices and antioxidant activity of peppermint (*Mentha piperita* L.) and basil (*Ocimum basilicum* L.) in conditions of hydroponics. *J. Aquac. Res. Dev* **2016**, *7*, 1–3. [CrossRef]
6. Saha, S.; Monroe, A.; Day, M.R. Growth, yield, plant quality and nutrition of basil (*Ocimum basilicum* L.) under soilless agricultural systems. *Ann. Agric. Sci.* **2016**, *61*, 181–186. [CrossRef]
7. Currey, C.J.; Flax, N.J.; Litvin, A.G.; Metz, V.C. Substrate Volumetric WATer Content Controls Growth and Development of Containerized Culinary Herbs. *Agronomy* **2019**, *9*, 667. [CrossRef]
8. Nobile, C.; Denier, J.; Houben, D. Linking biochar properties to biomass of basil, lettuce and pansy cultivated in growing media. *Sci. Hortic.* **2019**, *261*, 109001. [CrossRef]
9. Chen, J.; Wei, X. Controlled-Release Fertilizers as a Means to Reduce Nitrogen Leaching and Runoff in Container-Grown Plant Production. *Nitrogen Agric. Updates* **2018**, *33*. [CrossRef]
10. Sun, H.; Lu, H.; Chu, L.; Shao, H.; Shi, W. Biochar applied with appropriate rates can reduce N leaching, keep N retention and not increase NH3 volatilization in a coastal saline soil. *Sci. Total Environ.* **2017**, *575*, 820–825. [CrossRef]
11. Savci, S. An agricultural pollutant: Chemical fertilizer. *Int. J. Environ. Sci. Dev.* **2012**, *3*, 73. [CrossRef]
12. Bilderback, T.; Boyer, C.; Chappell, M.; Fain, G.; Fare, D.; Gilliam, C.; Jackson, B.; Lea-Cox, J.; LeBude, A.; Niemiera, A. *Best Management Practices: Guide for Producing Nursery Crops*; Southern Nursery Association: Acworth, Goregia, 2013.
13. Ngaatendwe, M.; Ernest, M.; Moses, M.; Tuarira, M.; Ngenzile, M.; Tanyaradzwa, Z. Use of vermicompost as supplement to pine bark for seedling production in nurseries. *World J. Agric. Res.* **2015**, *3*, 123–128.
14. El Sharkawi, H.M.; Ahmed, M.A.; Hassanein, M.K. Development of treated Rice Husk as an alternative substrate medium in cucumber soilless culture. *J. Agric. Environ. Sci.* **2014**, *3*, 131–149. [CrossRef]
15. Choi, H.-S.; Zhao, Y.; Dou, H.; Cai, X.; Gu, M.; Yu, F. Effects of biochar mixtures with pine-bark based substrates on growth and development of horticultural crops. *Hortic. Environ. Biotechnol.* **2018**, *59*, 345–354. [CrossRef]
16. Gu, M.; Li, Q.; Steele, P.H.; Niu, G.; Yu, F. Growth of 'Fireworks' gomphrena grown in substrates amended with biochar. *J. Food Agric. Environ.* **2013**, *11*, 819–821.
17. Buamscha, M.G.; Altland, J.E.; Sullivan, D.M.; Horneck, D.A. Micronutrient availability in fresh and aged Douglas fir bark. *HortScience* **2007**, *42*, 152–156. [CrossRef]
18. Torres-Quezada, E.A.; Santos, B.M.; Zotarelli, L.; Treadwell, D.A. Soilless Media and Containers for Bell Pepper Production. *Int. J. Veg. Sci.* **2015**, *21*, 177–187. [CrossRef]
19. Wright, R.D.; Jackson, B.E.; Barnes, M.C.; Browder, J.F. The landscape performance of annual bedding plants grown in pine tree substrate. *HortTechnology* **2009**, *19*, 78–82. [CrossRef]
20. Cole, D.M.; Sibley, J.L.; Blythe, E.K.; Eakes, D.J.; Tilt, K.M. Evaluation of cotton gin compost as a horticultural substrate. In *A Research Paper Presented at the Southern Nursery Association Researcher's Conference*; Department of Horticulture, Auburn University: Auburn, AL, USA, 2002; Volume 47, pp. 264–276.
21. Haynes, R.W. *An Analysis of the Timber Situation in the United States: 1952 to 2050*; Gen. Tech. Rep. PNW-GTR-560; US Department of Agriculture Forest Service Pacific Northwest Research Station: Corvallis, OR, USA, 2003; Volume 560, p. 254.

22. Lu, W.; Sibley, J.L.; Gilliam, C.H.; Bannon, J.S.; Zhang, Y. Estimation of US bark generation and implications for horticultural industries. *J. Environ. Hortic.* **2006**, *24*, 29–34.

23. Demirbas, A.; Arin, G. An overview of biomass pyrolysis. *Energy Sour.* **2002**, *24*, 471–482. [CrossRef]

24. Lehmann, J. A handful of carbon. *Nature* **2007**, *447*, 143–144. [CrossRef] [PubMed]

25. Nartey, O.D.; Zhao, B. Biochar preparation, characterization, and adsorptive capacity and its effect on bioavailability of contaminants: An overview. *Adv. Mate. Sci. Eng.* **2014**, *2014*, 715398. [CrossRef]

26. Guo, Y.; Niu, G.; Starman, T.; Volder, A.; Gu, M. Poinsettia Growth and Development Response to Container Root Substrate with Biochar. *Horticulturae* **2018**, *4*, 1. [CrossRef]

27. Guo, Y.; Niu, G.; Starman, T.; Gu, M. Growth and development of Easter lily in response to container substrate with biochar. *J. Hortic. Sci. Biotechnol.* **2018**, *94*, 80–86. [CrossRef]

28. Huang, L.; Gu, M. Effects of Biochar on Container Substrate Properties and Growth of Plants—A Review. *Horticulturae* **2019**, *5*, 14. [CrossRef]

29. Huang, L.; Niu, G.; Feagley, S.E.; Gu, M. Evaluation of a hardwood biochar and two composts mixes as replacements for a peat-based commercial substrate. *Ind. Crop Prod.* **2019**, *129*, 549–560. [CrossRef]

30. Dumroese, R.K.; Heiskanen, J.; Englund, K.; Tervahauta, A. Pelleted biochar: Chemical and physical properties show potential use as a substrate in container nurseries. *Biomass Bioenergy* **2011**, *35*, 2018–2027. [CrossRef]

31. Vaughn, S.F.; Kenar, J.A.; Thompson, A.R.; Peterson, S.C. Comparison of biochars derived from wood pellets and pelletized wheat straw as replacements for peat in potting substrates. *Ind. Crops Prod.* **2013**, *51*, 437–443. [CrossRef]

32. Zhang, L.; Sun, X.-Y.; Tian, Y.; Gong, X.-Q. Biochar and humic acid amendments improve the quality of composted green waste as a growth medium for the ornamental plant Calathea insignis. *Sci. Hortic.* **2014**, *176*, 70–78. [CrossRef]

33. Tian, Y.; Sun, X.; Li, S.; Wang, H.; Wang, L.; Cao, J.; Zhang, L. Biochar made from green waste as peat substitute in growth media for Calathea rotundifola cv. Fasciata. *Sci. Hortic.* **2012**, *143*, 15–18. [CrossRef]

34. Méndez, A.; Cárdenas-Aguiar, E.; Paz-Ferreiro, J.; Plaza, C.; Gascó, G. The effect of sewage sludge biochar on peat-based growing media. *Biol. Agric. Hortic.* **2017**, *33*, 40–51. [CrossRef]

35. Nieto, A.; Gascó, G.; Paz-Ferreiro, J.; Fernández, J.; Plaza, C.; Méndez, A. The effect of pruning waste and biochar addition on brown peat based growing media properties. *Sci. Hortic.* **2016**, *199*, 142–148. [CrossRef]

36. Headlee, W.L.; Brewer, C.E.; Hall, R.B. Biochar as a substitute for vermiculite in potting mix for hybrid poplar. *Bioenergy Res.* **2014**, *7*, 120–131. [CrossRef]

37. Hansen, V.; Hauggaard-Nielsen, H.; Petersen, C.T.; Mikkelsen, T.N.; Müller-Stöver, D. Effects of gasification biochar on plant-available WATer capacity and plant growth in two contrasting soil types. *Soil Tillage Res.* **2016**, *161*, 1–9. [CrossRef]

38. Spokas, K.; Koskinen, W.; Baker, J.; Reicosky, D. Impacts of woodchip biochar additions on greenhouse gas production and sorption/degradation of two herbicides in a Minnesota soil. *Chemosphere* **2009**, *77*, 574–581. [CrossRef]

39. Hansen, V.; Müller-Stöver, D.; Ahrenfeldt, J.; Holm, J.K.; Henriksen, U.B.; Hauggaard-Nielsen, H. Gasification biochar as a valuable by-product for carbon sequestration and soil amendment. *Biomass Bioenergy* **2015**, *72*, 300–308. [CrossRef]

40. Spokas, K.A.; Baker, J.M.; Reicosky, D.C. Ethylene: Potential key for biochar amendment impacts. *Plant Soil* **2010**, *333*, 443–452. [CrossRef]

41. Hina, K.; Bishop, P.; Arbestain, M.C.; Calvelo-Pereira, R.; Maciá-Agulló, J.A.; Hindmarsh, J.; Hanly, J.; Macìas, F.; Hedley, M. Producing biochars with enhanced surface activity through alkaline pretreatment of feedstocks. *Soil Res.* **2010**, *48*, 606–617. [CrossRef]

42. Locke, J.C.; Altland, J.E.; Ford, C.W. Gasified rice hull biochar affects nutrition and growth of horticultural crops in container substrates. *J. Environ. Hortic.* **2013**, *31*, 195–202.

43. Xu, G.; Zhang, Y.; Sun, J.; Shao, H. Negative interactive effects between biochar and phosphorus fertilization on phosphorus availability and plant yield in saline sodic soil. *Sci. Total Environ.* **2016**, *568*, 910–915. [CrossRef] [PubMed]

44. Webber, C.L., III; White, P.M., Jr.; Gu, M.; Spaunhorst, D.J.; Lima, I.M.; Petrie, E.C. Sugarcane and Pine Biochar as Amendments for Greenhouse Growing Media for the Production of Bean (*Phaseolus vulgaris* L.) Seedlings. *J. Agric. Sci.* **2018**, *10*, 58.

45. Yeager, T.; Fare, D.; Lea-Cox, J.; Ruter, J.; Bilderback, T.; Gilliam, C.; Niemiera, A.; Warren, S.; Whitewell, T.; White, R. *Best Management Practices: Guide for Producing Container-Grown Plants*; Southern Nursery Association: Marietta, Georgia, 2007.

46. Fonteno, W.; Hardin, C.; Brewster, J. *Procedures for Determining Physical Properties of Horticultural Substrates Using the NCSU Porometer*; North Carolina State University: Raleigh, NC, USA, 1995.

47. LeBude, A.; Bilderback, T. Pour-through extraction procedure: A nutrient management tool for nursery crops. *N.C. Coop. Ext.* **2009**, 1–8.

48. Gell, K.; van Groenigen, J.; Cayuela, M.L. Residues of bioenergy production chains as soil amendments: Immediate and temporal phytotoxicity. *J. Hazard. Mate.* **2011**, *186*, 2017–2025. [CrossRef]

49. Webber, C.L., III; White, P.M., Jr.; Spaunhorst, D.J.; Lima, I.M.; Petrie, E.C. Sugarcane Biochar as an Amendment for Greenhouse Growing Media for the Production of Cucurbit Seedlings. *J. Agric. Sci.* **2018**, *10*, 104. [CrossRef]

50. Xu, L.; Niu, H.; Xu, J.; Wang, X. Nitrate-nitrogen leaching and modeling in intensive agriculture farmland in China. *Sci. World J.* **2013**, *2013*. [CrossRef]

51. Luce, M.S.; Whalen, J.K.; Ziadi, N.; Zebarth, B.J. Nitrogen dynamics and indices to predict soil nitrogen supply in humid temperate soils. In *Advances in Agronomy*; Elsevier: Amsterdam, The Netherlands, 2011; Volume 112, pp. 55–102.

52. Wang, X.; Xing, Y. Evaluation of the effects of irrigation and fertilization on tomato fruit yield and quality: A principal component analysis. *Sci. Rep.* **2017**, *7*, 350. [CrossRef]

53. Nelson, P.V. *Greenhouse Operation and Management*; Prentice Hall: Upper Saddle River, NJ, USA, 2012.

54. Badr, M.; Hussein, S.A.; El-Tohamy, W.; Gruda, N. Nutrient uptake and yield of tomato under various methods of fertilizer application and levels of fertigation in arid lands. *Gesunde Pflanzen* **2010**, *62*, 11–19. [CrossRef]

55. Vinten, A.; Vivian, B.; Wright, F.; Howard, R. A comparative study of nitrate leaching from soils of differing textures under similar climatic and cropping conditions. *J. Hydrol.* **1994**, *159*, 197–213. [CrossRef]

56. Vaughn, S.F.; Eller, F.J.; Evangelista, R.L.; Moser, B.R.; Lee, E.; Wagner, R.E.; Peterson, S.C. Evaluation of biochar-anaerobic potato digestate mixtures as renewable components of horticultural potting media. *Ind. Crops Prod.* **2015**, *65*, 467–471. [CrossRef]

57. Dunlop, S.J.; Arbestain, M.C.; Bishop, P.A.; Wargent, J.J. Closing the loop: Use of biochar produced from tomato crop green waste as a substrate for soilless, hydroponic tomato production. *HortScience* **2015**, *50*, 1572–1581. [CrossRef]

58. Liu, R.; Gu, M.; Huang, L.; Yu, F.; Jung, S.-K.; Choi, H.-S. Effect of pine wood biochar mixed with two types of compost on growth of bell pepper (*Capsicum annuum* L.). *Hortic. Environ. Biotechnol.* **2019**, *60*, 313–319. [CrossRef]

59. Prendergast-Miller, M.; Duvall, M.; Sohi, S. Biochar–root interactions are mediated by biochar nutrient content and impacts on soil nutrient availability. *Eur. J. Soil Sci.* **2014**, *65*, 173–185. [CrossRef]

60. Rellán-Álvarez, R.; Lobet, G.; Dinneny, J.R. Environmental control of root system biology. *Ann. Rev. Plant Biol.* **2016**, *67*, 619–642. [CrossRef]

61. USDA-NASS. *Agricultural Statistics*; USDA, Ed.; United States Government Printing Office Washington: Washington, DC, USA, 2018; pp. 202–210.

Article

The Use of Dewpoint Hygrometry to Measure Low Water Potentials in Soilless Substrate Components and Composites

Jeb S. Fields [1], William C. Fonteno [2], Brian E. Jackson [2,*], Joshua L. Heitman [3] and James S. Owen Jr. [4]

[1] Hammond Research Station, Louisiana State University Agricultural Center, 21549 Old Covington Hwy, Hammond, LA 70403, USA; JFields@agcenter.lsu.edu

[2] Department of Horticultural Science, North Carolina State University, 152 Kilgore Hall, Campus Box 7609, Raleigh, NC 27695, USA; wcf@ncsu.edu

[3] Department of Crop and Soil Sciences, North Carolina State University, 3410 Williams Hall, Campus Box 7620, Raleigh, NC 27695, USA; jlheitma@ncsu.edu

[4] Application Technology Research Unit, Agricultural Research Service, United States Department of Agriculture, 1680 Madison Ave., Wooster, OH 44691, USA; jim.owen@usda.gov

* Correspondence: brian_jackson@ncsu.edu

Received: 23 July 2020; Accepted: 10 September 2020; Published: 15 September 2020

Abstract: Plant water availability in soilless substrates is an important management consideration to maximize water efficiency for containerized crops. Changes in the characteristics (i.e., shrink) of these substrates at low water potential (<-1.0 MPa) when using a conventional pressure plate-base can reduce hydraulic connectivity between the plate and the substrate sample resulting in inaccurate measures of water retention. Soilless substrate components *Sphagnum* peatmoss, coconut coir, aged pine bark, shredded pine wood, pine wood chips, and two substrate composites were tested to determine the range of volumetric water content (VWC) of surface-bound water at water potentials between -1.0 to -2.0 MPa. Substrate water potentials were measured utilizing dewpoint hygrometry. The VWC for all components or composites was between 5% and 14%. These results were considerably lower compared to previous research (25% to 35% VWC) utilizing conventional pressure plate extraction techniques. This suggests that pressure plate measurements may overestimate this surface-bound water which is generally considered unavailable for plant uptake. This would result in underestimating available water by as much as 50%.

Keywords: available water; coconut coir; dewpoint potentiometer; peat; pine bark; pine tree substrate; substrate processing; surface-bound water; unavailable water; wood substrate

---

1. Introduction

Traditionally, substrate scientists separate the water storage capacity of a soilless substrate into two categories, available water (AW; water that is available for plant uptake) and unavailable water (water that is bound tightly to soil surfaces and is unavailable for plant uptake). Soil and substrate scientists separate the availability of water as a function of water potential, as water within the substrate matrix is held at various tensions by a combination of matric and gravitational potentials. To absorb water from the substrate matrix, plants exert suction which must overcome the water tension. As the substrate volumetric water content (VWC) and water potential decreases (tension increases) the water becomes less available for plant uptake. The water potential at which the substrate transitions from AW to unavailable is not exact, but instead plant water availability is gradually reduced as the substrate dries and substrate water potential becomes more negative [1]. Water that is less available for plant uptake is most often tightly bound to particle surfaces, known herein as surface-bound water (SBW).

Water typically becomes less available for common agricultural crops in soil at potentials between −1.0 and −2.0 MPa [2]. Often, soils and substrate researchers use a water potential of −1.5 MPa as the potential at which water becomes plant unavailable for calculation. It is understood that many plant species can survive in soils with water potentials well below −1.5 MPa. However, the change in actual water content as water potential becomes more negative with drying beyond −1.5 MPa is typically negligible, with miniscule losses in water content accounting for substantial drops in water potential thereafter. Denmead and Shaw [3] reported that plants started to reduce transpiration levels at water potentials as high as −0.2 MPa, and Caron et al. [4] indicated that horticultural crops grown in containerized peat-based soilless substrates begin to show stress signals at substrate water potentials as high as −0.003 MPa. However, utilizing substrate VWC at substrate water potentials of −1.5 MPa as an estimated transition value, scientists can estimate substrate AW as a proxy for substrate water storage capacity, the water held at water potentials between container capacity (CC; the maximum volume occupied by water in a soilless substrate after drainage) and the water held at substrate water potentials of −1.5 MPa [5]. This calculation is useful for practical management considerations and to compare substrates.

To determine relationships between soil (or substrate) water potential and VWC, Bouyoucos [6] described an apparatus that introduces suction upon soil samples. This idea was refined by Richards and Fireman [7] who applied pressure and employed the use of porous plates that allowed water to be extracted at a given pressure representing tension. The most commonly used and described method of measuring soil water potential in situ is through the use of tensiometers [8]. However, water potentials below approximately −0.085 MPa cannot be measured with tensiometers due to vaporization and cavitation of water within the device [9]. More recent work has extended the range of tensiometers by employing a polymer in place of water [10]; however, the range remains limited to soil water potentials much greater than −1.5 MPa.

Presently, the most commonly cited method for measuring moisture content (MC) at low water potentials in soilless substrates is the pressure outflow apparatus method described by Cassel and Nielsen [11]. The pressure outflow apparatus is a modified version of Richards and Fireman's pressure plates [7]. This method determines soil or substrate MC at a specified water potential and can be conducted in a relatively short period of time. However, previous research has pointed out inaccuracies with this method due to a loss of hydraulic connectivity between the water in the sample and the water in the plate when the soil or substrate dries [12–15]. This occurs because water moves in porous media primarily by displacement. Therefore, when the water column is interrupted, water movement ceases, preventing equilibrium between the applied pressure and the soil or substrate water potential.

Dewpoint hygrometry has also been used to measure water potentials of porous substances [16]. A dewpoint potentiometer utilizes hygrometry to determine water potentials of porous media via a chilled mirror to measure the dew point temperature in the headspace above a sample [17]. Recent research has shown the effectiveness of the dewpoint potentiometer for determining water potentials below −0.1 MPa for soils [18,19] and soilless substrates [20]. Moreover, dewpoint potentiometer measurements have demonstrated inaccuracies in pressure plate measurements for mineral soils [21,22]. Curtis and Claassen [23] compared dewpoint hygrometry to pressure plate measurements for inorganic amendments at water potentials of −1.5 MPa, demonstrating more precision with dewpoint hygrometry than with the pressure plates. Fields et al. [15] used dewpoint hygrometry to describe inaccuracies in measuring water retention of highly porous organic soilless substrate components with a pressure plate set at −1.5 MPa.

As improved water management continues to be an imperative focus for horticultural crop production, refining the characterization of water storage and availability in horticultural substrates is critical. Therefore, the objective of this research was to utilize dewpoint hygrometry to assess VWC of traditional soilless substrates at water potentials <−1.0 MPa. Additionally, we compared the VWC at substrate water potential values near −1.5 MPa measured through dewpoint hygrometry with values attained through other accepted methodologies in the literature and make inferences upon the viability of utilizing dewpoint hygrometry in soilless substrate science.

2. Materials and Methods

Preparation of Substrate Components and Composites

Substrate components tested were coconut coir pith (Densu Coir, Toronto, ON, Canada), horticultural grade Sphagnum peat moss (Premier Tech, Riviere-du-Loup, Quebec Canada), aged pine bark (PB; Pacific Organics, Henderson, NC, USA), pine wood chips (PWC), and shredded pine wood (SPW) with examples shown in Figure 1. The coir was hydrated from compressed bricks with tap water and then fluffed by hand to reconstitute the material. The peat was removed from the compressed bale, fluffed by hand, hydrated, and screened by hand through a 1.25 cm screen to prevent any larger aggregates (foreign debris) from being included in the sampling. Pine bark derived from harvested loblolly pine (*Pinus taeda* L.) trees was processed in a hammer mill through a 16 mm screen, windrowed, and allowed to age for nine months.

Figure 1. Examples of the base materials used in this research, including (**A**) Sphagnum peat moss; (**B**) coconut coir pith; (**C**) shredded pine wood; (**D**) pine wood chips; and (**E**) aged pine bark.

For the pine wood materials, 12-year old loblolly pine trees were harvested at ground level, de-limbed, and processed through either a wood chipper or a wood shredder with bark intact. The pine trees used to create PWC were harvested on 9 December 2011 and processed through a DR Chipper (18 HP DR Power Equipment, Model 356447; Vergennes, VT, USA) on 3 January 2012 to produce the coarse wood chips that were then hammer-milled (Meadows Mills, North Wilkesboro, NC, USA) through a 6.35 mm screen on 5 January 2012 yielding the final PWC product (Figure 2). The pine trees used to create the SPW were harvested on 12 December 2011, shredded in a Wood Hog shredder (Morbark; Winn, MI, USA) on 9 January 2012 to create the coarsely shredded wood that was then hammer-milled through a 6.35 mm screen on 10 January 2012 yielding the final SPW product (Figure 2).

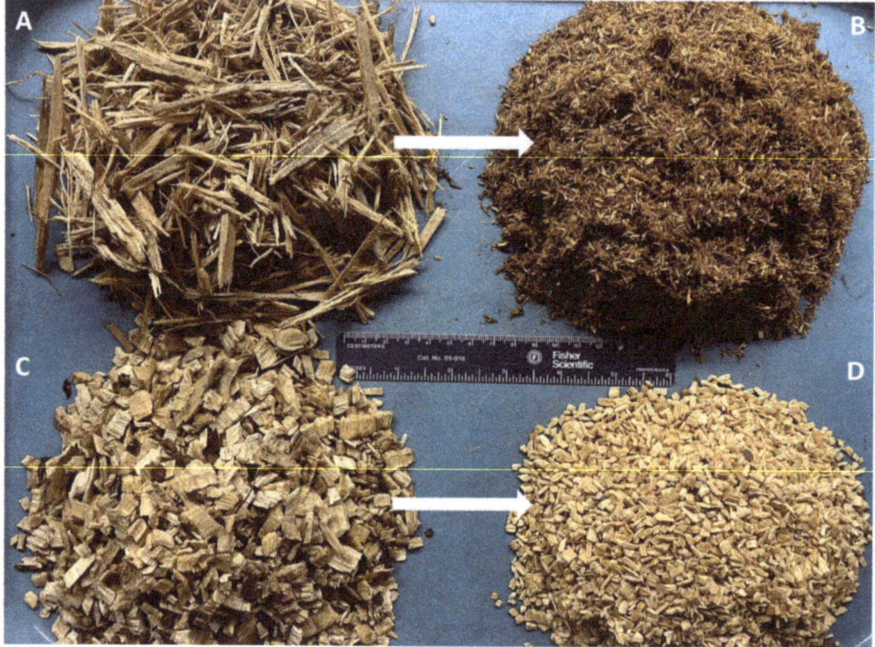

Figure 2. Pre and post hammer mill processing on the shredded pine wood (SPW) and pine wood chips (PWC). Shredded wood (**A**) passed through the hammer mill to produce SPW (**B**). Chipped pine wood (**C**) is passed through a hammer mill to produce PWC (**D**).

No additional screening was needed for the coir, PB, PWC, or the SPW. After acquiring, preconditioning, or creating the substrate materials all were placed in 60 L plastic bags, sealed, and stored in a controlled environment laboratory until experiment initiation.

On the day of sampling, bags were carefully turned upside down and mixed to ensure uniformity of the contents/materials, after which a representative sample of 14 L was collected. Moisture content (MC = mass of water/total mass) was measured for each material and adjusted to 55% by weight using procedures described by Fonteno and Harden [24]. Two substrate composites also tested in this experiment included a commercially available growing mix comprised of "Canadian sphagnum peat moss, pine bark, perlite, and vermiculite" (Fafard 4P; Sungro, Anderson, SC, USA) and an 80:20 (by vol.) peat: perlite substrate derived from peat (Berger Tourbe de Shaigne Blonde Golden; BP-P; Quebec Canada) that was taken from a compressed bale, loosened/fluffed by hand, and hydrated to 55% MC before being amended with horticultural grade perlite (Carolina Perlite Company, Gold Hill, NC, USA).

Measurements and Analysis. An initial test was conducted using a dewpoint potentiometer (WP4C, Decagon; Pullman, WA, USA) to determine water potentials for each substrate component and composite materials as they air dried. Based on these results, samples were prepared at target MC for each component and composite that fell within the water potential ranges of −1.0 to −2.0 MPa, and allowed to equilibrate for 24 h. Fifteen samples for each substrate component and twelve samples for each composite were evaluated. Five stainless steel sampling dishes (1.1 cm tall × 3.7 cm i.d.; Decagon; Pullman, WA, USA) were loosely filled to approximately half full (0.5 cm depth) from random locations in the prepared samples at each of the predetermined (through the initial test) MCs for each substrate. The dishes were immediately sealed with plastic lids and Parafilm® (American Can Co.; Greenwich, CT, USA) to prevent evaporative water loss. The samples were then individually analyzed for substrate water potential utilizing the dewpoint potentiometer in precision mode. Only samples that resulted in measures between −1.0 and −2.0 MPa were utilized, resulting in six to twelve utilized measures (12, 9, 8, 7, and 6 measurements for coir, peat, shredded wood, pine bark, and wood chips, respectively) for the components (see example; Figure 3) and four and five samples for the two composites. The reduction in measures within the −1.0 to −2.0 MPa range in the composites is due to the heterogeneous nature of these materials resulting in less uniformity in drying.

Table 1. Estimated substrate-bound water contents, container capacity and available water for substrate components and composites determined via dewpoint hygrometry measured between −1.0 and −2.0 MPa.

Substrate	Surface-Bound Water (% vol. ± SD) [z]	Container Capacity [y] (% vol.)	Bulk Density (g/cm³)	Available Water [x] (% vol.)
Coir	4.40 ± 0.30	75.2 b [w]	0.12 c	70.8
Peat	4.06 ± 0.36	80.1 a	0.09 d	76.0
Pine bark	7.35 ± 0.20	42.5 d	0.21 a	35.2
SPW [v]	4.77 ± 0.42	52.8 c	0.18 b	48.0
PWC [u]	4.60 ± 0.30	41.6 d	0.18 b	37.0
Mix 1 [t]	7.76 ± 1.62	59.3 c	0.10 d	51.5
Mix 2 [s]	8.42 ± 1.91	75.2 b	0.13 c	66.8

[z] Mean substrate-bound water content across substrate water potential between −1.0 and −2.0 MPa ± standard deviation (SD). [y] CC = container capacity values from NCSU porometer test. [x] AW = available water content calculated as difference between mean CC and SBW content. [w] Statistics preformed down columns using Tukey's HSD. Means with the same letter are not statistically different. [v] SPW = shredded pine wood made from loblolly pine (*Pinus taeda*) logs that were shredded prior to processing in a hammer mill through a 6.35 mm screen. [u] PWC = pine wood chips made from loblolly pine logs that were chipped prior to processing in a hammer mill through a 6.35 mm screen. [t] Mix 1 = Composite of Peat:perlite 80:20 (by vol.) [s] Mix 2 = Fafard 4P (Sungro,/Anderson, SC, USA).

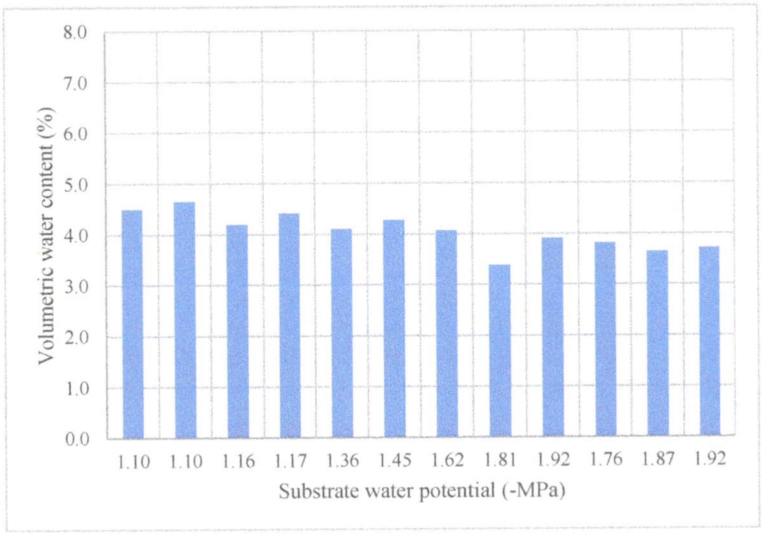

Figure 3. Example of individual sampling measurements of coconut coir. Variation of volumetric water content (VWC) was <1%. Data used to calculate values presented in Table 1.

Subsequently, mass wetness (MW = mass of water/mass of solid) was determined by placing the samples in a drying oven at 105 °C for 48 h to attain dry weights. Mass wetness for the samples was transformed to volumetric water content (VWC = volume of water/volume total) through: MW × Db of the material/density of water (1 g/cm³) = VWC. Since both VWC and water potential were measured (i.e., neither were precisely controlled), values for both are presented and discussed according to their range and average. Bulk density and container capacity (CC) values were obtained using the

NCSU porometer analysis following procedures of Fonteno and Harden [24] on three samples for each substrate. The values for VWC corresponding to the water potential range of −1.0 to −2.0 MPa were used as an estimate for soil-bound water and subtracted from the CC obtained from porometer analysis to obtain an estimate of available water holding capacity.

3. Results and Discussion

3.1. Estimating Water Availability at Low Water Potentials

Values for substrate-bound water obtained for all substrate components tested across the range of −1.0 MPa to −2.0 MPa were generally between 3% and 5% (Table 1), with PB providing the highest VWC at 7 to 8%. The PB had a higher VWC within the −1.0 to −2.0 MPa substrate water potential range likely resulting from reduced uniformity in the pore size distribution, as well as increased intraparticle porosity. As substrate water potential decreased, increased quantities of water became trapped within bark particles, thus limiting the water loss from the material. Furthermore, the majority of the accessible water present in the substrate at water potentials <−1.0 MPa exists primarily as hygroscopic water (water that is bound to particle surfaces). Previous research has demonstrated that at much higher substrate water potentials (i.e., −10 to −300 hPa or −0.0001 to −0.03 MPa) the VWC of PB is much lower than the other materials in this study [25]. This is likely from dual-porosity that is more evident in the PB than the other materials. Large pores created by irregular and large particles in PB readily drain at higher substrate water potentials, with smaller pores being either inaccessible or held more tightly at substrate water potentials between −1.0 and −2.0 MPa. The two composite substrates had similar VWCs (5 to 11%) at substrate water potentials in the range of −1.0 to −2.0 MPa (Table 1). These were only slightly larger than the other substrates, which indicates that the primary components in these substrates (i.e., peat or pine bark) dominate the hydraulic characteristics of the composites.

As expected, there was a large range in CC among the substrate materials and composites (Table 1). While many factors influence the CC of the substrate, the similarity between PB and PWC is likely a result of particle size and shape. Peat, coir, and SPW are more fibrous in structure, while PB and PWC had "plate-like" and "blockular" structure, respectively. The difference in particle size and shape can influence pore distribution and connectivity which has a great influence on water retention and CC, due to changes in the ratio of gravitational to capillary pores. Similar CC between the coir and Mix 2, as well as between SPW and Mix 1 highlight the similarities between fibrous materials and fiber-dominated mixtures (Table 1).

The estimated AW storage capacities for both peat and coir were >70% by volume. SPW, PWC, and PB had much lower AW (approx. 48%, 37%, and 35%, respectively). The differences in AW were primarily due to differences in CC, as there was little (<4%) difference observed in SBW among components. Moreover, by calculating the proportion of the water at CC that is AW (i.e., AW/CC from Table 1) coir, peat, SPW, and PWC are similar 94.1%, 94.9%, 90.9%, and 88.9%, respectively). However, the proportion of CC that is AW in PB is much lower at 82.7%. From data presented herein, PWC would appear to have similar properties to more traditionally used greenhouse substrate aggregates, such as PB and perlite. The SPW possessed a similar VWC at low substrate water potentials, yet a significantly greater CC, yielding increased AW (Table 1), which allows it to be incorporated into a substrate to increase drainage, while still retaining moisture needed for plant growth.

A review of the literature was performed and selected references associated with measuring soilless substrate VWC at substrate water potentials of −1.5 MPa were included in Table 2. The current accepted normal range of SBW is 23 to 35% by volume [26]. In fact, current best management practices for nursery growers recommend substrate SBW between 25 and 35% by vol. [27]. This acceptability range is further evidenced as much of the previous research utilizing ceramic pressure plates identifies commonly used substrates as having SBW within these ranges (Table 2). For example, Wright and Browder [28] reported SBW of PB as 26.6% and pine tree substrate at 23.6% (by vol.), respectively. This may be a significant overestimation of SBW (26.6% and 23.6% by vol. as compared to 7.5% and

4.8% by vol.), and therefore a large underestimation of AW (~20% by vol.) for these substrate materials. Water measurements (at −1.5 MPa) as high as 39.0% by volume have been reported in PB substrates using ceramic pressure plates [29]. With previous reports of miscalculations of soilless substrate SBW through ceramic pressure plate analysis at tensions <−1.0 MPa [12–15], it is entirely possible that many values within the literature are overestimating SBW.

Table 2. Survey of soilless substrate VWC at substrate water potentials of −1.5 MPa as measured through pressure plate extractors.

Publication	Material	Reported VMC (% vol.)
Altland and Krause [30]	Pine bark	29.9
	Pine bark: pine wood (1:1)	26.4
	Pine wood	24.2
Bilderback et al. [26]	Pine bark	33
	Pine bark: sand (4:1)	25
	Pine bark: peat moss (9:1)	32
	Pine bark: perlite (7:3)	33
	Pine bark: soil (9:1)	26
	Fir bark: peat: pumice (1:1:1)	25
Fonteno and Bilderback [13]	Pine bark	35.2
	Pine bark: sand (4:1)	35.6
Gabriel et al. [31]	Douglas fir bark	23
	Douglas fir bark: peat (7:3)	21
	Douglas fir bark: pumice (7:3)	25
Herring et al. [32]	Swine lagoon compost	28.7
	Pine bark fines-based potting mix	29.4
	Peat-based potting mix	27.5
Jackson et al. [33]	Pine bark	34.3
	Pine bark: clay (8:1)	31.6
	Pine bark: mortar sand (8:1)	27.3
Londra et al. [34]	Peat	13.1
	Peat: perlite (3:1)	12.3
	Coir	13
	Coir: perlite (3:1)	13
Milks et al. [35]	Peat: vermiculite	20
Niemiera et al. [29]	Pine bark	39
	Pine bark: sand (9:1)	27
	Pine bark: sand (5:1)	33
Owen, Jr. et al. [36]	Pine bark + clay	25
	Pine bark + sand	24
Owen, Jr. et al. [37]	Pine bark	38
	Pine bark + 12% mineral aggregate	36
Tyler et al. [38]	Pine bark	31.4
Warren et al. [39]	Pine bark	29
	Pine bark: cotton stalk/swine compost (85:15)	30
Wright et al. [40]	Peat:perlite:vermiculite: bark (45:15:15:25)	22
	Pine tree substrates	22
Write and Browder [28]	Pine bark	26.6
	Pine wood chips	23.6
	Pine wood chips: pine bark (3:1)	25

Previous research from the authors of this publication involved utilizing dewpoint hygrometry to assess the water potential of substrate components and field soils that had been squeezed to −1.5 MPa on ceramic plates [15]. The authors found that the water in the mineral soil samples did equilibrate at ~−1.5 MPa; however, the hydraulic connection between the coarse substrate components (bark, peat, and perlite) was broken at approx. −0.3 MPa, preventing additional water loss from the samples (Figure 4). Further investigation in that research showed that when peat and pine bark samples were squeezed at −0.1 and −0.3 MPa, the assessed water potential was close to the applied pressure. This leads the authors to believe that highly coarse substrate materials are not coming to equilibrium with pressures exceeding 0.3 MPa in traditional pressure plate extractors. This information further supports the hypothesis that the pressure plate analysis is overestimating water in samples at very low substrate water potentials.

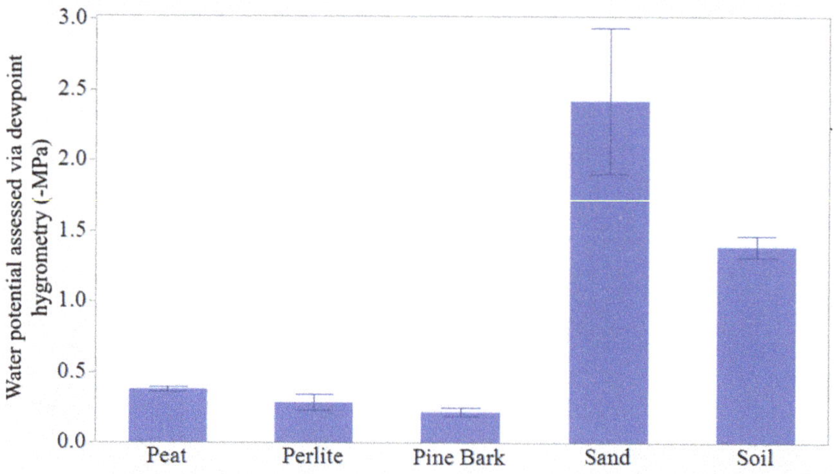

Figure 4. Substrate water potential of individual substrate and soil components assessed via dewpoint hygrometry, after being squeezed at −1.5 MPa on ceramic pressure plates in a volumetric pressure plate extractor. Data utilized from Fields et al., 2013 [15].

3.2. Gravimetric vs. Volumetric Water Contents

The authors also suggest a paradigm shift in discussing moisture contents that evolved during this work. The term used to describe the amount of moisture in a substrate had two forms: MC (expressed on a weight basis) and VWC (expressed on a volume basis). Initial moisture content for substrates is usually expressed as MC; however, almost all discussion of water content as a result of irrigation is expressed in terms of VWC. The moisture contents in these experiments were considered using both forms. In this case, MC was converted to VWC for comparisons in the table and figures (using yet another measure, MW). For example, coir at −1.5 MPa water potential has a resulting VWC of 7.62% (Table 1), which is equivalent to MC of 50% by weight. During the initial potting of most greenhouse crops, it is important to have adequate moisture in the substrate [27], and generally speaking, many growers tend to use ~50% MC substrates for planting. These results suggest that at this MC, coir is already at a water potential <−1.0 MPa, within the currently accepted range of plant unavailable water. Kiehl et al. [41] showed water stress symptoms occurring in plants at −16 kPa, much less negative than the −1.0 MPa of coir at 50% MC. These high (50%) MC values convert to much lower VWC values due to the very low bulk density of organic components. Moisture contents of 50% are considered to be heavy for transportation, in fact, coir is normally dried, compressed, and formed into blocks for shipping [42]. Peat is normally compressed two to three times and bailed at a MC of about 20–25% (personal observation) for shipping purposes, which is significantly lower than at substrate water potential of −1.5 MPa (37% MC). This establishes that not only is proper hydration of substrates important for potting/planting, but previously accepted MC levels are essential in the plant unavailable range. This also implies that recently potted plants should not be allowed to "sit" for prolonged periods of time before initial hydration (i.e., water) is applied.

4. Conclusions

The use of dewpoint hygrometry allowed estimates of soil-bound water in the water potential range (1.0 to 2.0 MPa) typically considered plant unavailable. These estimates are much lower than values previously reported for similar substrate components using pressure plates. The authors agree that problematic measures <−1.0 MPa can potentially overestimate SBW due to reported issues associated with highly porous organic materials in pressure plate analysis. As such, it is important that more efforts are utilized to investigate SBW from a substrate standpoint and identify more precise methods of analysis to truly identify substrate water relations at low water potentials. The use of dewpoint hygrometry has the potential to improve the estimation of SBW for substrate analysis. If further investigations find that dewpoint hygrometry measures are in fact accurate, best management practices and acceptable ranges should be updated.

Author Contributions: Conceptualization, J.S.F., W.C.F., B.E.J., J.L.H., J.S.O.J.; methodology, J.S.F., W.C.F., B.E.J., J.L.H., J.S.O.J.; formal analysis; J.S.F., W.C.F.; investigation, J.S.F.; writing—original draft preparation, J.S.F.; writing—review and editing, J.S.F., W.C.F., B.E.J., J.L.H., J.S.O.J.; supervision, W.C.F. and B.E.J. All authors have read and agreed to the published version of the manuscript.

Funding: This research received no external funding.

Conflicts of Interest: The authors declare no conflict of interest.

References

1. Hagan, R.M. Factors affecting soil moisture-plant growth relations. In Proceedings of the XIV International Horticultural Congress, The Hague, The Netherlands, 29 August–6 September 1955; pp. 82–98.

2. Richards, L.; Wadleigh, C. Soil water and plant growth. In *Soil Physical Conditions and Plant Growth*; Shaw, B.T., Ed.; Academic Press: New York, NY, USA, 1952; pp. 73–251.

3. Denmead, O.T.; Shaw, R.H. Availability of soil water to plants as affected by soil moisture content and meteorological conditions. *Agron. J.* **1962**, *45*, 385–390. [CrossRef]

4. Caron, J.; Xu, H.L.; Bernier, P.Y.; Duchesne, I.; Tardif, P. Water availability in three artificial substrates during *Prunus x cistena* growth: Variable threshold values. *J. Am. Soc. Hortic. Sci.* **1998**, *123*, 931–936. [CrossRef]

5. Veihmeyer, F.J.; Hendrickson, A.H. The relation of soil moisture to cultivation and plant growth. *Proc. First. Intern. Cong. Soil Sci.* **1927**, *3*, 498–513.

6. Bouyoucos, G.J. A new, simple, and rapid method for determining the moisture equivalent of soils, and the role of soil colloids in this moisture equivalent. *Soil Sci.* **1929**, *27*, 233–241. [CrossRef]

7. Richards, L.A.; Fireman, M. Pressure-plate apparatus for measuring moisture sorption and transmission by soils. *Soil Sci.* **1943**, *50*, 395–404. [CrossRef]

8. Cassel, D.K.; Klute, A. Water potential: Tensiometry. In *Methods of Soil Analysis Pt 1—Physical and Mineralogical Methods*, 2nd ed.; Klute, A., Ed.; American Society of Agronomy: Madison, WI, USA, 1986; pp. 563–596.

9. Klute, A. *Methods of Soil Analysis, Part I, Physical and Mineralogical Methods*, 2nd ed.; American Society of Agronomy: Madison, WI, USA, 1986.

10. Bakker, G.; van der Ploeg, M.J.; de Rooij, G.H.; Hoogendam, C.W.; Gooren, H.P.A.; Huiskes, C.; Koopal, L.K.; Kruidhof, H. New polymer tensiometers: Measuring matric pressures down to the wilting point. *Vadose Zone J.* **2007**, *6*, 196–202. [CrossRef]

11. Cassel, D.K.; Nielsen, D.R. Field capacity and available water capacity. In *Methods of Soil Analysis, Part 1—Physical and Minerological Methods*; Klute, A., Ed.; Agronomy Monographs 9; American Society of Agronomy: Madison, WI, USA, 1986; pp. 901–926.

12. Stevenson, D.S. Unreliability's of pressure plate 1500 kilopascal data in predicting soil water contents at which plants become wilted in soil-peat mixes. *Can. J. Soil Sci.* **1982**, *62*, 415–419. [CrossRef]

13. Fonteno, W.C.; Bilderback, T.E. Impact of hydrogel on physical properties of coarse-structured horticultural substrates. *J. Am. Soc. Hortic. Sci.* **1993**, *118*, 217–222. [CrossRef]

14. Gee, G.W.; Ward, A.L.; Zhang, Z.F.; Campbell, G.S.; Mathison, J. Influence of hydraulic nonequilibrium on pressure plate data. *Vadose Zone J.* **2002**, *1*, 172–178. [CrossRef]

15. Fields, J.S.; Fonteno, W.C.; Jackson, B.E. Plant available and unavailable water in greenhouse substrates: Assessment and considerations. *Acta Hortic.* **2013**, *1034*, 341–346. [CrossRef]

16. Campbell, E.C.; Campbell, G.S.; Barlow, W.K. A dewpoint hygrometer for water potential measurement. *Agric. Meteorol.* **1973**, *12*, 113–121. [CrossRef]

17. Campbell, G.S.; Smith, D.M.; Teare, B.L. Application of a dew point method to obtain the soil water characteristic. In *Experimental Unsaturated Soil Mechanics*; Schanz, T., Ed.; Springer: Berlin, Germany, 2007; Volume 112, pp. 71–77.

18. Durner, W.; Iden, S.C.; Schelle, H.; Peters, A. Determination of hydraulic properties of porous media across the whole moisture range. In Proceedings of the CMM Conference, Karlsruhe, Germany, 12–14 October 2011.

19. Schelle, H.; Heise, L.; Janicke, K.; Durner, W. Water retention characteristics of soils over the whole moisture range: A comparison of laboratory methods. *Eur. J. Soil Sci.* **2013**, *64*, 814–821. [CrossRef]

20. Wang, T. Chemical, Physical, and Moisture Release Characteristics of Compost Based Potting Media. Master's Thesis, Clemson University, Clemson, SC, USA, 2013.

21. Solone, R.; Bittelli, M.; Tomei, F.; Morari, F. Errors in water retention curves determined with pressure plates: Effects on the soil water balance. *J. Hydrol.* **2012**, *470*, 65–74. [CrossRef]

22. Gubiani, P.I.; Reichert, J.M.; Campbell, C.; Reinert, D.J.; Gelain, N.S. Assessing errors and accuracy in dew-point pontentiometer and pressure plate extractor measurements. *Soil Sci. Soc. Am. J.* **2013**, *77*, 19–24. [CrossRef]

23. Curtis, M.J.; Claassen, V.P. An alternative method for measuring plant available water in inorganic amendments. *Crop Sci.* **2008**, *48*, 2447–2452. [CrossRef]

24. Fonteno, W.C.; Harden, C.T. *North Carolina State University Horticultural Substrates Lab Manual*; North Carolina State University: Raleigh, NC, USA, 2010; Volume 49, pp. 827–832.

25. Fields, J.S.; Fonteno, W.C.; Jackson, B.E.; Heitman, J.L.; Owen, J.S.J. Hydrophysical properties, moisture retention, and drainage profiles of wood and traditional components for greenhouse substrates. *HortScience* **2014**, *49*, 827–832. [CrossRef]

26. Bilderback, T.E.; Warren, S.L.; Owen, J.S.J.; Albano, J.P. Healthy substrates need physicals too! *HortTechnology* **2005**, *15*, 747–751. [CrossRef]

27. Bilderback, T.E.; Boyer, C.B.; Chappell, M.; Fain, G.B.; Fare, D.C.; Gilliam, C.H.; Jackson, B.E.; Lea-Cox, J.D.; LeBude, A.V.; Niemiera, A.X.; et al. *Best Management Practices: Guide for Producing Container-Grown Plants*, 3rd ed.; Southern Nurserymen's Association: Acworth, GA, USA, 2013.

28. Wright, R.D.; Browder, J.F. Chipped pine logs: A potential substrate for greenhouse and nursery crops. *HortScience* **2005**, *40*, 1513–1515. [CrossRef]

29. Niemiera, A.X.; Bilderback, T.E.; Leda, C.E. Pine bark characteristics influence pour-through nitrogen concentrations. *HortScience* **1994**, *29*, 789–791. [CrossRef]

30. Altland, J.E.; Krause, C.R. Substituting pine wood for pine bark affects physical properties of nursery substrates. *HortScience* **2012**, *47*, 1499–1503. [CrossRef]

31. Gabriel, M.Z.; Altland, J.E.; Owen, J.S.J. The effect of physical and hydraulic properties of peatmoss and pumice on Douglas fir based soilless substrates. *HortScience* **2009**, *44*, 874–878. [CrossRef]

32. Herring, P.L.; Noah, A.C.; Kraus, H.T. Swine lagoon compost as transplant substrate for basil, chives, and dill. *HortTechnology* **2018**, *28*, 337–343. [CrossRef]

33. Jackson, B.; Kraus, H.; Bilderback, T. What to do when pine bark runs short: Physical properties of pine bark alternatives. *South. Nurs. Assoc. Res. Conf. Proc.* **2010**, *55*, 413–418.

34. Londra, P.; Paraskevopoulou, A.; Psychogiou, M. Hydrological behavior of peat- and coir-based substrates and their effect on begonia growth. *Water* **2018**, *10*, 722. [CrossRef]

35. Milks, R.R.; Fonteno, W.C.; Larson, R.A. Hydrology of horticultural substrates: III. Predicting air and water content in limited-volume plug cells. *J. Am. Soc. Hortic. Sci.* **1989**, *114*, 57–61.

36. Owen, J.S.J.; Warren, S.L.; Bilderback, T.E.; Albano, J.P. Phosphorus rate, leaching fraction, and substrate influence on influent quantity, effluent nutrient content, and response of a containerized woody ornamental crop. *HortScience* **2008**, *43*, 906–912. [CrossRef]

37. Owen, J.S.J.; Warren, S.L.; Bilderback, T.E.; Cassel, D.K.; Albano, J.P. Physical properties of pine bark substrate amended with industrial mineral aggregate. *Acta Hortic.* **2008**, *779*, 131–138. [CrossRef]

38. Tyler, H.T.; Warren, S.L.; Bilderback, T.E.; Fonteno, W.C. Composted turkey litter: I. Effect on chemical and physical properties of a pine bark substrate. *J. Environ. Hortic.* **1993**, *11*, 131–136. [CrossRef]

39. Warren, S.L.; Bilderback, T.E.; Owen, J.S.J. Growing media for the nursery industry: Use of amendments in traditional bark-based media. *Acta Hortic.* **2007**, *819*, 143–155. [CrossRef]

40. Wright, R.D.; Jackson, B.E.; Browder, J.F.; Latimer, J.G. Growth of chrysanthemum in a pine tree substrate requires additional fertilizer. *HortTechnology* **2008**, *18*, 111–115. [CrossRef]

41. Kiehl, P.A.; Lieth, J.H.; BurgeF, D.W. Growth response of Chyrsanthemum to various container medium moisture tension levels. *J. Am. Soc. Hortic. Sci.* **1992**, *117*, 224–229. [CrossRef]

42. Raviv, M.; Lieth, J.H. *Soilless Culture Theory and Practice*; Elsevier: San Diego, CA, USA, 2008.

Article

Comparison of Water Capture Efficiency through Two Irrigation Techniques of Three Common Greenhouse Soilless Substrate Components

Brian A. Schulker [1,*], Brian E. Jackson [1,*], William C. Fonteno [1], Joshua L. Heitman [2] and Joseph P. Albano [3]

1   Department of Horticultural Science, North Carolina State University, 21 Kilgore Hall, Raleigh, NC 27695, USA; wcf@ncsu.edu
2   Department of Crop and Soil Sciences, North Carolina State University, 3410 Williams Hall, Raleigh, NC 27695, USA; jlheitma@ncsu.edu
3   U.S. Horticultural Research Laboratory, USDA-ARS, 2001 S. Rock Rd., Fort Pierce, FL 34945, USA; joseph.albano@ars.usda.gov
*   Correspondence: baschulk@ncsu.edu (B.A.S.); brian_jackson@ncsu.edu (B.E.J.)

Received: 24 July 2020; Accepted: 10 September 2020; Published: 14 September 2020

Abstract: Substrate wettability is an important factor in determining effective and efficient irrigation techniques for container-grown crops. Reduced substrate wettability can lead to lower substrate water capture, excessive leaching and poor plant growth. This research examined substrate water capture using surface and subirrigation under three initial moisture contents (IMC). Sphagnum peat moss, coconut coir, and pine bark were tested at IMCs of 67% 50%, and 33%. Substrate water capture was influenced by both IMC and irrigation technique. Surface irrigation increased the water capture of coir and peat, regardless of IMC, whereas IMC influenced pine bark water capture more than irrigation method. Surface-irrigated coir at or above 50% IMC provided the greatest water capture across all treatments. The first irrigation had the highest capture rate compared to all other events combined. Container capacities of pine bark and coir were unaffected by IMC and irrigation type, but the CC of peat was less by ~ 40% volumetrically under low IMC conditions. Coir, had the greatest ability to capture water, followed by pine bark and peat, respectively. Moisture content, irrigation type and component selection all influence the water capture efficiency of a container substrate.

Keywords: irrigation; soilless substrates; water; coconut coir; initial moisture; mass wetness; peatmoss; pine bark; wettability; capillary rise; container capacity; capture rate

1. Introduction

Water use efficiency of horticultural soilless substrates represents one of the biggest variables in container plant production. With nearly 21,500 acres of land devoted to greenhouse operations in the U.S., representing a 148% increase since 1998, growers specializing in container plant production need to be able to understand how irrigation specifics impact water use efficiency of soilless substrates [1]. Greenhouse production uses less water and fewer nutrients than many agricultural resources [2], and decrease crop water requirements by up to 40% compared to open field cultivation [3–5]. As water quality, conservation, scarcity and operational costs increase, plant producers must adopt new strategies to improve the sustainable use of water to confront water-climate policies [6–8]. In order for growers to meet these increasing regulations in water use, we need to increase the overall understanding of irrigation techniques.

Two parameters affecting water efficiency in substrates are container capacity (CC) [9–12] and wettability [13–17]. Both wettability and CC are vital to the wetting of a substrate, however neither

Agronomy **2020**, *10*, 1389; doi:10.3390/agronomy10091389   www.mdpi.com/journal/agronomy

completely describes the effectiveness of water capture during irrigation [18]. Container capacity is the maximum amount of water a substrate can hold after wetting and drainage. Wettability includes a liquid's ability to spread laterally at and below the surface of a material [19]. In substrates, proper wettability helps to provide a uniform distribution of water throughout the rooting environment [20]. Moisture content in substrates affects both wettability and CC of a substrate. Fields et al., [21] showed the variability in CC based on substrate and initial moisture content (IMC), with coir and pine bark being less variable than peat. Initial moisture content in this context references the moisture content prior to an irrigation event. At low moisture conditions, peat can have a ~30% lower CC than at high moisture conditions [11]. In mineral soils, hydrophobicity is caused by organic residues coating the mineral materials from the breakdown of organic matter. In substrates, most components are composed almost entirely of organic materials which complicates the nature of hydrophobicity. As organic materials naturally break down, the intensity of hydrophobicity can change, which then alters the substrate's behavior during rewetting [22,23]. Hydrophobicity issues also arise as organic materials dry, and materials such as peat and pine bark begin to see reductions in water capture based on repellency [13,24]. Adequate substrate particle structure, stability, density, and CC are needed to allow water movement through the containers [25,26].

Most irrigation is applied to the top of the soil column. However, in containers, the irrigation delivery direction can be reversed and delivered from the bottom (subirrigation). Irrigating from below can require a finer textured, micro-pore abundant substrate to take up water mainly through capillary action [27]. Conversely, greater air space (AS) and pore size diversity favor surface irrigation methods. Capillary action, the movement of water from a saturated zone upward into an unsaturated zone through surface tension and soil matric potential, provides water and nutrients to the plant root [28]. Subirrigation is a combination of flooding from a perched water table and capillary rise. Ebb and flow subirrigation, was found to reduce water use by ~40% compared to hand watering [29,30]. The confluence of these factors combine to play a pivotal role in the effectiveness of water uptake in specific combinations of irrigation method and substrate components. Water transport research in mineral soils [31] provides the basis to understand soilless substrate systems, but the substantial differences in physical properties and their accompanying calculated values between soil and soilless substrates requires substrate-specific research.

Research has identified the impacts hydrophobicity can have on the wettability of some horticultural substrates [13,19,20]. However, few have studied the influence of irrigation delivery method on the ability of substrates to capture water or rehydrate. In soilless substrates, water distribution in the container can largely change due to a substrate's hydrophobicity, physical characteristics (texture/particle size), as well as the irrigation method used. The objective of this study was to characterize the water capture and retention of three substrate components based on irrigation technique and IMC.

2. Materials and Methods

2.1. Preparation of Substrates

On 11 April, 2019, sphagnum peat (Premier Pro-Moss, Quakertown, PA) was hydrated and fluffed to an initial IMC of 70% (by weight; ~2.5 w). To do that, peat was removed from the compressed bale and placed into a large tub, water was then added in 3 L increments after which peat was mixed/agitated by hand to allow water absorption. Moisture levels were then tested using an Ohaus MB27 soil moisture determination balance (Ohaus Corp., Parsippany, NJ) to determine further water additions needed to bring peat to an initial moisture of 70%. On 15 April, 2019, three compressed bricks of coconut coir (Densu Coir, Ontario, Canada) were hydrated individually by adding 14 L of water (in 1 L increments), by hand, until the coir was completely broken apart. Moisture levels were then tested using the soil moisture balance to determine further water additions needed to bring coir to an initial moisture of 70%. On 16 April, 2019, loblolly (*Pinus taeda* L.) pine bark (Pacific Organics, Henderson, NC) which had been aged in outdoor windrows for four months and specifically engineered (hammer milled and screened) to have a CC of 55% volumetric water content (VWC). The volume of pine bark was measured out, initial moisture level was tested and recorded before the bark was further hydrated to a moisture level of 70%.

Each substrate component was tested under three IMC treatments. The most common and recommended IMC for potting soils has a mass wetness of 1.0 g g^{-1}. To test effects of IMC, each component was also brought to half (0.5 g g^{-1}) and double (2.0 g g^{-1}) this normal level, which resulted in percent IMCs of 33%, 50%, and 67% by weight. To do this the wet weights and dry weights were determined by taking 500 g subsamples of each substrate, weighing, drying, and then reweighing. Substrate samples were wet to an initial IMC of 70% IM, before being air-dried down to initial IMC's of 67%, 50%, and 33% IMC. Initial IMC and total weight of each sample were used to calculate how much water needed to be lost to reach initial IMC's of 67% IM, 50% IM, and 33% IM. Using a 160 cm × 49 cm × 69 cm four-tier PVC-enclosed dehumidifying chamber, substrates were allowed to air dry to desired wetness. Once the target initial IMC was reached, samples were transferred to plastic bags and sealed to prevent further water loss, while allowing the substrate to reach moisture equilibrium.

2.2. Particle Size

Particle size distribution (Table 1) was performed on three 50 g oven dried samples of each substrate with 7 sieves. The sieve sizes used were 6.3 mm, 2.0 mm, 0.71 mm, 0.5 mm, 0.25 mm, 0.11 mm, and the bottom pan to collect fine particulates. The 7 sieves (6 sieves and the pan) were stacked together and substrate samples were poured into the top sieve, and placed into the RX-29 Ro-Tap sieve shaker (278 oscillations/min, 150 taps/min; W.S, Tyler, Mentor, OH). The sieves and pan were shaken for five min and the particle fractions retained on each sieve and the amount collected in the bottom pan (representing the smallest particle fractions) were weighed.

Table 1. Particle size distribution of three traditional greenhouse substrate components.

Particle Size Distribution (%) [z]			
Sieve (mm)	**Coir**	**Peat**	**Pine Bark**
6.3	0.2	2.0	8.6
2.0	6.0	17.0	45.0
0.71	40.2	29.0	30.6
0.5	19.8	11.2	7.2
0.25	26.0	25.0	5.8
0.11	5.4	11.4	1.8
<0.11 (pan)	2.4	4.4	1.0
Texture			
Coarse [y]	6.2 C c [v,u]	19.0 B b	53.6 A a
Medium [x]	60.0 A a	40.2 A b	37.8 B b
Fines [w]	33.8 B a	40.8 A a	8.6 C b

[z] Particle size distribution calculated on a dry weight scale using means of three oven-dried samples. [y] Coarse = particles that are greater than 2.0 mm in diameter. [x] Medium = particles that are less than 2.0 mm but greater than 0.5 mm in diameter. [w] Fines = particles that measure less than 0.5 mm in diameter. [v] Values are means of percentages of the total sample. [u] Statistics are determined down columns (denoted by an uppercase letter) and across rows (denoted by a lowercase letter) using Tukey's honestly significant difference to determine similarities and differences across all components.

2.3. Surface Irrigation

In order to determine the effects of IMC with surface applied irrigation, this experiment followed the procedures described by Fonteno et al. [18] and Fields et al. [20]. The equipment consisted of a transparent cylinder, 5 cm i.d. × 15 cm h^{-1}, with a mesh screen (mesh size 18 × 16; New York Wire, York, PA, USA) (Figure 1A), attached to one end, using rubber pressure plate rings (Soil moisture Equipment Corp., Santa Barbara, CA, USA); a 250-mL beaker; a 250-mL funnel; as well as a 10 mL plastic vial (4-cm diameter) with five evenly spaced 2.33 mm diameter holes in the base to act as a diffuser displayed in

Figure 1. This allowed the water dripping through the funnel to be evenly dispersed through the five holes onto the surface of the substrate in the cylinder.

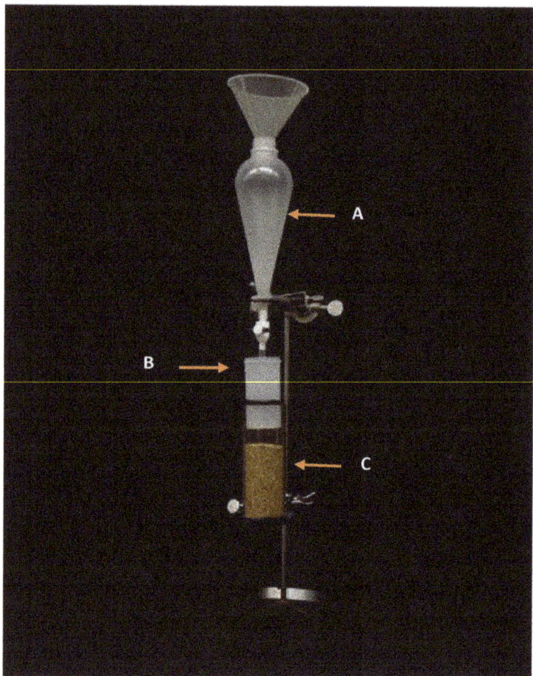

Figure 1. Surface irrigation apparatus. (**A**) Funnel, sparatory funnel with stopcock. (**B**) Water diffuser with O-ring above cylinder. (**C**) Sample cylinder with 200 mL of substrate.

The transparent cylinders were packed with each substrate component to have a weight within 5% of other samples of the same component. To achieve this, cylinders were filled (by weight) with substrate then raised 12 cm off a flat surface, then tapped four times to bring the top of all 4 replications to 10 cm from the base of the cylinder, representing 200 mL of substrate and providing similar Db across all replications. With three substrates, at three IMCs, and four replications there were a total of 36 experimental samples. After the cylinders were packed, each was fitted with a diffuser and placed in the clamps held up by a ring stand, just under the separatory funnel (Figure 1). Two hundred mls of water was poured into the separatory funnels and allowed to drip onto the surface of the substrate at an average rate of 40 mL min^{-1}, using the stopcock valve to control the flow. Water was applied in 10 individual hydration events. Substrates with an IMC of 33% required the rate of flow to be less to prevent ponding of water on the substrate surface which would have created a hydraulic head greater than 0.5 cm and alter the influence of any native hydrophobicity in the sample. Water was passed from the separatory funnel, through the diffuser, and onto the surface of the substrate. With the help of gravity, water was able to penetrate the surface of the substrate and percolate through the 10 cm depth. Some of the water volume was absorbed as it moved through the substrate, the rest was collected out of the bottom by a 250 mL beaker. After ~5 min, water flow ceased; the substrate was then held at equilibrium for two min. The effluent volume was measured and recorded while water retained was calculated by subtracting the amount of water applied (200 mL) from the amount of effluent captured. With the total event lasting ~7 min, 5 min time intervals were measured out in between events to keep treatments even. This procedure was repeated for each of the 10 hydration events.

2.4. Subirrigation

In order to understand how IMC influences substrate water capture through subirrigation, this experiment was conducted using materials and modified procedures described by Fonteno et al., [12]. The same transparent cylinders as described in surface irrigation above were prepared in the same way (Figure 2), The subirrigation method used an ebb and flood irrigation system (Hawthorn Hydroponics, Vancouver WA) 60.96 cm wide by 121.92 cm in length (Figure 2). Water was introduced into this system via a faucet and controlled through a series of gate valves connected to the system (Figure 3B). Water was maintained at a continuous height with a flow rate of ~21 L min^{-1}. To be able to control the height of the water while also having a steady flow into the bench, a standing copper pipe was cut to allow water to be held at 2.5 cm at a steady state (Figure 2C).

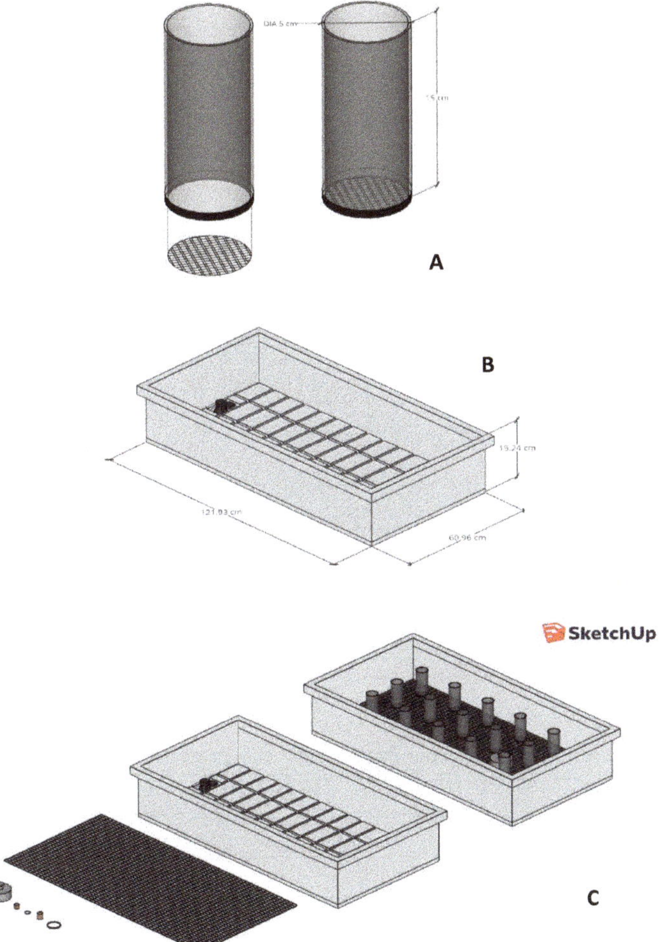

Figure 2. Subirrigation system. (Left, right, bottom) (**A**) Cylinder (15 cm × 5 cm) with rubber pressure plate ring at base with mesh screen with DIA representing the cylinder diameter. (**B**) Ebb and flood subirrigation system. (**C**) Separated full system (from left to right) with central weight, three copper leveling pipes, large steel ring (for raising wire screen off surface), black wire mesh screen, fully assembled system.

Figure 3. Ebb and flood subirrigation system. (**A**) Container capacity testing with 2 kg aluminum weights. (**B**) Partially constructed ebb and flood unit with aluminum rings with mesh screen. (**C**) Fully constructed system complete with packed substrate cylinders.

The transparent cylinders were packed in identical manner as the samples used in the surface irrigation system. Cylinders were then placed on an elevated mesh screen to optimize the lower surface area exposure to water. The unit was then filled with water. It took approximately one minute for the water to reach the bottom of the cylinders and another minute for the water to reach 2.5 cm above the base. At that time, water flow input equaled output, allowing constant flow of water without a change in water level. The substrate samples were kept in the unit for five minutes for each of the hydration events. Once an event was finished, water drained from the unit for one minute before each cylinder was weighed. The weights were used to calculate the amount of water captured by the substrate by subtracting it from the initial weight of the packed cylinders. This procedure was repeated 10 times (10 hydration events), with a total time of hydration equaling 50 min.

2.5. Container Capacity

After the final hydration event was complete and final weights were taken, CC was then determined for each cylinder. The cylinders were returned to the ebb and flood unit (Figure 2), and CC was determined using a modified version of the NC State University Porometer Method [32]. After placed in the subirrigation unit, an aluminum weight of approximately 2 kg was placed on the top of each cylinder to prevent tipping and buoyancy (Figure 3A). The samples were then saturated from below by allowing water to flow into the unit until it reached 1/3 of the height of the sample (three cm from the base of the sample). After two minutes, additional water was allowed to enter the unit until reaching 2/3 of the height of the cylinder, or six cm from the base of the cylinder. After an additional two minutes, the water was applied until reaching the top of the sample within the cylinder (Figure 3A), 10 cm from the base of the cylinder. After saturating in the system for an additional 30 min, the water was drained and samples were reweighed to record changes in weight (water captured and retained). Samples were then placed into a forced-air drying oven at 105 °C for 48 h to dry, after which each sample was weighed and the dry weights were used to determine total water retained and IMC.

2.6. Water Capture Rate

Water CR was calculated for subirrigated substrates using a modified version of the flow rate formula to account for variables in this experiment, the equation was written as:

$$CR = \frac{C_i - C_p}{t} \tag{1}$$

where CR is the mL/min of water captured by the substrate after one irrigation event, C_i [water capture (g) in the initial irrigation event] is the weight of the substrate after the present irrigation event (minus the weight of the cylinder), C_p (previous water capture in grams) is the weight of the substrate (minus the cylinder) taken after the previous irrigation (for the first irrigation event, C_p is equal to the pack weight of the cylinder (minus the weight of the cylinder), t is the amount of time per irrigation (in minutes). For surface irrigated samples that have a defined volume of water passing through the substrate, the equation was written as

$$CR = \frac{A_w - E}{t} \tag{2}$$

where CR is the amount of water captured by the substrate after one irrigation per unit time (in mL min^{-1}), A_w is the amount of water applied to the substrate per irrigation event (in this case, 200 mL), E is the effluent captured after the individual irrigation event (in mL), t is the amount of time per irrigation event (minutes).

2.7. Water Capture Curves

The IMCs of 33%, 50%, and 67% were all determined by weight. Wettability curves were determined by VWC to describe the amount of water contained within the substrate. These curves show a VWC reading at event zero, and represent the percent VWC at the IMC. Therefore, an IMC of 50% (by weight) was actually 12% to 15% $v\,v^{-1}$ (moisture) for peat. For coir, an IMC of 50% ranged from 9% to 11% $v\,v^{-1}$, and for pine bark (at 50% IMC) they were 16% to 18% $v\,v^{-1}$.

2.8. Capture Efficiency Values

In order to provide both statistical and numeric comparisons, water capture efficiency of the substrates was described in three ways: (1) first hydration, (2) final hydration and (3) CC. First hydration was the amount of water absorbed by the substrate after one irrigation event, and compared across all substrates and moisture levels. Final hydration was the amount of water absorbed by the substrate after the tenth irrigation event. Container capacity was the maximum water content the sample could hold after saturation and drainage. Physical properties of the substrates, including CC, AS, total porosity (TP), and bulk density (Db) were derived using the NC State University Porometer method [20] with three representative samples of each substrate, and CC is reported in Table 2.

Table 2. First hydration (H_1), container capacity (CC), and final hydration (H_{10}) of three substrate components analyzed at three different initial moisture contents (IMC) irrigated by either subirrigation or surface application.

	Coir				Peat				Pine Bark			
Surface	H_1[z]	H_{10}[y]	CC[x]	S*[w]	H_1	H_{10}	CC	S*	H_1	H_{10}	CC	S*
33% IMC	35.4 de	59.6 b	75.4 ab	L** Q*	17.8 cd	37.6 c	78.3 ab	L** Q*	24.1 c	42.1 bc	54.5 a	L* Q*
50% IMC	67.6 b[v]	73.5 a	73.5 b	L*Q *	21.1 c[v]	37.1 c	73.1 b	L** Q*	45.2 b	50.2 ab	59.6 a	L* Q*
67% IMC	75.2 a	75.2 a	76.0 ab	NS	67.6 a	70.8 a	81.8 a	NS	51.9 a	57.9 a	57.1 a	L* Q*
Subirrigation												
33% IMC	32.2 e	47.7 c	76.1 ab	L** Q*	8.6 d	15.7 d	49.3 d	L** Q*	28.1 c	36.6 c	56.2 a	L *** Q*
50% IMC	40.0 d	51.6 bc	74.7 b	L*Q *	15.7 cd	26.0 d	58.0 c	L* Q*	40.9 b	42.5 bc	60.5 a	L* Q*
67% IMC	48.7 c	54.7 bc	79.0 a	L*Q *	45.6 b	53.5 b	77.3 ab	L* Q*	46.5 ab	48.5 b	57.3 a	L* Q*
Significance[v]	L*Q *	L*Q *	L*Q *		L*Q *	L*Q *	L*Q *		L*Q*	L*Q *	NS	

[z] H_1 = the amount (by volume) of water that is absorbed by the substrate after one irrigation event. [y] H_{10} = the amount (by volume) of water that is absorbed by the substrate after the final hydration event. [x] CC = maximum volumetric moisture content attained by sample. [w] Significance: Linear (L) and Quadratic (Q) regression significance test, NS = nonsignificant, *** $p \leq 0.001$, ** $p \leq 0.01$ * $p \leq 0.05$ down all columns for peat, coir, and pine bark. S *: Linear (L) and Quadratic (Q) regression significance test, NS = nonsignificant, *** $p \leq 0.001$, ** $p \leq 0.01$ * $p \leq 0.05$ across rows for individual substrates, moisture contents, and irrigation techniques. [v] Statistics using Tukey's honestly significant difference with alpha = 0.05 are given down individual columns at a given initial moisture content. Means with the same letter are not statistically different.

Statistics were determined using SAS v. 9.4 (SAS Institute; Cary, NC, USA). A Tukey's HSD test with alpha = 0.05 was used to identify differences and similarities between substrates at individual IMCs and irrigation events. This test also determined the similarities and differences of CC, first hydration, and final hydration across substrates, IMCs, and irrigation techniques. Both linear and quadratic regression was performed and significance was determined using p values with significance ranging from > 0.001 to 0.05. An analysis of variance test was conducted to test the effects of initial IMC and irrigation technique on CC, first hydration and final hydration within individual substrate components.

3. Results

3.1. Particle Size

Substrate particle size analysis was performed on all three substrates, with the results displayed in Table 1. Coir represented the substrate with the highest percentage of particles smaller than 2.0 mm, representing 93.8% of all particles tested while pine bark showed the highest percentage of coarse particles with a value of 53.6%. Peat occupied a middle ground between coir and pine bark with 13% more coarse particles than coir, but still 34% less than that of pine bark.

3.2. Coir Water Capture

The VWC curves for coir (Figure 4A–C) indicated a pattern directly related to IMC. Regardless of IMC, the first hydration event had the most water absorbed by the substrate compared to all other irrigation events. The IMC affected the amount of water absorbed in the first hydration, and increased as the IMC increased. For surface irrigation, coir at 33% IMC (Figure 4A) needed four irrigation events to reach maximum absorption through irrigation. Coir at 50% IMC (Figure 4B) needed two events to reach maximum absorption and at 67% IMC (Figure 4C) needed just one. For subirrigation, IMC contributed to the ability of coir to absorb water across all 10 events, however, coir never reached a steady state or maximum absorption at any initial moisture level with subirrigation. At 50% IMC, coir reached a final hydration of 73.5% VWC through surface irrigation and 51.6% VWC with subirrigation (Table 2). However, at 33% IMC, coir was ~20% VWC below the CC after the final hydration in both irrigation techniques. Coir samples at 67% IMC reached near CC in one irrigation using the surface application technique, with a final hydration value < 1% below the CC.

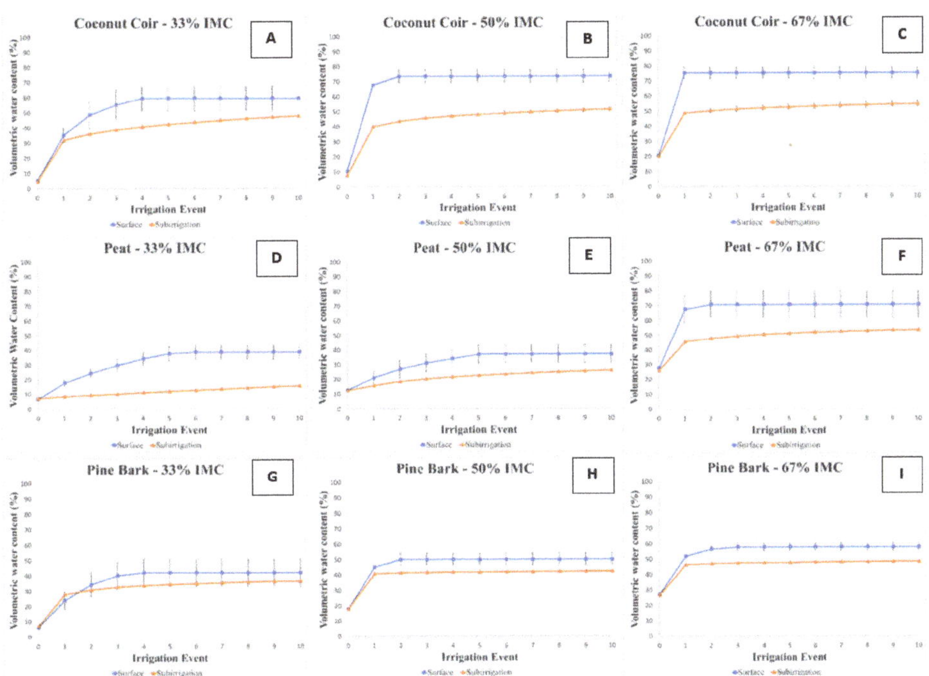

Figure 4. Substrate water capture volumetric water content curves for peat, coir, and pine bark over ten irrigation events at three moisture contents and two irrigation techniques. With (**A**) coir at 33% IMC, (**B**) coir at 50% IMC, (**C**) coir at 67% IMC, (**D**) representing peat at 33% IMC, (**E**) peat at 50% IMC, (**F**) peat at 67% IMC, (**G**) pine bark at 33% IMC, (**H**) pine bark at 50% IMC, and (**I**) pine bark at 67% IMC.

The capture rate of coir was affected by irrigation method. Capture rates for coir were greatest at 50% IMC (Figure 5), where 50% IMC captured ~60% VWC of water through one irrigation event whereas 67% IMC captured ~50% VWC of water in surface irrigation. However, with subirrigation the differences were smaller. Initial moisture contents of 33%, 50%, and 67% were within 8% of total water captured (volumetrically) between each increase in moisture. Also, as IMC increased, the difference between events one and 10 were smaller. With surface irrigation, as IMC increased, fewer events were needed to reach maximum capture. In subirrigation, the effect was similar, although 33% IMC and 50% IMC showed minimal differences in water capture.

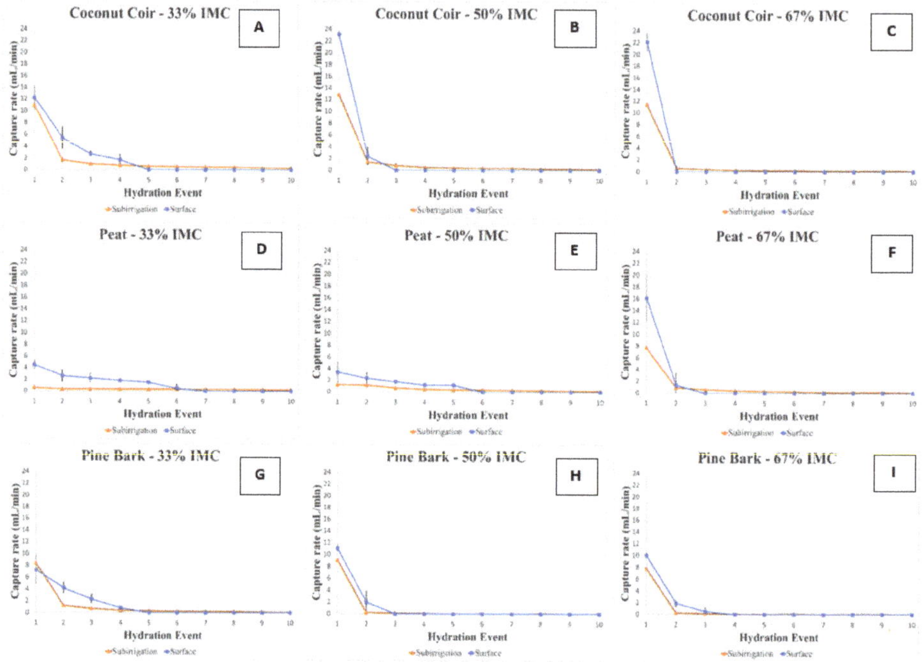

Figure 5. Water capture rate (CR) for peat, coir, and pine bark over ten irrigation events at three moisture contents and two irrigation techniques. With (**A**) coir at 33% IMC, (**B**) coir at 50% IMC, (**C**) coir at 67% IMC, (**D**) representing peat at 33% IMC, (**E**) peat at 50% IMC, (**F**) peat at 67% IMC, (**G**) pine bark at 33% IMC, (**H**) pine bark at 50% IMC, and (**I**) pine bark at 67% IMC.

3.3. Peat Water Capture

The VWC curves (Figure 4D–F) for peat, similar to coir, identified a pattern related to IMC. With surface irrigation, 33% IMC required six irrigation events to reach maximum absorption through irrigation while 50% IMC needed just five irrigation events (Figure 4D,E), showing very little difference between the two IMC levels. At 67% IMC, maximum absorption was reached in two surface irrigation events. Comparing the surface and subirrigated VWC curves (Figure 4D–F), there was ~20% less water taken up by capillary rise than from gravitational flow. Peat contained 81% of particles < 2 mm (coir had 92%) and nearly 17% of particles between 6.3 mm and 2 mm (Table 1). At both 50% IMC and 33% IMC, the final hydration values were less than 20% VWC, with subirrigated peat at 33% IMC reaching less than 20% VWC after 10 irrigations (Table 2). At 67% IMC, the first hydration of subirrigated peat increased by 30% VWC compared to that of 50% IMC.

The water capture of peat was affected by both irrigation method and IMC. Water CRs for peat show the effects of a low moisture condition on hydration with surface irrigation at 33% IMC reaching ~5 mL min^{-1} and 50% IMC reaching ~4 mL min^{-1} (Figure 5E). Subirrigated peat at 33% IMC and 50% IMC had much lower CRs than surface irrigation, with CRs at or below 1 mL min^{-1}. Peat did not begin to show a change in hydration until IMCs of 67% in both irrigation methods, with very minimal differences between 33% IMC and 50% IMC. Surface CR at 67% IMC, while more variable than subirrigation, reached 16 mL/min while subirrigation peaked ~8 mL min^{-1} (Figure 5). The strong interaction between water capture and IMC was clearly evident in peat.

3.4. Pine Bark Water Capture

Pine bark had a more consistent increase in VWC over the 10 irrigation events than either peat or coir. Of the three substrates, bark contained the highest percentage of coarse particles (Table 1), while also having a similar portion of medium (2.0–6.3 mm) sized particles compared to peat. The VWC curves (Figure 4G-I) identify a degree of consistency between irrigation techniques, regardless of initial moisture level or irrigation method. At 50% IMC and 67% IMC, subirrigation produced maximum irrigation absorption after one irrigation event, with less than 2% difference between first and final hydration. At all IMCs, surface irrigation had higher VWC after the final irrigation event compared to subirrigation, but the difference between surface irrigation and subirrigation after the final hydration was less than 10% VWC (Table 2). Similar to coir, pine bark achieved maximum capture within the first two to three irrigations at 50% IMC and 67% IMC in both irrigation methods. At 50% IMC, the difference in first hydration between surface and subirrigation was < 5% VWC. Similar to coir, the CC was not influenced by IMC or irrigation method. For bark, water capture differences are evident between 33% and 50% IMC. First hydrations increased from ~25% VWC to ~42% VWC (Figure 4G–I). Unlike coir and peat, pine bark showed less variability by irrigation method, with all first and final hydrations between 50% IMC and 67% IMC less than 8% VWC gain between all ten irrigation events. Pine bark water capture was the most consistent of the three substrate materials.

Water CRs also showed similarities across irrigation methods, with subirrigation having the higher CR at 33% IMC (Figure 5G). In just one hydration event, pine bark at 67% IMC reached 0.90 (90%) of its CC by surface irrigation and 0.81 (81%) of its CC by subirrigation. Water CRs for pine bark at 50% IMC and 67% IMC are within 2 mL/min of each other at the first hydration, with surface irrigation representing the maximum CR across all pine bark treatments. The low variability in water capture and high percentage of coarse and medium sized particles (Table 1) allowed pine bark to have a high capture efficiency, regardless of irrigation technique.

3.5. Capture Rate

Capture rates of surface-irrigated samples were highest in coir, regardless of IMC. At 50% IMC, coir CR was ~23 mL min^{-1} before falling to ~3 mL/min by the second irrigation event (Figure 5E). The steep drop in CR was attributed to the substrate's ability to absorb water at such a high rate during the first irrigation, nearly reaching the maximum it could absorb (through surface application) after one irrigation event. With first hydrations capturing 67.6% VWC for 50% IMC and 75.2% VWC for 67% IMC, coir had the highest absorption rate of all three substrates. The lower initial moisture conditions in peat at 33% IMC and 50% IMC impacted CR more so than irrigation method (Figure 5A,B), further reducing the wettability of low-moisture peat. Conversely, the similar responses of pine bark between IMC and irrigation method could be attributed to increasing particle size which might have decreased variability in uptake. The water volumes absorbed by pine bark were less than coir, however the CC of pine bark was ~20% lower volumetrically than that of coir, giving pine bark physically less capacity to hold water.

4. Discussion

From the data in Figures 4 and 5, it appears that initial moisture content prior to the first irrigation event had the overall greatest effect on the water capture and retention of peat, coir, and pine bark across both irrigation techniques. The container capacity of pine bark and coir were less affected by irrigation technique than peat. Surface irrigation provided the highest water capture in the first hydration across nearly all substrates and IMCs. Peat had higher initial and final hydration values with surface irrigation compared to subirrigation over all IMCs.

At all initial moisture levels, coir was able to take up water. However, IMC altered the intensity of imbibition. With surface irrigation, coir at 33% IMC needed four irrigation events to reach its maximum of 60% VWC. At 50% IMC, coir needed two irrigation events, and at 67% IMC it needed just one for water capture to equal CC. Coir is known to be very hydrophilic, having a sponge-like ability to capture and hold water [33]. Surface irrigation has the additional potential of gravity to draw water through the substrate, allowing droplets to travel a path of least resistance. This allows water to move through macro and mesopores to hydrate the substrate. Conversely, with subirrigation, water must travel via capillary action and against gravity, along particle surfaces and through mostly micro-pores [34]. Initial moisture content did not have an effect on coir CC. With 92% of coir particles ranging from 2 mm or less (Table 1), water retention was very high.

For peat, IMC had the greatest influence on the substrate's ability to capture water with surface irrigation. As is well documented, intensity of hydrophobicity of peat increases at lower substrate moisture contents. These hydrophobic intensities can influence rewetting and impair the physical properties of the substrate [35,36]. At 33% IMC and 50% IMC, peat had difficulty hydrating through the first five irrigation events (Figure 4A). In the case of peat at 33% IMC and irrigated from the surface, water delivery from the separatory funnel had to be slowed to reduce ponding of water on the surface and increasing the hydraulic head at the substrate surface. For perspective, the first hydration at 33% IMC and 50% IMC through surface irrigation for peat was 17.8% and 21.1% (Table 2) respectively, while the first hydration of coir at the same moisture levels reached 35.4% and 67.6% respectively (Table 2). It wasn't until 67% IMC that peat began to capture and retain water during surface irrigation. The large proportion of coarse particles may relate to a greater portion of macro-pores in peat than coir (Table 1). These larger pores allowed surface irrigated water to preferentially flow through peat, even at lower initial moisture levels. Water moved through the large pores with less wetting of the substrate matrix due to increased intensity of hydrophobicity of the peat at both 33% IMC and 50% IMC. Conversely, with subirrigation, at 33% IMC, peat was unable to eclipse 13% VWC after 10 irrigation events, representing the lowest first hydration of all treatments. At 50% IMC, peat reached 23% VWC with a final hydration of 37.1% VWC and a CC of 58.0 (Table 2). Compared to coir, irrigation method and IMC both impacted the CC of peat. However, at 67% IMC, the substrate absorbed water in the first irrigation event. The capture potential, or total volumetric water captured, of peat was nearly 40% less than that of coir.

In pine bark, an increase in fine (greater than 0.5 mm) particles has been shown to greatly influence the physical properties (AS and CC), while larger particle sizes had a minor influence on physical properties [9]. Larger particles, for surface irrigation, may provide water with more channels to move through the container, better hydrating the bark as pine bark just doesn't have as much surface area/microporosity for absorption. However, micro and meso-pores have higher abilities to capture and retain water. Larger pore sizes tend to drain more easily under gravitational potentials than smaller pores [37]. Pine bark can have variable properties based on processing, and it can have more AS, lower TP and easily available water than both peat and coir [38]. Pine bark had the most similar water capture and retention across all IMC and, aside from 33% IMC, reached their maximum VWC in the first two irrigation events.

5. Conclusions

Comparing first hydration, final hydration, and capture rate across all treatments (Table 2), there were varied effects among irrigation methods and IMCs. Peat was highly affected by IMC, the intensity of hydrophobicity was altered by IMC, and irrigation delivery. Coir and peat, through every IMC, captured less water through subirrigation than surface irrigation. Most notably, the higher the initial moisture content in the substrate prior to irrigation, the greater the overall water capture. These three substrate components demonstrated markedly different responses to water capture and retention in response to irrigation method and IMC.

Author Contributions: Conceptualization, B.A.S., B.E.J., W.C.F., J.L.H.; methodology, B.A.S., B.E.J., W.C.F., J.L.H., J.P.A.; formal analysis, B.A.S., B.E.J., W.C.F.; investigation, B.A.S.; writing-original draft preparation, B.A.S.; writing-review and editing, B.A.S., B.E.J., W.C.F., J.L.H.; supervision, B.E.J., W.C.F. All authors have read and agreed to the published version of the manuscript.

Funding: This research received no external funding.

Conflicts of Interest: The authors declare no conflict of interest.

References

1. U.S. Department of Agriculture. *Census of Horticultural Specialties*; U.S. Department of Agriculture: Washington, DC, USA, 2014.
2. Stanghellini, C. Horticultural production in greenhouses: Efficient use of water. *Acta Hortic.* **2014**, *1034*, 25–32. [CrossRef]
3. Fernandes, C.; Corá, J.; Araújo, J. Reference evapotranspiration estimation inside greenhouses. *Sci. Agric.* **2003**, *60*, 591–594. [CrossRef]
4. Kitta, E.; Bartzanas, T.; Katsoulas, N.; Kittas, C. Benchmark irrigated under cover agriculture crops. *Agric. Agric. Sci. Procedia* **2015**, *4*, 348–355. [CrossRef]
5. Levidow, L.; Zaccaria, D.; Maia, R.; Vivas, E.; Todorovic, M.; Scardigno, A. Improving water-efficient irrigation: Prospects and difficulties of innovative practices. *Agric. Water Manag.* **2014**, *146*, 84–94. [CrossRef]
6. Daccache, A.; Ciurana, J.S.; Rodriguez Diaz, J.A.; Knox, J.W. Water and energy footprint of irrigated agriculture in the Mediterranean region. *Environ. Res. Lett.* **2014**, *9*, 1–12. [CrossRef]
7. Egea, G.; Fernández, J.E.; Alcon, F. Financial assessment of adopting irrigation technology for plant-based regulated deficit irrigation scheduling in super high-density olive orchards. *Agric. Water Manag.* **2017**, *187*, 47–56. [CrossRef]
8. Montesano, F.F.; Van Iersel, M.W.; Boari, F.; Cantore, V.; D'Amato, G.; Parente, A. Sensor-based irrigation management of soilless basil using a new smart irrigation system: Effects of set-point on plant physiological responses and crop performance. *Agric. Water Manag.* **2018**, *203*, 20–29. [CrossRef]
9. Handreck, K.A.; Black, N.D. *Growing Media for Ornamental Plants and Turf*; New South Wales University Press: Kensington, Australia, 1984; pp. 115–117.
10. Puustjarvi, V. Physical properties of peat used in horticulture. *Acta Hortic.* **1974**, *37*, 1922–1929. [CrossRef]
11. Milks, R.R.; Fonteno, W.C.; Larson, R.A. Hydrology of horticultural substrates: II. Predicting physical properties of media in containers. *J. Am. Soc. Hortic. Sci.* **1989**, *114*, 53–56.
12. Wallach, R.; da Silva, F.F.; Chen, Y. Hydraulic characteristics of Tuff (Scoria) used as a container medium. *J. Am. Soc. Hortic. Sci.* **1992**, *117*, 415–421. [CrossRef]
13. Michel, J.C.; Riviere, L.M.; Bellon-Fontaine, M.N. Measurement of wettability of organic materials in relation to water content by the capillary rise method. *Eur. J. Soil Sci.* **2001**, *52*, 459–467. [CrossRef]
14. Urrestarazu, M.; Guillen, C.; Carolina Mazuela, P.; Carrasco, G. Wetting agent effect on physical properties of new and reused rockwool and coconut coir waste. *Sci. Hortic.* **2008**, *116*, 104–108. [CrossRef]
15. Levesque, M.P.; Dinel, H. Fiber content, particle-size distribution and some related properties of four peat materials in Eastern Canada. *Can. J. Soil Sci.* **1977**, *57*, 187–195. [CrossRef]
16. Bunt, A.C. *Media and Mixes for Container-Grown Plants*, 2nd ed.; Unwin Hyman Ltd.: London, UK, 1988.
17. Bilderback, T.E.; Lorscheider, M.R. Wetting agents used in container substrates are they BMP's? *Acta Hortic.* **1997**, *450*, 313–319. [CrossRef]
18. Fonteno, W.C.; Fields, J.S.; Jackson, B.E. A pragmatic approach to wettability and hydration of horticultural substrates. *Acta Hortic.* **2013**, *1013*, 139–146. [CrossRef]
19. Letey, J.; Osborn, J.; Pelishek, R.E. Measurement of liquid-solid contact angles in soil and sands. *Soil Sci.* **1962**, *93*, 149–153. [CrossRef]
20. Fields, J.S.; Fonteno, W.C.; Jackson, B.E. Hydration efficiency of traditional and alternative greenhouse substrate components. *HortScience* **2014**, *49*, 336–342. [CrossRef]
21. Fields, J.S.; Fonteno, W.C.; Jackson, B.E.; Heitman, J.L.; Owen, J.S., Jr. Hydrological properties, moisture retention, and draining profiles of wood and traditional components for greenhouse substrates. *HortScience* **2014**, *49*, 827–832. [CrossRef]
22. Jouany, C.; Chenu, C.; Chassin, P. Wettability of soil constituents from angle measurements: Bibliographical review. *Sci. Sol.* **1992**, *30*, 33–47.
23. Dekker, L.W.; Ritsema, C.J. Wetting patterns and moisture variability in water repellent Dutch soils. *J. Hydrol.* **2000**, *232*, 148–164. [CrossRef]

24. Beardsell, D.V.; Nichols, D.G. Wetting properties of dried-out nursery container media. *Sci. Hortic.* **1982**, *17*, 49–59. [CrossRef]

25. Elia, A.; Parente, A.; Serio, F.; Santamaria, P. Some aspects of trough benches system and its performances in cherry tomato production. *Acta Hortic.* **2003**, *614*, 161–166. [CrossRef]

26. Oh, M.M.; Cho, Y.Y.; Kim, K.S.; Son, J.E. Comparisons of water content of growing media and growth of potted kalanchoe among nutrient-flow wick culture and other irrigation systems. *HortTechnology* **2007**, *17*, 62–66. [CrossRef]

27. Biernbaum, J.A. Subirrigation could make environmental and economical sense for your greenhouse. *Prof. Plant Grow. Assn. Nwsl.* **1993**, *24*, 2–14.

28. Uva, W.L.; Weiler, T.C. Economic analysis of adopting zero runoff subirrigation systems in greenhouse operations in the northeast and north central United States. *HortScience* **2001**, *36*, 167–173. [CrossRef]

29. Holcomb, E.J.; Gamez, S.; Beattie, D.; Elliott, G.C. Efficiency of fertigation programs for baltic ivy and asiatic lily. *Hort. Technol.* **1992**, *2*, 43–46. [CrossRef]

30. Dumroese, R.K.; Pinto, J.R.; Jacobs, D.F.; Davis, A.S.; Horiuchi, B. Subirrigation reduces water use, nitrogen loss, and moss growth in a container nursery. *Nativ. Plants J.* **2006**, *7*, 253. [CrossRef]

31. Russo, D.D.G. *Water Flow and Solute Transport in Soils: Developments and Applications*; Springer: Berlin/Heidelberg, Germany; New York, NY, USA, 1993.

32. Fonteno, W.C.; Hardin, C.T.; Brewster, J.P. *Procedures for Determining Physical Properties of Horticultural Substrates Using the NCSU Porometer. Horticultural Substrates Laboratory*; North Carolina State University: Raleigh, NC, USA, 1995.

33. Abad, M.; Frones, F.; Carrion, C.; Noguero, V. Physical properties of various coconut coir dusts compared to peat. *HortScience* **2005**, *40*, 2138–2144. [CrossRef]

34. Biernbaum, J.A.; Versluys, N.B. Water Management. *HortTechnology* **1998**, *8*, 504–509. [CrossRef]

35. Valat, B.; Jouany, C.; Riviere, L.M. Characterization of the wetting properties of air dried peats and composts. *Soil Sci.* **1991**, *152*, 100–107. [CrossRef]

36. Michel, J.C. Study of the Wettability of Organic Materials Used as a Culture Support. Ph.D. Thesis, Agrocampus Ouest, Rennes, France, 1998.

37. Drzal, M.S.; Fonteno, W.C.; Cassel, D.K. Pore fraction analysis: A new tool for substrate testing. *Acta Hortic.* **1999**, *481*, 43–54. [CrossRef]

38. Bilderback, T.E.; Warren, S.L.; Owen, J.S.; Albano, J.P. Healthy substrates need physicals too! *HortTechnology* **2005**, *15*, 747–751. [CrossRef]

MDPI

St. Alban-Anlage 66

4052 Basel

Switzerland

www.mdpi.com

Agronomy Editorial Office

E-mail: agronomy@mdpi.com

www.mdpi.com/journal/agronomy